Modern Group Analysis:
Advanced Analytical and Computational Methods
in Mathematical Physics

Modern Group Analysis: Advanced Analytical and Computational Methods in Mathematical Physics

Proceedings of the International Workshop Acireale, Catania, Italy, October 27–31, 1992

Edited by

N. H. Ibragimov
M. Torrisi
and
A. Valenti

Department of Mathematics,
University of Catania,
Catania, Italy

SPRINGER SCIENCE+BUSINESS MEDIA, B.V.

Library of Congress Cataloging-in-Publication Data

Modern group analysis: advanced analytical and computational methods in mathematical physics : proceedings of the international workshop, Acireale, Catania, Italy, October 27-31, 1992 / edited by N.H. Ibragimov, M. Torrisi, and A. Valenti.

p. cm.

ISBN 978-94-010-4908-5 ISBN 978-94-011-2050-0 (eBook)

DOI 10.1007/978-94-011-2050-0

1. Mathematical physics--Congresses. 2. Numerical analysis--Congresses. I. Ibragimov, N. Kh. (Nail' Khaĭrullovich) II. Torrisi, M. III. Valenti, A.

QC19.2.M63 1993

530.1'5--dc20 93-20973

ISBN 978-94-010-4908-5

Printed on acid-free paper

The Workshop was dedicated to the 150th anniversary of Sophus Lie

PREFACE

On the occasion of the 150th anniversary of Sophus Lie, an International Workshop "Modern Group Analysis: advanced analytical and computational methods in mathematical physics" has been organized in Acireale (Catania, Sicily, October 27–31, 1992).

The Workshop was aimed to enlighten the present state of this rapidly expanding branch of applied mathematics.

Main topics of the Conference were:

- classical Lie groups applied for constructing invariant solutions and conservation laws;
- conditional (partial) symmetries;
- Bäcklund transformations;
- approximate symmetries;
- group analysis of finite-difference equations;
- problems of group classification;
- software packages in group analysis.

The success of the Workshop was due to the participation of many experts in Group Analysis from different countries.

This book consists of selected papers presented at the Workshop.

We would like to thank the Scientific Committee for the generous support of recommending invited lectures and selecting the papers for this volume, as well as the members of the Organizing Committee for their help.

The Workshop was made possible by the financial support of several sponsors that are listed below.

It is also a pleasure to thank our colleague Enrico Gregorio for his invaluable help during the preparation of this volume.

N. H. Ibragimov
M. Torrisi
A. Valenti

Scientific Committee

N. H. Ibragimov (Chairman)	M. Torrisi (Co-chairman)
W. F. Ames	R. L. Anderson
A. M. Anile	G. W. Bluman
V. Boffi	C. Cercignani
Y. Choquet-Bruhat	A. Donato
D. Fusco	G. Grioli
P. G. L. Leach	P. J. Olver

L. V. Ovsiannikov

Organizing Committee

M. Lalicata	S. Lombardo
A. Grasso	A. E. Pompei
M. A. Rigano	M. Torrisi (Chairman)
A. Valenti	G. Zappalà

Sponsors

Mathematical Committee of the Italian National Research Council (C.N.R.)

Gruppo Nazionale per la Fisica Matematica (G.N.F.M.-C.N.R.)

Università di Catania

Dipartimento di Matematica di Catania

Provincia Regionale di Catania

Assessorato Ecologia ed Ambiente della Provincia Regionale di Catania

A.A.P.I.T. di Catania

Azienda Turismo di Acireale

S.I.M.A.I.

Fondazione IBM Italia

TABLE OF CONTENTS

HIDDEN AND NONLOCAL SYMMETRIES OF NONLINEAR DIFFERENTIAL EQUATIONS

B. ABRAHAM-SHRAUNER and A. GUO*
Department of Electrical Engineering
Washington University St. Louis, Missouri, 63130 U.S.A.

Abstract. New results on hidden and nonlocal symmetries of nonlinear ordinary differential equations (NLODEs) are presented. Two types of hidden symmetries have been identified. A type I (II) hidden symmetry of an ODE occurs if a symmetry is lost (gained) when the order of the ODE is reduced. Both type I and type II hidden symmetries are found in the reduction of a third-order NLODE invariant under a three-parameter nonsolvable Lie group. Nonlocal group generators are determined of the exponential form and a new linear form. The ODEs can be reduced by the nonlocal group generators until first-order ODEs are obtained where the procedure fails because canonical coordinates cannot be calculated in that case. ODEs cannot be reduced by the linear nonlocal group generators.

1. Introduction

The pioneering research of Sophus Lie in the latter nineteenth century on the use of symmetries to solve differential equations has led to widely diverse applications of Lie groups. The results reported here relate to a further development of his original intent, the solution of differential equations by examining the symmetries of the equations. In the present day the Lie classical method for finding point symmetries of the differential equations is the most common method used. However, many symmetries are not found by the classical method and this has led to the investigation of contact symmetries [1,2], generalized symmetries [3], nonclassical symmetries [4,5]. All these methods share with the Lie classical method that they are direct methods which given the differential equations then determine the symmetries of the differential equations. The term direct method should not be confused with the method of that name [6]. Nonetheless, Sophus Lie, himself, frequently used an indirect method where he started with the group and then determined the general form of the differential equations under which the differential equations were invariant. Tables were compiled of the general from of ODEs [7].

Not all symmetries are found by the direct methods mentioned above. These symmetries are called hidden symmetries since for ODEs the type I (II) symmetry is lost (gained) when the order of the ODE is reduced. These symmetries are connected to nonlocal symmetries since hidden symmetries may be represented by nonlocal group generators and nonlocal transformations between ODEs of the same order occur in the presence of hidden symmetries. The significance of the loss or gain of these symmetries was first stressed by Olver [3] and development of their properties has been reported in our earlier work [8–11]. In this article the hidden symmetries

* Supported in part by a grant from the Southwestern Bell Corporation.

N. H. Ibragimov et al. (eds.),
Modern Group Analysis: Advanced Analytical and Computational Methods in Mathematical Physics, 1–5.

of a nonsolvable group of type $sl(2,R)$ are explored. In addition the properties of the nonlocal group generators are investigated.

2. Hidden Symmetries of 3-Parameter Group

We consider the three-parameter group of structure $sl(2,R)$ which has the group generators

$$U_1 = \frac{\partial}{\partial z}, \quad U_2 = z\frac{\partial}{\partial z}, \quad U_3 = z^2\frac{\partial}{\partial z}, \tag{1}$$

Two features are important for the Lie algebra of this set of group generators. First, if we transform z to $-1/z$, the form of the group generators remains the same; that is U_1 maps to U_3, U_3 maps to U_1 and U_2 remains the same. Second, the commutators are

$$[U_1, U_2] = U_1, \quad [U_2, U_3] = U_3, \quad [U_1, U_3] = 2U_2. \tag{2}$$

From Eq. (2) we conclude that this group is nonsolvable which implies that a third-order ODE cannot be reduced to quadratures by the symmetries of this group. That follows because no matter which subgroups we use to reduce the order of the ODE, at least one of the three symmetries needed to reduce the the third-order ODE is lost.

The form of the third-order ODE is determined by calculating the differential invariants associated with the three group generators and finding the overlap of these invariants. For this simple case that can be done by inspection; a more systematic method is to assume that a differential equation is a function of one set of invariants and apply the extended group generators of the other two subgroups in succession. The solved form of the ODE is

$$u_{zzz} = 1.5u_z^{-1}u_{zz}^2 + u_z^3 g(u) \tag{3}$$

where $g(u)$ is arbitrary function of u. We investigate the reduction of Eq. (3) by the differential invariants of the groups under which it is invariant. Since U_1 and U_3 are equivalent under the inverse transformation, we consider only the two possible orders of reduction. These are case A where we reduce the ODE by the invariants of U_1, then by those of U_2 and finally by those of U_3 and case B where we reduce the ODE by the invariants of U_2, then by those of U_1 and finally by those of U_3. The reduction order U_1, then U_3, and finally U_2 is not possible because the once-extended U_3 in the differential invariants of U_1 is a new type of nonlocal group generator whose differential invariants cannot be calculated.

The order of the Eq. (3) is reduced for case A by the invariants of the group generators in the order $U_1 \to U_2 \to U_3$. The variables of the second-order ODE are $Y = u_z^{1/2}$, $x = u$ in a modified set of differential invariants and path curves chosen such that the ODE is linear. The reduced ODE is

$$Y'' - \frac{g(x)}{2}Y = 0. \tag{4}$$

where $'$ denotes differentiation with respect to x. The Eq. (4) is invariant under an eight-parameter group as it is linear but the original third-order ODE is invariant under a three-parameter group. One symmetry, that of the group represented by U_1, was used to reduce the ODE and one associated with U_3 is lost. Consequently, seven new subgroups are gained. The groups generators U_2 and U_3 become in the once-extended extended form

$$\begin{aligned} U_{2a}^{(1)} &= \frac{Y}{2}\frac{\partial}{\partial Y} \\ U_{3a}^{(1)} &= Y\int\frac{dx}{Y^2}\frac{\partial}{\partial Y} \end{aligned} \tag{5}$$

The seven new local group generators are not listed here but they depend on the solutions of a linear ODE which contains the arbitrary function $g(x)$. Type II hidden symmetries occur in the third-order ODE, Eq. (3), since the reduced second-order ODE, Eq. (4), is invariant under seven new groups in addition to the symmetry group of $U_{2a}^{(1)}$. The second-order ODE has a type I hidden symmetry as the group associated with U_3 is lost; the nonlocal group generator $U_{3a}^{(1)}$ is a consequence of this lost symmetry. The new feature of this nonlocal group generator is that it is linear in the integral over the dependent variable. This property follows from the Lie algebra. Unlike an exponential nonlocal group generator this group generator can not be used to reduce the order of the of the ODE since the integral over the dependent variable does not factor out of the characteristic equations for the differential invariants.

The second-order ODE, Eq. (4) is reduced by the invariants of the once-extended group, generator $U_{2a}^{(1)}$. The resultant first-order ODE is a Riccati equation

$$\frac{dw}{dv} + w^2 = \frac{g(v)}{2} \tag{6}$$

This ODE is not invariant under any local group of those listed for the second-order ODE and under any obvious new groups. This ODE has type I hidden symmetries as the seven local group generators that appeared in the symmetry analysis of Eq. (4) are lost and the group of $U_{2a}^{(1)}$ has been used to reduce the order of the ODE. The remarkable feature is that the nonlocal group generator, $U_{3a}^{(1)}$ in Eq. (5), has changed from linear form in to exponential form in Eq. (7). The expression for $U_{3a}^{(1)}$ is

$$U_{3a}^{(1)} = \exp[-2\textstyle\int w\,dv]\dfrac{\partial}{\partial w} \tag{7}$$

However, the Riccati equation (6) cannot be reduced to quadratures even by this exponential nonlocal group generator because one cannot find the canonical coordinates which are needed to reduce a first-order ODE.

The same procedure can be tried for case B where the reduction is done in the order $U_2 \to U_1 \to U_3$. The new variables are $\bar{Y} = (zu_z)^{1/2}$, $\bar{x} = u$ from the group represented by U_2. The second-order ODE is the Pinney equation

$$\frac{d^2\bar{Y}}{dx^2} - \frac{g(x)}{2}\bar{Y} = -\frac{1}{4\bar{Y}^3}. \tag{8}$$

This nonlinear ODE is invariant under a new three-parameter group which has the group generators U_{scj} for $j = 1, 2, 3$ in Eq. (12) in a previous paper [11].

The group generators U_1 and U_3 when once-extended become exponential nonlocal group generators. These are

$$\begin{aligned} U_{1b}^{(1)} &= \exp[-\textstyle\int \bar{Y}^{-2}\, dx]\frac{Y}{2}\frac{\partial}{\partial \bar{Y}} \\ U_{3b}^{(1)} &= \exp[\textstyle\int \bar{Y}^{-2}\, dx]\frac{Y}{2}\frac{\partial}{\partial \bar{Y}} \end{aligned} \tag{9}$$

The Pinney equation does not retain the point symmetries of U_1 or U_3 as it was reduced from the third-order ODE by variables of a non-normal subgroup. The third-order ODE has type II hidden symmetries as new group invariances appear in the reduced second-order ODE, the Pinney equation. The Pinney equation has type II hidden symmetries as the symmetries of U_1 and U_3 were lost.

The Pinney equation can be reduced to a first-order ODE by using the invariants found from either of $U_{1b}^{(1)}$ or $U_{3b}^{(1)}$ as these are exponential nonlocal group generators. The local group generators for the three groups under which Eq. (8) is invariant were not used to reduce the order of the Pinney equation since we must find the solutions of a linear third-order ODE, which contains an arbitrary function $g(x)$, to write down the local group generators explicitly. This is an unusual case where the invariants of a nonlocal group generator rather than those of a local group generator are used to reduce the order of an ODE. The first-order ODE is found from Eq. (7) by letting

$$\bar{w} = \frac{\bar{Y}'}{\bar{Y}} + \frac{1}{2\bar{Y}'^2}, \qquad \bar{v} = x. \tag{10}$$

The reduced first-order ODE is identical in form to Eq. (6) if we let $\bar{w} = w$ and $\bar{v} = v$. The two paths of reduction give the same first-order ODEs. The difference is that this first-order ODE has lost three Lie point symmetries of the Pinney equation whereas in case A seven Lie point symmetries of the linear second-order ODE, Eq. (4), were lost.

3. Nonlocal Group Generator

The nonlocal group generator represents the nonlocal symmetries. Nonlocal transformations between the second-order Pinney equation, Eq. (7), and the linear second-order ODE, Eq. (4), also occur. The nonlocal transformation is expected between ODEs of the same order but invariant under different Lie groups since a local transformation would leave unchanged the dimension and structure of the Lie group under which the differential equation is invariant. Consequently, the presence of hidden symmetries is associated with nonlocal transformations between ODEs of the same order and nonlocal group generators of ODEs. The nonlocal transformations have been discussed previously [8]. Here we point out some curious properties of the nonlocal group generators. The differential invariants of the exponential nonlocal group generators but not those of the linear nonlocal group generator can be used to reduce the order of an ODE. This follows because the exponential factors out of each term in the characteristic equations from which the differential invariants are calculated. The differential invariants of the linear nonlinear group generators cannot be calculated explicitly since the integral term does not factor out of all terms in the characteristic equations and it cannot be found.

The exponential or linear form of the nonlocal group generator depends on the structure of the Lie algebra associated with the Lie group under which the ODE is invariant. An exponential nonlocal group generator is found if the nonlocal subgroup variables are used to reduce an ODE invariant under a two-parameter group. This can be shown from the commutator. However, also evident is that the exponential nonlocal group generator may not signify a lost symmetry of a non-normal subgroup. For our example we found a lost symmetry of a third-order ODE led to an exponential group generator of a first-order ODE.

References

1. R. L. Anderson and N. H. Ibragimov, *Lie-Bäcklund Transformations in Applications*, SIAM, Philadelphia, 1979.
2. G. W. Bluman and S. Kumei, *Symmetries and Differential Equations*, Appl. Math. Sci. No. 81, Springer-Verlag, New York, 1989.
3. P. J. Olver, *Applications of Lie Groups to Differential Equations*, Springer-Verlag, New York, 1986.
4. G. W. Bluman and J. D. Cole, *Similarity Methods for differential Equations*, Springer-Verlag, 1974.
5. P. J. Olver and P. Rosenau, SIAM J. Appl. Math. **47**, 263–275, 1987.
6. P.Clarkson and M. Kruskal, "New similarity reductions of the Boussinesq equation," J. Math. Phys. **30**, 2201–2213, 1989.
7. A. Cohen, *An Introduction to the Lie theory of One-Parameter Groups with Applications to the Solution of differential Equations*, D. C. Heath, New York, 1911.
8. B. Abraham-Shrauner and Ann Guo, "Hidden Symmetries Associated with the Projective Group of Nonlinear First-Order Ordinary Differential Equations," J. Phys. A. **25**, 5597, 1992.
9. A. Guo and B. Abraham-Shrauner, "Hidden Symmetries of Energy Conserving Differential Equations," IMA J. Appl. Math. (submitted for publication).
10. B. Abraham-Shrauner and Ann Guo, "Hidden symmetries of Differential Equations," Proceedings of the AMS March, 1992 meeting in Springfield, Missouri (submitted).
11. B. Abraham-Shrauner and P. G. L. Leach, "Hidden Symmetries of Nonlinear Ordinary Differential Equations," AMS-SIAM Summer Seminar Proceedings (submitted).

INTERNAL SYMMETRIES OF DIFFERENTIAL EQUATIONS

IAN ANDERSON
Department of Mathematics
Utah State University
Logan, Utah 84322-3900

NIKY KAMRAN*
Department of Mathematics
McGill University
Montréal, Québec
CANADA H3A 2K6

and

PETER J. OLVER† ‡
Department of Mathematics
University of Maryland
College Park, MD
U.S.A. 20742

Abstract. Bäcklund's Theorem, which characterizes contact transformations, is generalized to give an analogous characterization of "internal symmetries" of systems of differential equations. For a wide class of systems of differential equations, every internal symmetry comes from a first order generalized symmetry and, conversely, every first order generalized symmetry satisfying certain explicit contact conditions determines an internal symmetry. We analyze the contact conditions in detail, deducing powerful necessary conditions for a system of differential equations admit "genuine" internal symmetries, *i.e.*, ones which do not come from classical "external" symmetries. Applications include a direct proof that both the internal symmetry group and the first order generalized symmetries of a remarkable differential equation due to Hilbert and Cartan are the noncompact real form of the exceptional simple Lie group G_2.

The work we will survey in this paper, which will appear in [1], had its genesis in a series of lectures on the variational bicomplex given by the first author while visiting the University of North Carolina at Chapel Hill. Robert Bryant, who was in the audience, asked Ian to compute the symmetry group of the innocent looking underdetermined ordinary differential equation $u' = (v'')^2$. Robert knew well the history of this equation, which we have decided to call the Hilbert–Cartan equation; in particular, Elie Cartan had proved that the "symmetry group" of this equation is a realization of the non-compact real form of the exceptional simple Lie group G_2! Robert was suitably impressed when Ian came back with a fourteen dimensional symmetry algebra for the equation. There matters rested until, during a Conference on Symbolic Manipulation hosted by the Institute for Mathematics and Its Applications, Robby Gardner asked Fritz Schwarz to answer the same question using his

* *Supported in part by an NSERC Grant.*
† *Supported in part by NSF Grant DMS 92–04192.*
‡ On leave from School of Mathematics, University of Minnesota, Minneapolis, Minnesota, U.S.A. 55455

N. H. Ibragimov et al. (eds.),
Modern Group Analysis: Advanced Analytical and Computational Methods in Mathematical Physics, 7–21.

computer algebra package for computing symmetry groups in SCRATCHPAD (now renamed AXIOM). Fritz only found a six-dimensional symmetry group. After Ian sent the results of his earlier (hand!) computations, we realized that the discrepancy was due to the fact that Ian had computed the first order generalized symmetries of the equation, whereas Fritz' program was designed to compute classical point symmetries; this is why he failed to detect the eight remaining vector fields. However, upon reflection, it occurred to us that much more was at stake than merely the difference between point symmetries and generalized symmetries. Cartan was certainly not aware of the concept of a generalized symmetry, and all his symmetries were realized as geometrical transformations of some finite-dimensional space, which the generalized symmetries are not. Contact transformations fit into Cartan's framework, but these were not the objects Cartan had computed for this particular equation since, according to Bäcklund's Theorem, there are no contact transformations (beyond prolonged point transformations) if the number of dependent variables is greater than one. What Cartan had computed were what we will call "internal symmetries", which are transformations which preserve the contact ideal *only* when restricted to the equation submanifold. (These are also known as "dynamical symmetries" in the mathematical physics literature, and have also received mention in the abstract work of Vinogradov and his collaborators, *cf.* [9].) The restrictions of Bäcklund's Theorem no longer apply, and there are internal symmetries which depend explicitly on higher order derivatives. Thus, a new question arose: for the Hilbert–Cartan equation, why did the computed Lie algebra of generalized symmetries coincide with Cartan's Lie algebra of internal symmetries?

Our results answer this question in general, and can be summarized as follows. First, and obvious, is the fact that every external symmetry restricts to an internal symmetry. In many cases, all internal symmetries arise in this way, although the Hilbert–Cartan equation is a significant exception; in the final section we present some preliminary results in this direction. Second, under a certain condition on the systems, which we name the "descent property", we prove that every internal symmetry comes from a *first order* generalized symmetry, a result that significantly ameliorates the computation of these symmetries. The systems covered by this result include all second order systems of differential equations, all normal systems of partial differential equations, and a wide class of higher order underdetermined ordinary differential equations; the principal exceptional cases are the normal systems of ordinary differential equations of order three or more. This Theorem is a significant generalization of Bäcklund's Theorem for internal symmetries of differential equations. Finally, we prove that every first order generalized symmetry which satisfies additional contact conditions is equivalent to an internal symmetry. In certain cases, such as the "codimension 1" ordinary differential equations, of which the Hilbert–Cartan equation is a particular example, there are no contact restrictions, hence there is a one-to-one correspondence between internal symmetries and first order generalized symmetries. This explains the aforementioned calculations for the Hilbert–Cartan equation. More generally, in the case of systems of ordinary differential equations, the contact conditions naturally split into "tangential" and "normal" components. First order generalized symmetries which satisfy the tangential contact conditions give rise to internal symmetries. In the case of systems of partial differential equations, the contact conditions are much more restrictive, and, in many cases, preclude the existence of any "genuine" internal symmetries, meaning ones that do not come from restriction of an external symmetry. In particular, we will prove that every internal symmetry of a normal system of partial differential equations (meaning a system that can be placed into Cauchy–Kovalevskaya form) of order at least two extends to an external symmetry, hence only for first order normal systems of partial differential equations can interesting new internal symmetries arise. Further results based on analysis of the characteristic variety of the system for the existence of non-extendable internal symmetries are discussed, including a few examples. However, the complete analysis of the contact conditions remains a significant open problem.

In order to keep the exposition as brief as possible, we will assume that the reader is reasonably familiar with the standard theory of symmetry groups of differential equations as presented, for instance, in [11]. We will work with local coordinates throughout, although all of these results have analogous, more general, statements for arbitrary fiber bundles over smooth manifolds. Consider a system of differential equations in p independent variables $x = (x^1, \ldots, x^p)$, and q dependent variables $u = (u^1, \ldots, u^q)$

$$\mathcal{R}: \qquad \Delta_\kappa(x, u^{(n)}) = 0, \qquad \kappa = 1, \ldots, r. \tag{1}$$

The derivatives of the dependent variables are denoted by $u^\alpha_J = \partial^J u^\alpha / \partial x^J$, where $J = (j_1, \ldots, j_k)$, $1 \le j_\nu \le p$, is a symmetric multi-index, of order $k = \#J$. We let $u^{(n)}$ denote the collection of all such derivatives of orders $k \le n$, which provide coordinates on the associated jet space J^n. We will assume that the system 1 satisfies the nondegeneracy conditions of being both maximal rank and locally solvable, *cf.* [10; §2.6], and can identify it with the corresponding implicitly defined submanifold $\mathcal{R} \subset \mathrm{J}^n$. (These nondegeneracy conditions are quite mild and are satisfied by virtually every system of differential equations arising in applications.)

In general, by a symmetry of the system of differential equations 1 we mean a transformation which maps solutions to solutions. The most basic type of symmetry is a point transformation, meaning a local diffeomeorphism of the space of independent and dependent variables:

$$\Phi: (x, u) \longmapsto (\bar{x}, \bar{u}).$$

Such transformations act on solutions $u = f(x)$ by pointwise transforming their graphs. Let G denote a local group of point transformations. We will always assume that our transformation group G is connected, thereby consciously omitting discrete symmetry groups, which, while also of great interest for differential equations, are unfortunately not amenable to Lie's techniques. Connectivity implies that it is sufficient to work with the associated infinitesimal generators, which, in the case of groups of point transformations, form a Lie algebra of vector fields of the form

$$\mathbf{v} = \sum_{i=1}^{p} \xi^i(x, u) \frac{\partial}{\partial x^i} + \sum_{\alpha=1}^{q} \varphi^\alpha(x, u) \frac{\partial}{\partial u^\alpha}, \tag{2}$$

on the space of independent and dependent variables. The group transformations in G are recovered from the infinitesimal generators by the usual process of exponentiation.

Since the transformations in G act on functions $u = f(x)$, they also act on their derivatives, and so induce so-called prolonged transformations

$$\mathrm{pr}^{(n)}\,\Phi\colon (x, u^{(n)}) \longmapsto (\bar{x}, \bar{u}^{(n)}),$$

which is defined on an appropriate open subset of J^n. The explicit formula for the prolonged group transformations is very complicated; however the corresponding prolonged infinitesimal generators have a rather simple "prolongation formula". Explicitly, the nth prolongation of the vector field 2, which is the infinitesimal generator of its prolonged action of the associated one-parameter group, is the vector field

$$\mathrm{pr}^{(n)}\,\mathbf{v} = \sum_{i=1}^{p} \xi^i(x,u)\,\frac{\partial}{\partial x^i} + \sum_{\alpha=1}^{q}\ \sum_{\#J=j=0}^{n} \varphi_J^\alpha(x,u^{(j)})\,\frac{\partial}{\partial u_J^\alpha}, \tag{3}$$

on J^n. The coefficients φ_J^α are determined recursively via the well-known formula

$$\varphi_{J,i}^\alpha = D_i\varphi_J^\alpha - \sum_{j=1}^{p} D_i\xi^j\, u_{J,j}^\alpha, \tag{4}$$

where D_i denotes the total derivative with respect to x^i.

Theorem 1 *Assume that the system of partial differential equations 1 is nondegenerate. Then the vector field* $\mathbf{v}$ *in 2 will generate a one-parameter symmetry group of the system 1 if and only if the classical infinitesimal symmetry criterion holds:*

$$\mathrm{pr}^{(n)}\,\mathbf{v}(\Delta_\nu) = 0, \qquad \nu = 1,\ldots,r, \qquad \text{whenever} \qquad \Delta = 0, \tag{5}$$

The "determining equations" 5 form a large over-determined linear system of partial differential equations for the coefficients ξ^i, φ^α of $\mathbf{v}$, and can, in practice, be explicitly solved to determine the complete (connected) symmetry group of the system 1. There are now a wide variety of computer algebra packages available which will automate most of the routine steps in the calculation of the symmetry group of a given system of partial differential equations. See [4] for a good survey of the different packages available as of 1991, and a discussion of their strengths and weaknesses.

The theory of point symmetries of differential equations is classical, and, in more or less the same form, dates back to the original work of Sophus Lie. After this theory is well understood, a number of possible generalizations come to mind. The first direction, originally taken by E. Noether, [10], is to allow *generalized vector fields*

$$\mathbf{v} = \sum_{i=1}^{p} \xi^i(x,u^{(k)})\,\frac{\partial}{\partial x^i} + \sum_{\alpha=1}^{q} \varphi^\alpha(x,u^{(k)})\,\frac{\partial}{\partial u^\alpha}, \tag{6}$$

whose coefficients can also depend on derivatives of u. The condition that $\mathbf{v}$ be a generalized symmetry of the system of differential equations 1 is the same as before, 5, although now one must also take into account the derivatives (prolongations) of the system:

$$\mathrm{pr}^{(k)}\,\mathcal{R}: \qquad D_J\Delta_\kappa(x,u^{(n+k)}) = 0, \qquad \kappa = 1,\ldots,r, \quad \#J \le k, \tag{7}$$

with $D_J = D_{j_1} \cdots D_{j_l}$ denoting the total derivative of order $l = \#J$. Every generalized symmetry is equivalent to one in *evolutionary form*

$$\mathbf{v}_Q = \sum_{\alpha=1}^{q} Q^\alpha(x, u^{(1)}) \frac{\partial}{\partial u^\alpha}, \tag{8}$$

where the q-tuple of functions $Q = (Q^1, \ldots, Q^q)$, known as the *characteristic* of $\mathbf{v}$, has entries

$$Q^\alpha(x, u^{(k)}) = \varphi^\alpha(x, u^{(k)}) - \sum_{i=1}^{p} \xi^i(x, u^{(k)}) \frac{\partial u^\alpha}{\partial x^i}, \qquad \alpha = 1, \ldots, q. \tag{9}$$

Replacing the generalized vector field $\mathbf{v}$ by its evolutionary form $\mathbf{v}_Q$ leads to a simpler set of determining equations in that they only involve the q unknown functions Q^α rather than the $p+q$ unknown coefficients ξ^i, φ^α of $\mathbf{v}$. (This technique even works for point symmetries, where the associated characteristic depends linearly on first order derivatives.) An evolutionary vector field $\mathbf{v}_Q$ is a *trivial symmetry* of the system 1 if the characteristic Q vanishes on all solutions. Two generalized symmetries $\mathbf{v}$ and $\mathbf{w}$ are *equivalent* if their respective evolutionary forms differ by a trivial evolutionary symmetry.

A kth order generalized vector field is will not usually prolong to a well-defined vector field on any jet bundle J^n since its nth prolongation will involve derivatives of orders up to $k+n$. Beyond point transformations, the only exceptions to this are the infinitesimal contact transformations, which correspond to first order generalized symmetries in the case of just one dependent variable. In general, recall that a contact transformation is a map on J^n which preserves the *contact ideal* $\mathcal{I}^{(n)}$. In local coordinates, $\mathcal{I}^{(n)}$ is generated by the basic *contact one-forms*

$$\theta^\alpha_J = du^\alpha_J - u^\alpha_{J,i}\, dx^i, \qquad \alpha = 1, \ldots, q, \quad 0 \le \#J < n. \tag{10}$$

Therefore a (locally defined) transformation $\Psi: \mathrm{J}^n \to \mathrm{J}^n$ on the jet space will determine a contact transformation provided its pull-back Ψ^* maps every contact form to a linear combination of contact forms, which means that it preserves the contact ideal:

$$\Psi^*(\mathcal{I}^{(n)}) \subset \mathcal{I}^{(n)}. \tag{11}$$

A contact transformation acts on a function $u = f(x)$ by pointwise transforming the graph of its n-jet or prolongation $u^{(n)} = \mathrm{pr}^{(n)} f(x)$; the contact condition 11 ensures that the transformed graph is (locally) the n-jet of some function. The infinitesimal version of this criterion is that a vector field

$$\mathbf{X} = \sum_{i=1}^{p} \xi^i(x, u^{(n)}) \frac{\partial}{\partial x^i} + \sum_{\alpha=1}^{q} \sum_{\#J=0}^{n} \varphi^\alpha_J(x, u^{(n)}) \frac{\partial}{\partial u^\alpha_J}, \tag{12}$$

on J^n generates a one-parameter group of contact transformations provided the Lie derivative of any contact form is contained in the contact ideal, *i.e.*, for each α, K,

$$\mathbf{X}[\theta^\alpha_K] = \sum_{\beta=1}^{q} \sum_{\#J=0}^{\#K} \mu^{\alpha,J}_{K,\beta}\, \theta^\beta_J, \qquad \alpha = 1, \ldots, q, \quad \#K < n, \tag{13}$$

for some functions $\mu^{\alpha,J}_{K,\beta}\colon \mathrm{J}^n \to \mathbb{R}$. These conditions are quite restrictive, as Bäcklund's Theorem, *cf.* [6], shows.

Theorem 2 *If the number of dependent variables is more than one, $q > 1$, then every contact transformation on J^n is the nth prolongation of a point transformation. If there is a single dependent variable, $q = 1$, then every contact transformation on J^n is the $(n-1)^{\text{st}}$ prolongation of a first order contact transformation on J^1.*

The projection

$$\pi(\mathbf{X}) = \sum_{i=1}^{p} \xi^i(x, u^{(1)}) \frac{\partial}{\partial x^i} + \sum_{\alpha=1}^{q} \varphi^\alpha(x, u^{(1)}) \frac{\partial}{\partial u^\alpha}, \tag{14}$$

of any contact vector field gives a first order generalized vector field, or, if $q > 1$, of a point vector field, as in 2. Conversely, the contact conditions 13 imply that $\mathbf{X}$ will coincide with the nth prolongation of its projection $\pi(\mathbf{X})$. The next lemma is utilized to provide a characterization of which generalized vector fields produce contact transformations. As such, it plays a key role in the standard infinitesimal proof of Bäcklund's 2, [6].

Lemma 3 *An evolutionary vector field $\mathbf{v}_Q$ is equivalent to an infinitesimal contact transformation if and only if its characteristic $Q(x, u^{(1)})$ depends on at most first order derivatives, and there exist functions $\xi^i(x, u^{(1)})$, $i = 1, \ldots, p$, such that the following* contact conditions *hold:*

$$\frac{\partial Q^\alpha}{\partial u^\beta_i} + \xi^i \delta^\alpha_\beta = 0, \qquad \alpha, \beta = 1, \ldots, q, \quad j = 1, \ldots, p. \tag{15}$$

Indeed, in this case, the ξ^i's will be the coefficients of the $\partial/\partial x^i$ in the generator and the coefficients of the $\partial/\partial u^\alpha$ will be defined by

$$\varphi^\alpha = Q^\alpha + \sum_{i=1}^{p} \xi^i u^\alpha_i, \qquad \alpha = 1, \ldots, q. \tag{16}$$

The contact vector field $\mathbf{X}$ is then just the nth prolongation of

$$\mathbf{v} = \sum_{i=1}^{p} \xi^i(x, u^{(1)}) \frac{\partial}{\partial x^i} + \sum_{\alpha=1}^{q} \varphi^\alpha(x, u^{(1)}) \frac{\partial}{\partial u^\alpha}, \tag{17}$$

cf. 9. Note that left hand sides of the contact conditions 15 appear in the prolongation formula as the coefficients of the terms in $\mathrm{pr}^{(n)}\,\mathbf{v}$ which depend on derivatives of order $n+1$, hence their vanishing is a necessary and sufficient condition that the prolongation $\mathrm{pr}^{(n)}\,\mathbf{v}$ of the first order generalized vector field 17 define a genuine vector field on J^n.

In the case of one dependent variable, $q = 1$, there are no Greek indices in the contact conditions 15, and so these equations serve to define the coefficients ξ^i. Thus, any first order generalized symmetry will give rise to a contact transformation.

Indeed, the characteristic $Q(x, u^{(1)})$ can be identified with the negative of Lie's characteristic function (hence the name), which is the Hamiltonian generating the one-parameter group of contact transformations. For more than one dependent variable, $q > 1$, the integrability conditions for the system of partial differential equations 15 will require that ξ^i, φ^α depend only on x, u, and so every contact transformation reduces to a point transformation.

We shall call a group of contact transformations which preserves a given system of differential equations an *external symmetry group* as the transformations are defined on open subsets of the the jet space J^n, and can thereby be used to transform arbitrary functions $u = f(x)$. Thus any external symmetry group of a system of differential equations is characterized by two conditions:

- It maps the equation manifold $\mathcal{R}$ to itself.
- It preserves the contact ideal on J^n.

Bäcklund's 2 implies that the second condition is very restrictive and severely limits the possible geometrical symmetries beyond point transformations. However, since we are only really interested in what the symmetry group does to solutions of the system of differential equations, and thus in its restriction to the equation submanifold $\mathcal{R}$, it makes sense to relax the second condition and only require that the group transformations preserve the contact ideal on $\mathcal{R}$, rather than all of J^n. Taking 15 into account, we are naturally led to the definition of an internal symmetry of a system of differential equations.

Definition 4 Let $\mathcal{R} \subset \mathrm{J}^n$ be a system of differential equations. An *internal symmetry* of the system is an invertible transformation $\Psi: \mathcal{R} \to \mathcal{R}$ which maps $\mathcal{R}$ to itself and preserves the restriction of the contact ideal on $\mathcal{R}$:

$$\Psi^*(\mathcal{I}^{(n)} \mid \mathcal{R}) \subset \mathcal{I}^{(n)} \mid \mathcal{R}, \tag{18}$$

where $\mid \mathcal{R}$ denotes the pull-back to the submanifold $\mathcal{R}$, *i.e.*, if $\iota: \mathcal{R} \to \mathrm{J}^n$ is the natural embedding, then $\mathcal{I}^{(n)} \mid \mathcal{R} = \iota^* \mathcal{I}^{(n)}$.

Note that, as is the case with external symmetries, internal symmetries form groups of geometrical transformations, now only well-defined on the equation submanifold, which map solutions of the system to solutions. Clearly any external symmetry restricts to an internal symmetry, but it is not necessarily true that an internal symmetry can be extended off the solution manifold to a genuine contact transformation. Indeed, Bäcklund's Theorem in its original form no longer applies to internal symmetries, and, as we shall see, there are n^{th} order internal symmetries which are not the prolongation of any lower order contact map. However, every internal symmetry can be viewed as a particular type of generalized symmetry, and so internal symmetries are seen to occupy a position intermediate to external and generalized symmetries. They form the widest possible class of symmetries which can be realized as local geometrical transformations on some finite dimensional submanifold of jet space, and which map solutions of the system to solutions.

In the case of continuous groups of internal symmetries, we can again work infinitesimally. Let $\mathbf{X}$ be a vector field on the equation submanifold $\mathcal{R}$, which, in local coordinates, takes the form 12 above, where the coefficients ξ^i, φ_J^α are now only need be defined on $\mathcal{R}$, although we may always assume, without essential loss of generality, that we have extended the vector field off the submanifold, the precise extension not being important. The infinitesimal symmetry condition is that $\mathbf{X}$ is tangent to $\mathcal{R}$, which, in local coordinates, says

$$\mathbf{X}(\Delta_\nu) = 0, \qquad \nu = 1, \ldots, r, \quad \text{whenever} \quad \Delta = 0, \tag{19}$$

in direct analogy with 5. In addition, $\mathbf{X}$ must preserve the contact ideal on $\mathcal{R}$:

$$\mathbf{X}(\mathcal{I}^{(n)} \mid \mathcal{R}) \subset \mathcal{I}^{(n)} \mid \mathcal{R}. \tag{20}$$

Note that the projection $\mathbf{v} = \pi(\mathbf{X})$, *cf.* 14, of any internal symmetry determines an nth order generalized vector field. (The coefficients ξ^i, φ^α are *a priori* only defined on $\mathcal{R}$, but the projections of two different extensions of $\mathbf{X}$ will differ only by a trivial generalized symmetry.) It is not difficult to see that $\mathbf{v}$ is a generalized symmetry of the system whose prolongation agrees with $\mathbf{X}$ when restricted to the system. Now, in general, $\mathbf{X}$ and $\mathbf{v}$ will depend on n^{th} order derivatives of the u's. The crucial new result of our work is that, under certain conditions on the system of differential equations, the characteristic Q of any internal symmetry $\mathbf{X}$ depends on at most *first* order derivatives! The technical condition is the following:

Definition 5 Let $n \geq 2$. An nth order system of differential equations $\mathcal{R}$ is said to have the *descent property* if the only (smooth) functions $Q(x, u^{(n-1)})$ of order $n-1$ *all* of whose total derivatives , $i = 1, \ldots, p$, restricted to the system, have order $n-1$, are functions $Q(x, u^{(n-2)})$ of order $n-2$.

In other words, if the system $\mathcal{R}$ has the descent property, and $Q(x, u^{(n-1)})$ is such that its total derivatives $D_i Q \mid \mathcal{R}$ does not depend on derivatives of order n, then $Q = Q(x, u^{(n-2)})$ cannot depend on derivatives of order $n-1$. First order systems are said to have the descent property without any restrictions. The following systems can be shown to have the descent property:

- Any open subset of J^n.
- Any second order system of differential equations.
- Any normal system of partial differential equations in $p > 1$ independent variables.

The main source of examples of systems which do not have the descent property are higher order systems of ordinary differential equations, and severely overdetermined system of partial differential equations. The established inter-connections among internal, external and generalized symmetries of differential equation can now be summarized in the following fundamental theorem.

Theorem 6 *Let $\mathcal{R}$ be a nondegenerate system of differential equations having the descent property.*

- *Every external symmetry restricts to an internal symmetry.*
- *Every internal symmetry is equivalent to a first order generalized symmetry in evolutionary form.*
- *Conversely, every first order generalized symmetry which satisfies certain contact conditions (equations 25 below) is equivalent to an internal symmetry.*

In local coordinates, the contact conditions 20 take the form

$$\mathbf{X}[\theta_K^\alpha] = \sum_{\beta,J} \mu_{K,\beta}^{\alpha,J}\,\theta_J^\alpha + \sum_\kappa \lambda_K^{\alpha,\kappa}\, d\Delta_\kappa, \qquad \text{on} \quad \mathcal{R}, \tag{21}$$

in analogy with 13. Here the $\mu_{K,\beta}^{\alpha,J}$, $\lambda_K^{\alpha,\kappa}$ are functions on $\mathcal{R}$, and by the phrase "on $\mathcal{R}$" we mean that the individual coefficients of the basis one-forms dx^i, du_K^α must agree when restricted to the submanifold $\mathcal{R}$. (The fact that the pull-backs of these one-forms to $\mathcal{R}$ are no longer linearly independent has been taken care of by the $\lambda_K^{\alpha,\kappa}$, which play the role of Lagrange multipliers.) Detailed analysis of 21 shows that the coefficients of $\mathbf{X}$ must satisfy the prolongation formula 4 modulo the system and its derivatives:

$$\varphi_{J,i}^\alpha = D_i\varphi_J^\alpha - \sum_{j=1}^p D_i\xi^j\, u_{J,j}^\alpha \qquad \text{on} \quad \mathrm{pr}^{(n)}\mathcal{R}. \tag{22}$$

Moreover, there is an additional set of contact conditions analogous to 15, arising from the fact that, on the equation submanifold, the nth order terms arising from the restricted prolongation formula 22 cannot depend on $(n+1)^{\text{st}}$ order derivatives. In order to write these in a reasonably compact form, we introduce some additional auxiliary variables $\zeta = (\zeta_1, \ldots, \zeta_p)$, and two important matrices, which depend both on the point $(x, u^{(n)}) \in \mathcal{R}$, and, as homogeneous polynomials, on the auxiliary variables ζ. First, the $r \times q$ matrix of homogeneous polynomials of degree n in ζ given by

$$\mathbf{D}(\zeta) = \left(\sum_{\#K=n} \frac{\partial \Delta_\kappa}{\partial u_K^\beta}\,\zeta_K \right), \tag{23}$$

where, for a symmetric multi-index $K = (k_1, \ldots, k_n)$, we set $\zeta_K = \zeta_{k_1}\zeta_{k_2}\cdots\zeta_{k_n}$, plays a key role in the definition of the classical characteristic directions (not to be confused with the characteristic Q!) for the system of partial differential equations 1. For example, assume that $r = q$, so we have the same number of equations as unknowns. A complex direction ζ, which should be thought of as defining coordinates in the complexified cotangent bundle $T^*_{\mathbb{C}}X = T^*X \otimes \mathbb{C}$ of the independent variable space X, determines a characteristic direction if and only if $\det \mathbf{D}(\zeta) = 0$. A system is called *normal* if not every direction is characteristic, *i.e.*, $\det \mathbf{D}(\zeta) \not\equiv 0$. This is equivalent to the existence of local coordinates in which the system assumes a form amenable to the application of the Cauchy-Kovalevskaya existence theorem, *cf.* [11], Theorem 2.79. Second, given the characteristic $Q(x, u^{(1)})$ of a first order generalized symmetry, define the $q \times q$ matrix of homogeneous linear polynomials

$$\mathbf{R}(\zeta) = \left(\sum_{j=1}^p \frac{\partial Q^\alpha}{\partial u_j^\beta}\,\zeta_j \right), \tag{24}$$

of ζ, where $u_j^\beta = \partial u^\beta / \partial x^j$. The relevant contact conditions can then be cast into the general form:

Theorem 7 *Let $\mathbf{v}_Q$ be a first order generalized symmetry of a nondegenerate system of partial differential equations $\mathcal{R}$. Then there is an internal symmetry $\mathbf{X}$ with evolutionary form $\mathbf{v}_Q$ if and only if there exist functions $\xi^1(x, u^{(n)}), \ldots, \xi^p(x, u^{(n)})$ defined on $\mathcal{R}$, such that, for* every *homogeneous scalar polynomial $P(\zeta)$ of degree n, there exists an $q \times r$ matrix of linear polynomials $\mathbf{L}_P(\zeta)$ (which can depend on both the polynomial P and the point $(x, u^{(n)}) \in \mathcal{R}$) satisfying the* internal contact conditions

$$P(\zeta)\mathbf{R}(\zeta) + (\xi \cdot \zeta)\,\mathbf{I} = \mathbf{L}_P(\zeta) \cdot \mathbf{D}(\zeta), \qquad \text{on} \quad \mathcal{R}. \tag{25}$$

In this matrix equation, $\mathbf{I}$ denotes the $q \times q$ identity matrix, and $\xi \cdot \zeta = \sum_{i=1}^{p} \xi^i \zeta_i$.

If the internal contact conditions 25 are satisfied, then the internal symmetry $\mathbf{X} = \mathrm{pr}^{(n)}\,\mathbf{v}$ associated with the evolutionary symmetry $\mathbf{v}_Q$ is the nth prolongation of the equivalent generalized vector field 6 whose coefficients ξ^i, φ^α are related to Q via 16. The internal contact conditions 25 guarantee that, on the equation submanifold $\mathcal{R}$, the nth prolongation of $\mathbf{v}$ does not depend on $(n+1)^{\text{st}}$ order derivatives, and so defines a genuine internal symmetry; see the remarks following 3. Also note that, even though $\mathbf{v}_Q$ is a first order generalized vector field, the equivalent generalized vector field $\mathbf{v}$ can have order n since the functions ξ^i which satisfy 25 may depend on higher order derivatives.

In order to understand what these conditions mean more concretely, we discuss some particular examples. First consider the extreme case in which there are no differential equations, *i.e.*, the equation submanifold $\mathcal{R}$ is an open subset of J^n. In this degenerate case, the right hand side of the contact conditions 25 is automatically zero, and so the polynomial $P(\zeta)$ can be ignored. The resulting condition

$$\mathbf{R}(\zeta) + (\xi \cdot \zeta)\mathbf{I} = 0, \tag{26}$$

are easily seen to be the same as the contact conditions 15 for ordinary contact transformations — an "internal symmetry of J^n" just means an ordinary contact transformation. Thus, in this case, 6 reduces to the classical Theorem of Bäcklund that every contact transformation comes from a first order contact transformation, and we are justified in labelling 6 as a generalization of Bäcklund's 2 to systems of differential equations.

In many cases, the contact conditions 25 will be so restrictive as to automatically imply that the left hand side must vanish. Indeed, the $q \times q$ matrix

$$\mathbf{M}(\zeta) = \mathbf{R}(\zeta) + (\xi \cdot \zeta)\,\mathbf{I}, \tag{27}$$

of linear functions of ζ measures, in a sense, the "degree of internalness" of the symmetry $\mathbf{v}_Q$. More specifically, according to 3, an internal symmetry will extend to an external (contact) symmetry if and only if the corresponding matrix $\mathbf{M}(\zeta)$ is identically zero for some choice of functions ξ^i. Internal symmetries which do not extend to external symmetries, *i.e.*, ones for which $\mathbf{M}(\zeta) \not\equiv 0$ for all ξ, will be called *non-extendable*, and these are, in a sense, the only "true" internal symmetries. Many systems of partial differential equations do not have any non-extendable internal symmetries, and the contact conditions 25 are an effective means of detecting this. For instance, this is the case for a normal system of partial differential equations of order at least 2.

Theorem 8 *If $\mathcal{R}$ is a normal system of partial differential equations in $p > 1$ independent variables, of order $n \geq 2$, then every first order internal symmetry extends to an external symmetry.*

For an nth order system of ordinary differential equations,

$$\Delta_\kappa(x, u^{(n)}) = 0, \qquad \kappa = 1, \ldots, r, \tag{28}$$

the general internal contact conditions 25 dramatically simplify. First, there is just a single parameter $\zeta = \zeta_1$. Moreover, $P(\zeta)$ is a multiple of ζ_1^n, so ζ_1 can be eliminated entirely. Let

$$\mathbf{D} = \left(\frac{\partial \Delta_\kappa}{\partial u_n^\beta}\right), \qquad \mathbf{R} = \left(\frac{\partial Q^\alpha}{\partial u_x^\beta}\right), \tag{29}$$

where $u_n^\beta = d^n u^\beta / dx^n$, be, respectively, $r \times q$ and $q \times q$ matrices depending on $(x, u^{(n)})$ and $(x, u^{(1)})$. The internal contact conditions 25 reduce to

$$\mathbf{R} + \xi \mathbf{I} = \mathbf{L} \cdot \mathbf{D}, \qquad \text{on } \mathcal{R}, \tag{30}$$

for some $q \times r$ matrix $\mathbf{L}$, which, in components, is

$$\frac{\partial Q^\alpha}{\partial u_x^\beta} + \xi\, \delta_\beta^\alpha = \sum_{\kappa=1}^{r} \lambda_\kappa^\alpha \frac{\partial \Delta_\kappa}{\partial u_n^\beta}, \qquad \alpha, \beta = 1, \ldots, q, \qquad \text{on } \mathcal{R}, \tag{31}$$

for some unspecified functions ξ, λ_κ^α. As before, the internal symmetry associated with $\mathbf{v}_Q$ is given by $\mathbf{X} = \mathrm{pr}^{(n)}\, \mathbf{v}$, where coefficients of $\mathbf{v}$ are related to Q via 16.

It is not always necessary to verify all of the contact conditions 31, as some of them are direct consequences of the symmetry conditions. Assume that the $q \times r$ Jacobian matrix $\mathbf{D}$ has maximal rank r. (In particular, we assume that the system is not over-determined, *i.e.*, $r \leq q$.) Such a system will be said to be of *codimension* $c = q - r$. The implicit function theorem assures us that we can locally solve for r of the top order derivatives, say $u_n^1, \ldots, u_n^r$. This results in a system of ordinary differential equations of the form

$$u_n^\kappa = \Gamma^\kappa(x, u^{(n-1)}, u_n^{r+1}, \ldots, u_n^q), \qquad \kappa = 1, \ldots, r. \tag{32}$$

With this choice, we will refer to the variables $u^1, \ldots, u^r$, as *normal* directions, and the variables $u^{r+1}, \ldots, u^q$, as *tangential* directions. Although a normal system of ordinary differential equations of order $n \geq 3$ does not satisfy the descent property, most of the underdetermined systems 32 do.

Proposition 9 *If the $\frac{1}{2}(q - r)(q - r + 1) \times r$* tangential Hessian matrix *with entries $\partial^2 \Gamma^\kappa / \partial u_n^\lambda \partial u_n^\mu$, with rows indexed by $\lambda, \mu = r + 1, \ldots, q$ and columns indexed by $\kappa = 1, \ldots, r$, has (maximal) rank r, then the system 32 has the descent property. For such systems, a key simplification is that we only need verify the internal contact conditions in the tangential directions.*

Theorem 10 *Let $\mathcal{R}$ be an nth order system of ordinary differential equations 28 whose Jacobian matrix $\mathbf{D}$, cf. 29, has maximal rank r. Let $\mathbf{v}_Q$ be a first order* *generalized symmetry of* $\mathcal{R}$. *Then there is an internal symmetry* $\mathbf{X}$ *with evolutionary form* $\mathbf{v}_Q$ *if and only if there exist functions* $\xi(x,u^{(n)})$, $\lambda^\alpha_\kappa(x,u^{(n)})$, $\kappa=1,\ldots,r$, $\alpha=r+1,\ldots,q$, *defined on* $\mathcal{R}$, *satisfying the* tangential contact conditions

$$\frac{\partial Q^\alpha}{\partial u^\beta_x}+\xi\delta^\alpha_\beta=\sum_{\kappa=1}^{r}\lambda^\alpha_\kappa\,\frac{\partial\Delta_\kappa}{\partial u^\beta_n},\qquad \beta=1,\ldots,q,\quad \alpha=r+1,\ldots,r,\qquad \text{on}\quad \mathcal{R}. \tag{33}$$

Indeed, the remaining normal contact conditions, *i.e.*, 31 for $\alpha=1,\ldots,r$, $\beta=1,\ldots,q$, are found to be direct consequences of the tangential contact conditions and the symmetry conditions. In particular, if the system satisfies the hypotheses of 9, then 10 provides a one-to-one correspondence between internal symmetries and first order generalized symmetries which satisfy the contact conditions in the tangential directions.

Let us look at a few special cases of interest. First consider a normal system of ordinary differential equations, which is one of the standard form

$$u^\alpha_n=F^\alpha(x,u^{(n-1)}),\qquad \alpha=1,\ldots,q, \tag{34}$$

in which there are the same number of equations as unknowns, and we have solved for the top order derivatives. Here $r=q$, the codimension is 0, and there are no tangential directions. Therefore *every* first order generalized symmetry of a normal system of ordinary differential equations determines an internal symmetry. Indeed, the internal contact conditions (31) form a system of q^2 equations, with q^2+1 undetermined functions $\xi,\lambda^\alpha_\kappa$, $\alpha,\kappa=1,\ldots,q$. Therefore, for any given function ξ, we can determine q^2 functions λ^α_κ so as to satisfy the contact conditions. This implies that the correspondence between internal symmetries and generalized symmetries is not one-to-one for normal system of ordinary differential equations. Indeed, any multiple $\xi(x,u^{(n-1)})\mathbf{d}_x$ of the vector field

$$\mathbf{d}_x=\frac{\partial}{\partial x}+\sum_{\alpha=1}^{q}\sum_{j=1}^{n-2}u^\alpha_{j+1}\frac{\partial}{\partial u^\alpha_j}+\sum_{\alpha=1}^{q}F^\alpha(x,u^{(n-1)})\,\frac{\partial}{\partial u^\alpha_{n-1}}+\sum_{\alpha=1}^{q}\widehat{F}^\alpha(x,u^{(n-1)})\,\frac{\partial}{\partial u^\alpha_n}$$

is trivially an internal symmetry. The vector field $\mathbf{d}_x$ is just the restriction of the total derivative D_x to the equation submanifold, and $\widehat{F}^\alpha=\mathbf{d}_xF^\alpha$ is the function which agrees with the derivative u^α_{n+1} of a solution to the system. Geometrically, a trivial internal symmetry is just the (reparametrized) flow along the solution curves of the system. Each first order generalized symmetry determines an infinite number of internal symmetries, each of which differ by such a trivial internal symmetry.

Next consider a system of codimension 1. In this case, we have $r=q-1$ and $\mathbf{D}$ has rank $q-1$. There is just one tangential direction, say u^q, and so the tangential contact conditions 33 for $\alpha=q$ form a system of q equations with precisely q undetermined functions ξ,λ^q_κ, $\kappa=1,\ldots,q-1$. Therefore, for each first order generalized symmetry, we can uniquely determine the functions ξ,λ^q_κ, $\kappa=1,\ldots,q-1$, so as to satisfy the tangential contact conditions; the remaining normal contact conditions will then follow automatically from the symmetry conditions. Thus for codimension 1 systems, there is a one-to-one correspondence between first order generalized symmetries and internal symmetries.

The most important example of such a system is the under-determined ordinary differential equation

$$v_x = u_{xx}^2. \tag{35}$$

Equation 35 was introduced by Hilbert, [5], as an example of an equation whose general solution could not be expressed in terms of an arbitrary function and a finite number of its derivatives. Subsequently, Cartan, [2], [3], proved that this equation has the real non-compact form of the 14 dimensional exceptional Lie group G_2 as an internal symmetry group. Cartan's result can be verified directly using the following result.

Theorem 11 *Every first order generalized symmetry of the Hilbert–Cartan equation is a linear constant coefficient combination of the following fourteen generalized vector fields*

$$\begin{aligned}
\mathbf{v}_1 &= \partial_u, \\
\mathbf{v}_2 &= \partial_v, \\
\mathbf{v}_3 &= x\partial_u, \\
\mathbf{v}_4 &= x^2\partial_u + 4u_x\partial_v, \\
\mathbf{v}_5 &= x^3\partial_u + 12(xu_x - u)\partial_v, \\
\mathbf{v}_6 &= u\partial_u + 2v\partial_v, \\
\mathbf{v}_7 &= 3v\partial_u + 4v_x^{3/2}\partial_v, \\
\mathbf{v}_8 &= u_x\partial_u + v_x\partial_v, \\
\mathbf{v}_9 &= (2xu_x - 3u)\partial_u + 2xv_x\partial_v, \\
\mathbf{v}_{10} &= (x^2u_x - 3xu)\partial_u + (x^2v_x - 4u_x^2)\partial_v, \\
\mathbf{v}_{11} &= (3xv - 4u_x^2)\partial_u + (4xv_x^{3/2} - 8u_xv_x)\partial_v, \\
\mathbf{v}_{12} &= (3x^2v - 8xu_x^2 + 12uu_x)\partial_u + (12u_xv + 12uv_x + 4x^2v_x^{3/2} - 16xu_xv_x)\partial_v, \\
\mathbf{v}_{13} &= (3x^3v - 12x^2u_x^2 + 36xuu_x - 36u^2)\partial_u \; + \\
&\qquad +(36xu_xv - 36uv + 4x^3v_x^{3/2} - 24x^2u_xv_x + 36xuv_x - 16u_x^3)\partial_v, \\
\mathbf{v}_{14} &= (9uv - 4u_x^3)\partial_u + (9v^2 - 12u_x^2v_x + 12uv_x^{3/2})\partial_v.
\end{aligned} \tag{36}$$

According to 16, 31, any first order generalized symmetry

$$\mathbf{v}_Q = Q(x,u,u_x,v,v_x)\partial_u + R(x,u,u_x,v,v_x)\partial_v \tag{37}$$

of a codimension one system of the special form

$$u_{xx} = F(x,u,u_x,v,v_x,v_{xx}), \tag{38}$$

is equivalent to the internal symmetry $\mathbf{X} = \mathrm{pr}^{(2)}\,\mathbf{v} \mid \mathcal{R}$, where

$$\mathbf{v} = \xi\partial_x + \varphi\partial_u + \psi\partial_v, \quad \xi = -R_{u_x}F_{v_{xx}} - R_{v_x}, \quad \varphi = Q + \xi u_x, \quad \psi = R + \xi v_x. \tag{39}$$

Thus, for example, the internal symmetry equivalent to $\mathbf{v}_7$ is given by

$$\mathbf{X}_7 = -6v_x^{1/2}\partial_x + (3v - 6u_xv_x^{1/2})\partial_u - 2v_x^{3/2}\partial_v - 3v_x\partial_{u_x} + 3v_x^{-1/2}v_{xx}^2\partial_{v_{xx}}.$$

Note that, according to 39, the six symmetries $\mathbf{v}_1$, $\mathbf{v}_2$, $\mathbf{v}_3$, $\mathbf{v}_6$, $\mathbf{v}_8$, $\mathbf{v}_9$ are found to be equivalent to point symmetries, while the remaining eight are true internal symmetries. Since each of the vector fields in 11 corresponds to a unique internal symmetry, we deduce that these vector fields close to form a Lie algebra when restricted to the equation. Using standard Lie-algebraic techniques (Killing form, Cartan subalgebra, root diagrams, etc.), it can be proven that this Lie algebra is isomorphic to the non-compact real form of Lie algebra for the exceptional simple Lie group G_2. Therefore, we obtain Cartan's explicit realization of G_2 as the group of internal symmetry transformations of the six dimensional manifold defined by the Hilbert–Cartan equation.

Interestingly, there are additional higher order generalized symmetries of the Hilbert–Cartan equation. An explicit example is the third order symmetry

$$\mathbf{v} = u_{xxx}\partial_u + (2u_{xx}u_{xxxx} - u_{xxx}^2)\partial_v.$$

The full structure of the generalized symmetries of the Hilbert–Cartan equation and various generalizations has been determined by P. Kersten, [7], [8].

For underdetermined systems of ordinary differential equations having higher codimension, the tangential contact conditions impose additional constraints for a first order generalized symmetry to give an internal symmetry. For a system of codimension c, the tangential contact conditions 33 are a system of qc equations involving the $c(q-c)+1$ undetermined functions ξ, λ_κ^α, $\alpha = 1, \ldots, q-c$, $\kappa = 1, \ldots, c$. Therefore there will be $qc - c(q-c) - 1 = c^2 - 1$ additional equations a first order generalized symmetry must satisfy in order that it correspond to an internal symmetry. For instance, any first order generalized symmetry of a codimension 2 system must satisfy 3 additional constraints for it to be an internal symmetry. For example, the equation $v_x u_{xx} = w$ has the first order generalized symmetry $\mathbf{v} = x^2\partial_u + 2v_x\partial_w$, but there is no internal counterpart, since it does not satisfy the tangential contact conditions 33.

References

1. Anderson, I.M., Kamran, N., and and Olver, P.J., Internal, external and generalized symmetries, *Adv. in Math.*, to appear.
2. Cartan, É., Sur l'équivalence absolue de certains systèmes d'équations différentielles et sur certaines familles de courbes, *in: Oeuvres Complétes*, part. II, vol. 2, Gauthiers–Villars, Paris, 1953, pp. 1133–1168 .
3. Cartan, É., Sur l'intégration de certains systèmes indéterminés d'équations différentielles, *in: Oeuvres Complétes*, part. II, vol. 2, Gauthiers–Villars, Paris, 1953, pp. 1169–1174 .
4. Champagne, B., Hereman, W., and Winternitz, P., The computer calculation of Lie point symmetries of large systems of differential equations, *Comp. Phys. Comm.* **66** (1991), 319–340 .
5. Hilbert, D., Über den Begriff der Klasse von Differentialgleichungen, *in: Gesammelte Abhandlungen*, vol. 3, Springer–Verlag, Berlin, 1935, pp. 81–93 .
6. Ibragimov, N.H., *Transformation Groups Applied to Mathematical Physics*, D. Reidel, Boston, 1985 .
7. Kersten, P.H.M., The general symmetry algebra structure of the underdetermined equation $u_x = (v_{xx})^2$, *J. Math. Phys.* **32** (1991), 2043–2050 .
8. Kersten, P.H.M., The Lie-Bäcklund algebra structure for the general underdetermined equation $u_r = F(x, u, \ldots, u_{r-1}, v, \ldots, v_k)$,, *Nonlinearity* **5** (1992), 763–770
9. Krasil'shchik, I.S., Lychagin, V.V., and Vinogradov, A.M., *Geometry of Jet Spaces and Nonlinear Partial Differential Equations*, Gordon and Breach, New York, 1986 ..

10. Noether, E., Invariante Variationsprobleme, *Nachr. Konig. Gesell. Wissen. Göttingen, Math.-Phys. Kl.* (1918), 235–257 .
11. Olver, P.J., *Applications of Lie Groups to Differential Equations*, Graduate Texts in Mathematics, vol. 107, Springer–Verlag, New York, 1986 .

EXAMPLES OF COMPLETELY INTEGRABLE BATEMAN PAIRS

ROBERT L. ANDERSON
Department of Physics and Astronomy, UGA,
Athens, GA, 30602

PIOTR W. HEBDA
Math and Science Division, Reinhardt College,
Waleska, GA, 30183

and

GUY RIDEAU
Laboratoire de Physique Théorique et Mathématique
Université Paris VII,
75251 Paris Cedex 05

Abstract. Recently, a Hamiltonian formalism was presented which treats the singular nature of Bateman Lagrangians describing quasi-linear integro-differential equations. This is in exchange for the pairing of a given system of equations with another system via Bateman's Lagrangian prescription. One possible important consequence of this approach in the case of a system of ordinary differential equations is that a particular Bateman pair may form a completely integrable Hamiltonian system even though the original system is not one. It is the main purpose of this paper to exhibit concrete detailed examples of such systems.

1. Introduction

Recently, we have extended Hamiltonian techniques to include all systems described by systems of quasi-linear integro-differential equations [1]. This is in exchange for pairing the system of equations describing a given system with another system via Bateman's Lagrangian prescription [2] (see §2). This pair of systems of equations we call a Bateman pair. The Bateman Lagrangian prescription is singular, intrinsically so for equations first-order in time derivatives, hence the need for a formalism to deal with this fact. In §2, the Bateman-Hamiltonian formalism reported in Ref. [1] is described and illustrated with a Bateman pair that includes the Lorentz system [3].

One possible important consequence of this approach in the case of systems described by ordinary differential equations is that a particular Bateman pair may form a completely integrable Hamiltonian system even though the original system is not one. It is the main purpose of this paper to exhibit concrete detailed examples of such systems. Complete integrability is reviewed in §2.

One class of examples is provided by Bateman pairs which consist of a nonlinear oscillator governed by a restoring force described by an arbitrary odd-degree polynomial in its position whose gradient is nonnegative along with the set of all linear oscillators determined by setting their spring constants equal to this gradient evaluated on all solutions of this nonlinear oscillator. One linear oscillator for

N. H. Ibragimov et al. (eds.),
Modern Group Analysis: Advanced Analytical and Computational Methods in Mathematical Physics, 23–33.

each solution of the nonlinear oscillator. In §3 this Bateman pair is shown to be a completely integrable Hamiltonian system.

Another example is the linear Bateman-Morse-Feshbach pair [2], [4] which includes the damped oscillator. In §4, we double the dimensions to obtain more structure and a Bateman pair which includes a two-dimensional isotropic damped oscillator. The results reported here include explicit angle-action type variables and two quad-Hamiltonian structures. These results make extensive use of our prior treatment of the Bateman-Morse-Feshbach pair which includes the one-dimensional damped oscillator (see [5]) and reduce to our prior results when restricted to an appropriate four-dimensional subspace.

2. Bateman-Hamiltonian Formalism — Preliminaries

We begin with a description of the most general context for the Bateman-Hamiltonian formalism presented in reference [1]. Let $x = (x_1, \ldots, x_s) \in \mathbf{R}^s$, $s \geq 0$, $t \in \mathbf{R}$, $u(x,t) = (u_1(x,t), \ldots, u_m(x,t)) \in \mathbf{R}^m$, $m \geq 1$, $\partial_k u_\alpha = \partial u_\alpha/\partial x_k$, $\partial_1^{r_1} \ldots \partial_s^{r_s} u_\alpha = \partial^r u_\alpha$, $u_\alpha^{(n)} = \partial^n u_\alpha/\partial t^n$, and $\dot{w} = \partial w/\partial t$. Consider the following class of systems of quasi-linear (linear in the highest derivative with respect to time $u_\alpha^{(n_\alpha)}$, $n_\alpha \geq 1$), in general, nonautonomous system of nonlinear intro-differential (partial) equations defined on m functions in $s+1$ "space-time"

$$u_\alpha^{(n_\alpha)} + \Lambda_\alpha(t, x, \ldots, \partial^r u_\beta^{(j)}, \ldots; 0 \leq j \leq n_\beta - 1, \beta = 1, \ldots, m, r \in \mathbf{Z}_+^s) = 0, \quad (2.1)$$

where Λ_α is, in general, a nonlinear space-time dependent space-functional of the variables $\partial^r u_\beta^{(j)}$. In our applications we restrict ourselves to the case where the multi-indices r as well as the n_α take their values in finite sets.

Following Bateman [2], we define the Lagrangian density, up to an overall minus sign, by

$$L_B = \sum_{\alpha=1}^{m} u_\alpha^*(u_\alpha^{(n_\alpha)} + \Lambda_\alpha). \quad (2.2)$$

Then the Euler-Lagrange equation $\delta L_B/\delta u_\alpha^* = 0$ yields eq.(2.1) and $\delta L_B/\delta u_\alpha = 0$ yields

$$(-1)^{n_\alpha} u_\alpha^{*(n_\alpha)} + \delta\Big(\sum_{\beta=1}^{m} u_\beta^* \Lambda_\beta\Big) \Big/ \delta u_\alpha = 0, \quad (2.3)$$

where the variational derivative $\delta/\delta w$ is given in our notation by

$$(\delta/\delta w)A = \sum_{n=0} (-1)^n (\partial^n/\partial t^n)(\delta/\delta w^{(n)})A, \quad (2.4)$$

where $\delta/\delta w^{(n)}$ indicates variation of only $w^{(n)}$ and its spatial derivatives and $\partial/\partial t$ is total partial differentiation with respect to t. Eqs.(2.1) and (2.3) constitute what we call a Bateman pair. The following proposition characterizes the Bateman-Hamiltonian formalism (This is Proposition 1 in reference [1]).

Proposition. [1] With the identification for $\alpha = 1, \ldots, m$, $j = 1, \ldots, n_\alpha$

$$q_{\alpha,j} = u_\alpha^{(j-1)}, \qquad p_{\alpha,j} = \sum_{k=j} (-1)^{k-j} (\partial^{k-j}/\partial t^{k-j})(\delta L_B/\delta u_\alpha^{(k)}), \tag{2.5}$$

the Euler-Lagrange equations $\delta L_B/\delta u_\alpha^* = 0$ (eq.(2.1)) and $\delta L_B/\delta u_\alpha = 0$ (eq.(2.3)) for the Bateman Lagrangian density (2.2) are equivalent to Hamilton's equations

$$\dot{q}_{\alpha,i} = \delta H/\delta p_{\alpha,i}, \qquad \dot{p}_{\alpha,i} = -\delta H/\delta q_{\alpha,i}, \tag{2.6}$$

where the Hamiltonian density H is given by

$$\begin{aligned} H = & \sum_{\alpha=1}^{m} \Big(\Big(\sum_{j=1}^{n_\alpha - 1} p_{\alpha,j} q_{\alpha,j+1} \Big) \\ & - p_{\alpha,n_\alpha} \Lambda_\alpha(t, x, \ldots, \partial^r q_{\beta,j}, \ldots; 1 \le j \le n_\beta, \beta = 1, \ldots, m, r \in \mathbf{Z}_+^s) \Big), \end{aligned} \tag{2.7}$$

and the Poisson bracket is given by

$$\{A, B\} = \sum_{\alpha=1}^{m} \sum_{j=1}^{n} ((\delta A/\delta q_{\alpha,j})(\delta B/\delta p_{\alpha,j}) - (\delta A/\delta p_{\alpha,j})(\delta B/\delta q_{\alpha,j})). \tag{2.8}$$

The above Proposition allowed us, for example, to formulate a time-independent Hamiltonian treatment for a Bateman pair which includes the Navier-Stokes equation for viscous, incompressible, external force-free, three-dimensional flow in a fluid of uniform density [1]. This restricted Navier-Stokes equation in momentum space is a first order integro-differential (partial) equation. Another example of a first-order system is given by the following example. The Bateman Lagrangian (2.2) for the Bateman pair which includes the Lorentz system [3] is given explicitly by

$$L_B = u_1^*(\dot{u}_1 + \sigma\, u_1 - \sigma\, u_2) + u_2^*(\dot{u}_2 - r\, u_1 + u_1 u_3) + u_3^*(\dot{u}_3 - b\, u_3 + u_1 u_2). \tag{2.9}$$

The identification (2.5) is given explicitly by

$$q_i = u_i, \quad p_i = u_i^*, \qquad i = 1, 2, 3, \tag{2.10}$$

and the Hamiltonian (2.7) for the pair is given explicitly by

$$H = -p_1(\sigma\, q_1 - \sigma\, q_2) - p_2(-r q_1 + q_2 + q_1 q_3) - p_3(b\, q_3 - q_1 q_2). \tag{2.11}$$

(The form (2.11) for first-order ordinary differential equations was known to P.A. Lagerstrom [private communication].)

In the rest of this paper, we discuss only ordinary differential equations. So for the convenience of the reader, the x-dependence in the Proposition may be suppressed, and the following substitutions may be made: $\delta w^{(k)} \to \partial/\partial w^{(k)}$ in eq.(2.4), $\delta/\delta u^{(k)} \to \partial/\partial u^{(k)}$ in eq.(2.5).

Turning to the characterization of complete integrability, we will use the standard definitions. It is sufficient for the results presented in the subsequent paragraphs, to adapted them for Hamiltonian systems describing flows in a $2n$-dimensional phase space M (actually, $n = 4$ or 8 in our examples). With our conventions, described in more detail in §4, the fundamental Poisson bracket $\{.,.\}$ is given in coordinates $(v_1, \ldots, v_{2n})$ by

$$\begin{aligned}\{A, B\}_{(v_1,\ldots,v_{2n})} = \partial A/\partial v_{n+1}\partial B/\partial v_1 - \partial A/\partial v_1 \partial B/\partial v_{n+1} + \ldots \\ + \partial A/\partial v_{2n}\partial B/\partial v_n - \partial A/\partial v_n \partial B/\partial v_{2n},\end{aligned} \tag{2.12}$$

for arbitrary differentiable functions A and B on M. A set $(I_1, \ldots, I_n, \varphi_1, \ldots \varphi_n)$ is a set of Darboux (canonical) coordinates which linearize the flows on a neighborhood $\mathcal{O}$ of a point $v' \in \mathcal{O} \subset M$, if for every $v \in \mathcal{O}$ there exists a time-independent transformation $(p_1, \ldots, p_n, q_1, \ldots, q_n) \to (I_1, \ldots, I_n, \varphi_1, \ldots \varphi_n)$ such that

$$\{A, B\}_{(p_1,\ldots,p_n,q_1,\ldots,q_n)} = \{A, B\}_{(I_1,\ldots,I_n,\varphi_1,\ldots,\varphi_n)}, \tag{2.13}$$

$$H(p_1, \ldots, p_n, q_1, \ldots, q_n) = H(I_1, \ldots, I_n). \tag{2.14}$$

3. Nonlinear oscillator Bateman pairs

Consider a linear oscillator with position coordinate u^* subject to a linear restoring force governed by a time-dependent spring constant $k(t)$, *i.e.*,

$$\ddot{u}^*(t) + k(t)u^*(t) = 0, \quad k(t) \geq 0 \text{ for all } t \in \mathbf{R}^+, \tag{3.1}$$

where

$$k(t) = V''(y(t)), \tag{3.2}$$

and we require the fixed function $V(\xi)$ to be an even degree polynomial in ξ, $V'(\xi)/\xi \geq 0$, $V''(\xi) \geq 0$ for all $\xi \in \mathbf{R}$, and the fixed function $y(t)$ satisfies

$$\ddot{y}(t) + V'(y(t)) = 0. \tag{3.3}$$

Here we use the notation $' = d/dy$.

Now embed the particular oscillator system (3.1)–(3.3) in the coupled system

$$\ddot{u}^*(t) + V''(u(t))u^*(t) = 0, \tag{3.4}$$

$$\ddot{u}(t) + V'(u(t)) = 0, \tag{3.5}$$

i.e., consider the set of all linear oscillators with their spring constants determined by all solutions of Eq.(3.5).

The advantage of this embedding is that the coupled system Eqs.(3.4)–(3.5) is, in fact, a Bateman pair. The Bateman Lagrangian (2.2) is given explicitly by

$$L_B = u^*(\ddot{u} + V'(u)). \tag{3.6}$$

The identification (2.5) is given explicitly by

$$q_1 = u, \qquad q_2 = \dot{u}, \qquad p_1 = -\dot{u}^*, \qquad p_2 = u^*, \tag{3.7}$$

and the Hamiltonian (2.7) for the pair is given explicitly by

$$H = p_1 q_2 - p_2 . V'(q_1). \tag{3.8}$$

The Hamiltonian (3.8) is time-independent, hence a constant of the motion. There is an obvious second constant of the motion corresponding to the subsystem (3.5) given by

$$h = q_2^2/2 + V(q_1). \tag{3.9}$$

The constants of the motion H and h are obviously functionally independent. What was a pleasant surprise is that they are in involution. In particular, it is easy to verify that

$$\{H, h\}_{(p_1,p_2,q_1,q_2)} = 0, \tag{3.10}$$

and $\{A, B\}_{(p_1,p_2,q_1,q_2)}$ is given by Eq.(2.12). Hence, since our Hamiltonian system is four-dimensional, it is completely integrable. Thus the Bateman pair (3.4)–(3.5) corresponds to a completely integrable Hamiltonian system.

4. Eight-Dimensional Bateman-Morse-Feshbach Pair

The Bateman [2] and Morse and Feshbach [4] Lagrangian, extended to include a two-dimensional isotropic damped linear oscillator is given by

$$\begin{aligned} L(\dot q, \dot q^*, q, q^*) &= \\ &= -\sum [q_i^*(m\ddot q_i + R\dot q_i + kq_i) - d(mq_i^*\dot q_i + Rq_i^* q_i)/dt] \\ &= \sum [m\dot q_i \dot q_i^* + Rq_i \dot q_i^*/2 - R\dot q_i q_i^*/2 - kq_i q_i^*], \end{aligned} \tag{4.1}$$

where R, k, m are nonnegative physical constants and $\sum = \sum_{i=1}^{2}$ everywhere in this paragraph. For L given by eq.(4.1), eqs.(2.1) and (2.3) become

$$m\ddot q_i + R\dot q_i + kq_i = 0, \quad i = 1,2, \qquad m\ddot q_i^* - R\dot q_i^* + kq_i^* = 0, \quad i = 1,2, \tag{4.2}$$

respectively. The Hamiltonian eq.(2.7) becomes

$$H(p, p^*, q, q^*) = \sum [(p_i + R\, q_i^*/2) \times (p_i^* - R\, q_i/2)/m + k\, q_i q_i^*]. \tag{4.3}$$

Hereafter, we shall restrict ourselves to the so-called underdamped case $k/m - R^2/4m^2 > 0$. If, for convenience, we make the canonical transformation

$$P_i = p_i/\sqrt{m\Omega}, \quad Q_i = \sqrt{m\Omega} q_i, \quad P_i^* = p_i^*/\sqrt{m\Omega}, \quad Q_i^* = \sqrt{m\Omega} q_i^*, \qquad i = 1,2 \tag{4.4}$$

where $m\Omega^2 = k - R^2/4m > 0$, then the corresponding Hamiltonian H, is given by

$$H = \mathcal{H}_{12} + \mathcal{H}_{22}, \tag{4.5a}$$

$$\mathcal{H}_{12} = \Omega\, I_1 + R/2m\, I_2, \qquad \mathcal{H}_{22} = \Omega\, I_3 + R/2m\, I_4, \tag{4.5b}$$

$$I_1 = P_1 P_1^* + Q_1 Q_1^*, \qquad I_2 = P_1^* Q_1^* - P_1 Q_1, \tag{4.5c}$$

$$I_3 = P_2 P_2^* + Q_2 Q_2^*, \qquad I_4 = P_2^* Q_2^* - P_2 Q_2, \tag{4.5d}$$

Direct computation yields that the I_i's are in involution, *i.e.*,

$$\{I_i, I_j\} = 0, \qquad i,j = 1,\ldots,4, \tag{4.6}$$

and independent on $\mathbf{R}^8 \setminus (0,0,0,0,0,0,0,0)$.

Let us introduce the planes

$$\begin{aligned}
\mathcal{P}_1 &= \{(0, P_1^*, P_2, P_2^*, Q_1, 0, Q_2, Q_2^*) \mid P_1^*, P_2, P_2^*, Q_1, Q_2, Q_2^* \in \mathbf{R}\},\\
\mathcal{P}_2 &= \{(P_1 0, P_2, P_2^*, 0, Q_1^*, Q_2, Q_2^*) \mid P_1, P_2, P_2^*, Q_1^*, Q_2, Q_2^* \in \mathbf{R}\},\\
\mathcal{P}_3 &= \{(P_1, P_1^*, 0, P_2^*, Q_1, Q_1^*, Q_2, 0) \mid P_1, P_1^*, P_2^*, Q_1, Q_1^*, Q_2 \in \mathbf{R}\}, \text{ and}\\
\mathcal{P}_4 &= \{(P_1, P_1^*, P_2, 0, Q_1, Q_1^*, 0, Q_2^*) \mid P_1, P_1^*, P_2, Q_1, Q_1^*, Q_2^* \in \mathbf{R}\}.
\end{aligned}$$

Then direct computation verifies that $(P_1, P_1^*, P_2, P_2^*, Q_1, Q_1^*, Q_2, Q_2^*) \to (I_1, \ldots, I_4, \varphi_1, \ldots, \varphi_4)$, where the I's are given by eqs.(4.5c)–(4.5d) and the φ's are given by

$$\begin{aligned}
\varphi_1 &= -[\tan^{-1}(P_1/Q_1) + \tan^{-1}(P_1/Q_1^*)]/2,\\
\varphi_2 &= -(\ln[(Q_1^2 + P_1^{*2})/(Q_1^{*2} + P_1^2)])/4,
\end{aligned} \tag{4.7a}$$

$$\begin{aligned}
\varphi_3 &= -[\tan^{-1}(P_2^*/Q_2) + \tan^{-1}(P_2/Q_2^*)]/2,\\
\varphi_4 &= -(\ln[(Q_2^2 + P_2^{*2})/(Q_2^{*2} + P_2^2)])/4,
\end{aligned} \tag{4.7b}$$

satisfy eq.(2.12) on $\mathbf{R}^8/(\mathcal{P}_1 \cup \mathcal{P}_2 \cup \mathcal{P}_3 \cup \mathcal{P}_4)$. It follows from the preceding and the fact that eq.(4.5a) is of the form given by eq.(2.13) that $(I_1, \ldots, I_4, \varphi_1, \ldots, \varphi_4)$ is a set of linearizing Darboux coordinates on $\mathbf{R}^8 \setminus (\mathcal{P}_1 \cup \mathcal{P}_2 \cup \mathcal{P}_3 \cup \mathcal{P}_4)$.

In order to illustrate another technique for establishing complete integrability, we turn to a discussion of a multi-Hamiltonian structure for this eight-dimensional Bateman pair. A system whose time evolution is described by a vector field K which is Hamiltonian with respect to two symplectic structures is said to be bi-Hamiltonian. Such structures were first identified and studied within the context of completely integrable Hamiltonian systems by Magri [9,10] (See also ref.[11]).

The phase space M of interest here is a subspace of $\mathbf{R}^8$. Points v in M are described locally by coordinates $(p_1, p_1^*, p_2, p_2^*, q_1, q_1^*, q_2, q_2^*)$ with respect to the basis $\{e_1, \ldots, e_8\}$ where $e_j = (\delta_{1j}, \ldots, \delta_{8j})$. The flows on M are governed by an eight-component evolution equation in Hamiltonian form, *i.e.*,

$$\dot{v}(t) = K(v(t)), \qquad t \in \mathbf{R}, v(t) \in M, \tag{4.8}$$

$$\begin{aligned}
K = (&-\partial H/\partial q_1, -\partial H/\partial q_1^*, -\partial H/\partial q_2, -\partial H/\partial q_2^*,\\
&\partial H/\partial p_1, \partial H/\partial p_1^*, \partial H/\partial p_2, \partial H/\partial p_2^*).
\end{aligned} \tag{4.9}$$

We write $T_v M$ for the tangent space to M at the point $v \in M$ and $TM = \bigcup_{v \in M} T_v M$, the disjoint union of the $T_v M$, for the tangent bundle on M. The tangent bundle TM is modelled on $\mathbf{R}^8 \times \mathbf{R}^8$ with coordinates $(p_1, p_1^*, p_2, p_2^*, q_1, q_1^*, q_2, q_2^*, \xi_1, \ldots, \xi_8)$ where the ξ's are with respect to the basis $(\partial/\partial p_1, \partial/\partial p_1^*, \partial/\partial p_2, \partial/\partial p_2^*, \partial/\partial q_1, \partial/\partial q_1^*, \partial/\partial q_2, \partial/\partial q_2^*)$. In this basis K is given by the vector field

$$\begin{aligned}
K \;=\; &-(\partial H/\partial q_1)\partial/\partial p_1 - (\partial H/\partial q_1^*)\partial/\partial p_1^*\\
&-(\partial H/\partial q_2)\partial/\partial p_2 - (\partial H/\partial q_2^*)\partial/\partial p_2^*\\
&+(\partial H/\partial p_1)\partial/\partial q_1 + (\partial H/\partial p_1^*)\partial/\partial q_1^*\\
&+(\partial H/\partial p_2)\partial/\partial q_2 + (\partial H/\partial p_2^*)\partial/\partial q_2^*,
\end{aligned} \tag{4.10}$$

where for each point $v \in M$, $K(v) \in T_v M$. The operator commutator or Lie bracket of two vector fields X and Y with respect to the basis $(\partial/\partial p_1, \partial/\partial p_1^*, \partial/\partial p_2, \partial/\partial p_2^*, \partial/\partial q_1, \partial/\partial q_1^*, \partial/\partial q_2, \partial/\partial q_2^*)$ is defined in the usual way to be the Lie bracket $[X, Y] = XY - YX$. The infinitely differentiable vector fields $\chi(M)$ on M form a Lie algebra with respect to this bracket. In order to work within this Lie structure, we require that $K \in \chi(M)$, *i.e.*, we require that our H is infinitely differentiable.

The cotangent bundle T^*M of one-forms is modelled on $\mathbf{R}^8 \times \mathbf{R}^8$ with coordinates $(p_1, p_1^*, p_2, p_2^*, q_1, q_1^*, q_2, q_2^*, \eta_1, \ldots, \eta_4)$ where the η's are with respect to the basis $(dp_1, dp_1^*, dp_2, dp_2^*, dq_1, dq_1^*, dq_2, dq_2^*)$. The fundamental symplectic two-form $\omega : TM \times TM \to \mathbf{R}$ is given in these coordinates by,

$$\omega = \sum (dp_i \wedge dq_i + dp_i^* \wedge dq_i^*), \tag{4.11}$$

where $\alpha \wedge \beta(X, Y) = \alpha(X)\beta(Y) - \alpha(Y)\beta(X)$, for $\alpha, \beta \in T^*M$ and $X, Y \in TM$. The two-form ω is, by definition, antisymmetric, closed (*i.e.*, $d\omega = 0$), and nondegenerate (*i.e.*, $\omega(X, Y) = 0$, for fixed X and all Y, $X, Y \in TM$ implies $X = 0$). Therefore using ω, one constructs in the usual way (up to sign conventions) an invertible map ω^b between TM and T^*M, which connects vector fields and one-forms. In particular, it connects a subalgebra of $\chi(M)$, the subalgebra of Hamiltonian vector fields with one forms which are differentials of functions on M. Our conventions are those of Arnold [12], $\omega^b : TM \to T^*M$ given by $TM \in X \to \alpha_X = \omega(\,.\,, X) \in T^*M$, and $\omega^\# : T^*M \to TM$ is given by $\omega^\# = (\omega^b)^{-1}$. In this language, $K = \omega^\# \, dH$. With respect to the ordering of coordinates above, ω and ω^b are represented by the matrix $\begin{bmatrix} 0 & -I \\ I & 0 \end{bmatrix}$, where I, 0 are the 4×4 identity and zero matrices respectively. For vector fields X_A and X_B, Hamiltonian with respect to ω, *i.e.*, $X_A = \omega^\# \, dA$ and $X_B = \omega^\# \, dB$, where A and B are differentiable functions on M, we have

$$\omega(X_B, X_A) = \{A, B\}, \tag{4.12}$$

where $\{A, B\}$ is given by the usual Poisson bracket (see eq.(4.11′) and $\{A, B\}_2$ below it in the following paragraph). For the purposes of the discussion here, the above structure is what is meant by a symplectic structure and we denote it by (M, ω). A Hamiltonian system is then a triple (M, ω, H) such that the vector field K which governs the evolution of the system is given by $K = \omega^\# \, dH$.

It follows from the above discussion that for this system, one Hamiltonian structure is given by (M, ω_2, H_2), where for ω_2 we take ω given by eq.(4.11) and subject it to the canonical transformation given by eq.(4.4), *i.e.*,

$$\omega_2 = \sum (dP_i \wedge dQ_i + dP_i^* \wedge dQ_i^*), \tag{4.11′}$$

and for H_2 we accordingly take H given by eqs.(4.5a)–(4.5d). Eq.(4.11′) implies the Poisson structure

$$\begin{aligned} \{A, B\}_2 = \sum (&\partial A/\partial Q_i \partial B/\partial P_i - \partial A/\partial P_i \partial B/\partial Q_i \\ &+ \partial A/\partial Q_i^* \partial B/\partial P_i^* - \partial A/\partial P_i^* \partial B/\partial Q_i^*). \end{aligned}$$

Now, we turn to the identification of another Hamiltonian structure for the damped oscillator. Direct computation shows that there exists a second symplectic two-form

$$\omega_1 = dP_1 \wedge dP_1^* - dQ_1 \wedge dQ_1^* + dP_2 \wedge dQ_2 + dP_2^* \wedge dQ_2^*, \tag{4.13}$$

which implies a second Poisson bracket

$$\begin{aligned} \{A, B\}_1 \;=\;& \partial A/\partial P_1^* \partial B/\partial P_1 - \partial A/\partial P_1 \partial B/\partial P_1^* \\ & -\partial A/\partial Q_1^* \partial B/\partial Q_1 + \partial A/\partial Q_1 \partial B/\partial Q_1^* \\ & +\partial A/\partial Q_i^* \partial B/\partial P_i^* - \partial A/\partial P_i^* \partial B/\partial Q_i^*, \end{aligned} \tag{4.14}$$

and a second Hamiltonian H_1

$$H_1 = \mathcal{H}_{11} + \mathcal{H}_{22}, \tag{4.15}$$

$$\mathcal{H}_{11} = \Omega\, I_2 - R/2m\, I_1, \tag{4.16}$$

such that

$$\omega_1^{\#}\, dH_1 = \omega_2^{\#}\, dH_2. \tag{4.17}$$

Eq.(4.17) is a statement that the pair of Hamiltonian structures (M, ω_1, H_1) and (M, ω_2, H_2) are equivalent in the sense that they both yield the same Hamiltonian vector field. Note we have interchanged here what was called ω_1 and ω_2 in ref.[5].

Direct computation shows that the two-form ω_1 given by Eq.(4.13) can be related to the two-form ω_1 given by (4.11′) via

$$\omega_2(X, Y) = \omega_1(X, \Phi Y), \tag{4.18}$$

where $\Phi \in End(TM)$ is an operator whose matrix representative $[\Phi]$ is

$$[\Phi] = \begin{bmatrix} 0 & 0 & 0 & 0 & 0 & -1 & 0 & 0 \\ 0 & 0 & 0 & 0 & 1 & 0 & 0 & 0 \\ 0 & 0 & 1 & 0 & 0 & 0 & 0 & 0 \\ 0 & 0 & 0 & 1 & 0 & 0 & 0 & 0 \\ 0 & -1 & 0 & 0 & 0 & 0 & 0 & 0 \\ 1 & 0 & 0 & 0 & 0 & 0 & 0 & 0 \\ 0 & 0 & 0 & 0 & 0 & 0 & 1 & 0 \\ 0 & 0 & 0 & 0 & 0 & 0 & 0 & 1 \end{bmatrix}, \tag{4.19}$$

where $[\Phi]^4 = id$. The operator appearing in eq.(4.18) is called a recursion operator and the structure given by the triple $(\Phi, K_{\omega_2,H_1}, \Phi\, K_{\omega_2,H_1})$ is called a bi-Hamiltonian structure, where we have introduced the following extended notation for the vector field associated with H_j through ω_i; $K_{\omega_i,H_j} = \omega_i^{\#}\, dH_j$.

Now, we shall use the following theorem [8] which describes hierarchies $\Phi^{j-1}X_1 = X_j$, $j = 1, \ldots, n$ of commuting Hamiltonian vector fields whose Hamiltonians are in involution to obtain more equivalent Hamiltonian structures.

Theorem. [8] Let (M, ω) be a symplectic manifold. Consider $K \in \chi(M)$ and $\Phi \in End(T(M))$ and define the "hierarchy" $K_j = \Phi^{j-1}K$, $j = 1, 2, \ldots$, of vector fields on M. Assume

1. a) Φ is symplectically self-adjoint, *i.e.*, $\omega(\Phi X, Y) = \omega(X, \Phi Y)$ for all $X, Y \in \chi(M)$.

b) Φ is Nijenhuis, *i.e.*, $\Phi^2[X,Y] = \Phi[X,\Phi Y] + \Phi[\Phi X, Y] - [\Phi X, \Phi Y]$ for all $X, Y \in \chi(M)$.

2. K and ΦK are symplectic.

Then, there exist locally on M (and globally if the first de Rham cohomology group of M is 0), functions $\{h_j\}$ in involution such that $K_j = X_{h_j}$ (the symplectic vector field corresponding to the 1-form dh_j). In particular, $\{K_j\}$ is an isotropic set of commuting vector fields on M, *i.e.*, for any pair K_i, K_j $(i,j = 1,2,\ldots)$ of the vector fields, $\omega(K_i, K_j) = 0$ and $[K_i, K_j] = 0$.

We now identify ω in the above Theorem with ω_2 given by Eq.(4.11′) and K_1 in the Theorem with K_{ω_2,H_1}. The triple $(\Phi, K_{\omega_2,H_1}, \Phi K_{\omega_2,H_1})$ where Φ is given by Eq.(4.19) satisfies the three tests of the Theorem given above, namely, Φ is self-adjoint with respect to ω_2 (*i.e.*, $\omega_2(X, \Phi Y) = \omega_2(\Phi X, Y)$ by direct computation), Φ is Nijenhuis because Φ is a constant operator, and both K_{ω_2,H_1} and $\Phi K_{\omega_2,H_1}$ are symplectic with respect to ω_2 (*i.e.*, $K_{\omega_2,H_j} = \omega_1^{\#}\, dH_j$, $j = 1,2$, by definition and direct calculation yields $\Phi\, K_{\omega_2,H_1} = K_{\omega_2,H_2}$). Hence, it follows from the Theorem and the relation $\Phi^4 = id$ (Eq.(4.19)), that the vector fields $\Phi^2 K_{\omega_2,H_1} = K_{\omega_2,H_3}$, $\Phi^3 K_{\omega_2,H_1} = K_{\omega_2,H_4}$ are Hamiltonian for some H_3, H_4 respectively. Computation yields

$$H_3 = -\mathcal{H}_{11} + \mathcal{H}_{22}, \quad H_4 = -\mathcal{H}_{12} + \mathcal{H}_{22} \tag{4.20}$$

Hence, it follows from the theorem that the hierarchy of Hamiltonian vector fields

$$\{K_{\omega_2,H_1}, \Phi K_{\omega_2,H_1} = K_{\omega_2,H_2}, \Phi^2 K_{\omega_2,H_1} = K_{\omega_2,H_3}, \Phi^3 K_{\omega_2,H_1} = K_{\omega_2,H_4}\} \tag{4.21}$$

satisfies

$$\omega_2(K_{\omega_2,H_i}, K_{\omega_2,H_j}) = \{H_j, H_i\}_2 = 0, \qquad i,j = 1,\ldots,4 \tag{4.22}$$

i.e., H_i's are in involution with respect to the Poisson bracket $\{\ ,\ \}_2$. The hierarchy (4.21) truncates because $\Phi^4 = id$ in this case.

It also follows from the theorem that

$$[K_{\omega_2,H_i}, K_{\omega_2,H_j}] = 0, \qquad i,j = 1,\ldots,4. \tag{4.23}$$

Thus, K_{ω_2,H_i} $i = 1,3,4$ generates a local symmetry in the sense of Lie for the system described by K_{ω_2,H_2}, *i.e.*, each K_{ω_2,H_i} generates a one parameter transformation group which takes solutions of

$$\dot{v} = K_{\omega_2,H_2}(v), \tag{4.24}$$

into solutions of itself.

This is called a quad-Hamiltonian hierarchy because

$$\omega_1(X, \Phi Y) = \omega_2(X, Y) = \omega_3(X, \Phi^{-1} Y) = \omega_4(X, \Phi^{-2} Y), \tag{4.25}$$

which implies

$$\omega_1^{\#}\, dH_1 = \omega_2^{\#}\, dH_2 = \omega_3^{\#}\, dH_3 = \omega_4^{\#}\, dH_4, \tag{4.26}$$

where

$$\omega_1 = dP_1 \wedge dP_1^* - dQ_1 \wedge dQ_1^* + dP_2 \wedge dQ_2 + dP_2^* \wedge dQ_2^*, \quad (4.27a)$$

$$\omega_2 = dP_1 \wedge dQ_1 + dP_1^* \wedge dQ_1^* + dP_2 \wedge dQ_2 + dP_2^* \wedge dQ_2^*, \quad (4.27b)$$

$$\omega_3 = -dP_1 \wedge dP_1^* + dQ_1 \wedge dQ_1^* + dP_2 \wedge dQ_2 + dP_2^* \wedge dQ_2^*, \quad (4.27c)$$

$$\omega_4 = -dP_1 \wedge dQ_1 - dP_1^* \wedge dQ_1^* + dP_2 \wedge dQ_2 + dP_2^* \wedge dQ_2^*, \quad (4.27d)$$

$$H_1 = \Omega I_2 - (R/2m)I_1 + \Omega I_3 + (R/2m)I_4, \quad (4.28a)$$

$$H_2 = \Omega I_1 + (R/2m)I_2 + \Omega I_3 + (R/2m)I_4, \quad (4.28b)$$

$$H_3 = -\Omega I_2 + (R/2m)I_1 + \Omega I_3 + (R/2m)I_4, \quad (4.28c)$$

$$H_4 = -\Omega I_1 - (R/2m)I_2 + \Omega I_3 + (R/2m)I_4, \quad (4.28d)$$

and

$$I_1 = P_1 P_1^* + Q_1 Q_1^*, \qquad I_2 = P_1^* Q_1^* - P_1 Q_1, \quad (4.29)$$

$$I_3 = P_2 P_2^* + Q_2 Q_2^*, \qquad I_4 = P_2^* Q_2^* - P_2 Q_2. \quad (4.30)$$

Technically, in this hierarchy, there are only three functionally independent constants of the motion in the hierarchy (4.21), namely, $\mathcal{H}_{11}$, $\mathcal{H}_{12}$, $\mathcal{H}_{22}$. A fourth one $\mathcal{H}_{21}$

$$\mathcal{H}_{21} = \Omega I_4 - (R/2m)I_3, \quad (4.31)$$

is identified by interchanging the the subscripts. This yields a second quad-Hamiltonian hierarchy generated by a recursion operator. Because our phase space is eight dimensional, there are at most four independent constants of the motion which are in involution. Therefore, the existence of these four constants of the motion which are in involution is also a statement that our system is completely integrable with respect to the symplectic structure (M, ω_2).

References

1. Anderson R. L., Hebda, P. W., and Rideau, G., "Hamiltonian Treatment of Quasi-Linear Systems via a Bateman Lagrangian," LMP 22, 335–343 (1991).
2. Bateman, H., "On Dissipative Systems and Related Principles," Phys. Rev. 38, 815–819 (1931).
3. Lorentz, E. N., "Deterministic Nonperiodic Flow," J. Atmospheric Sciences 20, 130–141 (1963).
4. Morse, P. M. and Feshbach,H., Methods of Theoretical Physics, McGraw-Hill Book Co., Chapter 3, New York, (1953).
5. Anderson R. L., Hebda, P. W., and Rideau, G., "Bateman-Morse-Feshbach Treatment of a Damped Linear Oscillator — Linearizing Canonical Coordinates and Bi-Hamiltonian Structures," preprint, (October, 1990).
6. Lanczos, C., The Variational Principles of Mechanics, University of Toronto Press, Toronto (1949).
7. Ostrogradsky, M., "Les Equations Differentielles Relatives Au Probleme Des Isoperimetres," Mem. Acad. St. Petersburg 6 (4), 385–517 (1850); Whittaker, E. T., Treatise on the Analytical Dynamics of Particles and Rigid Bodies, Cambridge Press, Cambridge (1927).
8. Adams, M., Anderson, R. L., and Varley, R., "Remarks On Integrable Hierarchies In Finite Dimensions," Proceedings of the CRM Workshop on "Hamiltonian Systems, Transformation Groups and Spectral Transform Methods," Oct 20–26, 1989, CRM, Université de Montréal, ed. J. Harnad and J. E. Marsden.

9. Magri, F., "A simple model of the integrable Hamiltonian equation," J. Math. Phys. 19, 1156 (1978); "A geometrical approach to the nonlinear solvable equations," Lecture notes in Physics (Springer) 120, 233 (1980) (Eds. M. Boiti, F. Pempinelli, G. Soliani).
10. Magri F., and Morosi, C., "A geometrical characterization of integrable integrable Hamiltonian systems through the theory of Poisson-Nijenhuis manifolds," Univ. of Milano preprint (1984).
11. Gel'fand, I. M. and Dorfman, I. Y., "Hamiltonian Operators and Algebraic Structures Related to Them," Funct. Analysis Appl. 13, 248 (1979); "The Shouten Bracket and Hamiltonian Operators," Funct. Analysis Appl. 14, 223 (1980).
12. Arnold, V. I., Mathematical Methods of Classical Mechanics, Springer-Verlag, New York (1978).
13. Anderson, R. L. and Kupershmidt, B. A., "Multi-Hamiltonian Structures for N^{th}-Order Linear Matrix Differential Equations," work in progress.

GROUP METHOD ANALYSIS OF THE DISPERSION OF GASEOUS POLLUTANTS IN THE PRESENCE OF A TEMPERATURE INVERSION

N. A. BADRAN and M. B. ABD-EL-MALEK*
Department of Engineering, Mathematics and Physics,
Faculty of Engineering, Alexandria University,
Alexandria, Egypt

Abstract. The group transformation theoretic approach is applied to present an analytical study of the dispersion of gaseous pollutants in the presence of a temperature inversion. The pollutants are assumed to be spread in the layer bounded by the ground surface and the inversion level as well as they are driven by wind in the horizontal direction. The analytical study indicates that the cross-wind effects cannot be ignored during dispersion calculations and depends on the height of the inversion layer.

1. Introduction

In the last decade, increased efforts have been made to gain an understanding of the dispersion of gaseous pollutants to achieve more effective methods for detection and control. The emission of pollutants into the atmosphere is due to free burning fire, large firestorms produced by nuclear bombing, liquid droplets from chimneys and motor vehicles.

In 1986, Kumar [7] studied the effects of cross-wind shear on horizontal dispersion using a numerical model, based on K-theory, of a steady state diffusion equation and boundary layer equation. In 1987, Ayad [6] studied a turbulence model for wind flows above fire areas using the finite difference method. In 1992, Toson [11] considered the same problem analytically by the method of separation of variables under the assumption that the pollutant diffusion in the horizontal direction is minimal.

The mathematical technique used in the present analysis is the one-parameter group transformation, which leads to a similarity representation of the problem. Similarity analysis has been applied intensively by Ames [3, 4, 5], Abd-el-Malek *et al.* [1, 2], Moran and Gaggioli [8, 9, 10].

In this work the analytical study has been carried out for two limiting cases, namely, if the ground does not absorb any gaseous pollutants and if the ground absorbs all the pollutants. It is hoped that the information in this paper will confirm our existing knowledge of cross-wind effects.

* Presently on leave at: Department of Science, Mathematics Unit, The American University in Cairo, P.O. Box 2511, Cairo, Egypt

N. H. Ibragimov et al. (eds.),
Modern Group Analysis: Advanced Analytical and Computational Methods in Mathematical Physics, 35–41.

2. Formulation of the problem and the governing equations

The gaseous pollutant is bounded from below by the ground surface and from above by the inversion layer, which is at height h from the ground surface.

Following Toson [11], the pollution, with concentration $C(x, y)$, is assumed to be evenly distributed throughout the layer and the mean concentration of the pollutant at $x = 0$ averaged over $0 \le y \le h$ is constant and equal to C_0. The diffusion of the pollutants takes place due to the wind that has a constant mean velocity $u = u(x)$ in the x-direction, and the eddy diffusivities k_1 and k_2 in the x and y-directions, respectively, are also independent of y.

The normalized steady state diffusion equation, that governs the dispersion of the gaseous pollutants is

$$u(x)C_x = k_1(x)C_{xx} + k_2(x)C_{yy} \tag{2.1}$$

with the boundary conditions

$$k_2C_y = \lambda\sqrt{(h/u_0)}\,C \qquad \text{at } y = 0, \tag{2.2.1}$$

$$C_y = 0 \qquad \text{at } y = 1, \tag{2.2.2}$$

$$C = 1 \qquad \text{at } x = 0 \text{ and } 0 < y < 1, \tag{2.2.3}$$

$$C_x = 0 \qquad \text{as } x \to \infty, \tag{2.2.4}$$

where all x and y are scaled with respect to h, C with respect to C_0, u with respect to u_0, k_1 and k_2 with respect to u_0h, u_0 is a reference velocity; $\lambda \ll 1$ corresponds to the case where no pollutant is absorbed by the ground, while $\lambda \gg 1$ corresponds to the case where all pollutant is absorbed by the ground. A schematic diagram of the problem with its normalized boundary conditions is illustrated in Figure 1.

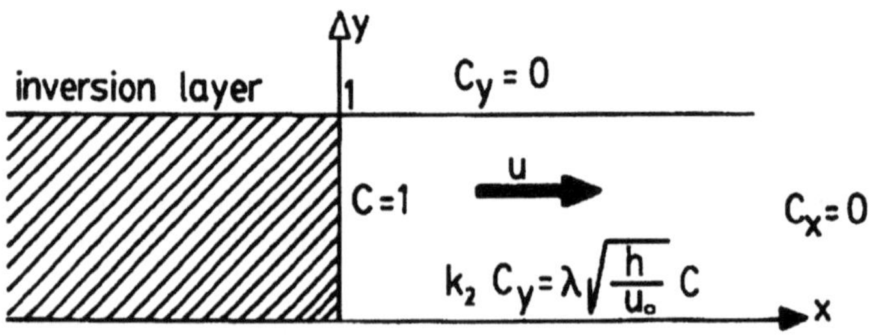

FIGURE 1. Physical model of the problem in normalized form.

If we introduce the nondimensional function $\theta(x, y)$ and $C^*(x)$ such that

$$C(x, y) = \theta(x, y)C^*(x), \tag{2.3}$$

equation (2.1) becomes

$$u(C^*\theta_x + \theta C_x^*) = k_1(C^*\theta_{xx} + 2\theta_x C_x^* + \theta C_{xx}^*) + k_2C^*\theta_{yy}. \tag{2.4}$$

3. Solution of the problem

The number of independent variables is reduced by one by applying a one-parameter transformation group.

3.1. The group systematic formulation

The procedure is initiated with the group G, a class of one-parameter group of the form

$$G: \quad \bar{S} = E^S(a)S + K^S(a), \tag{3.1}$$

where S stands for x, y; k_1, k_2, u, C^*, θ and the E^S and K^S are real-valued functions and at least differentiable in their argument.

3.2. The invariance analysis

To transform the differential equation, transformations of the derivatives are obtained from G via chain rule operations

$$\left.\begin{aligned} \bar{S}_i &= (E^S/E^i)S_i \\ \bar{S}_{ij} &= (E^S/E^iE^j)S_{ij} \end{aligned}\right\} \quad i = x, y, \quad j = x, y, \tag{3.2}$$

where S stands for C^* and θ.

Following Abd-el-Malek *et al.* [1] we get the group G_1 which transforms invariantly the equation (2.4) and the boundary conditions (2.2). The group G_1 is of the form

$$G_1 \left\{ \begin{aligned} G_s & \left\{ \begin{aligned} \bar{x} &= E^x(a)x, \\ \bar{y} &= y, \end{aligned}\right. \\ \bar{u} &= E^x(a)u, \\ \bar{k}_1 &= (E^x(a))^2 k_1, \\ \bar{k}_2 &= k_2, \\ \bar{C}^* &= E^{C^*}(a)C^*, \\ \bar{\theta} &= 0. \end{aligned}\right. \tag{3.3}$$

3.3. The complete set of absolute invariants

The complete set of absolute invariants consists of the absolute invariant of the independent variables (similarity variable) and the invariants corresponding to the dependent variables.

If $\eta \equiv \eta(x, y)$ is the absolute invariant of the independent variables, then

$$g_j(x, y, u, k_1, k_2, C^*, \theta) = F_j(\eta(x, y)), \qquad j = 1, 2, 3, \dots \tag{3.4}$$

are the dependent absolute invariants. From the group theory, the function g of equation (3.4) is an absolute invariant of a one-parameter group if it satisfies the first order linear differential equation

$$\sum_{i=1}^{6} (\alpha_{2i-1}S_i + \alpha_{2i}) \frac{\partial g}{\partial S_i} = 0, \tag{3.5}$$

where S_1, S_2, S_3, S_4, S_5 and S_6 stand for x, y, u, k_1, k_2 and θ, respectively, and $\alpha_1 = (\partial E^x/\partial a)(a^0)$, $\alpha_2 = (\partial k^x/\partial a)(a^0)$, etc.; a^0 is the identity element of the group G. Following Abd-el-Malek *et al.* [1], we get: The absolute invariant of the independent variable $\eta(x, y)$ will be

$$\eta \equiv \eta(y), \tag{3.6}$$

and the absolute invariants of the dependent variables will be

$$\begin{aligned} F(x,u) = u/x, \quad G(x,k_1) = k_1/x^2, \\ \theta = \theta(y), \quad k_2 = k_2(x), \quad C^* = C^*(x). \end{aligned} \tag{3.7}$$

4. The reduction to ordinary differential equations

With no loss of generality the independent absolute invariant $\eta(y)$ in (3.6) may assume the form

$$\eta = y. \tag{4.1}$$

Substitution for the dependent variables in terms of their corresponding absolute invariants into equation (2.4) gives

$$[k_2 C^*]\theta_{yy} + [x^2 G C^*_{xx} - x F C^*_x]\theta = 0. \tag{4.2}$$

The requirement of reducing equation (4.2) of two ordinary differential equations implies that

$$[x^2 G C^*_{xx} - x F C^*_x] = p^2[k_2 C^*], \tag{4.3}$$

for some constant p^2.

Under this assumption, equation (4.2) gives the ordinary differential equation of $\theta(y)$,

$$\theta_{yy} + p^2\theta = 0,$$

which has the solution

$$\theta(y) = A \cos p(y + \varepsilon). \tag{4.4}$$

Rearranging equation (4.3), we get

$$\alpha C^*_{xx} - 2\beta C^*_x - p^2 C^* = 0 \tag{4.5}$$

where, using equation (3.7), the constants α and β are

$$\alpha = (x^2 G)/k_2 = k_1/k_2 \qquad \text{and} \qquad \beta = (xF)/2k_2 = u/2k_2. \tag{4.6}$$

The ordinary differential equation (4.5) has the general solution

$$C^*(x) = A_1 e^{m_1 x} + A_2 e^{m_2 x}. \tag{4.7}$$

To satisfy the boundary conditions (2.2.4) as $x \to \infty$, only the negative exponent may be considered. Taking $\alpha > 0$, we get the solution

$$C^*(x) = A_1 e^{mx}, \tag{4.8}$$

where

$$m = \beta - \sqrt{\beta^2 + \alpha p^2}. \tag{4.9}$$

Substituting (4.4) and (4.8) in (2.3), we get

$$C(x,y) = Be^{mx}\cos p(y+\varepsilon). \tag{4.10}$$

Application of the boundary condition (2.2.2) at the inversion level determines the value of ε to be -1. Hence

$$C(x,y) = Be^{mx}\cos p(y-1). \tag{4.11}$$

Application of the boundary condition (2.2.1) at the ground surface yields

$$\tan p = \frac{\lambda}{k_2}\sqrt{(h/u_0)}(1/p). \tag{4.12}$$

The constants B and p will be determined corresponding to the two limiting cases of λ.

CASE (1): No pollutant is absorbed by the ground ($\lambda \ll 1$).
Assuming $\frac{\lambda}{k_2}\sqrt{(h/u_0)}(1/p)$ to be a very small quantity leads to

$$p^2 = \frac{\lambda}{k_2}\sqrt{(h/u_0)}, \tag{4.13}$$

and hence, by applying the boundary condition (2.2.3), we get $B = 1$.
The concentration distribution for the case (1) is:

$$C = e^{mx}, \tag{4.14}$$

where m is given by (4.9).

CASE (2): All pollutant is absorbed by the ground ($\lambda \gg 1$).
In this case, from equation (4.12), we get

$$p = (2n-1)\frac{\pi}{2}; \qquad n = 1,2,3,\dots.$$

The concentration distribution model will be

$$C = \sum_{n=1}^{\infty} B_n e^{mx}\cos\left[\frac{(2n-1)\pi}{2}(y-1)\right], \tag{4.15}$$

where

$$m = \beta - \sqrt{\beta^2 + \frac{(2n-1)^2\pi^2}{4}\alpha}. \tag{4.16}$$

The constants B_n will be determined as the cosine's Fourier coefficients of the expansion of the function $C = 1$ in $0 < y < 1$ and $x = 0$; we get

$$B_n = 2\left(\frac{\sin\frac{(2n-1)\pi}{2}}{\frac{(2n-1)\pi}{2}}\right).$$

Hence

$$C(x,y) = 2\sum_{n=1}^{\infty}\left(\frac{\sin\frac{(2n-1)\pi}{2}}{\frac{(2n-1)\pi}{2}}\right)e^{mx}\cos\left[\frac{(2n-1)\pi}{2}(y-1)\right] \tag{4.17}$$

where m is given by (4.16).

5. Results and discussion

We shall deal with each case of the two derived cases individually.

CASE (1): No pollutant is absorbed by the ground surface.
We found that the concentration distribution has the form:

$$C = \exp\left\{\frac{1}{2k_2}\left[u - \sqrt{u^2 + 4\lambda k_1 \sqrt{(h/u_0)}}\right]\right\} x \tag{5.1}$$

which is independent of y and mainly depends on u, k_1, k_2, h and λ. For very small λ or very large u, it is clear that no pollutant will be absorbed by the ground surface. As it is clear in Figure 2, as u increases the absorbed pollutant by the ground surface will be less. Also as the eddy diffusivity ratio, $\alpha = k_1/k_2$, increases the absorbed pollutant by the ground surface will be more.

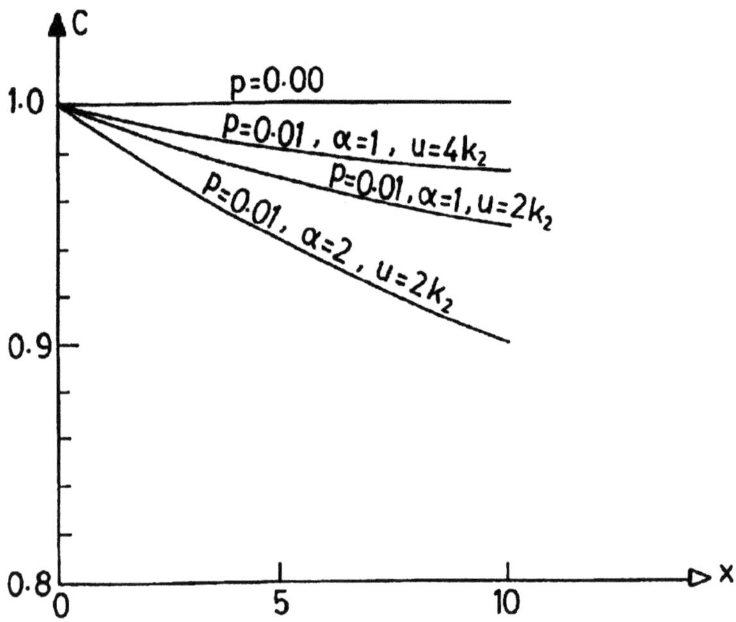

FIGURE 2. Effect of u, the eddy diffusivity ratio α, and parameter p on C for case (1).

CASE (2): All pollutant is absorbed by the ground surface.
We found that the concentration distibution has the form

$$C(x,y) = 2\sum_{n=1}^{\infty}\left(\frac{\sin\frac{(2n-1)\pi}{2}}{\frac{(2n-1)\pi}{2}}\right)\left\{\exp\left[\frac{1}{2k_2}\left(u - \sqrt{u^2 + (2n-1)^2\pi^2 k_1}\right)\right] x\right\}$$
$$\times \cos\left[\frac{(2n-1)\pi}{2}(y-1)\right]. \tag{5.2}$$

It is found that at the inversion layer:

$$C(0) = 5.073 \times 10^{-2}, \quad C(1) = 1.122 \times 10^{-18}, \quad C(2) = 2.481 \times 10^{-35},$$

i.e., starting from distance equal to the height of the inversion layer all pollutant will be absorbed by the ground surface. At the mid way between the ground surface and the inversion layer:

$$C(0) = 3.464 \times 10^{-2}, \quad C(1) = 0.766 \times 10^{-18}, \quad C(2) = 1.694 \times 10^{-35},$$

i.e., the rate of absorption will be more than the previuos level. It is clear from the derived formula that, at the ground surface, all pollutants will be absorbed.

Acknowledgements

We acknowledge with thanks Mrs. Soraya Garas, The American University in Cairo, for her tireless efforts in typing the manuscript and for an excellent job.

References

1. Abd-el-Malek, M. B., Boutros, Y. Z. and Badran, N. A. (1990) "Group method analysis of unsteady free-convective boundary-layer flow on a nonisothermal vertical flat plate", J. Engineering Mathematics **24**, 343–368.
2. Abd-el-Malek, M. B. and Badran, N. A. (1990) "Group method analysis of unsteady free-convective laminar boundary-layer flow on a nonisothermal vertical circular cylinder", Acta Mechanica **85**, 193–206.
3. Ames, W. F. (1965) "Similarity for the nonlinear diffusion equation", I & EC Fundamentals **4**, 72–76.
4. Ames, W. F. (1972) *Nonlinear partial differential equations in engineering*, Academic Press, New York.
5. Ames, W. F. and Nucci, M. C. (1985) "Analysis of fluid equations by group methods", J. Engineering Mathematics **20**, 181–187.
6. Ayad, S. S. (1987) "A turbulence model for wind flows above fire areas", Proc. second conf. on Aeronautical Sciences and Aviation Technology, Military Technical College, Cairo, Egypt, 29–39.
7. Kumar, A. (1986) "Effects of cross-wind shear on horizontal dispersion", J. Environmental Engineering, ASCE, **112**, 1–12.
8. Moran, M. J. and Gaggioli, R. A. (1968) "Similarity analysis via group theory", AIAA J. **6**, 2014–2016.
9. Moran, M. J. and Gaggioli, R. A. (1968) "A new systematic formalism for similarity analysis with application to boundary layer flows", U.S. Army Math. Research Center, Tech. Summary Report 918.
10. Moran, M. J. and Gaggioli, R. A. (1968) "Reduction of the number of variables in systems of partial differential equations with auxiliary conditions", SIAM J. Applied Mathematics **16**, 202–215.
11. Toson, S. (1992) "Dispersion predictions of air pollutants with an occurrence of temperature inversion", submitted for publication in J. Engineering and Applied Science, Cairo University, Egypt.

APPLICATIONS OF THE GENERALIZED SYMMETRY METHOD

G. BAUMANN
Abteilung für Mathematische Physik,
Universität Ulm
Albert-Einstein-Allee 11,
D-7900 Ulm, Germany
e-mail: bau@theophys.physik.uni-ulm.de

Abstract. In the following we shall demonstrate the applications of the genaralized symmetry method on two ions in a trap, the Hénon Heiles model, and a system of nonlinear quartic oscillators. The method of deriving generalized symmetries is reviewed and the construction of integrals of motion is demonstrated. The conditions of integrability follow as side conditions from the determining equations for the generalized characteristics in which cases the corresponding integrals of motion are derivable. The generalized symmetry analysis is supported by a *Mathematica* program called **LieBaecklund**.

1. Introduction

At the end of the last century, Lie considered the invariance of differential equations under the transformation of dependent and independent variables [1]. Lie was able to classify and solve some types of ordinary as well as partial differential equations. Although very general, his original work mainly considered the geometric aspects of the transformations related to the equations. Essentially, he considered invariance under point transformations and demonstrated that some sort of contact transformations are related to point transformations in case of first order partial differential equations. However, it turns out that much wider classes of transformations leave differential equations invariant including those considered by Lie. Emmi Noether [2] and Felix Klein [3] demonstrated that a system of differential equations derivable from a variational principle allows a much greater set of transformations under which a given system of differential equations is invariant. They significantly extended the application of symmetry group methods by including derivatives of the dependent variables in the transformations. Moreover, they were able to offer a regular procedure to construct the related conservation laws which are based on the investigation of the invariance properties of the variational integral under the action of transformation groups. Since Noether published her work, it is well known that Lie-type point symmetries cannot provide for the complete symmetry group underlying a given system of differential equations. Consequently, the required integrals of motion and the solutions are not available for many systems.

In recent years, the symmetry methods have become more attractive, especially in the field of nonlinear dynamics [4, 5]. In this article, we will consider a generalized method to determine the symmetries for nonlinear ordinary differential equations (ODE) of second order. We will demonstrate that the method is applicable on models such as the Hénon-Heiles model, two dimensional quartic oscillators relevant

N. H. Ibragimov et al. (eds.),
Modern Group Analysis: Advanced Analytical and Computational Methods in Mathematical Physics, 43–53.

in the description of light propagation in a fiber, and the motion of two ions in a Paul or Penning trap. In section 2, we will outline the mathematical procedure used in *Mathematica* programs based on the Lie group approach. In section 3, we will give some results for the above models, generalized symmetries and integrals of motion. Finally, some concluding remarks are given in section 4.

2. Generalized symmetries and consrvation laws

A great number of problems in physics can be described by second order ordinary differential equations

$$\Delta_k(t, \{q_j\}, \{\dot{q}_j\}, \{\ddot{q}_j\}) = 0 \qquad k, j = 1, 2, \ldots, N \tag{1}$$

where $\{q_j\} = \{q_j(t)\}$ are the set of the dependent variables. Dots denote differentiation with respect to the independent variable t. Δ_k are given functions of dependent and independent variables which for example can be considered as Newton's or Lagrange's equations.

Let us assume that (1) can be derived from a generating function of the S action by applying the variational calculus and by making a variation of the individual paths q_j's in

$$S = \int_{t_1}^{t_2} \mathcal{L}(t, \{q_j\}, \{\dot{q}_j\})dt. \tag{2}$$

We well know that under these assumptions the system of equations (1) are the Euler-Lagrange equations $E_k(\mathcal{L}) = \Delta_k = \frac{\delta S}{\delta q_k} = 0$, $k = 1, 2, \ldots, N$ derivable by the variation of the path q_k. The symbol $\frac{\delta}{\delta q_k} = \sum_{n=0}^{\infty}(-1)^n \frac{d^n}{dt^n} \frac{\partial}{\partial q_k^{(n)}}$ denotes the variational derivative or the Euler operator. At this preliminary stage of our examination, we know the equations of motion and probably the physical interpretation contained in the equations. It is, however, unknown to us under which conditions solutions for the system (1) can be found.

2.1. Generalized symmetries

As mentioned above, Lie's original procedure $pr^{(2)}\vec{v}\Delta_k)|_{\Delta_k=0} = 0$, $k = 1, 2, \ldots, N$ to determine symmetries of ODE's does not provide all symmetries of the equations. In turn, we have to use a generalized method to uncover the 'hidden' symmetries. A generalization of the point transformations can be obtained, if the infinitesimals τ and ϕ_j do not only depend on the dependent and independent variables but also on derivatives of the q_j's. In case of second order differential equations, only the space of coordinates by the first derivatives needs to be extended. All higher derivatives can be eliminated by the differential equations and its derivatives. To denote the extended dependencies including derivatives, we use square brackets $[q] = (t, \{q_j\}, \{\dot{q}_j\})$ in the infinitesimals $\tau = \tau[q]$, and $\phi_j = \phi_j[q]$.

The corresponding generalized vector field for such transformations can be formally given by

$$\vec{v} = \tau[q]\frac{\partial}{\partial t} + \sum_{j=1}^{N} \phi_j[q]\frac{\partial}{\partial q_j}. \tag{3}$$

In a second step, we can introduce so-called evolutionary symmetries which are related to the generalized symmetries by $Q_j[q] = \phi_j[q] - \tau[q]\dot{q}_j$. The corresponding vector field has the representation

$$\vec{v}_Q = \sum_{j=1}^{N} Q_j[q]\frac{\partial}{\partial q_j}. \tag{4}$$

The symmetries contained in (3) and (4) are essentially the same [5]. The only difference is that (4) considers transformations of the dependent variables and derivatives in the extended form while (3) additionally takes into account a transformation of the independent variables. The main adventage of the evolutionary representation is its simplicity in the representation of the prolongation

$$\mathbf{pr}\vec{v}_Q = \sum_{j=1}^{N}\sum_{J=0}^{\infty}\frac{d^J}{dt^J}\left(Q_j[q]\right)\frac{\partial}{\partial q_j^{(J)}}. \tag{5}$$

The assumption in the calculation of the related symmetries of (1) is the same as in the case of point symmetries. Thus, the computation of the generalized symmetries proceeds in essentially the same way as the computation of Lie point symmetries. Since we consider the variables and derivatives as independent up to a fixed order, we not only take equations (1) for themselves into account but also consider the derivatives of (1) as concerns the invariance condition. The invariance condition remains nearly the same and reads

$$\mathbf{pr}\vec{v}_Q\left(\Delta_k\right)|_{\frac{d^J}{dt^J}\Delta_k=0} = 0 \qquad \text{with} \qquad J = 0,1,2,\ldots \quad \text{and} \quad k = 1,2,\ldots,N. \tag{6}$$

Contrary to point symmetries, differential equations can be used to eliminate higher derivatives and also the differential consequences of equations (1). Equation (6) provides the determining equations for the characteristics Q_j. Though linear, it is very complicated to solve this system of equations as it yields a large overdetermined system of equations. With some skill, solutions of (6) can be found by some ad hoc ansatz for the Q_j's; for example, by assuming that Q_j has a special dependence (powers, polynomials, etc.) on one or some of its arguments. By this method, equation (6) will split into several equations that may or may not have a nonzero solution. In our *Mathematica* program **LieBaecklund**, we use for second order ODEs a polynomial ansatz in the coordinates and momenta. If we can solve the huge system of linear equations in the coefficients introduced by the ansatz, we will have determined the characteristics up to some power in the variables q_j and momenta $\dot{q}_j$. This information can in turn, be used to construct conservation laws and invariant solutions. For partial differential equations, **LieBaecklund** delivers a simplified system of linear partial differential equations determining the characteristics Q_j's.

2.2. Conservation laws

With equation (2), we assumed that the system of equations (1) is derivable by a variation of the path q_j. In close analogy to the question of invariance of (1), we can ask for the invariance of the action itself. This means that we have to assume

$$\int_{t_1}^{t_2}\mathcal{L}(t,\{q_j\},\{\dot{q}_j\})dt = \int_{t_1^*}^{t_2^*}\mathcal{L}(t^*,\{q_j^*\},\{\dot{q}_j^*\})dt^*. \tag{7}$$

Two questions arise at this points **i)** Are the group elements $g \in G$ of (7) derivable from infinitesimal conditions? **ii)** How are the symmetries of the Euler-Lagrange equations related to these variational symmetries? It is obvious that local transformations preserving the invariance of a functional transform one solution of the variational problem into another. This must be the case if relation (7) is correct. Thus, we can use the Lie procedure applied to the Euler-Lagrange equations corresponding to (2), taking into account that not all symmetries of the equations (1) are symmetries of (2). To uncover the variational symmetries, we can use the methods discussed so far but need to check in addition the relation [5]

$$\mathbf{pr}\vec{v}_Q(\Delta_k) + D_Q^*(\Delta_k) = 0 \qquad k = 1, 2, \ldots, N \tag{8}$$

where D_Q^* denotes the adjoint Fréchet derivative of the characteristic Q. The evolutionary vector field $\vec{v}_Q$ with its characteristics Q is a variational symmetry if equation (8) holds for all t and q_j's. We now have a criterion singling out these symmetries from the generalized symmetries which are also variational symmetries. Noethers theorem establishes the connection between the variational symmetries with their characteristics Q_j and the conserved quantities. The case of one independent variable can be summarized in the formula

$$\frac{d}{dt}I = \sum_{j=1}^{N} Q_j \Delta_j = 0 \tag{9}$$

which is a special case of the more general divergence formula given by Olver [5]. The essential points in determining integrals of motion for a given Euler-Lagrange system are i) to find the characteristics Q_j for the variational symmetries, and ii) to integrate equation (9). These two, are steps among other implemented in the *Mathematica* function **LieBaecklund**. To show how the outlined analysis can be used in practical applications, we will give some examples demonstrating these *Mathematica* functions.

3. Models of second order

In this section, we will deal with the application of the generalized symmetry method previously introduced. We discuss the two ion problem in a Paul trap and present results for the Hénon–Heiles model, and a quartic anharmonic oscillator. We will describe how the method works and where its shortcommings come into play. All calculations are supported by the two *Mathematica* programs **Lie** [6] and **LieBaecklund**.

3.1. Two ions in a trap

The two-ion motion confined to a Paul trap has been frequently discussed in different contexts. Experiments were carried out by Blümel et al. [9] and Hoffnagle et al. [10]. The experimental settings of a dynamical ion trap were first described by Paul et al. [11] in the late fifties. The motion of two ions in such a trap is governed mainly by two forces, the quadrupole force $F^{(trap)}$ and the Coulomb forces $F^{(coul)}$.

The quadrupole force in the center of the trap is derivable from the time dependent potential $\Phi^{(trap)} = (U_0 + V_0 \cos(\Omega t))(x^2 + y^2 - 2z^2)/(r_0^2 + 2z_0^2)$, where U_0, V_0 and Ω denote amplitudes of *dc* and *ac* voltage, and Ω is the corresponding driving frequency. r_0 and z_0 give the geometric extensions of the trap in radial and axial directions. The basic equations which we will consider here are derivable in a secular approximation from the generalized Mathieu equations for the relative coordinates. Applying Landau's fast field averaging [12], we can write down the equations of motion in cylindrical coordinates to read

$$\Delta = \begin{pmatrix} \ddot{\rho} - \frac{\nu^2}{\rho^3} + \rho - \dfrac{\rho}{(\rho^2 + \zeta^2)^{3/2}} \\ \ddot{\zeta} + \lambda^2 \zeta - \dfrac{\zeta}{(\rho^2 + \zeta^2)^{3/2}} \end{pmatrix} = 0. \tag{10}$$

λ is the ratio of secular frequencies and ν is the ratio of angular momentum ℓ_z and of the radial secular frequency. The coordinates of relative motion ρ, ζ and time τ are scaled by an appropriate length and the driving frequency. The second order equations (10) are derivable from the Lagrangian

$$\mathcal{L} = \frac{1}{2}(\dot{\rho}^2 + \dot{\zeta}^2) - \{\frac{1}{2}(\rho^2 + \lambda^2\zeta^2) + \frac{1}{(\rho^2 + \zeta^2)^{1/2}} + \frac{\nu^2}{2\rho^2}\}. \tag{11}$$

It is evident that equations (10) are highly nonlinear. With $\lambda = 0$, they have some resemblance to a two dimensional Kepler problem. In the language of classical mechanics, the equations describe the motion of one particle in a two-dimensional potential. The second ion in this picture is fixed at the origin of coordinates. The phase space of this Hamiltonian system is four-dimensional and thus is a potential candidate for chaotic motion. The question now is under which parameter conditions regular motion can be found.

We will consider transformations of dependent and independent variables (ρ,-$\zeta, \dot{\rho}, \dot{\zeta}, t$) to carry over solutions of (10) into solutions. We take invariance into account which is represented by equation (6). The vector field $\vec{v}$ in an evolutionary representation is here given by (4). The characteristics Q_i are functions of the dependent and independent variables $Q_i = Q_i(\rho, \zeta, \dot{\rho}, \dot{\zeta}, t)$ $(i = 1, 2)$. Expanding equation (6), we obtain a linear system of partial differential equations determining the characteristics Q_i.

Preliminary to the general procedure, we examined point symmetries of equations (10) by using the *Mathematica* program **Lie**. From a set of 18 equations, we obtained the characteristics Q_1 and Q_2 to be

$$Q_1 = -k_2\zeta - k_1\dot{\rho} \qquad Q_2 = k_2\rho - k_1\dot{\zeta}, \tag{12}$$

where the group constants k_1 and k_2 are arbitrary. The geometrical interpretation of k_1 and k_2 correspond to translational invariance in time and rotational invariance in the (ρ, ζ)-plane, respectively. We find that the symmetry given by k_1 exists independent of the choice of parameters λ und ν. Contrary to this global existence, rotational invariance exists only for $\lambda = \pm 1$ and $\nu = 0$. It is easy to check that the point symmetries of (10) are also symmetries of the Lagrangian (11). In cases in which symmetries are also variational symmetries, Noether's theorem provides the corresponding integrals of motion. Thus, we obtain the total energy E and the angular momentum ℓ corresponding to k_1 and k_2

$$E = \frac{1}{2}(\dot{\rho}^2 + \dot{\zeta}^2) + \frac{1}{2}(\rho^2 + \lambda^2\zeta^2) + \frac{1}{(\rho^2 + \zeta^2)^{1/2}} + \frac{\nu^2}{2\rho^2} \tag{13}$$

$$\ell = \rho\dot{\zeta} - \zeta\dot{\rho}. \tag{14}$$

as integrals of motion, respectively. Provided that the z-component of the angular momentum ℓ_z equals zero and that secular frequencies in axial and radial directions are identical, we find that the motion of the two ions is completely determined by the integrals E and ℓ. These two conserved quantities restrict the motion of the ions to a two-dimensional surface in the phase space. Thus, the motion of the ions is completely regular for $\lambda = \pm 1$ and $\nu = 0$.

Enlarging the symmetry class from point symmetries to generalized symmetries, we find four different parameter sets (λ, ν) for which integrals of motion exist. The generalized symmetries are provided if we use (6) in its explicit form

$$\tfrac{d^2}{dt^2}Q_1 + 3\nu^2\tfrac{1}{\rho^4}Q_1 + Q_1 - \tfrac{Q_1}{(\rho^2+\zeta^2)^{3/2}} + \tfrac{3Q_1\rho}{(\rho^2+\zeta^2)^{5/2}} + \tfrac{3Q_2\zeta}{(\rho^2+\zeta^2)^{5/2}} = 0 \tag{15}$$

$$\tfrac{d^2}{dt^2}Q_2 + \lambda^2 Q_2 - \tfrac{Q_2}{(\rho^2+\zeta^2)^{3/2}} + \tfrac{3Q_2\zeta}{(\rho^2+\zeta^2)^{5/2}} + \tfrac{3Q_1\rho}{(\rho^2+\zeta^2)^{5/2}} = 0. \tag{16}$$

This system of equations delivers a linear system of equations determining the unknown functions Q_1 and Q_2. Carrying out the solution, we get five cases of parameter combinations (λ, ν):

i) λ and ν arbitrary (conservation of energy),

ii) $\lambda = \pm 1$, $\nu =$ arbitrary (conservation of angular momentum),

iii) $\lambda = \pm 2$, $\nu =$ arbitrary,

iv) $\lambda = \pm 1/2$ and $\nu = 0$, and

v) $\lambda = \pm 1/2$ and $\nu =$ arbitrary.

For each of these cases we can determine a pair of characteristics Q_1 and Q_2 given by

case	Q_1	Q_2
i)	$\dot{\rho}$	$\dot{\zeta}$
ii)	$\rho\zeta\dot{\zeta} - \zeta^2\dot{\rho}$	$\rho\zeta\dot{\rho} - \rho^2\dot{\zeta}$
iii)	$\dot{\zeta}\rho - 2\zeta\dot{\rho}$	$\rho\dot{\rho}$
iv)	$\zeta\dot{\zeta}$	$\zeta\dot{\rho} - 2\rho\dot{\zeta}$
v)	$2\zeta\dot{\zeta}^2\left(\zeta\dot{\rho} - \rho\dot{\zeta}\right) + \frac{1}{2}\rho\zeta^3\dot{\zeta} - - \frac{2\rho\zeta\dot{\zeta}}{(\rho^2+\zeta^2)^{(1/2)}} + 2\nu^2\dot{\rho}$	$\rho\zeta^2\left(\frac{1}{2}\zeta\dot{\rho} - \rho\dot{\zeta}\right) + + 4\rho^2\dot{\zeta}^3 + \frac{2\rho}{(\rho^2+\zeta^2)^{(1/2)}}\left(2\rho\dot{\zeta} - \zeta\dot{\rho}\right) + 2\zeta\dot{\rho}\dot{\zeta}\left(\zeta\dot{\rho} - 3\rho\dot{\zeta}\right) + 2\nu^2\left(2\dot{\zeta} + \frac{\zeta^2\dot{\zeta}}{\rho^2}\right).$

A comparison of the characteristics **i)** and **ii)** with those obtained from point symmetries shows that time translation and rotational invariance are also generalized symmetries. Based upon the aforegoing, we know the invariance properties of equations (10) but not the related integrals.

To calculate the corresponding integrals of motion, we use

$$Q \cdot \Delta = \frac{d}{dt} I. \tag{17}$$

I is an invariant, $Q = (Q_1, Q_2)^T$ the characteristic, and Δ the left hand side of the Euler-Lagrange equations (10). In order to demonstrate that (17) essentially contains the information to determine integrals of motion, we will consider as an example the general case where λ and ν are arbitrary. With respect to case **i)**, we find from (17)

$$\begin{pmatrix} \dot{\rho} \\ \dot{\zeta} \end{pmatrix} \begin{pmatrix} \ddot{\rho} - \frac{\nu^2}{\rho^3} + \rho - \dfrac{\rho}{(\rho^2+\zeta^2)^{3/2}} \\ \ddot{\zeta} + \lambda^2\zeta - \dfrac{\zeta}{(\rho^2+\zeta^2)^{3/2}} \end{pmatrix} = \frac{d}{dt}E = 0, \tag{18}$$

where E can be calculated by quadrature to be

$$E = \frac{1}{2}(\dot{\rho}^2 + \dot{\zeta}^2) + \frac{1}{2}(\rho^2 + \lambda^2\zeta^2) + \frac{1}{(\rho^2+\zeta^2)^{1/2}} + \frac{1}{2}\frac{\nu^2}{\rho^2}. \tag{19}$$

The definition of momenta by $p_\rho = \dot{\rho}$ and $p_\zeta = \dot{\zeta}$ shows that E is nothing but the energy in the relative coordinates of the two ions in the Paul trap. This result is more or less trivial since we started from a Lagrangian.

With respect to case ii) with $\lambda = \pm 1$ and $\nu = 0$, we already know that ℓ is an invariant of motion. The question now is what happens if the z-component of ℓ is different from zero? If we insert the characteristics into (17) for this case, we get

$$\begin{pmatrix} \zeta(\rho\dot{\zeta}-\zeta\dot{\rho}) \\ \rho(\zeta\dot{\rho}-\rho\dot{\zeta}) \end{pmatrix} \begin{pmatrix} \ddot{\rho}-\frac{\nu^2}{\rho^3}+\rho-\dfrac{\rho}{(\rho^2+\zeta^2)^{3/2}} \\ \ddot{\zeta}+\zeta-\dfrac{\zeta}{(\rho^2+\zeta^2)^{3/2}} \end{pmatrix} = \frac{d}{dt}I^{(ii)} = 0. \tag{20}$$

The corresponding integral of motion $I^{(ii)}$ reads

$$I^{(ii)} = -\left[\frac{1}{2}(\rho\dot{\zeta}-\zeta\dot{\rho})^2 + \frac{1}{2}\frac{\nu^2}{\rho^2}\zeta^2\right]. \tag{21}$$

The expression (21) contains the square of the angular momentum ℓ and a term depending on the square of ℓ_z. At this point, we derive the first new result from the generalized symmetry analysis. In (21), we have a quadratic dependence in ζ/ρ generalizing the invariant ℓ known from point analysis.

With respect to case iii) with $\lambda = \pm 2$ and ν arbitrary, we can determine the second integral of motion in the same fashion to be

$$I^{(iii)} = \rho\dot{\zeta}\dot{\rho} - \zeta\dot{\rho}^2 + \rho^2\zeta - \frac{\zeta}{(\rho^2+\zeta^2)^{1/2}} - \frac{\nu^2\zeta}{\rho^2}. \tag{22}$$

Since ν is arbitrary in the calculation of the characteristics, we can choose $\nu = 0$ in order to find correspondence with results published by Blümel et al. [8]

$$I^{(iii)}\Big|_{\nu=0} = \rho\dot{\zeta}\dot{\rho} - \zeta\dot{\rho}^2 + \rho^2\zeta - \frac{\zeta}{(\rho^2+\zeta^2)^{1/2}}. \tag{23}$$

With respect to cases iv) and v), a different situation appears. Here, the symmetry analysis allows the combinations $\lambda = \pm 1/2$, $\nu = 0$, and $\lambda = \pm 1/2$, $\nu =$ arbitrary. The generalized symmetry for $\nu = 0$ gives the second integral of motion in the form

$$I^{(iv)} = \dot{\zeta}\dot{\rho}\zeta - \rho\dot{\zeta}^2 + \frac{1}{4}\zeta^2\rho - \frac{\rho}{(\rho^2+\zeta^2)^{1/2}}. \tag{24}$$

This is the second case where integrals of motion given in [8] coincide with our results. Contrary to the relations in [8], we find from the generalized symmetry v) with $\nu \neq 0$ and $\lambda = \pm 1/2$ the following representation of the second integral

$$\begin{aligned} I^{(v)} = {} & \frac{\nu^4}{\rho^2} + \nu^2\rho^2 + \frac{\nu^2\zeta^2}{2} + \frac{\rho^2\zeta^4}{16} + \frac{\rho^2}{\rho^2+\zeta^2} + \frac{2\nu^2}{\sqrt{\rho^2+\zeta^2}} \\ & - \frac{\rho^2\zeta^2}{2\sqrt{\rho^2+\zeta^2}} + \nu^2\dot{\rho}^2 + \frac{\rho\zeta^3\dot{\rho}\dot{\zeta}}{2} - \frac{2\rho\zeta\dot{\rho}\dot{\zeta}}{\sqrt{\rho^2+\zeta^2}} \\ & + 2\nu^2\dot{\zeta}^2 + \frac{\nu^2\zeta^2\dot{\zeta}^2}{\rho^2} - \frac{\rho^2\zeta^2\dot{\zeta}^2}{2} + \frac{2\rho^2\dot{\zeta}^2}{\sqrt{\rho^2+\zeta^2}} + \zeta^2\dot{\rho}^2\dot{\zeta}^2 \\ & - 2\rho\zeta\dot{\rho}\dot{\zeta}^3 + \rho^2\dot{\zeta}^4. \end{aligned} \tag{25}$$

Thus, we succeeded in an analytic construction of the second integral of motion for this case.

However, we note that in cases **ii)** - **v)**, there exists beside of the given second integrals the integral of energy. In all cases for which generalized symmetries are known, we find that the two ions are confined to a two-dimensional surface in phase space and are thus integrable. The reason for the integrability is the fact that the ratios of the two secular frequencies have specific values. With respect to case **ii)**, both frequencies are equal; in case **iii)**, the axial frequency is twice as large as the radial frequency ($\lambda = \pm 2$); and with respect to $\lambda = \pm 1/2$, the axial frequency is one half of the radial frequency. We note that the characteristics in cases **iii)** and **iv)** are well known from the classical Kepler problem corresponding to the Runge-Lenz vector. The difference between the classical Kepler problem with its $1/r$ dependence on the potential and the Paul trap is an additional harmonic part in the potential and its two-dimensionality. However, the symmetry of the Runge-Lenz vector of the Kepler problem is identical to the symmetry of the Paul trap. What has changed are the corresponding integrals of motion for the Paul trap.

Similar calculations for the Hénon Heiles model and a quartic anharmonic oscillator can be easily done. Using the *Mathematica* program **LieBaecklund** the integrals of motion and the special parameter combinations are derivable in a straightforward way. The results of such calculations are given in the Appendix.

4. Concluding remarks

We demonstrated the application of the generalized symmetry method (Lie–Bäcklund method) on some second order ordinary differential equation. We find that the application of this method will give us a straightforward way to determine the related integrals of motion and to decide the question of integrability. For the Hénon–Heiles model, the quartic anharmonic oszillators, and the two ions in a Paul trap, we calculated integrals of motion by applying a generalization of Lie's method. The calculation for all models was supported by the *Mathematica* programs **Lie** and **LieBaecklund**. These programs allow us to find the simplified determining equations of the characteristics and make an ansatz of polynomial form in case of the **LieBaecklund** program for second order ODEs. However, this ansatz of polynomials in the coordinates and derivatives does not give us the most general solutions for the characteristics but instead provides us with a form which is managable on a computer. The Lie group approach in its generalized form is particularly effective since it furnishes both the generalized Lie symmetries and all their invariants. For the models under discussion, it was possible to identify all generalized symmetries with variational symmetries and thus, the integrals of motion are available. The relations among the parameters follow as side conditions from a linear system of determining equations. By using the polynomial ansatz for the characteristics, we have an effective scheme to determine the symmetries. The corresponding parameter combinations are consistent with those derived from Painlevé tests. The advantage of the method under consideration is that it furnishes the parameters and the integrals of motion in a constructive way.

5. Appendix

5.1. The Hénon-Heiles model

The Hénon-Heiles model for gravitating stars in a cylindrical galaxy [13] is given by

$$\Delta = \begin{pmatrix} \ddot{q}_1 + Aq_1 + 2Dq_1q_2 \\ \ddot{q}_2 + Bq_2 + Dq_1^2 - Cq_2^2 \end{pmatrix} = 0 \tag{26}$$

where A, B, C and D are constants which determine the dynamical behavior of equation (26). Given an arbitrary choice of parameters, chaotic evolution occurs for q_1 and q_2 above a critical energy level [13, 14]. The non-chaotic cases are given in table I where also the generalized symmetries and integrals of motion are listed.

TABLE I
The Hénon-Heiles model

Parameters	Q_1	Q_2	Integral
A, B, C, D	$\dot{q}_1$	$\dot{q}_2$	$\frac{1}{2}(p_1^2+p_2^2)+\frac{A}{2}q_1^2+\frac{B}{2}q_2^2+Dq_1^2q_2-\frac{C}{3}q_2^3$
$A = B$, $D = -C$	$\dot{q}_2$	$\dot{q}_1$	$\dot{q}_1\dot{q}_2 + \frac{D}{3}q_1^3 + Dq_1q_2^2 + Aq_1q_2$
$A, B, C = -6D$	$-2q_2\dot{q}_1 + \frac{1}{2D}(4A-B)\dot{q}_1 + q_1\dot{q}_2$	$q_1\dot{q}_1$	$\dot{q}_1(q_1\dot{q}_2 - q_2\dot{q}_1) + \frac{4A-B}{4D}\dot{q}_1^2 + \frac{D}{4}q_1^4 + Dq_1^2q_2^2 + Aq_1^2q_2 + \frac{A(4A-B)}{4D}q_1^2$
$B = 16A$, $C = -16D$	$(2Dq_2q_1^2 + Aq_1^2)\dot{q}_1 - \frac{1}{3}Dq_1^3\dot{q}_2 + \dot{q}_1^3$	$-\frac{1}{3}Dq_1^3\dot{q}_1$	$\frac{1}{4}\dot{q}_1^4 + \left(\frac{A}{2}q_1^2 + Dq_1^2q_2\right)\dot{q}_1^2 - \frac{D}{3}q_1^3\dot{q}_1\dot{q}_2 - \frac{D^2}{18}q_1^6 - \frac{D^2}{3}q_1^4q_2^2 - \frac{AD}{3}q_1^4q_2 + \frac{A^2}{4}q_1^4$

5.2. Two dimensional quartic oscillators

Some physical models from nonlinear optics [7] and accelerator physics, for example, can be formulated in a standard form for a coupled system of anharmonic oscillators

$$\Delta = \begin{pmatrix} \ddot{q}_1 + \frac{\alpha_1}{2}q_1 - 2\Gamma_1 q_1^3 - 2q_1q_2^2 \\ \ddot{q}_2 + \frac{\alpha_2}{2}q_2 - 2\Gamma_2 q_2^3 - 2q_2q_1^2 \end{pmatrix} = 0 \tag{27}$$

where α_i and Γ_i determine the strength of quadratic and quartic terms in the harmonic and anharmonic part of the potential, respectively. The integrals of motion, generalized symmetries, and parameter combinations are given in table II.

TABLE II
Two dimensional quartic oscillators

Parameters	Q_1	Q_2	Integral
$\alpha_1, \alpha_2, \Gamma_1, \Gamma_2$	$\dot{q}_1$	$\dot{q}_2$	$\frac{1}{2}\sum_{i=1}^{2}\{\dot{q}_i^2 + \frac{\alpha_i}{2}q_i^2 - \Gamma_i q_i^4\} - q_1^2 q_2^2$
$\alpha_1 \neq \alpha_2$, $\Gamma_1, \Gamma_2 = 1$	$\frac{1}{2}(\alpha_1 - \alpha_2)\dot{q}_1 +$ $+ q_2(q_2\dot{q}_1 - q_1\dot{q}_2)$	$q_1(q_1\dot{q}_2 -$ $- q_2\dot{q}_1)$	$\frac{1}{2}(\alpha_1 - \alpha_2)(q_1^4 + q_1^2 q_2^2 - \frac{\alpha_1}{2}q_1^2 - \dot{q}_1^2) -$ $- (q_1\dot{q}_2 - q_2\dot{q}_1)^2$
$\alpha_1 = \alpha_2$, $\Gamma_1 = 1/3$, $\Gamma_2 = 1/3$	$\dot{q}_2$	$\dot{q}_1$	$\dot{q}_1\dot{q}_2 + q_1 q_2\left(\frac{\alpha_1}{2} - \frac{2}{3}(q_1^2 + q_2^2)\right)$
$4\alpha_1 = \alpha_2$, $\Gamma_1 = 1/6$, $\Gamma_2 = 8/3$	$q_1\dot{q}_2 - 2q_2\dot{q}_1$	$q_1\dot{q}_1$	$\dot{q}_1(q_1\dot{q}_2 - q_2\dot{q}_1) - \frac{2}{3}q_1^2 q_2^3 - \frac{1}{3}q_1^4 q_2 +$ $+ \frac{\alpha_1}{2}q_1^2 q_2$
$\alpha_2 = 4\alpha_1$, $\Gamma_1 = 1/3$, $\Gamma_2 = 8/3$	$\frac{8}{3}q_1^3 q_2\dot{q}_2 - 8q_1^2 q_2^2\dot{q}_1 +$ $+ 4\dot{q}_1^3 - \frac{4}{3}q_1^4\dot{q}_1 +$ $+ 2\alpha_1 q_1^2\dot{q}_1$	$\frac{8}{3}q_1^3 q_2\dot{q}_1 -$ $- \frac{4}{3}q_1^4\dot{q}_2$	$\dot{q}_1^4 + 4\left(\frac{\alpha_1}{4} - \frac{1}{6}q_1^2 - q_2^2\right)q_1^2\dot{q}_1^2 +$ $+\frac{8}{3}q_1^3 q_2\dot{q}_1\dot{q}_2 + \frac{1}{9}(q_1^2 + 2q_2^2)^2 q_1^4 - \frac{2}{3}q_1^4\dot{q}_2^2 +$ $+ \alpha_1\left(\frac{\alpha_1}{4} - \frac{1}{3}q_1^2 - \frac{2}{3}q_2^2\right)q_1^4$

References

1. S. Lie, Math. Ann. **32**, 313 (1988).
2. E. Noether, Nachr. Ges. Wiss. Göttingen Math. Phys. **2**, 235 (1918).
3. F. Klein, Nachr. Ges. Wiss. Göttingen Math. Phys. **2**, 171 (1918).
4. G.W. Bluman and S. Kumei, Symmetries and differential equations (Springer, New York, 1989).
5. P.J. Olver, Applications of Lie groups to differential equations (Springer, Berlin, 1986).
6. G. Baumann, Lie symmetries of differential equations: A *Mathematica* program to determine Lie symmetries. (Wolfram Research Inc., Champaign, Illinois, MathSource 0202-622 1992).
7. G. Baumann and T.F. Nonnenmacher, in: Nonlinear evolution equations: integrability and spectral methods, p. 229. Eds. A. Degasperis, A.P. Fordy and M. Lakshmanan, (Manchester University Press, Manchester 1990).
8. R. Blümel, C. Kappler, W. Quint and H. Walther, Phys. Rev. A **40**, 808 (1989).
9. R. Blümel, J.M. Chen, E. Peik, W. Quint, W. Schleich, Y.R. Shen and H. Walther, Nature **334**, 309 (1988).
10. J. Hoffnagle, R.G. DeVoe, L. Reyna and R.G. Brewer, Phys. Rev. Lett. **61**, 255 (1988).
11. W. Paul, O. Osberghaus and E. Fischer, Ein Ionenkäfig, Forschungsberichte des Wissenschafts- und Verkehrsministeriums Nordrhein-Westfalen 415 (1958).
12. L.D. Landau and E.M. Lifschitz, Mechanik (Akademie-Verlag, Berlin, 1981)
13. M. Hénon and C. Heiles, Astrophys. **69**, 73 (1964).
14. R.H.G. Helleman, In: Fundamental problems in statistical mechanics Ed. E.G.D. Cohen (North-Holland, Amsterdam, 1980).

CONFORMAL INVARIANCE, HUYGENS PRINCIPLE AND FUNDAMENTAL SOLUTIONS FOR SCALAR SECOND ORDER HYPERBOLIC EQUATIONS

YU. YU. BEREST and N. H. IBRAGIMOV
Institute of Mathematical Modeling of Russian Ac. Sci.
Miusskaya Sq. 4,
Moscow 125047, Russia

and

A. O. OGANESYAN
Yerevan State University, Department of Physics
Alek Manukyan St. 1,
Yerevan 375049, Armenia

Abstract. New results on group theoretical aspects of the Huygens principle are presented. An extension of Darboux-Lagnese-Stellmacher's transformation to the wave equations in spaces with non-trivial conformal group is considered. Fundamental solutions of selected Huygens' equations are constructed by using a new adaptation of the group theoretic technique for distributions. It is observed that the hierarchy of Huygens' equations is closely related to that of rational solutions for higher Korteweg-de Vries equations.

1. Huygens principle

The history of the principle goes back to the famous memoir of Christian Huygens [1]. Being put forward as a geometrical method for constructing wave fronts this principle was succeeded in describing various light phenomena. Its adequate mathematical formulation was given subsequently [2], [3], [4]. But Hadamard [5] was the first who raised the problem on Huygens principle, emphasized its importance and advanced essentially towards its solution. Hadamard formulated the problem within the framework of his integration theory of Cauchy problem for hyperbolic second order differential equations

$$\mathbb{L}[u] \equiv \sum_{i,j=1}^{n} g^{ij}(x)u_{x^ix^j} + \sum_{i=1}^{n} b^i(x)u_{x^i} + c(x) = 0, \tag{1}$$

$$u\Big|_S = f, \qquad \frac{\partial u}{\partial \nu}\Big|_S = g. \tag{2}$$

Here S is a space-like manifold of the dimension $n-1$, and $\frac{\partial u}{\partial \nu}$ is a derivative with respect to the normal ν to S.

Definition 1. *Eq.* (1) *satisfies the Huygens principle if the value of the solution of the Cauchy problem* (1), (2) *at any point* $P = (x^1, \ldots, x^n)$ *is determined by the Cauchy data given on the intersection of the initial manifold and the characteristic*

N. H. Ibragimov et al. (eds.),
Modern Group Analysis: Advanced Analytical and Computational Methods in Mathematical Physics, 55–69.

conoid with vertex at P (the Huygens principle in the sense of "Hadamard's minor premise").

In other words the Huygens principle states that the wave, carrying a disturbance initially limited in space and time, does not leave any trace behind.

The property of an operator $\mathbb{L}$ to be a Huygens' operator is not disturbed, if $\mathbb{L}$ is subject to so-called elementary transformations:

a) non-degenerate changes of coordinates x^i;

b) linear homogeneous transformation of the unknown function $u \to \lambda(x)u$, $\lambda(x) \neq 0$;

c) transition from $\mathbb{L}$ to $\mu(x)\mathbb{L}$, $\mu(x) \neq 0$.

Two equations are called equivalent when one of them is derived from another by means of elementary transformations.

Hadamard's problem is to single out the Huygens' operators from all hyperbolic operators. It was Hadamard who gave the first description of these operators. He proved that for $n = 2$ and odd n the Huygens' principle never holds, whereas for even $n \geq 4$ it is valid if and only if there is no logarithmic term in elementary solution of the adjoint equation

$$\mathbb{L}^*[u] = 0. \tag{3}$$

Let $\Gamma(x, x_0)$ be a square of the geodesic distance in Riemannian space V_n ($n \geq 4$ even) with tensor $g_{ij}(x)$, where $\sum_{j=1}^{n} g_{ij}g^{jk} = \delta_i^k$. The function [5]

$$u = V(x, x_0)\Gamma^{-\frac{n-2}{2}} - W(x, x_0)\ln\Gamma, \tag{4}$$

where

$$V(x, x_0) = \sum_{\nu=0}^{\frac{n-4}{2}} U_\nu \Gamma^\nu, \qquad W(x, x_0) = \sum_{\nu=\frac{n-2}{2}}^{\infty} U_\nu \Gamma^\nu$$

is called an elementary solution (in Hadamard's terms) of Eq. (3), if it satisfies Eq. (3) outside the characteristic conoid given by

$$\Gamma(x, x_0) = 0. \tag{5}$$

Hadamard's criterion states: Eq. (1) satisfies Huygens principle iff

$$W(x, x_0) = 0 \tag{6}$$

for each $x_0 \in V_n$ and each x on the conoid (5). According to the construction of the elementary solution (4), $W(x, x_0)$ is the solution of a Goursat problem with the data on the conoid (5). Consequently, the condition (6) is equivalent to

$$W(x, x_0)\Big|_{\Gamma=0} = U_{\frac{n-2}{2}}\Big|_{\Gamma=0} = 0 \tag{7}$$

The classical wave equation

$$u_{tt} - \sum_{i=1}^{n-1} u_{x^i x^i} = 0 \qquad (n \geq 4 \text{ even}) \tag{8}$$

is the simplest example of a Huygens' equation. It follows from Poisson's [6] and Tédone's [7] formulas and can be proved by Hadamard's criterion, too. These are the so-called trivial Huygens' equations.

There arises the question: do there exist non-trivial Huygens' equations, i.e., different from and non-equivalent to Eq. (8). Mathisson [8] and Asgeirsson [9] proved the following negative result:

Theorem 1. *Any Huygens' equation in a flat Lorentzian space V_4 is equivalent to the wave equation* (8).

The first examples of non-trivial Huygens' equations were constructed by Stellmacher [10], [11]. In [12]–[15] all Huygens' equations of the form

$$u_{tt} - \sum_{i=1}^{n-1} u_{x^i x^i} + c(t)u = 0 \tag{8'}$$

were described.

The class of Huygens' equations with variable principal part was discovered by Günther [16] and independently by Ibragimov and Mamontov [17]. These are the following equations

$$u_{tt} - u_{xx} - \sum_{i,j=1}^{n-2} a^{ij}(x-t)u_{y^i y^j} = 0. \tag{9}$$

They were investigated from group theoretic point of view in [18].

The special case $n = 4$ was considered in [19]. The Riemannian space V_4 with the metrics

$$ds^2 = dt^2 - dx^2 - \sum_{i,j=1}^{n-2} a_{ij}(x-t)dy^i dy^j \quad \left(\sum_{k=1}^{n-2} a_{ik}a^{kj} = \delta_i^j\right)$$

is an example of the space with a non-trivial conformal group.

Definition 2. ([20]) *The group of conformal transformations in a space V_n is called trivial if it coincides with the group of isometric motions in a space, conformal to V_n. If the conformal group is not trivial the corresponding space V_n is called a space with non-trivial conformal group.*

In the case $n = 4$, a complete description of Lorentzian spaces V_4 (i.e., Riemannian spaces of signature $(- - - +)$) with non-trivial conformal group was given by Bilialov (see, e.g., [20], sect. 41, and [18], sect. 8.5). This result can be formulated as follows:

Lemma 1. *Any Lorentzian space V_4 with a non-trivial conformal group G can be reduced by changing coordinates and mapping to a conformally equivalent space, to a space with the following metric tensor g^{ij}:*

$$(g^{ij}) = \begin{pmatrix} -1 & 0 & 0 & 0 \\ 0 & -a^{11}(x-t) & -a^{12}(x-t) & 0 \\ 0 & -a^{12}(x-t) & -1 & 0 \\ 0 & 0 & 0 & 1 \end{pmatrix},$$

where $\Delta = (a^{12})^2 - a^{11} < 0$. *The order of the conformal group* G *can be* 6, 7, *and* 15 *in the case of a conformally flat space.*

This result can be extended to any dimension n (Chupakhin [21], see also [18], sect. 8.5).

Let G be a transformation group in the space of variables (x, u) generated by

$$X = \sum_{i=1}^{n} \xi^i(x)\frac{\partial}{\partial x^i} + \sigma(x)u\frac{\partial}{\partial u}$$

and G_x be a restriction of G on x-space with the generator

$$X_1 = \sum_{i=1}^{n} \xi^i(x)\frac{\partial}{\partial x^i}.$$

Suppose that G is a symmetry group of Eq. (1). Then its restriction G_x is a subgroup of the conformal group of the space V_n, associated with Eq. (1).

Definition 3. *Eq.* (1) *for which the group* G_x *coincides with the group of all conformal transformations in* V_n *is called a conformally invariant equation.*

Lemma 2. ([18], sect. 10.5) *Any conformally invariant equation* (1) *in the Lorentzian space* V_n *with non-trivial conformal group is equivalent to* (9).

Let $n = 4$. Consider the validity of Huygens principle for equations obtained from Eq. (9) by adding lower order terms to the left-hand side (these equations are surely expected not to be equivalent to Eq. (9)). We are interested in the question if there exist any equations of the form (1) in the spaces V_4 with a non-trivial conformal group being not conformally invariant but those of Huygens' type. For $n = 4$, the answer proved to be negative according to the following

Theorem 2. ([18], [19]) *Let a Lorentzian space* V_4 *have a non-trivial conformal group. Then Eq.* (1) *in this space satisfies Huygens principle if and only if this equation is conformally invariant.*

This theorem completely solves the Hadamard's problem for spaces V_4 with non-trivial conformal group. Notwithstanding, the answer on the question above is affirmative for $n > 4$. The proper examples were constructed in [22], [23]. They are concerned with the following class of Huygens' equations:

$$\mathbb{L}_n^{(k)}[u] \equiv u_{tt} - u_{xx} - a^k(x-t)u_{y^k y^k} - \sum_{\substack{i,j=1 \\ i,j\neq k}}^{n-2} a^{ij}(x-t)u_{y^i y^j} + \\ + a^k(x-t)b_k(y^k)u = 0, \qquad k = 1, 2, \ldots, n-2 \tag{10}$$

which has been studied in [24]–[26].

Let $k = 1$ and

$$l_\mu = \frac{\partial}{\partial y^1} - \frac{\mu'(y^1)}{\mu(y^1)}, \qquad l^*_\mu = -\frac{\partial}{\partial y^1} - \frac{\mu'(y^1)}{\mu(y^1)}.$$

Here $\mu(y^1)$ is an arbitrary solution of the equation

$$\mu'' - b_1(y^1)\mu = 0. \tag{11}$$

Then the operator $\mathbb{L}^{(1)}_n$ may be written in the form

$$\mathbb{L}^{(1)}_n = \mathbb{P} + a^1(x-t)l^*_\mu l_\mu$$

where

$$\mathbb{P} = \frac{\partial^2}{\partial t^2} - \frac{\partial^2}{\partial x^2} - \sum_{i,j=2}^{n-2} a^{ij}(x-t)\frac{\partial^2}{\partial y^i \partial y^j}.$$

We define the operator

$$\tilde{\mathbb{L}}^{(1)}_n = \mathbb{P} + a^1(x-t)l_\mu l^*_\mu$$

which is referred to as l_μ-transformation or Darboux-Lagnese-Stellmacher's transformation. It is evident that

$$l_\mu \mathbb{L}^{(1)}_n = \tilde{\mathbb{L}}^{(1)}_n l_\mu. \tag{12}$$

Theorem 3. *Let $\mathbb{L}^{(1)}_n$ be a Huygens' operator and $\tilde{\mathbb{L}}^{(1)}_n$ be a l_μ-transformation of $\mathbb{L}^{(1)}_n$. Then $\tilde{\mathbb{L}}^{(1)}_{n+2}$ is a Huygens' operator, too.*

Denote by $\mathbb{L}^{(1)}_{0n}$ the operator $\mathbb{L}^{(1)}_n$ with $b^1 = 0$. Given a Huygens' operator $\mathbb{L}^{(1)}_{04}$, Theorem 3 provides other Huygens' operators with $n \geq 6$ variables. It follows that the following equations

$$u_{tt} - u_{xx} - a^1(x-t)u_{y^1y^1} - \sum_{i,j=2}^{n-2} a^{ij}(x-t)u_{y^iy^j} + \frac{a^1(x-t)q(q+1)}{(y^1)^2}u = 0$$

satisfy Huygens principle for $q = 0, 1, \ldots, (n-4)/2$.

Theorem 4. ([25], [27]) *Let the operator $\mathbb{L}^{(1)}_n$ satisfy Huygens principle. Then it can be obtained from the operator $\mathbb{L}^{(1)}_{0n}$ by l_μ-transformations applied for no more than $(n-4)/2$ times.*

It follows from this theorem and Eqs. (11), (12) that there exist operators l_{μ_0}, $l_{\mu_1}, \ldots, l_{\mu_q}$ with $q \leq (n-4)/2 + 1$ such that

$$l_{\mu_q} \cdots l_{\mu_0} \mathbb{L}^{(1)}_{0n} = \mathbb{L}^{(1)}_n l_{\mu_q} \cdots l_{\mu_0}$$

where

$$\mu_0 = \alpha y^1 + \beta, \qquad (\alpha, \beta = \text{const}),$$
$$(l_{\mu_k} l^*_{\mu_k})\mu_{k+1} = 0, \qquad k = 0, \ldots, q.$$

If m is a minimum number of transformations turning the operator $\mathbb{L}_{0n}^{(1)}$ into $\mathbb{L}_n^{(1)}$, then the functions μ_k $(k = 0, \ldots, m-1)$ are determined uniquely up to a constant factor [15], and moreover:

a) $$b_1(y') = -2\frac{d^2}{d(y^1)^2}\left[\ln \prod_{k=0}^{m-1} \mu_k(y^1)\right];$$

b) $$\prod_{\substack{k=0\\ 0\le l\le m-1}}^{l} \mu_k(y^1) = \mathbb{P}_l(y^1) \text{ is a polynomial of degree } \frac{1}{2}(l+1)(l+2);$$

c) each of $\mu_k(y^1)$ is a rational function.

Using these results one can prove the following

Theorem 5. *Let Eq. (10) satisfy Huygens principle and let $b_1(y^1) \neq 0$. Then*

a) singularities of the function $b_1(y^1)$ at points $a_1, \ldots, a_s$ are poles of the second order, and $s \le \frac{1}{2}m(m+1) \le \frac{(n-2)(n-4)}{8}$;

b) if the function $b_1(y^1)$ can be represented as

$$b_1(y^1) = \frac{\beta_j(y^1)}{(y^1 - a_j)^2}$$

in a neighbourhood of a singular point a_j, where $1 \le j \le s$, then $\beta_j(a_j)$ can assume only one of the following values:

$$\beta_j(a_j) = q_j(q_j+1) \qquad q_j = 0, 1, \ldots$$

where

$$\sum_{j=1}^{s} q_j(q_j+1) \le \frac{(n-4)(n-2)}{4}$$

and $\beta'(a_j) = 0$.

c) the equation

$$\mu_m'' - b_1(y^1)\mu_m = 0$$

has the following solution:

$$\mu_m = \frac{T_r(y^1)}{(y^1-a_1)^{q_1}\cdots(y^1-a_s)^{q_s}}$$

where $T_r(y^1)$ is a polynomial of order $r = m + 1 + \sum_{k=1}^{s} q_k$.

2. Fundamental solutions as group invariant solutions

Being properly extended to distributions group analysis provides an effective and unified method for finding fundamental solutions of linear differential equations. Having been first illustrated by the heat and Laplace equations [36] this approach was applied in [37] to the wave equation (9) in the spaces with non-trivial conformal group (see also [38], [39]). In this section a brief theoretical sketch of the method is followed by detailed examples on construction of fundamental solutions for a few hyperbolic Huygens' equations.

Let $\mathcal{D}(\mathbb{R}^n)$ be a space of test functions over $\mathbb{R}^n$ (i.e., the set of all C^∞-functions with compact support in $\mathbb{R}^n$). Let $\mathcal{D}'(\mathbb{R}^n)$ be a corresponding space of distributions (i.e., linear continuous functionals over $\mathcal{D}(\mathbb{R}^n)$). Consider a system of linear differential equations:

$$\mathbb{L}(x, D)[u] \equiv \sum_{i_1,\ldots,i_n} a_{i_1\ldots i_n}(x) D_1^{i_1} \ldots D_n^{i_n} u = F(x), \tag{13}$$

where $x = (x^1, \ldots, x^n)$, $u = (u^1, \ldots, u^m)$, $D_i = \frac{\partial}{\partial x^i}$ and $F(x) = (F_1(x), \ldots, F_m(x))$. The coefficients $a_{i_1\ldots i_n}(x)$ belong to the class C^∞ of infinitely differentiable functions while the components $F_k(x)$ are arbitrary distributions from $\mathcal{D}'(\mathbb{R}^n)$.

A symmetry group of Eq. (13) (in a view of linearity) should be looked for in the class of smooth invertible transformations of the following form

$$\begin{cases} \bar{x}^\mu = f^\mu(x, a) \\ \bar{u}^k = \varphi_l^k(x, a) u^l \end{cases} \tag{14}$$

where $f^\mu(x, 0) = x^\mu$, $\mu = 1, \ldots, n$; $\varphi_l^k(x, 0) = \delta_l^k$, $l, k = 1, \ldots, m$; and a is a group parameter.

It is convenient to express the infinitesimal representation of the transformations (14) in terms of Q-operators (see, e.g., [40]) which act on the space of the dependent variable u:

$$Q = \xi^\mu(x) \frac{\partial}{\partial x^\mu} + \eta(x) \tag{15}$$

The operator (15) corresponds to the following generator of the group (14):

$$X = \xi^\mu(x) \frac{\partial}{\partial x^\mu} - (\eta(x) u)^k \frac{\partial}{\partial u^k}, \tag{15$'$}$$

where

$$\xi^\mu(x) = \left. \frac{\partial f^\mu}{\partial a} \right|_{a=0}, \qquad \eta(x) = \left. \frac{\partial \varphi}{\partial a} \right|_{a=0}.$$

In this notation the determining equations are written

$$\begin{cases} [\mathbb{L}, Q] = \lambda(x) \mathbb{L} & (16) \\ \\ QF + \lambda(x) F = 0, & (17) \end{cases}$$

where $\lambda(x)$ is a $(m \times m)$-matrix of C^∞-functions to be a priori unknown.

Eq. (16) determines all symmetries of Eq. (13) with $F = 0$ in the class of transformations (14). Whereas Eq. (17), considered in the space of distributions, singles out the operators admitted by Eq. (13). In particular, if $m = 1$ and $F(x) = \delta(x - x_0)$ then Eq. (17) is equivalent to

$$\begin{aligned} &\lambda(x_0) + \eta(x_0) - \sum_{\mu=1}^{n} \frac{\partial \xi^\mu(x_0)}{\partial x^\mu} = 0 \\ &\xi(x_0) = 0, \qquad \mu = 1, 2, \ldots, n. \end{aligned} \tag{17$'$}$$

We summarize: it is necessary, for calculating the symmetry group of Eq. (13) in the space of distributions:

i) to find a full group of point symmetries of the homogeneous equation (i.e., Eq. (13) with $F = 0$) in the class of transformations (14) by solving Eq. (16);

ii) to single out a subspace of such operators in the space of solutions of Eq. (16) that also satisfy Eq. (17) or, in particular, Eqs. (17′). The algebra of these operators will be a searched symmetry algebra for Eq. (13).

Given a fixed symmetry algebra with a basis of operators $\{Q_i\}$, Q_i being that of the form (15), the procedure of construction of invariant solutions for Eq. (13) in the class of distributions, i.e., weak solutions of the form $u = f(x)$, $f \in \mathcal{D}'$, consists in finding so-called weak invariants of the algebra, i.e., in solving the system of linear differential equations

$$Q_i f(x) = 0 \tag{18}$$

in the space $\mathcal{D}'$ of distributions. Weak invariant solutions of Eq. (1) are those from the space of common solutions of Eq. (18) and Eq. (13).

Now we apply our approach to Eq. (9). The operator

$$\mathbb{L}[u] = \left(D_t^2 - D_x^2 - \sum_{i,j=1}^{n-1} a^{ij}(x-t) D_{y^i} D_{y^j}\right) u = 0 \tag{19}$$

is self-adjoint. Therefore its fundamental solution is defined as a (weak) solution of the following problem:

$$\mathbb{L}[u] = \delta(t - t_0, x - x_0, y - y_0), \qquad u\Big|_{t<t_0} = 0 \tag{19′}$$

where $y - y_0 = (y_0^1 - y_0^1, \ldots, y^{n-1} - y_0^{n-1})$. This solution is to be of the following form

$$u = f(t, x, y, t_0, x_0, y_0)\theta(t - t_0), \tag{20}$$

t_0, x_0, y_0 being considered as parameters.

To find the group of point symmetries for (19′) one should first solve the determining equation (16). Eq. (19) is known to be conformly invariant (see lemma 1). Hence its symmetry group can be easily obtained as a proper extension of the conformal group of the space with a plane wave metrics. In non-special case we have (see [18]) a $(2n+1)$-dimensional algebra generated by the following operators

$$\begin{gathered} X_1 = \frac{\partial}{\partial t} + \frac{\partial}{\partial x}, \qquad X_2 = \frac{\partial}{\partial y}, \\ Q_i = -y^1\left(\frac{\partial}{\partial t} + \frac{\partial}{\partial x}\right) + \sum_{j=1}^{n-1} A^{ij}(x-t)\frac{\partial}{\partial y^j}, \\ Q_n = (t+x)\left(\frac{\partial}{\partial t} + \frac{\partial}{\partial x}\right) + \sum_{j=1}^{n-1} y^j \frac{\partial}{\partial y^j}, \quad Q_{n+1} = 1, \end{gathered} \tag{21}$$

where $i = 1, \ldots, n-1$, and $\|A^{ij}(x-t)\|$ is a matrix of functions depending on the one variable and being primitive for a^{ij}, i.e.,

$$A^{ij}(\xi) = \int a^{ij}(\xi)\, d\xi.$$

The subalgebra admitted by Eq. (19′) can be found by means of Eqs. (17′). It appears to be an n-dimensional algebra with the following basis:

$$\begin{aligned} \bar{Q}_i &= -(y^i - y_0^i)\left(\tfrac{\partial}{\partial t} + \tfrac{\partial}{\partial x}\right) + \textstyle\sum_{j=1}^{n-1}[A^{ij}(x-t) - A^{ij}(x_0 - t_0)]\tfrac{\partial}{\partial y}, \\ \bar{Q}_n &= -(t - t_0 + x - x_0)\left(\tfrac{\partial}{\partial t} + \tfrac{\partial}{\partial x}\right) + \textstyle\sum_{j=1}^{n-1}(y^j - y_0^j)\tfrac{\partial}{\partial y_j} + n - 1. \end{aligned} \tag{22}$$

The set of operators $\bar{Q}_i$ has exactly two common classical invariants:

$$I_1 = \Gamma(t, x, y, t_0, x_0, y_0), \qquad I_2 \equiv \sigma = x - t,$$

the latter being an invariant of the operator $\bar{Q}_n$ as well. Here Γ is a square of geodesic distance between the points (t, x, y) and (t_0, x_0, y_0) in the Lorentzian space with plane wave metrics:

$$\Gamma = (t-t_0)^2 - (x-x_0)^2 - (t - t_0 + x - x_0)\sum_{i,j=1}^{n-1} \widetilde{A}_{ij}(y^i - y_0^i)(y^j - y_0^j), \tag{23}$$

where $\|\widetilde{A}_{ij}\| = \|A^{ij}(x-t) - A^{ij}(x_0 - t_0)\|^{-1}$.

To be weak invariant under the operators $\bar{Q}_i$, $\bar{Q}_n$, the solution of the problem (19′) has to be expressed as follows,

$$u = h(\sigma)F(\Gamma)\theta(t - t_0), \tag{24}$$

where F is, in general, a distribution determined on the conoid and $h(\sigma)$ is a smooth function of the one variable σ. In addition, the invariance conditions (18) should be valid, i.e.,

$$\begin{aligned} &\bar{Q}_i u = -h(\sigma)(y^i - y_0^i)F(\Gamma)\delta(t - t_0) = 0, \qquad i = 1, \ldots, n-1, \\ &\bar{Q}_n u = h(\sigma)\theta(t-t_0)[2\Gamma F'(\Gamma) + (n-1)F(\Gamma)] + h(\sigma)(x - x_0)F(\Gamma)\delta(t - t_0) = 0, \end{aligned}$$

whence

$$\begin{cases} F(\Gamma)\delta(t - t_0) = 0, \\ 2\Gamma F'(\Gamma) + (n-1)F(\Gamma) = 0. \end{cases} \tag{25}$$

The latter equation has the following solution in distributions [41]:

$$F(\Gamma) = \begin{cases} C_1\delta^{(\frac{n-3}{2})}(\Gamma) + C_3\Gamma^{\frac{1-n}{2}}, & n \geq 3,\ n \text{ is odd} \\ C_2\left(\dfrac{\theta(\Gamma)}{\sqrt{\Gamma}}\right)^{(\frac{n-2}{2})} + C_4\Gamma^{\frac{1-n}{2}}, & n \geq 2,\ n \text{ is even} \end{cases} \tag{26}$$

Since (see [41], p. 289)

$$\lim_{t\to t_0} \delta^{(\frac{n-3}{2})}(\Gamma) = 0, \qquad \lim_{t \to t_0}\left(\frac{\theta(\Gamma)}{\sqrt{\Gamma}}\right)^{(\frac{n-2}{2})} = 0,$$

one can obtain that (26) satisfies the system (25) iff $C_3 = C_4 = 0$. Upon substituting (24) (with $F(\Gamma)$ given by (26)) into Eq. (19′) one can derive first the following equation:

$$\frac{d}{d\sigma}\ln h(\sigma) = \frac{n-1}{2\sigma} - \frac{1}{2}\frac{d}{d\sigma}\ln\left(\det \|\widetilde{A}^{ij}(\sigma)\|\right),$$

which is solved by

$$h(\sigma) = \left[\frac{|\sigma|^{n-1}}{|\det \|\widetilde{A}^{ij}(\sigma)\||}\right]^{\frac{1}{2}},$$

and, second, find the values $C_1 = \frac{1}{2\pi^{(n-1)/2}}$, $C_2 = \frac{1}{2\pi^{n/2}}$.

So, an explicit formula for the fundamental solution of Eq. (19) can be written as follows:

$$u = \begin{cases} \dfrac{\theta(t-t_0)}{2\pi^{(n-1)/2}}\left[\dfrac{|\sigma|^{n-1}}{|\det \|\widetilde{A}^{ij}(\sigma)\||}\right]^{\frac{1}{2}} \delta^{(\frac{n-3}{2})}(\Gamma) & n \geq 3,\ n \text{ is odd}, \\ \dfrac{\theta(t-t_0)}{2\pi^{n/2}}\left[\dfrac{|\sigma|^{n-1}}{|\det \|\widetilde{A}^{ij}(\sigma)\||}\right]^{\frac{1}{2}} \left(\dfrac{\theta(\Gamma)}{\sqrt{\Gamma}}\right)^{(\frac{n-2}{2})} & n \geq 2,\ n \text{ is even}. \end{cases} \tag{27}$$

Here $\|\widetilde{A}^{ij}(\sigma)\| = \|A^{ij}(x-t) - A^{ij}(x_0 - t_0)\|$ and Γ is given by (23). The distributions $\delta(\Gamma)$ and $\theta(\Gamma)$ are specially determined on the characteristic conoid (see, e.g., [41]):

$$\Gamma(t, x, y, t_0, x_0, y_0) = 0.$$

Remark. The fundamental solution of the classical wave equation (8) can be readily obtained from (27) if chosen: $a^{ij}(x-t) = \delta^{ij}$.

Another illustration of the method is provided by the Euler-Poisson-Darboux equation

$$\mathbb{L}[u] = u_{tt} - \sum_{i=1}^{n} u_{x^i x^i} - \frac{k}{t^2}u = 0 \tag{28}$$

which is known to satisfy Huygens principle iff n is odd, $n \geq 3$, $k = p(p+1)$, $p = 0, 1, \ldots, \frac{n-3}{2}$.

Solving the determining equations (16) for $\mathbb{L}$ given by (28), one can obtain the following basis generators of the symmetry algebra of Eq. (28):

$$\begin{gathered} Q_i = \frac{\partial}{\partial x_i}, \qquad Q_{ij} = x^i \frac{\partial}{\partial x_j} - x^j \frac{\partial}{\partial x_i} \quad (i < j) \\ Q_{n+k} = (t^2 - |x|^2)\frac{\partial}{\partial x_k} + 2x^k\left(t\frac{\partial}{\partial t} + \sum_{j=1}^{n} x^j \frac{\partial}{\partial x_j}\right) + (n-1)x^k \\ Q_{2n+1} = t\frac{\partial}{\partial t} + \sum_{j=1}^{n} x^j \frac{\partial}{\partial x_j}, \qquad Q_{2n+2} = 1, \end{gathered} \tag{29}$$

where $i, j, k = 1, \ldots, n$, $|x|^2 = \sum_1^n (x^i)^2$.

To construct the fundamental solution, we look for a weak solution of the associated problem

$$\mathbb{L}[u] = \delta(t - t_0, x - x_0), \qquad u\Big|_{t<t_0} = 0, \tag{28'}$$

where $x_0 = (x_0^1, \dots, x_0^n)$. Eq. (17′) singles out among (29) operators admitted by Eq. (28′). These are the following operators:

$$\bar{Q}^i = [(t^2 - |x|^2) - (t_0^2 - |x_0|^2)]\frac{\partial}{\partial x_i} + \\ + 2(x^i - x_0^i)\left(t\frac{\partial}{\partial t} + \sum_{j=1}^{n} x^j \frac{\partial}{\partial x_j}\right) + (n-1)(x^i - x_0^i) \tag{30}$$

$$\bar{Q}_{ij} = (x^i - x_0^i)\frac{\partial}{\partial x_j} - (x^j - x_0^j)\frac{\partial}{\partial x_i} \quad (i < j),$$

As discussed above we look for the fundamental solution of the given $\mathbb{L}$-operator as a weak invariant solution of Eq. (28′) under the operators (30). At first one should find the classical invariants of the operators $\bar{Q}_{ij}$ and $\bar{Q}_i$, the latter being restricted before on (x,t)-space. We have $I_1 = \Gamma$, $I_2 = \frac{\Gamma}{t}$, where $\Gamma = (t-t_0)^2 - |x-x_0|^2$. Hence a weak invariant solution of Eq. (28′) is that of the form $u = \theta(t-t_0)f(\Gamma/t)F(\Gamma)$, the left hand side satisfying the condition (18):

$$\bar{Q}_i u = 2t_0(x^i - x_0^i)f(\Gamma/t)F(\Gamma)\delta(t-t_0) + \\ + \theta(t-t_0)f(\Gamma/t)[2\Gamma F'(\Gamma) + (n-1)F(\Gamma)]. \tag{31}$$

Eq. (31) gives rise to the system (25), and the general form of weak invariant solutions can be represented as

$$u = C\theta(t-t_0)f(\Gamma/t)\delta^{(\frac{n-3}{2})}(\Gamma), \quad \text{when } n \text{ is odd.} \tag{32}$$

To find $f(\Gamma/t)$, we substitute (32) into (28′) and derive after simple computations the following o.d.e.:

$$H\left(\frac{1-\sqrt{4k+1}}{2}, \frac{1+\sqrt{4k+1}}{2}, \frac{n+1}{2}, (4t_0)^{\frac{1-n}{2}} f, z\right) = 0. \tag{33}$$

Here $z \equiv \frac{\Gamma}{4tt_0} + 1$ and

$$H(\alpha, \beta, \gamma, y, x) \equiv x(x-1)y'' + [(\alpha+\beta)x - \gamma]y' + \alpha\beta y = 0$$

is a hypergeometric Gauss' equation, the solution of which can be expressed in terms of the hypergeometric function:

$$F(\alpha, \beta, \gamma, x) = 1 + \sum_{k=1}^{\infty} \frac{\alpha(\alpha+1)\dots(\alpha+k-1)\beta(\beta+1)\dots(\beta+k-1)}{\gamma(\gamma+1)\dots(\gamma+k-1)}\frac{x^k}{k!}, \tag{34}$$

For Eq. (28) to be of Huygens type, the hypergeometric series (34) must be truncated, the degree of a polynomial obtained being no more than $\frac{n-3}{2}$ (i.e., the order of differentiation of δ-function in (32)). It is sure to be valid iff $\sqrt{4k+1} = 2p+1$, i.e., $k = p(p+1)$ and $p \le \frac{n-3}{2}$. In this case the fundamental solution of Eq. (28) can be represented in the explicit form

$$u = \frac{\theta(t-t_0)}{2\pi^{(n-1)/2}} F\left(-p, p(p+1), \frac{n+1}{2}, 1 + \frac{\Gamma}{4tt_0}\right) \delta^{(\frac{n-3}{2})}(\Gamma), \tag{35}$$

the constant $C = \frac{1}{2\pi^{(n-1)/2}}$ having been easily found by reducing (35) to the wave equation with $p = 0$.

3. Huygens' complex. Deformations preserving Huygens principle

The recent results in the soliton theory hint an unexpected link between a well-known KdV-hierarchy and a hierarchy of Huygens' equations in the spaces with non-trivial conformal group. For the sake of simplicity we'll concern here only to the hierarchy of Lagnese-Stellmacher equations ($8'$) in Minkowski space. An extension to the general class of spaces with non-trivial conformal group is immediate.

As mentioned in section 1, Lagnese and Stellmacher put forward a procedure for constructing all Huygens' operators of the form ($8'$). Let us put in Eq. (10) $a^{ij}(x-t) = \delta^{ij}$, $a^k(x-t) \equiv 1$ and consider $c(t)$ instead of $b(y^k)$ reducing thereby Eq. (10) to Eq. ($8'$). The Lagnese's method is based upon permutation formula (12). In this way any Huygens' equation ($8'$) with the coefficient $c(t)$ generates a one-parameter family of Huygens' equations with coefficients $\tilde{c}(t,\gamma)$:

$$\tilde{c} = c - 2\frac{d^2}{dt^2}\ln\mu \tag{36}$$

where $\mu(t)$ is a solution of the equation

$$\mu'' + c(t)\mu = 0. \tag{37}$$

Eqs. (36), (37) define a rational Darboux transformation which generates a hierarchy of rational solutions for KdV-equations (i.e., so-called Adler-Moser potentials; see [42], [43] and references therein). This simple observation as well as main results of the work [42] adapted properly give us an opportunity to geometrically describe the hierarchy of Huygens' operators in a given space with non-trivial conformal group.

Let V_n be an n-dimensional Minkowski space, $n \geq 4$ being even. The set of all Huygens' operators of the form ($8'$) represents a following finite sequence $\{\mathbb{L}_{nk}\}$ of operators depending on complex parameter γ:

$$\begin{aligned}
\mathbb{L}_{n0} &\equiv \Box_n = 0\\
\mathbb{L}_{n1} &\equiv \Box_n + c(t,\gamma_1) = 0\\
&\dots\dots\dots\dots\dots\dots\dots\dots\dots\\
\mathbb{L}_{nk} &\equiv \Box_n + c(t,\gamma_1,\dots,\gamma_k) = 0\\
&\dots\dots\dots\dots\dots\dots\dots\dots\dots\\
\mathbb{L}_{nd} &\equiv \Box_n + c(t,\gamma_1,\dots,\gamma_d) = 0
\end{aligned} \tag{38}$$

$c(t,\gamma_1,\dots,\gamma_{k+1})$ being derived from $c(t,\gamma_1,\dots,\gamma_k)$ by means of rational Darboux transformation (36) and $d = \frac{n-4}{2}$. Moreover, each element $\mathbb{L}_{nk}$ in the sequence (38) can be considered as a k-dimensional complex space $M_k(V_n)$, where the parameters $(\gamma_1,\dots,\gamma_k)$ define a global coordinate system therein. Not any coordinate change, however, is admitted but that to be so-called birational, i.e.,

$$\gamma_i^* = a_i\gamma_i + g_i(\gamma_1,\dots,\gamma_{i-1}), \qquad i = 1,\dots,k, \quad a_i \neq 0 \tag{39}$$

where g_i are certain rational functions such that γ_i can be expressed in terms of γ_i^* just in the same rational manner [42]. So one can refer to the space $M_k(V_n)$ as a (complex) cell of dimension k, any two such cells being not intersected. That's why a space $M(V_n)$ of all Huygens' operators (38) admits a finite cell partition, i.e., it represents a cell complex the dimension of which coincides, by definition, with the dimension of maximal cell: $\dim M(V_n) \equiv d = \frac{n-4}{2}$. We call the space $M(V_n)$ a Huygens (cell) complex over the Riemannian space V_n.

Now we outline in a heuristic manner the algebraic structure connected with Huygens hierarchy. The Huygens principle being invariant under elementary transformations (see sect. 1) means that the family of all Huygens' equations (1) admits an infinite equivalence group generated by

$$X = \xi^0(x,t)\frac{\partial}{\partial t} + \xi^i(x,t)\frac{\partial}{\partial x^i} + \eta(x,t)u\frac{\partial}{\partial u} \tag{40}$$

It elucidates, in particular, a link between the Huygens principle and conformal invariance in the case of the wave equations because the symmetry groups of the latter equations are generated by operators of the form (40) and coincide with conformal groups.

It is natural to calculate an equivalence group for the family (8′) taking a slightly more extended form of the operator

$$X = \xi^0(x,t)\frac{\partial}{\partial t} + \xi^i(x,t)\frac{\partial}{\partial x^i} + \eta(x,t)u\frac{\partial}{\partial u} + \mu(x,t,c)\frac{\partial}{\partial c} \tag{40′}$$

To find it, one can use the infinitesimal invariance criterion for Eq. (8′):

$$\begin{cases} \Box_n u + cu = 0 \\ \dfrac{\partial c}{\partial x_i} = 0 \end{cases} \tag{41}$$

The equivalence algebra obtained in this way is spanned by

$$\begin{gathered} X_0 = \frac{\partial}{\partial t}, \quad X_i = \frac{\partial}{\partial x_i}, \quad X_{n+2} = u\frac{\partial}{\partial u}, \\ X = t\frac{\partial}{\partial t} + x^j\frac{\partial}{\partial x_j} - 2c\frac{\partial}{\partial c}, \quad X_{ij} = x^i\frac{\partial}{\partial x_j} - x^j\frac{\partial}{\partial x_i} \quad (i<j). \end{gathered} \tag{42}$$

It is quite evident that the algebra with the basis (42) converts any Huygens' equation of the form (8′) into another Huygens' equation of the same form. So the Huygens complex described above remains invariant. Since we are interested in group deformations of the operator $\mathbb{L} = \Box_n + c(t)$ it is convenient to restrict the equivalence algebra to the (x,t)-space and rewrite the remainder operators in the canonical form:

$$\bar{X}_0 = c_t\frac{\partial}{\partial c}, \qquad \bar{X} = (tc_t + 2c)\frac{\partial}{\partial c}. \tag{42′}$$

It is remarkable that this algebra of deformations preserving Huygens principle admits nontrivial extension by means of the recursion operator of the KdV-hierarchy

$$R = -D_t^2 + 4c + 2c_t D_t^{-1}. \tag{43}$$

More precisely, we consider a Lie-Bäcklund algebra of operators

$$\{R^k(\bar{X}_0), R^j(\bar{X})\}_{k,j=0}^{\infty}$$

which includes respectively KdV-vector fields $R^k(\bar{X}_0)$ and so-called master symmetries fields $R^j(\bar{X})$. Being interpreted within the framework of the KdV-theory (see, e.g., [43]) this algebra revealed a few of remarkable properties. In particular, it is of Kac-Moody-Virasoro type, the KdV fields constituting a commutative ideal therein:

$$[R^k(\bar{X}_0), R^j(\bar{X}_0)] = 0,$$
$$[R^k(\bar{X}), R^j(\bar{X})] = (j-k)R^{j+k}(\bar{X}),$$
$$[R^k(\bar{X}_0), R^j(\bar{X})] = -\left(k+\frac{1}{2}\right) R^{j+k}(\bar{X}_0).$$

Moreover, it is proved in the work [43] that both KdV-flows and master symmetry ones preserve the hierarchy of rational solutions of KdV equations. So recalling the observation above we can conclude that the obtained algebra leaves invariant the Huygens complex. More specifically, this algebra leaves invariant any cell of the complex involved. It means that the Huygens complex represents a set of orbits of the deformation algebra.

One of the authors (Yu. Yu. Berest) would like to express his appreciation to T. Esirkepov and E. Ivashkevich for encouragement and helpful discussion.

References

1. Christian Huygens, *Traité de lumière ...*, Pierre van der Aa, The Hague, 1690; latest reprint: Dawsons of Pall Mall, London, 1966.
2. G. R. Kirchhoff, *Zur Theorie der Lichtstrahlen*, Sitzungsber K. Preuss Akad. Wiss, Berlin, 1882, 641–669.
3. E. Beltrami, *Sul principio di Huygens*, Ist. Lombardo Sci. Lett. Rend. (2) 22 (1889), 428–438.
4. V. Volterra, *Sur les vibrations des corps élastiques isotropes*, Acta Math. 18 (1894), 161–232.
5. J. Hadamard, *Lectures on Cauchy's problem in linear partial differential equations*, Yale Univ. Press, New Haven, Conn., 1923.
6. D. Poisson *Mémoire sur la théorie du son*, J. de l'École Polytéchn., 7 (1807), 319–392.
7. O. Tédone, *Sull'integrazione dell'equazione* $\frac{\partial^2\varphi}{\partial t^2} - \sum_{i=1}^m \frac{\partial^2\varphi}{\partial x_i^2} = 0$, Ann. di Mat. Pura Appl., 1 Ser. III (1898) 1–23.
8. M. Mathisson, *Le problème de Hadamard rélatif a la diffusion des ondes*, Acta Math., 71 (1939), 249–282.
9. L. Asgeirsson *Some Hints on Huygens' Principle and Hadamard's Conjecture*, Comm. Pure and Appl. Math., 9, N3, (1956), 307–327.
10. K. L. Stellmacher, *Ein Beispiel einer Huygensschen Differentialgleichung*, Nachr. Acad. Wiss. Göttingen Math.-Phys. 10, K1 11a, (1953), 133–138.
11. K. L. Stellmacher, *Eine Klasse Huygensscher Differentialgleichungen und ihre Integration*, Math Ann., 130 (1955), 219–233.
12. J. E. Lagnese, K. L. Stellmacher, *A method of generating classes of Huygens' operators*, J. Math. Mech., 17, N5, (1967), 461–472.
13. J. E. Lagnese, *A solution of Hadamard's problem for a restricted class of operators*, Proc. Amer. Math. Soc. 19 (1968), 981–988.
14. J. E. Lagnese, *A New Differential Operator of the Pure Wave Type*, J. Diff. Eq. 1 (1965), 171–187.
15. J. E. Lagnese, *The Structure of a Class of Huygens' Operators*, J. Math. Mech. 18, N12, (1969), 1195–1201.
16. P. Günther, *Ein Beispiel einer nichttrivialen Huygensschen Differentialgleichung mit vier unabhanginen variablen*, Arch. Rat. Mech. Anal. 18, N2, (1965), 103–106.
17. N. H. Ibragimov, E. V. Mamontov, *Sur le problème de J. Hadamard relatif à la diffusion des ondes*, C. R. Acad. Sci. Paris, 270 (1970), 456–458.

18. N. H. Ibragimov, *Transformation Groups Applied to Mathematical Physics*, Nauka, Moscow, 1983, (English transl.: Reidel, Dordrecht, 1985).
19. N. H. Ibragimov, *Huygens Principle*, in: *Some problems of Mathematics and Mechanics*, Nauka, Leningrad, 1970, (English transl.: Amer. Math. Soc. Transl. (2), 104 (1976), pp. 141–152).
20. A. Z. Petrov, *New Methods in the General Theory of Relativity*, Nauka, Moscow, 1966 (English transl.: *Einstein Spaces*, Pergamon Press, Oxford, New York, 1969).
21. A. P. Chupakhin, *Non-trivial conformal groups in Riemannian spaces*, Dokl. Akad. Nauk SSSR 246, N5, (1979) (English transl.: Soviet Math. Dokl. 20, N3, 596–599).
22. G. M. Kagan, *On a class of singular problems satisfying Huygens principle.* Dokl. Acad. Nauk SSSR, 256, N6, (1981), 1307–1311.
23. K. Nishiwada, *Huygens' principle for a generalized Euler-Poisson-Darboux equation*, Proc. Japan. Acad., 60, Ser. A, (1984), 197–200.
24. A. O. Oganesyan, *Generalization of Lagnese-Stellmacher's method on a class of equations satisfying Huygens' principle*, IAN Armenii, Ser. Math., 25, N5, (1990), 462–473. (Russian).
25. N. H. Ibragimov, A. O. Oganesyan, *Hierarchy of Huygens' equations in the spaces with non-trivial conformal group*, Uspechi Math. Nauk, 46, N3, (1991), 111–146 (Russian).
26. P. Günther, *Zur Gültigkeit des Huygensschen Prinzips bei partielle Differentialgleichungen vom normalen hyperbolischen Typus*, Berichte über Verh. Sachs. Akad. Wiss. Leipzig, Math.-Naturwiss. K1, 100, Heft 2, (1952).
27. A. O. Oganesyan, *Solution of Hadamard's Problem for Certain Class of Hyperbolic Equations with Variable Coefficients*, IAN Armenii, Ser. Math., 26, N2, (1991).
28. R. G. McLenaghan, *An explicit determination of the empty space-time on which the wave equation satisfies Huygens' principle*, Proc. Camb. Philos. Soc. 65, (1969), 139–155.
29. R. G. McLenaghan, *On the validity of Huygens' principle for second order partial differential equations with four independent variables. Part 1: derivation of necessary conditions*, Ann. Inst. Henri Poincaré, Sect. A. 20, N2, (1974), 153–188.
30. V. Wünsch, *Über selbstadjungierte Huygenssche Differentialgleichungen*, Math. Nachr., 47 (1970), 131–154.
31. B. Rinke, V. Wünsch, *Zum Huygensschen Prinzip bei der skalaren Wellengleichung*, Beitrage zur Analysis, 18 (1981), 43–75.
32. J. Hadamard, *The problem of diffusion of waves*, Ann. Math., 2 Ser. 43 (1942), 510–522.
33. R. Penrose, *Conformal Treatment of Infinity*, in: Relativité, Groupes et Topologie Les Houches Lectures, 1963 Summer School of Theor. Phys., Univ. Grénoble, New York: Gordon & Breach, (1964), 565–584.
34. N. H. Ibragimov, *On the group classification of second order differential equations*, Soviet Math. Dokl. 9 (1968), 1365–1369.
35. P. Günther, *Huygens' Principle and Hyperbolic Equations*, New York, Academic Press, (1988).
36. N. H. Ibragimov, *Primer of the Group Analysis*, Znanie, Moscow, 1989 (Russian).
37. Yu. Yu. Berest, *Construction of Fundamental Solutions for Huygens' Equations as Invariant Solutions*, Dokl. Akad. Nauk SSSR 317, N4, (1991), 786–789.
38. Yu. Yu. Berest, *Weak Invariants of Symmetry Group Applied for Constructing Fundamental Solutions*, Proc. of the International Workshop "Modern Group Analysis", Bashkiria, USSR, June, 1991.
39. N. H. Ibragimov, *Group Theoretic Treatment of Fundamental Solutions*, in: "Analysis, Manifolds and Physics", Kluwer, Dordrecht, to appear.
40. W. I. Fushchych, W. M. Shtelen, N. I. Serov, *Symmetry Analysis and Exact Solutions of Nonlinear Equations of Mathematical Physics*, Kiev, Naukova Dumka, 1986 (Russian).
41. I. M. Gel'fand, G. E. Shilov, *Generalized Functions*, vol. 1, Academic Press, New York, 1964.
42. M. Adler, J. Moser, *On a class of polynomials connected with the KdV equations*, Commun. Math. Phys. 61 (1978), 1–30.
43. J. Zubelli, F. Magri, *Differential equations in the spectral parameter, Darboux transformations and a hierarchy of master symmetries for KdV equations*, Commun. Math. Phys. 141 (1991), 329–351.

POTENTIAL SYMMETRIES AND EQUIVALENT CONSERVATION LAWS

GEORGE BLUMAN
Department of Mathematics
University of British Columbia
Vancouver, B.C., CANADA, V6T 1Z2

Abstract. If one equation of a given system of PDEs is a conservation law, the system can be embedded in an auxiliary system. A point symmetry of an auxiliary system yields a nonlocal symmetry of the given system if it does not project onto a point symmetry.

A conservation law of a given system of PDEs is an equivalent conservation law if the replacement of a PDE by the conservation law yields a system of PDEs with the same complex solution set as the given system. The problem of finding equivalent conservation laws is considered.

1. Introduction

A symmetry admitted by a system of differential equations $R\{x,u\}$ with n independent variables x and m dependent variables u is a transformation which maps any solution of $R\{x,u\}$ to another solution of $R\{x,u\}$. In principle "any" system $R\{x,u\}$ has "continuous" symmetries which map solutions into neighbouring solutions. The problem is to find explicit symmetries of $R\{x,u\}$ and make use of them to simplify boundary value problems (BVPs) posed for $R\{x,u\}$ by reducing the number of independent variables (Bluman and Cole [1], Bluman and Kumei [2]), to find conservation laws for $R\{x,u\}$ (Noether [3], Olver [4], Ibragimov [5], Bluman and Kumei [2]), to discover whether or not $R\{x,u\}$ can be linearized and construct an explicit linearization when one exists (Kumei and Bluman [6], Bluman and Kumei [7], [2]), to discover PDEs equivalent to $R\{x,u\}$ (Ovsiannikov [8], Lisle [9], Ibragimov et al. [10]), etc.

An explicit symmetry can only be realized as a coordinate-dependent transformation. Sophus Lie gave an algorithm which yields "determining equations" to find infinitesimal generators of point symmetries admitted by a given $R\{x,u\}$. Such local symmetries are point transformations acting on (x,u)-space. Various symbolic manipulation programs now exist to implement Lie's algorithm: setting up determining equations, finding the dimension of their solution space without explicitly solving them, and/or yielding explicit infinitesimal generators (Head [11], Schwarz [12], Wolf and Brand [13], Champagne et al. [14], Kersten [15], Reid [16,17], Baumann [18]).

If $R\{x,u\}$ does not admit local symmetries, local symmetries yielding its linearization by an invertible mapping, or local symmetries of a posed BVP, one could still find useful nonlocal symmetries admitted by $R\{x,u\}$ through embedding $R\{x,u\}$ in an auxiliary "covering system" $S\{x,u,v\}$ with auxiliary dependent variables v (Krasil'shchik and Vinogradov [19,20]). A point symmetry of $S\{x,u,v\}$, acting on (x,u,v)-space, yields a nonlocal symmetry of $R\{x,u\}$ if it does not project onto a point symmetry acting on (x,u)-space. Such an auxiliary system is realized

N. H. Ibragimov et al. (eds.),
Modern Group Analysis: Advanced Analytical and Computational Methods in Mathematical Physics, 71–84.

by replacement of a PDE of $R\{x,u\}$ by an equivalent conservation law, yielding $S\{x,u,v\}$ with $n-1$ augmented potential variables v. Corresponding nonlocal symmetries are called *potential symmetries* of $R\{x,u\}$ (Bluman and Kumei [21,2], Bluman et al. [22], Bluman [23–25]). Consequently one could obtain more symmetry reductions for $R\{x,u\}$ (see [21,22,2]), its linearization by a non-invertible mapping (Bluman and Kumei [26,2], Bluman [27]), etc.

A point symmetry of $R\{x,u\}$ always yields a symmetry of any of its potential systems $S\{x,u,v\}$. However such a symmetry could be "lost" in the sense that it does not correspond to a local symmetry of $S\{x,u,v\}$, i.e., the symmetry does not arise as a projection to (x,u)-space of any local symmetry admitted by $S\{x,u,v\}$.

For a given system $R\{x,u\}$ we consider the problems of finding all equivalent conservation laws, all possible potential systems as further potential variables are introduced, and most importantly, "grand" potential systems yielding all potential symmetries and recovering all "lost" symmetries. Two conjectures are presented as well as several examples.

2. Symmetries of Differential Equations

Consider a system of m PDEs, $R\{x,u\}$, given by

$$G^{\sigma}(x,u,\underset{1}{u},\underset{2}{u},\ldots,\underset{k}{u})=0, \qquad \sigma=1,2,\ldots,m \tag{1}$$

with n independent variables $x=(x_1,x_2,\ldots,x_n)$ and m dependent variables $u=(u_1,u_2,\ldots,u_m)$; $\underset{1}{u}$ denotes first order partial derivatives (a coordinate in $\underset{1}{u}$ is denoted by $u^i_j=\frac{\partial u^i}{\partial x_j}$; $\underset{k}{u}$ denotes k-th order partial derivatives (a coordinate in $\underset{k}{u}$ is denoted by $u^i_{j_1j_2\ldots j_k}=\frac{\partial^k u^i}{\partial x_{j_1}\partial x_{j_2}\ldots\partial x_{j_k}}$).

Definition 2.1: A *symmetry* admitted by $R\{x,u\}$ is a transformation mapping any solution of $R\{x,u\}$ into another solution of $R\{x,u\}$.

Note that the definition of a symmetry admitted by a system of DEs is topological. In particular a symmetry need not be coordinate dependent, i.e., a symmetry admitted by $R\{x,u\}$ does not have to be a transformation acting on (x,u)-space, or even $(x,u,\underset{1}{u},\underset{2}{u},\ldots,\underset{p}{u},\ldots)$-space. However, to be "useful", a symmetry transformation must be defined in terms of some coordinate frame.

2.1. Point symmetries

Definition 2.2: A (Lie) point symmetry admitted by $R\{x,u\}$ is a symmetry of $R\{x,u\}$ characterized by an infinitesimal generator of the form

$$\mathbf{X}=\sum_{\mu=1}^{m}\eta^{\mu}(x,u,\underset{1}{u})\frac{\partial}{\partial u^{\mu}} \tag{2}$$

with $\eta=(\eta^1,\eta^2,\ldots,\eta^m)$ linear in the coordinates of $\underset{1}{u}$:

$$\eta^{\mu}=\alpha^{\mu}(x,u)-\sum_{i=1}^{n}\xi_i(x,u)\frac{\partial u^{\mu}}{\partial x_i}. \tag{3}$$

The infinitesimal generator (2), (3) corresponds to a one-parameter (ε) Lie group of point transformations

$$\begin{aligned} x_i^* &= x_i + \varepsilon\xi_i(x,u) + O(\varepsilon^2), && i = 1,2,\dots,n \\ (u^\mu)^* &= u^\mu + \varepsilon\alpha^\mu(x,u) + O(\varepsilon^2), && i = 1,2,\dots,m \end{aligned} \tag{4}$$

The prolongation (extension) of (2) to an infinitesimal generator acting on all derivatives of u is given by

$$\mathbf{U} = \mathbf{X} + (D_i\eta^\mu)\frac{\partial}{\partial u_i^\mu} + \dots + (D_{i_1}D_{i_2}\dots D_{i_j}\eta^\mu)\frac{\partial}{\partial u_{i_1i_2\dots i_j}} + \dots \tag{5}$$

where summation is assumed over a repeated index and

$$\frac{D}{Dx_i} = D_i = \frac{\partial}{\partial x_i} + u_i^\mu\frac{\partial}{\partial u^\mu} + \dots + u_{ii_1i_2\dots i_j}\frac{\partial}{\partial u^\mu_{i_1i_2\dots i_j}} + \dots \tag{6}$$

is a total derivative operator, $i = 1,2,\dots,n$. Under the action of (5) a solution $u = \Theta(x)$ of $R\{x,u\}$ is mapped into solutions

$$u = \Phi(x) = e^{\varepsilon\mathbf{U}}u\Big|_{u=\Theta(x)}. \tag{7}$$

For details see [2], [4], or [5].

2.1.1. How to Find Point Symmetrics.
If $u^* = u + \varepsilon\eta(x,u,\underset{1}{u}) + O(\varepsilon^2)$, then

$$G^\sigma(x,u^*,\underset{1}{u}^*,\underset{2}{u}^*,\dots,\underset{k}{u}^*) = G^\sigma(x,u^*,\underset{1}{u},\underset{2}{u},\dots,\underset{k}{u}) + \varepsilon\sum_{\rho=1}^{m}\mathcal{L}_\rho^\sigma[u]\eta^\rho + O(\varepsilon^2), \tag{8}$$

($\sigma = 1,2,\dots,m$) where the differential operator

$$\mathcal{L}_\rho^\sigma[u] = \frac{\partial G^\sigma}{\partial u^\rho} + \frac{\partial G^\sigma}{\partial u_i^\rho}D_i + \dots + \frac{\partial G^\sigma}{\partial u^\rho_{i_1i_2\dots i_k}}D_{i_1}D_{i_2}\dots D_{i_k}, \quad \sigma,\rho = 1,2,\dots,m \tag{9}$$

is the Fréchet derivative of $R\{x,u\}$. The following theorem is easily proved (see [2], [4]):

Theorem 2.3: Point symmetry (2) is admitted by $R\{x,u\}$ if and only if

$$\sum_{\rho=1}^{m}\mathcal{L}_\rho^\sigma[u]\eta^\rho = 0, \qquad \sigma = 1,2,\dots,m \tag{10}$$

holds for any solution of $R\{x,u\}$, including all differential consequences of $G^\sigma = 0$, $\sigma = 1,2,\dots,m$.

It is convenient to express (10) in the form

$$\mathcal{L}[u]\eta = 0. \tag{11}$$

Equations (10) lead to an overdetermined linear system of PDEs (determining equations) with $m+n$ unknowns $(\alpha^1(x,u),\alpha^2(x,u),\ldots,\alpha^m(x,u);\xi_1(x,u),\xi_2(x,u),\ldots,\xi_n(x,u))$. Various symbolic manipulation programs exist in MAPLE, REDUCE, MATHEMATICA, MACSYMA, SCRATCHPAD, muMATH, etc. (see [11–18]), which set up these determining equations, simplify them, solve them and/or find the dimension (if finite) of their solution space automatically and/or interactively.

Definition 2.4: The *adjoint* of the differential operator $\mathcal{L}[u]$ is the differential operator $\mathcal{L}^*[u]$ which satisfies

$$\int_\Omega v\cdot\mathcal{L}[u]w\,dx=\int_\Omega w\cdot\mathcal{L}^*[u]v\,dx \tag{12}$$

on every domain $\Omega\subset\mathbf{R}^n$, for every pair of k times differentiable functions $v(x)$, $w(x)$ with compact support in Ω.

In particular $\mathcal{L}^*[u]$ is the adjoint of $\mathcal{L}[u]$ if $v\cdot\mathcal{L}[u]w-w\cdot\mathcal{L}[u]v$ is a divergence expression.

Definition 2.5: The differential operator $\mathcal{L}[u]$ is *self-adjoint* if $\mathcal{L}^*[u]=\mathcal{L}[u]$.

2.1.2. Prototype Example.

Let $R\{x,u\}$ be the scalar nonlinear diffusion equation

$$u_t=(Lu))_{xx}=(K(u)u_x)_x \tag{13}$$

with conductivity $K(u)=L'(u)\neq$ const. The Fréchet derivative of $R\{x,u\}$ is given by $\mathcal{L}[u]=K(u)\frac{D^2}{Dx^2}+2K'(u)u_x\frac{D}{Dx}-\frac{D}{Dt}+[K''(u)u_x^2+K'(u)u_{xx}]=\frac{D^2}{Dx^2}\cdot K(u)-\frac{D}{Dt}$. Then $v\cdot\mathcal{L}[u]w=w\cdot\mathcal{L}^*[u]v+\frac{D}{Dx}(v^2\frac{D}{Dx}(\frac{K(u)w}{v}))-\frac{D}{Dt}(vw)$ with $\mathcal{L}^*[u]=K(u)\frac{D^2}{Dx^2}+\frac{D}{Dt}\neq\mathcal{L}[u]$.

To within scalings and translations of u, the infinitesimal generators of $R\{x,u\}$ are found to be (see [1], [2], [8]):

1. $K(u)$ arbitrary: $\mathbf{X}_1^R=u_x\frac{\partial}{\partial u}$, $\mathbf{X}_2^R=u_t\frac{\partial}{\partial u}$, $\mathbf{X}_3^R=(xu_x+2tu_t)\frac{\partial}{\partial u}$.
2. $K(u)=u^\lambda$: $\mathbf{X}_1^R$, $\mathbf{X}_2^R$, $\mathbf{X}_3^R$, $\mathbf{X}_4^R=(2u-\lambda xu_x)\frac{\partial}{\partial u}$.
3. $K(u)=u^{-4/3}$: $\mathbf{X}_1^R$, $\mathbf{X}_2^R$, $\mathbf{X}_3^R$, $\mathbf{X}_4^R$ $(\lambda=-4/3)$, $\mathbf{X}_5^R=(3xu_x+x^2u_x)\frac{\partial}{\partial u}$.

2.2. Local and nonlocal symmetries

Definition 2.6: A *local symmetry* admitted by $R\{x,u\}$ is a symmetry of $R\{x,u\}$ characterized by an infinitesimal generator of the form

$$\mathbf{X}=\sum_{\mu=1}^m\eta^\mu(x,u,\underset{1}{u},\underset{2}{u},\ldots,\underset{p}{u})\frac{\partial}{\partial u^\mu}. \tag{14}$$

Theorem 2.3 for point symmetries generalizes to (see [2], [4]):

Theorem 2.7: Local symmetry (14) is admitted by $R\{x,u\}$ if and only if $\mathcal{L}[u]\eta=0$ holds for every solution of $R\{x,u\}$, including differential consequences of $G^\sigma=0$, $\sigma=1,2,\ldots,m$.

When $m = p = 1$, a local symmetry (14) is a (proper) *contact symmetry* if it is not a point symmetry; a local symmetry (14) is a *Lie-Bäcklund* (*higher, higher order, generalized*) *symmetry* if it is not a point or contact symmetry. The transformational and geometrical properties of local symmetries are discussed in [2], [4], [5].

Definition 2.8: A *nonlocal symmetry* admitted by $R\{x,u\}$ is a symmetry of $R\{x,u\}$ characterized by an infinitesimal generator which is not of local type (14).

An infinitesimal generator depending on integrals of u or depending on an infinite number of derivatives of u would yield a nonlocal symmetry. In principle "all" systems of PDEs admit nonlocal symmetries. The problem is to find algorithms yielding nonlocal symmetries which are "useful".

2.3. Uses of local symmetries

If $R\{x,u\}$ admits a local symmetry (14), the following applications are possible:

1. Construction of a one-parameter family of solutions $u = e^{\varepsilon \mathbf{U}} u\Big|_{u=\Theta(x)}$ from any known solution $u = \Theta(x)$.
2. Construction of invariant solutions: If $\eta^\mu(x,u,\underset{1}{u},\underset{2}{u},\ldots,\underset{p}{u}) = 0$ (with $\xi \not\equiv 0$ in the case of a point symmetry), $\mu = 1,2,\ldots,m$, is substituted into $R\{x,u\}$ one obtains a reduced system of m PDEs with $n-1$ independent variables. In the case of a point symmetry such a reduction holds for a BVP posed for $R\{x,u\}$ if the BVP has a unique solution, its domain is an invariant of the symmetry, and its boundary conditions are also invariant. (See [1], [2] for details.) In turn one can generalize this ansatz to generate so-called "nonclassical solutions" which have been discovered for many nonlinear PDEs by Clarkson and co-workers. The nonclassical method is discussed in Nucci and Clarkson [28] and Pucci [29].
3. From the nature of point or contact symmetries admitted by $R\{x,u\}$ one can determine algorithmically whether or not $R\{x,u\}$ can be linearized by an *invertible* point or contact transformation and construct an explicit linearization when one exists (see [2], [6], [7]).
4. Construction of conservation laws (see Olver [4]): If $\mathcal{L}[u]$ is self-adjoint and a set of *factors* (*multipliers, characteristics*) $\{\lambda^\sigma(x,u,\underset{1}{u},\underset{2}{u},\ldots,\underset{p}{u})\}$ exists so that $\sum_{\sigma=1}^m \lambda^\sigma(x,u,\underset{1}{u},\underset{2}{u},\ldots,\underset{p}{u})G^\sigma = \sum_{i=1}^n D_i f^i$ for some $\{f^i(x,u,\underset{1}{u},\underset{2}{u},\ldots,\underset{q}{u})\}$ then $\eta = \lambda$ must be a local symmetry of $R\{x,u\}$.

The self-adjointness of $\mathcal{L}[u]$ means that $R\{x,u\}$ can be derived from a variational principle. If a local symmetry is admitted by a corresponding action functional to within a divergence (variational symmetry) then the local symmetry is admitted by $R\{x,u\}$, and one can obtain an explicit formula for the resulting conservation law of $R\{x,u\}$ (see Noether [3], Bessel-Hagen [30]; also [2], [4], [5]).

3. Potential Symmetries

Suppose one PDE of $R\{x,u\}$, without loss of generality, $G^m = 0$ is a *conserved form* (conservation law) $\sum_{i=1}^n D_i f^i(x, u, \underset{1}{u}, \underset{2}{u}, \ldots, \underset{k-1}{u}) = 0$. Then $R\{x,u\}$ is the system:

$$G^\sigma(x, u, \underset{1}{u}, \underset{2}{u}, \ldots, \underset{k}{u}) = 0 \qquad \sigma = 1, 2, \ldots, m-1, \tag{15}$$

$$\sum_{i=1}^{n} D_i f^i(x, u, \underset{1}{u}, \underset{2}{u}, \ldots, \underset{k-1}{u}) = 0. \tag{16}$$

Through (16) one can introduce $n-1$ auxiliary dependent variables $v = (v^1, v^2, \ldots, v^{n-1})$ and form auxiliary system $S\{x,u,v\}$ of $m+n-1$ PDEs given by:

$$\begin{aligned}
f^1(x, u, \underset{1}{u}, \underset{2}{u}, \ldots, \underset{k-1}{u}) &= \frac{\partial v^1}{\partial x_2}, \\
f^\ell(x, u, \underset{1}{u}, \underset{2}{u}, \ldots, \underset{k-1}{u}) &= (-1)^{\ell-1}\left[\frac{\partial v^\ell}{\partial x_{\ell+1}} + \frac{\partial v^{\ell-1}}{\partial x_{\ell-1}}\right], \qquad 1 < \ell < n, \\
f^n(x, u, \underset{1}{u}, \underset{2}{u}, \ldots, \underset{k-1}{u}) &= (-1)^{n-1}\frac{\partial v^{n-1}}{\partial x_{n-1}}, \\
G^\sigma(x, u, \underset{1}{u}, \underset{2}{u}, \ldots, \underset{k}{u}) &= 0, \qquad \sigma = 1, 2, \ldots, m-1
\end{aligned}$$

3.1. Relationship between $R\{x,u\}$, $S\{x,u,v\}$

If $(u(x), v(x))$ solves $S\{x,u,v\}$ then $u(x)$ solves $R\{x,u\}$; if $u(x)$ solves $R\{x,u\}$ then there is some $v(x)$ (not unique) such that $(u(x), v(x))$ solves $S\{x,u,v\}$. (A BVP posed for $R\{x,u\}$ leads to a BVP posed for $S\{x,u,v\}$ and vice versa.) Consequently any symmetry of $S\{x,u,v\}$ yields a symmetry of $R\{x,u\}$ and, conversely, any symmetry of $R\{x,u\}$ yields a symmetry of $S\{x,u,v\}$. Most importantly, due to the non-invertible relationship between $R\{x,u\}$ and $S\{x,u,v\}$, a point symmetry of $S\{x,u,v\}$ could yield a nonlocal symmetry of $R\{x,u\}$ and, conversely, a point symmetry of $R\{x,u\}$ could yield a nonlocal symmetry of $S\{x,u,v\}$.

Definition 3.1: A *potential symmetry* of $R\{x,u\}$, related to auxiliary system $S\{x,u,v\}$, is a point symmetry of $S\{x,u,v\}$ which does not project onto a point symmetry of $R\{x,u\}$.

Due to the relationship between $R\{x,u\}$ and $S\{x,u,v\}$, it follows that a potential symmetry of $R\{x,u\}$ must be a nonlocal symmetry of $R\{x,u\}$. The following theorem obviously holds:

Theorem 3.2: Suppose

$$\begin{aligned}
\mathbf{X}^S = &\sum_{\mu=1}^{m}\left[\alpha^\mu(x,u,v) - \sum_{i=1}^{n}\xi_i^S(x,u,v)u_i^\mu\right]\frac{\partial}{\partial u^\mu} + \\
&\sum_{\nu=1}^{n-1}\left[\beta^\nu(x,u,v) - \sum_{i=1}^{n}\xi_i^S(x,u,v)v_i^\nu\right]\frac{\partial}{\partial v^\nu}
\end{aligned}$$

is a point symmetry of $S\{x,u,v\}$. Then $\mathbf{X}^S$ induces a potential symmetry of $R\{x,u\}$ if and only if $(\alpha(x,u,v), \xi^S(x,u,v))$ depends essentially on v; otherwise $\mathbf{X}^S$ projects onto a point symmetry of $R\{x,u\}$, in particular

$$\mathbf{X} = \sum_{\mu=1}^{m} \left[\alpha^\mu(x,u) - \sum_{i=1}^{n} \xi_i^S(x,u) u_i^\mu \right] \frac{\partial}{\partial u^\mu}.$$

Conversely, suppose $\mathbf{X}^R = \sum_{\mu=1}^{m} \left[\alpha^\mu(x,u) - \sum_{i=1}^{n} \xi_i^R(x,u) u_i^\mu\right] \frac{\partial}{\partial u^\mu}$ is a point symmetry of $R\{x,u\}$. Then $\mathbf{X}^R$ yields a symmetry of $S\{x,u,v\}$ which is not a point symmetry of $S\{x,u,v\}$ if and only if $\mathbf{X}^S = \mathbf{X}^R + \sum_{\nu=1}^{n-1} \zeta^\nu(x,u,\underset{1}{u},\underset{1}{v}) \frac{\partial}{\partial v^\nu}$ does not yield a point symmetry of $S\{x,u,v\}$ for any choice of $\zeta = (\zeta^1, \zeta^2, \ldots, \zeta^{n-1})$.

3.2. Calculation and uses of potential symmetries

Since potential symmetries of $R\{x,u\}$ arise as point symmetries of auxiliary systems, existing symbolic manipulation programs yield information about point symmetries of any auxiliary system. In turn, from Theorem 3.2 one can automatically determine whether such point symmetries yield potential symmetries of $R\{x,u\}$. Using the work of Reid [16,17], one can determine whether or not a given auxiliary system $S\{x,u,v\}$ can yield potential symmetries of $R\{x,u\}$ without calculating explicit point symmetries of $S\{x,u,v\}$. Hence the main problem in finding potential symmetries of $R\{x,u\}$ is the discovery of useful auxiliary systems. This problem will be considered in Section 4.

The nature of potential symmetries allows one to extend the uses of point symmetries to such nonlocal symmetries. In particular:

1. One can construct a one-parameter family of solutions of $R\{x,u\}$ from any known solution $(u,v) = (\Theta(x), \Phi(x))$ of $S\{x,u,v\}$.
2. An invariant solution of $S\{x,u,v\}$ could yield a solution of $R\{x,u\}$ which is not an invariant solution of $R\{x,u\}$ for any local symmetry admitted by $R\{x,u\}$. More generally, the functional form for invariant solutions of $S\{x,u,v\}$ could yield solutions of $R\{x,u\}$ which are not invariant solutions of $S\{x,u,v\}$ (Pucci and Saccomandi [31]). If a BVP for $R\{x,u\}$ embedded in a BVP for $S\{x,u,v\}$ admits a potential symmetry then the BVP for $R\{x,u\}$ is reduced. (See Bluman and Kumei [2], [32] for an example.) Moreover the ansatz to generate nonclassical solutions of $S\{x,u,v\}$ could yield solutions of $R\{x,u\}$ which are neither nonclassical solutions of $R\{x,u\}$ nor solutions arising from potential symmetries.
3. A potential symmetry of $R\{x,u\}$ yielding the linearization of $S\{x,u,v\}$ by an invertible point transformation results in the linearization of $R\{x,u\}$ by a non-invertible mapping ([2], [26], [27]).
4. Suppose a variational principle for $S\{x,u,v\}$ admits a potential symmetry yielding a conservation law. Such a conservation law is new for $R\{x,u\}$ since it depends on nonlocal potential variables v.

3.3. Prototype example

If $R\{x,u\}$ is PDE (13) then it is already a conserved form which yields system $S\{x,u,v\}$ given by

$$v_x = u, \qquad v_t = K(u)u_x. \tag{17}$$

To within scalings and translations of u, the infinitesimal generators of point symmetries of (17) are found to be (see [2], [22]):

1. $K(u)$ arbitrary: $\mathbf{X}_1^S = \mathbf{X}_1^R + v_x\frac{\partial}{\partial v}$, $\mathbf{X}_2^S = \mathbf{X}_2^R + v_t\frac{\partial}{\partial v}$, $\mathbf{X}_3^S = \mathbf{X}_3^R + (xv_x + 2tv_t - v)\frac{\partial}{\partial v}$, (no potential symmetries).
2. $K(u) = u^\lambda$: $\mathbf{X}_1^S, \mathbf{X}_2^S, \mathbf{X}_3^S, \mathbf{X}_4^S = \mathbf{X}_4^R + (2v - \lambda x v_x)\frac{\partial}{\partial v}$ (no potential symmetries).
3. $K(u) = u^{-2}$: $\mathbf{X}_1^S$, $\mathbf{X}_2^S$, $\mathbf{X}_3^S$, $\mathbf{X}_4^S$ $(\lambda = -2)$, and potential symmetry

   $$\mathbf{X}_\infty^S = [F^1(v,t)u_x + u^2F^2(v,t)]\frac{\partial}{\partial u} + F^1(v,t)\frac{\partial}{\partial v}$$

   where $\{F^1(v,t), F^2(v,t)\}$ are arbitrary solutions of $\frac{\partial F^1}{\partial t} = \frac{\partial F^2}{\partial v}$, $\frac{\partial F^1}{\partial v} = F^2$. This potential symmetry yields an explicit linearization of $R\{x,u\}$ by a non-invertible mapping (see [2], [26]).
4. $K(u) = \frac{1}{1+u^2}e^{a\arctan u}$, $a = \text{const}$: $\mathbf{X}_1^S$, $\mathbf{X}_2^S$, $\mathbf{X}_3^S$, and potential symmetry

   $$\mathbf{X}_5^S = [(u^2+1) + vu_x + atu_t]\frac{\partial}{\partial u} + [x + vv_x + atv_t]\frac{\partial}{\partial v}.$$

Note that for $K(u) = u^{-4/3}$, the point symmetry $\mathbf{X}_5^R$ is "lost" in the sense that it does not induce a point symmetry of $S\{x,u,v\}$; $\mathbf{X}_5^R$ yields a nonlocal symmetry of $S\{x,u,v\}$ since $3xv_x + x^2v_{xx} = (x^2v_x + xv_x)_x - v$. Thus $\mathbf{X}_5^S = \mathbf{X}_5^R + (x^2v_x + xv - D_x^{-1}v)\frac{\partial}{\partial v}$, which is a nonlocal symmetry of $S\{x,u,v\}$.

4. Further Potential Symmetries; Construction of Auxiliary Systems; Equivalent Conservation Laws

Now consider the problem of finding auxiliary systems whose point symmetries yield nonlocal symmetries of $R\{x,u\}$. In principle any conservation law of $R\{x,u\}$ yields an auxiliary system and any conservation law of an auxiliary system yields another auxiliary system. From topological considerations we will see that a very restricted class of conservation laws can be expected to yield auxiliary systems leading to the possibility of discovering nonlocal symmetries of $R\{x,u\}$. In particular, it will be seen that "physical" conservation laws (corresponding to point symmetries for which $\xi \not\equiv 0$ and conservation laws arising from Lie-Bäcklund symmetries cannot be expected to yield nonlocal symmetries.

4.1. Potential symmetries of $S\{x,u,v\}$

Let $v^{(1)} = v$, $S^{(1)} = S\{x,u,v^{(1)}\}$. Suppose a PDE of $S^{(1)}$ is a conservation law. Then one can introduce $n-1$ more potential variables $v^{(2)}$ and form auxiliary system $S^{(2)}\{x,u,v^{(1)},v^{(2)}\}$ of $m + 2(n-1)$ PDEs with $m + 2(n-1)$ dependent variables. A point symmetry of $S^{(2)}$ could yield a potential symmetry of $S^{(1)}$ which is either a point symmetry or nonlocal symmetry of $R\{x,u\}$. If this potential symmetry of $S^{(1)}$ yields a point symmetry of $R\{x,u\}$ then a "lost" point symmetry of $R\{x,u\}$ is "recovered".

One can continue this process to auxiliary system $S^{(J)}\{x,u,v^{(1)},v^{(2)},\ldots,v^{(J)}\}$ provided a PDE of $S^{(I)}\{x,u,v^{(1)},v^{(2)},\ldots,v^{(I)}\}$ is a conservation law, $I=1,2,\ldots,$ $J-1$.

4.1.1. Prototype Example.
Suppose $R\{x,u\}$ is PDE (13) and $S^{(1)}$ is system (17). Then PDE $v_t = K(u)u_x = (L(u))_x$ is a conservation law. Let $w = v^{(2)}$. Consequently $S^{(2)}\{x,u,v^{(1)},v^{(2)}\} = T\{x,u,v,w\}$ is given by

$$\begin{aligned} v_x &= u, \\ w_x &= v, \\ w_t &= L(u). \end{aligned} \tag{18}$$

To within scalings and translations of u, the infinitesimal generators of point symmetries of (18) are found to be:

1. $K(u)$ arbitrary: $\mathbf{X}_1^T = \mathbf{X}_1^S + w_x\frac{\partial}{\partial w}$, $\mathbf{X}_2^T = \mathbf{X}_2^S + w_t\frac{\partial}{\partial w}$, $\mathbf{X}_3^T = \mathbf{X}_3^S + (xw_x + 2tw_t - 2w)\frac{\partial}{\partial w}$, (no potential symmetries of $S\{x,u,v\}$).
2. $K(u) = u^\lambda$: $\mathbf{X}_1^T, \mathbf{X}_2^T, \mathbf{X}_3^T, \mathbf{X}_4^T = \mathbf{X}_4^S + (2w - \lambda xw_x)\frac{\partial}{\partial w}$ (no potential symmetries of $S\{x,u,v\}$).
3. $K(u) = u^{-2}$: $\mathbf{X}_1^T, \mathbf{X}_2^T, \mathbf{X}_3^T, \mathbf{X}_4^T$ $(\lambda=-2)$, $\mathbf{X}_\infty^T = \mathbf{X}_\infty^S + [F^3(v,t) - vF^1(v,t)]\frac{\partial}{\partial w}$ where $\{F^1(v,t), F^2(v,t), F^3(v,t)\}$ are arbitrary solutions of $\frac{\partial F^3}{\partial v} = F^1$, $\frac{\partial F^3}{\partial t} = F^2$, $\frac{\partial F^1}{\partial v} = F^2$. Infinitesimal generator $\mathbf{X}_\infty^T$ projects onto point symmetry $\mathbf{X}_\infty^S$ of $S\{x,u,v\}$, and hence yields no further nonlocal symmetry of $R\{x,u\}$.
4. $K(u) = \frac{1}{1+u^2}e^{a\arctan u}$: $\mathbf{X}_1^T, \mathbf{X}_2^T, \mathbf{X}_3^T, \mathbf{X}_5^T = \mathbf{X}_5^S + [\frac{1}{2}(x^2 - v^2) + vw_x + atw_t]\frac{\partial}{\partial w}$. Thus $\mathbf{X}_5^T$ yields no extra nonlocal symmetry of $R\{x,u\}$.
5. $K(u) = u^{-4/3}$: $\mathbf{X}_1^T, \mathbf{X}_2^T, \mathbf{X}_3^T, \mathbf{X}_4^T$ $(\lambda = -4/3)$, $\mathbf{X}_6^T = \mathbf{X}_5^R + (xv - w + x^2v_x)\frac{\partial}{\partial v} + (x^2w_x - xw)\frac{\partial}{\partial w}$. Consequently $\mathbf{X}_6^T$ is a potential symmetry of $S\{x,u,v\}$ which projects onto the "lost" point symmetry of $R\{x,u\}$ which is now "recovered".
6. $K(u) = u^{-2/3}$: $\mathbf{X}_1^T, \mathbf{X}_2^T, \mathbf{X}_3^T, \mathbf{X}_4^T$ $(\lambda = -2/3)$, $\mathbf{X}_7^T = (3uv + wu_x)\frac{\partial}{\partial u} + (v^2 + wv_x)\frac{\partial}{\partial v} + (x^2w_x - xw)\frac{\partial}{\partial w}$. From the form of $\mathbf{X}_7^T$ we see that it is a potential symmetry of $S\{x,u,v\}$ which does not project onto a local symmetry of $R\{x,u\}$, and hence $\mathbf{X}_7^T$ yields a new nonlocal symmetry of $R\{x,u\}$,

Most importantly, $T\{x,u,v,w\}$ covers $R\{x,u\}$ and $S\{x,u,v\}$ (see [19], [20]) in the sense that the point symmetries of $T\{x,u,v,w\}$ project onto all point symmetries of both $R\{x,u\}$ and $S\{x,u,v\}$ for any conductivity $K(u)$.

4.2. Equivalent conservation laws

Suppose a set of factors $\{\lambda^\sigma(x,u,\underset{1}{u},\underset{2}{u},\ldots,\underset{p}{u})\}$ exists with $\lambda^\rho \not\equiv 0$, so that

$$\sum_{\sigma=1}^{m} \lambda^\sigma(x,u,\underset{1}{u},\underset{2}{u},\ldots,\underset{p}{u})G^\sigma = \sum_{i-1}^{n} D_i f^i \tag{19}$$

for some $\{f^i(x,u,\underset{1}{u},\underset{2}{u},\ldots,\underset{q}{u})\}$. Consider system $\widehat{R}_\rho\{x,u\}$ obtained by replacing PDE $G^\rho = 0$ of $R\{x,u\}$ by conservation law $\sum_{i=1}^n D_i f^i = 0$. Then $\widehat{R}_\rho\{x,u\}$ is system

$$\begin{aligned} &G^\sigma = 0, \qquad \sigma = 1,2,\ldots,\rho-1,\rho+1,\ldots,m,\\ &\sum_{i=1}^n D_i f^i = 0. \end{aligned} \tag{20}$$

Clearly any solution of $R\{x,u\}$ is a solution of $\widehat{R}_\rho\{x,u\}$. But any function $u=\Psi(x)$ satisfying system $\widetilde{R}_\rho\{x,u\}$ given by

$$\begin{aligned} &G^\sigma = 0, \qquad \sigma = 1,2,\ldots,\rho-1,\rho+1,\ldots,m,\\ &\lambda^\rho(x,u,\underset{1}{u},\underset{2}{u},\ldots,\underset{p}{u}) = 0. \end{aligned} \tag{21}$$

also solves $\widehat{R}_\rho\{x,u\}$. The union of the sets of solutions of $R\{x,u\}$ and $\widetilde{R}_\rho\{x,u\}$ yields all solutions of $\widehat{R}_\rho\{x,u\}$.

Suppose some complex solution of $\widetilde{R}_\rho\{x,u\}$ is not a complex solution of $R\{x,u\}$. Then the solutions of $R\{x,u\}$ and $\widetilde{R}_\rho\{x,u\}$ are topologically distinct in the sense that not all solutions of $\widetilde{R}_\rho\{x,u\}$ can be continuously deformed into solutions of $R\{x,u\}$. Consequently any symmetry of $\widehat{R}_\rho\{x,u\}$ will be a symmetry of $R\{x,u\}$ whereas, conversely, there will be symmetries of $R\{x,u\}$ which are not symmetries of $\widehat{R}_\rho\{x,u\}$.

Thus auxiliary system $\widehat{S}_\rho\{x,u,v\}$ resulting from the displayed conservation law in (20) of $\widehat{R}_\rho\{x,u\}$ is "potentially" useful for obtaining potential symmetries of $R\{x,u\}$ provided $\widehat{R}_\rho\{x,u\}$ and $R\{x,u\}$ are equivalent in the sense of having the same sets of complex solutions, i.e., any complex solution of $\widetilde{R}_\rho\{x,u\}$ is also a solution of $R\{x,u\}$. This leads to the following definition:

Definition 4.1: A set of factors $\{\lambda^\sigma(x,u,\underset{1}{u},\underset{2}{u},\ldots,\underset{p}{u})\}$ yields a conservation law which is an *equivalent conservation law* for $R\{x,u\}$ if and only if for some ρ, $1\le\rho\le m$, $\widehat{R}_\rho\{x,u\}$ and $R\{x,u\}$ have the same sets of complex solutions (or, equivalently, each complex solution of $\widetilde{R}_\rho\{x,u\}$ is a complex solution of $R\{x,u\}$ for some ρ, $1\le\rho\le m$.

It follows that for a given system of PDEs $R\{x,u\}$ it is only worthwhile to determine equivalent conservation laws when seeking potential symmetries. In particular, if each factor λ^σ depends on coordinates in $\underset{k}{u}$ one would not expect the resulting conservation law to be an equivalent conservation law. Thus, from topological considerations, if one expects to have a chance of success in finding potential symmetries for a given system $R\{x,u\}$, it is recommended that one only seek conservation laws of $R\{x,u\}$ with factors of the form $\lambda^\sigma=\lambda^\sigma(x,u)$, $\sigma=1,2,\ldots,m$.

As an example, consider the linear wave equation $R\{x,u\}$ given by

$$L[x]u = u_{xx} - \frac{1}{c^2(x)}u_{tt} \tag{22}$$

where $L[x] = \frac{\partial^2}{\partial x^2} - \frac{1}{c^2(x)}\frac{\partial^2}{\partial t^2}$. The Fréchet derivative of (22) is $\mathcal{L}[u] = L[x] = \mathcal{L}^*[u]$. Consequently if a factor λ yields a conservation law of (22) then it must satisfy $L[x]\lambda = 0$ for any solution of (22) and its differential consequences. The factor $\lambda = 2u_t$ satisfies this condition. Moreover $\lambda = 2u_t$ yields a conservation law of (22) since $2u_t[u_{xx} - \frac{1}{c^2(x)}u_{tt}] = \frac{D}{Dx}(2u_xu_t) - \frac{D}{Dt}(u_x^2 + \frac{1}{c^2(x)}u_t^2)$. Then $\widetilde{R}\{x,u\}$ is the PDE $u_t = 0$. Clearly, for any wave speed $c(x)$, almost all solutions of $u_t = 0$ are not solutions of wave equation (22). Thus $\widehat{R}\{x,u\}$ given by $\frac{D}{Dx}(2u_xu_t) - \frac{D}{Dt}(u_x^2 + \frac{1}{c^2(x)}u_t^2) = 0$ is a conservation law of $R\{x,u\}$ which is not an equivalent conservation law. Correspondingly $\widehat{S}\{x,u,v\}$ is given by

$$v_t = 2u_xu_t, \qquad v_x = \frac{1}{c^2(x)}u_t^2 + u_x^2. \tag{23}$$

One can show that for any wave speed $c(x)$ a point symmetry of $\widehat{S}\{x,u,v\}$ is a point symmetry of both (22) and $u_t = 0$ (Ma [33]). It is easy to check that for many wave speeds $c(x)$ there exist point symmetries of (22) which are not point symmetries of $u_t = 0$ (see [2], [21]).

4.2.1. Change of Coordinates.

Suppose one seeks conserved forms of $R\{x,u\}$ through a change of coordinates defined by a contact transformation (which must be a point transformation when $m \geq 2$):

$$\begin{aligned} X_i &= X_i(x,u,\underset{1}{u}), \qquad i = 1,2,\ldots,n,\\ U_\nu &= U_\nu(x,u,\underset{1}{u}), \qquad \nu = 1,2,\ldots,m. \end{aligned}$$

Then the following theorem holds (Bluman [34]):

Theorem 4.2: $\sum_{\sigma=1}^{m} \lambda^\sigma(x,u,\underset{1}{u},\underset{2}{u},\ldots,\underset{p}{u})G^\sigma = \sum_{i=1}^{n} \frac{Df^i}{Dx_i}$ for some $\{f^i(x,u,\underset{1}{u},\underset{2}{u},\ldots,\underset{q}{u})\}$ if and only if $\sum_{\sigma=1}^{m} \lambda^\sigma G^\sigma = \sum_{i=1}^{n} \frac{DF^i}{DX_i}$ for some $\{F^i(x,u,\underset{1}{u},\underset{2}{u},\ldots,\underset{q}{u})\}$.

Consequently a conserved form is invariant under a change of coordinates and hence all conserved forms are conservation laws. Thus the search for equivalent conservation laws for a given system $R\{x,u\}$ is coordinate independent.

4.2.2. Factors for Linear Systems.

The following theorem holds for linear systems $R\{x,u\}$:

Theorem 4.3: Suppose $R\{x,u\}$ is a linear system of PDEs

$$G^\sigma = \sum_{\rho=1}^{m} L_\rho^\sigma[x]u^\rho = 0, \qquad \sigma = 1,2,\ldots,m \tag{24}$$

(In short form (24) can be expressed as $L[x]u = 0$.) Then $\sum_{\sigma=1}^{m} \lambda^\sigma(x)G^\sigma = \sum_{i=1}^{n} D_if^i$ for some $\{f^i(x,u,\underset{1}{u},\underset{2}{u},\ldots,\underset{k-1}{u})\}$, yielding an equivalent conservation law for $R\{x,u\}$, if and only if $L^*[x]\lambda(x) = 0$, where $\lambda(x) = (\lambda^1(x), \lambda^2(x), \ldots, \lambda^m(x))$ and $L^*[x]$ is the adjoint of linear operator $L[x]$.

The proof of Theorem 4.3 follows immediately from the fact that the Fréchet derivative of (24) is $\mathcal{L}[u] = L[x]$. Thus every linear system of PDEs has an infinite number of equivalent conservation laws. As a consequence one can discover whether or not a given nonlinear system can be linearized by an invertible mapping during the course of seeking its equivalent conservation laws (see Bluman [27]).

4.2.3. Factors for Nonlinear Systems.
If $R\{x,u\}$ is self-adjoint, i.e., its Fréchet derivative $\mathcal{L}[u]$ is self-adjoint, then factors $\lambda = (\lambda^1(x,u), \lambda^2(x,u), \ldots, \lambda^m(x,u))$, yielding equivalent conservation laws, satisfy $\mathcal{L}[u]\lambda = 0$ for any solution of $R\{x,u\}$. In particular $\sum_{\sigma=1}^{m} \lambda^\sigma(x,u)\frac{\partial}{\partial u^\sigma}$ is the infinitesimal generator of a point symmetry of $R\{x,u\}$ (Olver [4]).

If $R\{x,u\}$ is not self-adjoint, suppose $R\{x,u\}$ can be expressed in the form $\overline{\mathcal{L}}[u]u = 0$, i.e.,

$$\overline{\mathcal{L}}^\sigma_\rho[u]u^\rho = 0, \qquad \sigma = 1, 2, \ldots, m, \tag{25}$$

in terms of some linear operator $\overline{\mathcal{L}}[u]$. Let $\overline{\mathcal{L}}^*[u]$ be the adjoint of $\overline{\mathcal{L}}[u]$. Then the proof of the next theorem follows directly from Definition 2.4:

Theorem 4.4 Suppose $R\{x,u\}$ can be represented in the form (25). Factors $\lambda = (\lambda^1(x,u), \lambda^2(x,u), \ldots, \lambda^m(x,u))$ yield a conservation law of $R\{x,u\}$ if

$$\sum_{\rho,\sigma=1}^{m} u^\sigma \cdot \overline{\mathcal{L}}^{*\sigma}_{\ \rho}\lambda^\rho = 0 \tag{26}$$

holds for any solution of $R\{x,u\}$.

As an example, consider system $R\{x,u\}$ given by

$$\frac{\partial u^2}{\partial x} = u^1, \qquad \frac{\partial u^2}{\partial t} = \frac{1}{(u^1)^2}\frac{\partial u^1}{\partial x}, \tag{27}$$

$u = (u^1, u^2)$. The Fréchet derivative of (27) is the operator

$$\mathcal{L}[u] = \begin{bmatrix} 1 & \frac{D}{Dx} \\ \frac{2}{(u^1)^3}\frac{\partial u^1}{\partial x} - \frac{1}{(u^1)^2}\frac{D}{Dx} & \frac{D}{Dt} \end{bmatrix}.$$

Then

$$\mathcal{L}^*[u] = \begin{bmatrix} 1 & -\frac{1}{(u^1)^2}\frac{D}{Dx} \\ \frac{D}{Dx} & \frac{D}{Dt} \end{bmatrix} \neq \mathcal{L}[u].$$

Expressing (27) in the form (25), one finds that

$$\overline{\mathcal{L}}[u] = \begin{bmatrix} 1 & -\frac{D}{Dx} \\ \frac{1}{(u^1)^2}\frac{D}{Dx} & -\frac{D}{Dt} \end{bmatrix}, \qquad \overline{\mathcal{L}}^*[u] = \begin{bmatrix} 1 & \frac{1}{(u^1)^2}\frac{D}{Dx} \\ \frac{D}{Dx} & \frac{D}{Dt} \end{bmatrix}.$$

After solving the corresponding equations (26), one obtains $\lambda^1 = \frac{1}{u^1}F^1(u^2,t)$, $\lambda^2 = F^2(u^2,t)$ where $\{F^1, F^2\}$ is an arbitrary solution of the linear system $\frac{\partial F^2}{\partial u^2} = F^1$, $\frac{\partial F^1}{\partial u^2} + \frac{\partial F^2}{\partial t} + \frac{2}{u^2}F^1 = 0$. This indicates that (27) can be linearized by an invertible point transformation (see [27]).

4.2.4. Two Conjectures.

After considering many examples, including analysis of all nontrivial examples of nonlocal symmetries, obtained through heuristic procedures by Akhatov et al. [35], in the framework of potential symmetries, the following conjectures are proposed:

Conjecture 4.5: A process of obtaining a chain of auxiliary systems $S^{(1)}$, $S^{(2)}$, ..., $S^{(N)}$, through use of equivalent conservation laws, terminates at some finite N when either

(I) $S^{(N)}$ has no equivalent conservation laws; or

(II) $S^{(N)}$ has an infinite number of equivalent conservation laws. In this case $S^{(N)}$ is likely to be linearizable.

Conjecture 4.6: If $S^{(N)}$ has no equivalent conservation laws then the point symmetries of $S^{(N)}$ yield, through projections, all point symmetries of *any* subsystem of $S^{(N)}$ including $R\{x,u\}$, $S^{(1)}$, $S^{(2)}$, ..., $S^{(N-1)}$. (For a discussion of subsystems see Bluman [24].)

These two conjectures suggest that symbolic computation programs should be developed for determining symmetries of nonlinear systems of PDEs which:

1. Find $\overline{\mathcal{L}}[u]$, $\overline{\mathcal{L}}^*[u]$ for a given $R\{x,u\}$.
2. Find all sets of factors $\lambda(x,u)$ satisfying (26).
3. If the number of sets of factors is finite, then for each set $\lambda = (\lambda^1, \lambda^2, \dots, \lambda^m)$ find all systems $\widehat{R}_\rho\{x,u\}$ with the same solution set as $R\{x,u\}$.
4. Repeat steps (1) to (3) for each $\widehat{S}_\rho^{(1)} = \widehat{S}_\rho^{(1)}\{x,u,v\}$ obtained from (3), ...
5. Stop when any system $\widehat{S}_\rho^{(N)}$ has no equivalent conservation laws. Determine point symmetries for any such $\widehat{S}_\rho^{(N)}$.
6. For any system $\widehat{S}_\rho^{(N)}$, which has an infinite number of equivalent conservation laws, use the results in [27] and [2], [6], [7] to seek its linearization.

References

1. Bluman, G.W. and Cole, J.D. (1974) *Similarity Methods for Differential Equations.* Appl. Math. Sci. No. 13, Springer, New York.
2. Bluman, G.W. and Kumei, S. (1989). *Symmetries and Differential Equations.* Appl. Math.Sci. No. 81, Springer, New York.
3. Noether, E. (1918). Invariante Variationsprobleme. Nachr. König. Gesell. Wissen. Göttingen, Math-Phys. Kl., 235–257.
4. Olver, P.J. (1986). *Applications of Lie Groups to Differential Equations.* GTM, No. 107,Springer, New York.
5. Ibragimov, N.H. (1985). *Transformation Groups Applied to Mathematical Physics.* Reidel, Boston.
6. Kumei, S. and Bluman, G.W. (1982). When nonlinear differential equations are equivalent to linear differential equations. SIAM J. Appl. Math. **42**, 1157–1173.
7. Bluman, G.W. and Kumei, S. (1990). Symmetry-based algorithms to relate partial differential equations: I. Local symmetries. Euro. J. Appl. Math **1**, 189–216.
8. Ovsiannikov, L.V. (1982). *Group Analysis of Differential Equations.* Academic Press, New York.
9. Lisle, I. (1992). Equivalence Transformations for Classes of Differential Equations. Ph.D. Thesis, University of British Columbia.
10. Ibragimov, N.H., Torrisi, M. and Valenti, A. (1991). Preliminary group classification of equations $v_tt = f(x,v_x)v_xx + g(x,v_x)$. J. Math. Phys. **32**, 2988–2995.

11. Head, A.K. (1990, 1992). LIE: A muMATH Program for the Calculation of the LIE Algebra of Differential Equations. CSIRO Division of Material Sciences. Clayton, Australia.
12. Schwarz, F. (1985). Automatically determining symmetries of partial differential equations. Computing **34**, 91–106.
13. Wolf, T. and Brand, A. (1992). The computer algebra package CRACK for investigating PDEs, in *ERCIM Advanced Course on Partial Differential Equations and Group Theory.* Bonn.
14. Champagne, B., Hereman, W. and Winternitz, P. (1991). The computer calculation of Lie point symmetries of large systems of differential equations. Comp. Phys. Comm. **66**, 319–340.
15. Kersten, P.H.M. (1987). *Infinitesimal Symmetries: a Computational Approach.* CWI Tract No. 34, Centrum voor Wiskunde en Informatica, Amsterdam.
16. Reid, G.J. (1991). Algorithms for reducing a system of PDEs to standard form, determining the dimension of its solution space and calculating its Taylor series solution. Euro. J. Appl. Math. **2**, 293–318.
17. Reid, G.J. (1991). Finding abstract Lie symmetry algebras of differential equations without integrating determining equations. Euro. J. Appl. Math. **2**, 319–340.
18. Baumann, G. (1992). Lie symmetries of differential equations: A *Mathematica* program to determine Lie symmetries. Wolfram Research Inc., Champaign, Ill. Math-Source 0202-622.
19. Krasil'shchik, I.S. and Vinogradov, A.M. (1984). Nonlocal symmetries and the theory of coverings: an addendum to A.M. Vinogradov's Local symmetries and conservation laws. Acta Applic. Math. **2**, 79–96.
20. Krasil'shchik, I.S. and Vinogradov, A.M. (1989). Nonlocal trends in the geometry of differential equations: symmetries, conservation laws, and Bäcklund transformations. Acta Applic. Math. **15**, 161–209.
21. Bluman, G.W. and Kumei, S. (1987). On invariance properties of the wave equation. J. Math. Phys. **28**, 307–318.
22. Bluman, G.W., Kumei, S. and Reid, G.J. (1988). New classes of symmetries for partial differential equations. J. Math. Phys. **29**, 806–811, 2320.
23. Bluman, G.W. (1990). Potential symmetries, in *Proceedings of the Annual Seminar of the Canadian Mathematical Society on Lie Theory, Differential Equations and Representation Theory.* (V. Hussin, ed.). Centre de Recherches Mathématiques, Montréal, 85–100.
24. Bluman, G.W. (1992). Use and construction of potential symmetries. To appear in J. Math. Comp. Modelling.
25. Bluman, G.W. (1992). An overview of potential symmetries. To appear in Proceedings of AMS-SIAM 1992 Summer *Seminar Exploiting Symmetry in Applied and Numerical Analysis.*
26. Bluman, G.W. and Kumei, S. (1990). Symmetry-based algorithms to relate partial differential equations: II. Linearization by nonlocal symmetries. Euro. J. Appl. Math. **1**, 217–223.
27. Bluman, G.W. (1992). Potential symmetries and linearization. To appear in Proceedings of NATO Advanced Research Workshop *Applications of analytic and geometric methods to nonlinear differential equations.* Exeter, U.K., July 14–19, 1992, Kluwer.
28. Nucci, M.C. and Clarkson, P.A. (1992). The nonclassical method is more general than the direct method for symmetry reductions. An example of the Fitzhugh-Nagumo equation. Phys. Lett. A **164**, 49–56.
29. Pucci, E. (1992). Similarity reductions of partial differential equations. J. Phys. A **25**, 2631–2640.
30. Bessel-Hagen, E. (1921). Über die Erhaltungssätze der Elektrodynamik. Math. Ann. **84**, 258–276.
31. Pucci, E. and Saccomandi, G. (1992). Potential symmetries and solutions by reduction of partial differential equations. Preprint. To appear in J. Phys. A.
32. Bluman, G.W. and Kumei, S. (1989). Exact solutions for wave equations of two-layered media with smooth transition. J. Math. Phys. **29**, 86–96.
33. Ma, A. (1990). Potential Symmetries of the Wave Equation. M.Sc. Thesis, University of British Columbia.
34. Bluman, G.W. Invariance of conserved forms under contact transformations. Preprint.
35. Akhatov, I.S., Gazizov, R.K. and Ibragimov, N.K. (1991). Nonlocal symmetries. Heuristic approach. J. Sov. Math. **55**, 1401–1450.

SOME REMARKS ON A CLASS OF ORDINARY DIFFERENTIAL EQUATIONS: THE RICCATI PROPERTY*

SANDRA CARILLO
Dipartimento Metodi e Modelli Matematici Scienze Applicate
Università "La Sapienza"
I 00161 Roma

and

BENNO FUCHSSTEINER
Mathematics Department
University of Paderborn
D 4790 Paderborn

Abstract. Here ordinary differential equations of third and higher order are considered; in particular, a class of equations which can be solved by quadratures is exploited.

Indeed, crucial to obtain our result is the property of the Riccati equation, according to which, given one particular solution, then its general solution can be determined explicitly.

Thus, what we term the "Riccati" Property is introduced to point out that the members of such a class are differential equations which are of a generalized form of Riccati equation. Trivial examples of differential equations which enjoy the Riccati Property are all linear second order ordinary differential equations.

Here some further examples of ordinary differential equations which enjoy the same Property are considered. In particular, on the basis of group invariance requirements, a method to construct ordinary differential equations which enjoy the Riccati Property is given.

Remarkably, it follows that ordinary differential equations enjoying the Riccati Property are related to nonlinear evolution equations which admit a hereditary recursion operator.

Finally, further connections with nonlinear evolution equations are mentioned.

1. Introduction

Here the Riccati equation [7] [10] is briefly reconsidered recalling its link with linear ordinary differential equations of the second order. Subsequently, third and higher order ordinary differential equations are considered; in particular, a class of equations which can be solved by quadratures is exploited.

Accordingly, the Riccati equation enjoys the property that given one particular solution, its general solution can be determined explicitly. Hence, in [3], the "Riccati" Property is introduced referring to a class of differential equations which share the same property with the Riccati equation [7]. By definition, thus, such a class of differential equations may comprise equations linear and nonlinear of any order. The reason why we termed such a property "Riccati" is that the best known example of an ordinary differential equation which enjoys it is the Riccati equation ([10], [7], [9]). Trivial examples of differential equations enjoying the Riccati Property

* Work partially supported by the Gruppo Nazionale Fisica Matematica of the Italian National Research Council (C.N.R.)

N. H. Ibragimov et al. (eds.),
Modern Group Analysis: Advanced Analytical and Computational Methods in Mathematical Physics, 85–92.

are all linear second order ordinary differential equations. Indeed, the *variation of constants' method* implies that, known one particular solution, the general one can be obtained in explicit form.

The method to construct ordinary differential equations which enjoy the Riccati Property, proposed in [3], is reconsidered.

Subsequently, following the results obtained in [3], some linear ordinary differential equations of third order are proved to enjoy the Riccati property and, in particular, given one solution, other two independent ones are constructed.

Notably, see [3], given a third order nonlinear evolution equation which admits a hereditary recursion operator, an ordinary differential equation enjoying the Riccati Property can be constructed.

The paper closes providing further remarks concerning how ordinary differential equations which enjoy the Riccati Property are related to nonlinear evolution equations [3].

2. The "Riccati" Property

In this section some properties of the Riccati equation which turn out to be crucial for our study are briefly recalled [7], [9], [10]. The nonlinear Riccati equation reads:

$$\frac{dw}{dx} = Q(x) - P(x)w - w^2 \quad , \quad Q(x) \not\equiv 0$$

wherein $P(x)$ and $Q(x)$ are given real functions. It is transformed into

$$\frac{d^2u}{dx^2} + P(x)\frac{du}{dx} + Q(x)u = 0$$

by the Riccati transformation

$$w = \frac{u'}{u}.$$

Thus, it can be transformed into a second order **linear** ordinary differential equation, and, therefore, whenever one solution of the Riccati equation is known, then, the general one can be obtained by quadratures [7], [9].

However, typically, the general solution of the Riccati equation cannot be obtained by quadratures.

Definition 2.1: *An ordinary differential equation is said to possess the* **Riccati Property** *if, given one particular solution, it is possible to find explicitly its general solution by quadratures.*

Trivially, all linear second order ordinary differential equations enjoy the Riccati Property since, given a particular solution of any second order linear ordinary differential equation, the related general solution can be easily constructed on application of the *variation of constants' method*. Moreover, given a second order linear ordinary differential equation it can be transformed into a family of nonlinear first order Riccati equations [10] which depend on a parameter: the integration constant.

Further examples of equations which enjoy such a property are represented by the following linear third order ordinary differential equations in the unknown function $s(x)$:

$$s''' + 4us' + 2u's = 0 \tag{2.1}$$

and

$$u'(s'' - \lambda s) - u(s''' - \lambda s') - 4u^3 s' = 0 \tag{2.2}$$

wherein u is an arbitrary function of x and λ a real parameter.

Remarkably, the equation (2.1) reduces to an Euler linear equation on letting $u = -\frac{3}{4}x^{-2}$, namely:

$$x^3 s''' - 3xs' - 3s = 0. \tag{2.3}$$

This particular problem can be studied on use of classical techniques (see for instance [9]). However, even in this special case, our result turns out to be helpful since it allows to write immediately, given any particular solution of (2.3), its general one.

Indeed, since we proved [3] that if $v_1 = s$ is a solution of (2.1), then both:

$$v_2 = sD_{x_0}^{-1}(s^{-1}) \tag{2.4}$$

and

$$v_3 = \frac{1}{2}s\left(D_{x_0}^{-1}(s^{-1})\right)^2 \tag{2.5}$$

wherein

$$D_{x_0}^{-1} = \int_{x_0}^{x} \cdot \; d\xi$$

are again solutions of (2.1), then the construction of the general solution of (2.3) is straightforward. As an example, one solution is readily checked to be $s = x^3$; the other two independent ones can be evaluated on use of (2.4) and (2.5) wherein $x_0 \in \mathbb{R}\backslash\{0\}$. They, in turn, are:

$$v_2 = -\frac{1}{2}x + \frac{1}{2{x_0}^2}x^3 \quad \text{and} \quad v_3 = -\frac{1}{4{x_0}^2}x + \frac{1}{8{x_0}^4}x^3 + \frac{1}{8x} \tag{2.6}$$

and, thus, the general solution reads:

$$s(x) = c_1 x^3 + c_2 x + c_3 \frac{1}{x} \tag{2.7}$$

where c_1, c_2 and c_3 are arbitrary constants.

The latter provides only a trivial example of application of equations enjoying the Riccati Property; more general is the case represented by (2.1), wherein u is an arbitrarily given function of x: its general solution can be constructed via our procedure whenever one particular solution is known. The advantages of our procedure, that induces to write, from a single known solution, a set of independent solutions, from which the general solution follows, show up clearly in the case of (2.2). Furthermore, it can be remarked that (2.2) does not reduce to an Euler linear equation for any choice of the parameter λ and of the function u. In particular, we proved [3] that (2.2) on admitting a solution $v_1 = s$, admits also the solutions:

$$v_2 = sD_{x_0}^{-1}\left(s^{-2}\sqrt{\lambda\, s^2 - s'^2}\right) \tag{2.8}$$

and

$$v_3 = \frac{1}{2}s\left\{D_{x_0}^{-1}\left(s^{-2}\sqrt{\lambda\, s^2 - s'^2}\right)\right\}^2 - s\left(D_{x_0}^{-1}(s's^{-3})\right) . \tag{2.9}$$

3. How to construct equations possessing the Riccati Property

Here the method to construct linear equations which enjoy the Riccati Property [3] is retrieved in detail, furthermore links between the examples presented and well known ordinary differential equations are supplied. For sake of simplicity, we restrict ourselves to illustrate the method in the case of linear third order ordinary differential equations.

Method:

1) Consider a third order linear equation in the unknown function s such that:
   - the coefficients depend on a given function u;
   - if the equation is considered as an equation in the unknown function u, instead of s, then it is possible to determine explicitly its solution u as a function of s and its derivatives up to second order, say

   $$u = F(c_1, \ldots, c_n, s, s', s'')$$

   where the $c_1, \ldots, c_n$ are suitable constants of integration.
2) In the representation $u = F(c_1, \ldots, c_n, s, s', s'')$, choose the integration constants such that F is invariant under the substitution $s \to \alpha s$, $\alpha \in \mathbb{R}$. Such a choice is possible since the original equation is linear in s and, hence, invariant under this substitution.
3) Then, evaluate the variational derivative of F

   $$F'(s)[v] = \frac{\partial}{\partial \epsilon} F(s + \epsilon v, s' + \epsilon v', s'' + \epsilon v'')_{|\epsilon=0} \tag{3.1}$$

   and consider $F'(s)[v] = 0$ as a second order ordinary differential equation in the unknown function v. The homogeneity immediately implies that one solution is given by $v = s$ and a second independent solution

   $$v = G(s) \tag{3.2}$$

   can be computed on application of the variation of constants' method. This $G(s)$ represents also a solution of the original linear problem.
4) Compute, now, the third solution of the original linear problem via the known two solutions by further application of the variation of constants' method.

The method, we just presented, can be applied to the examples we already gave (2.1) and (2.2). Let us, at first, concentrate our attention on (2.1) and consider it as an ordinary differential equation in the unknown function u. This, solved with respect to u, gives:

$$u = \frac{1}{4s^2}(\lambda s^2 - 2ss'' + s'^2) + \frac{c_1}{s^2} \tag{3.3}$$

where c_1 is a constant of integration. According to step 2), let us require the invariance of u under the substitution $s \to \alpha s, \quad \alpha \in \mathbb{R}$. This implies $c_1 = 0$ and, thus, (3.3) reads:

$$u = \frac{1}{4s^2}(\lambda s^2 - 2ss'' + s'^2) \tag{3.4}$$

Then, step 3), we look for a one-parameter group which transforms the s-variable so that the corresponding u-variable is left unchanged. Let us denote by τ the group parameter and by $v = s_\tau$ its infinitesimal generator. Thus, on derivation with respect to τ of (3.4), it follows that v satisfies the linear differential equation:

$$-s^2v'' + ss'v' + (ss'' - s'^2)v = 0 \tag{3.5}$$

The homogeneity of (3.4) implies that the special one-parameter group $s \to \tau s$ leaves u unchanged. The infinitesimal symmetry generator of this special group is s itself, thus, $v_1 = s$ is a particular solution of (3.5), which, interpreted as a differential equation in the unknown v, enjoys the Riccati Property and, therefore, its general solution can be obtained by quadratures. Applying the variation of constants' method, a second solution of (3.5), independent from the previous one, is obtained:

$$v_2 = \frac{1}{2}s\left(D_{x_0}^{-1}(s^{-1})\right). \tag{3.6}$$

Now, since any solution s of (3.4) is a solution of (3.3), the group $s \to s(\tau)$, for which the generator is determined, leaves invariant also (3.3). Furthermore, the linearity of (2.1) implies that the related solution space is a linear space; consequently, any infinitesimal generator of a group acting on its solution space is itself a vector which belongs to the same space of solutions. Thus, any solution v of (3.5), whenever s in (3.5) is a solution of (2.1), satisfies (2.1) as well. This implies that (3.6) represents another solution of (2.1).

Finally, given the two independent solutions $v_1 = s$ and (3.6) of the third order equation (2.1), a third solution, independent from both the previous ones, can be obtained by the method of variation of constants. It follows that, if s is a solution of (2.1), then both (2.4) as well as (2.5) are again solutions of (2.1). Equation (2.1), thus, is proved to enjoy the Riccati Property. □

The same method allows to prove that also (2.2) enjoys the Riccati Property; indeed, solution of (2.2) with respect to u, gives:

$$2u = \frac{\lambda\, s - s''}{\sqrt{\lambda\, s^2 - s'^2 + c_1}}. \tag{3.7}$$

wherein $c_1 = 0$ is the only choice of such a constant which implies that $u(s)$ is homogeneous with respect to s, namely:

$$2u = \frac{\lambda\, s - s_{xx}}{\sqrt{\lambda\, s^2 - s_x{}^2}}. \tag{3.8}$$

The subsequent analysis of the group transformations under which u, expressed by (3.8), is left unchanged, implies that if s is a solution of (2.2) then

$$sD_{x_0}^{-1}\left(s^{-2}\sqrt{\lambda\, s^2 - s'^2}\right) \tag{3.9}$$

is again a solution of such equation. The third independent solution can be obtained considering (3.9) as an infinitesimal generator of a one-parameter group acting on the solution space of (3.9) itself. Namely, on use of the group

$$s_\tau = sD_{x_0}^{-1}\left(s^{-2}\sqrt{\lambda\, s^2 - s'^2}\right) \tag{3.10}$$

and of the linearity of the solution space, it follows that also $s_{\tau\tau}$ represents a solution of (2.2). Explicit computation delivers:

$$s_{\tau\tau} = \frac{1}{2}s\left\{D_{x_0}^{-1}\left(s^{-2}\sqrt{\lambda\, s^2 - s'^2}\right)\right\}^2 - s\left(D_{x_0}^{-1}(s's^{-3})\right) . \tag{3.11}$$

Thus, equation (2.2) is proved to admit, further to the solution $v_1 = s$, the solutions $v_2 = s_\tau$ and $v_3 = s_{\tau\tau}$, given, respectively, by (3.10) and (3.11), that is, in turn, (2.8) and (2.9). □

4. Connections with Nonlinear Evolution Equations

Here, the connection between ordinary differential equations enjoying the Riccati Property and related nonlinear evolution equations is briefly pointed out. In particular, we show how the examples (2.1) and (2.2) are related to nonlinear evolution equations which admit a hereditary recursion operator. On the other hand, we remark the important role which ordinary differential equations enjoying the Riccati Property may have in connection with the Cauchy problem for nonlinear evolution equations.

Most nonlinear evolution equations admitting a hereditary recursion operator [2] can be written in the form

$$u_t = K(u) \quad , \quad K(u) = \Phi(u)u_x \tag{4.1}$$

wherein Φ represents the hereditary recursion operator ([2], [4]). In particular, the Korteweg-deVries (KdV) equation reads:

$$u_t \;=\; u_{xxx} + 6uu_x \tag{4.2}$$

which can be written in the form (4.1) wherein [1] the hereditary recursion operator is the Lenard operator [1], [2], [4]:

$$\Phi(u) \;=\; D^2 + 2DuD^{-1} + 2u \quad ; \quad D^{-1} = \int_{-\infty}^{x} \cdot \; d\xi . \tag{4.3}$$

[1] Note, that both u in (4.2) as well as s in (4.4) are C^∞-functions with respect to $x \in \mathbf{R}$ such that they, together with their partial derivatives with respect to x, vanish at $\pm\infty$ faster than any rational function of x.

The Korteweg-deVries equation (4.2) is linked [6] to the KdV zero-interacting soliton equation [5]:

$$s^2 s_t = s^2 s_{xxx} - 3 s s_x s_{xx} + \frac{3}{2} {s_x}^3 \tag{4.4}$$

by the Bäcklund transformation:

$$u = \frac{1}{2}\left(\frac{s_x}{s}\right)_x - \frac{1}{4}\left(\frac{s_x}{s}\right)^2 = -\sqrt{s}\left(\frac{1}{\sqrt{s}}\right)_{xx} . \tag{4.5}$$

This, when we consider u and s as functions of the variable x only, can be read as an ordinary differential equation which, also, enjoys the Riccati property.

Moreover, (2.1) and (2.2) can be obtained, on application of the results in [5], again considering u and s as functions of the variable x only, from the eigenvalue problem

$$\Phi(u)s_x = \lambda s_x$$

related to the hereditary recursion operator of the Korteweg-deVries and modified Korteweg-deVries equation, respectively.

Further aspects of the connection between evolution equations and ordinary differential equations enjoying the Riccati Property have been studied in [3]; therein the Liouville equation, in the following form[2]

$$h_{xt} = e^{2h} . \tag{4.6}$$

is considered. Indeed, such equation enjoys many interesting properties and, in particular, its "general" solution ([12], [1]) is well known since there is a Bäcklund transformation between this equation and the wave equation ([11], [8]). However, it is mostly impossible to adapt this "general" solution to boundary conditions arising in applications ([13]). In these cases, the natural connection between the Liouville equation and the Riccati nonlinear ordinary differential equation, established in [3], allows to solve, by quadratures, the characteristic Cauchy problem.

Furthermore, according to [3], equations which enjoy the Riccati Property are naturally connected with a class of equations sharing with the Liouville equation the property that initial value problems can be solved by quadratures: thus, we termed them *generalized Liouville* equations. In [3] the third order problems enjoying the Riccati Property, (2.1) and (2.2), have been shown to be connected with these generalized Liouville equations and a method to obtain further examples of such generalized Liouville equations has been suggested.

References

1. F. Calogero and A. Degasperis: *Spectral Transform and Solitons I*, Studies in Mathematics and its Applications Vol. 13, North Holland Publishing Co., Amsterdam - New York - Oxford, 1982
2. B. Fuchssteiner: *Application of Hereditary Symmetries to Nonlinear Evolution Equations*, Nonlinear Analysis TMA, 3, p.849-862, 1979

[2] Note, however, that usually in applications the relevant form of the Liouville equation is obtained by a linear transform in the complex plane of independent variables.

3. B. Fuchssteiner and S. Carillo: *Generalized Liouville Equations and the Riccati Property*, in: Analysis and Geometry, Wissenschaftsverlag, Mannheim-Leipzig-Wien-Zürich, B.Fuchssteiner and W.A.J. Luxemburg, eds. , p.73-85, 1992
4. B. Fuchssteiner: *The Lie Algebra Structure of Nonlinear Evolution Equations admitting Infinite Dimensional Abelian Symmetry Groups*, Progr. Theor. Phys., 65, p.861-876, 1981
5. B. Fuchssteiner: *Solitons in Interaction*, Progress of Theoretical Physics, 78, p.1022-1050, 1987
6. B. Fuchssteiner and S. Carillo: *Soliton Structure versus Singularity Analysis: Third Order Completely Integrable Nonlinear Equations in $1+1$ Dimensions*, Physica, A 152, p.467-510, 1989
7. E. L. Ince: *Ordinary Differential Equations*, Dover Publications, New York, 1926
8. G. L. Lamb: *Elements of Soliton Theory*, Wiley Interscience Publ., New York-Chichester-Brisbane-Toronto, 1980
9. G. M. Murphy: *Ordinary Differential Equations and their Solutions*, Van Nostrand Reinhold Company, New York-Cincinnati-Toronto-London-Melbourne, 1960
10. W.T. Reid: *Riccati Differential Equations*, Academic Press, Boston - San Diego - New York - Berkeley - London - Sydney - Tokyo - Toronto, 1972
11. C. Rogers and W. F. Ames: *Nonlinear Boundary Value Problems in Science and Engineering*, Academic Press, Boston - San Diego - New York - Berkeley - London - Sydney - Tokyo - Toronto, 1989
12. C. Rogers and W. F. Shadwick: *Bäcklund Transformations and their Applications*, Mathematics in Science and Engineering Vol. 161, Academic Press, New York - London - Paris - Sydney - Tokyo - Toronto, 1982
13. E.R. Tracy and A.J. Neil and H.H. Chen and C.H. Chin: *Investigation of the Periodic Liouville equation*, in: Topics in Soliton Theory and Exactly solvable Nonlinear equations (eds: M. Ablowitz, B. Fuchssteiner, M. Kruskal) World Scientific Publ., Singapore, 1987 p.263-276,

DIFFERENTIAL-ALGEBRAIC AND DIFFERENTIAL-GEOMETRIC APPROACH TO THE STUDY OF INVOLUTIVE SYMBOLS

GIUSEPPA CARRÀ-FERRO*
Dipartimento di Matematica
Università di Catania, Italy

and

SERGEI V. DUZHIN †
Program Systems Institute,
Pereslavl-Zalessky, Russia

Abstract. Any system of partial differential equations after a finite number of prolongations becomes either inconsistent or involutive. The Hilbert function of any homogeneous ideal becomes polynomial beginning from a certain value of its argument. Any system of generators of a polynomial ideal becomes its Gröbner basis when multiplied by all monomials of sufficiently big degree. We discuss the interrelations between the number of steps required in these three procedures, for systems of partial differential equations with constant coefficients.

1. Introduction

Since multilinear algebra is basically a much simpler mathematical subject than the theory of partial differential equations, it has always been a concern of many mathematicians to reduce at least some problems of the latter to algebra.

There are several ways to study differential equations by algebraic methods.

The first one is the approach of differential algebra (F. Ritt, E. R. Kolchin), where only polynomial systems of differential equations are considered and, as such, studied with the techniques of differential fields.

The second one is the approach of E. Cartan who rewrites differential equations as exterior differential systems and then undertakes an infinitesimal investigation of the latter within the framework of multilinear exterior algebra.

The third approach, due to D. C. Spencer, H. Goldshmidt, J. F. Pommaret and others, is a modern development of the basic ideas of Janet with the machinery of homological algebra. Here, the study of a most general system of partial differential equations is performed through the study of its symbol, i.e. the leading homogeneous part of the linearized equations. In this way, investigations are carried over to the realm of symmetric polynomial algebra.

One of the most important notions inherent in all the mentioned approaches is that of an involutive system. According to Cartan, a system is involutive if there are no intrinsic obstacles for the existence of the general integral element of maximal dimension. According to Pommaret, involutivity constitutes a prerequisite

* Supported by M.U.R.S.T. 40% and CEC, ESPRIT BRA contract 6846 POSSO.

† Paper written during the author's stay at the Department of Mathematics, University of Catania, as a visiting professor with the support of C. N. R.

N. H. Ibragimov et al. (eds.),
Modern Group Analysis: Advanced Analytical and Computational Methods in Mathematical Physics, 93–99.

for formal integrability and is closely connected with the regularity of the series of prolongations of the system.

In the present paper, we restrict ourselves to the simplest case of systems of linear homogeneous differential equations with constant coefficients, trying to answer the basic questions about the link between the different notions of involutivity and the interrelations between regularity bound, involutivity bound and Gröbner basis bound.

2. Gröbner bases and Hilbert functions

Let K be a field of characteristic zero and let $R = K[X_1, \ldots, X_n]$ the algebra of polynomials over K. Denote by $T_R = \{m = X_1^{a_1} \cdots X_n^{a_n} : (a_1, \ldots, a_n) \in \mathbf{N}^n\}$ be the *set of terms of* R. A *term ordering* σ is a total order on T_R such that $1 <_\sigma m$ for all $m \in T_R \setminus \{1\}$ and $m_1 <_\sigma m_2$ implies $m_1 p <_\sigma m_2 p$ for all $p \in T_R$.

Any polynomial $f \in R^* = R \setminus \{0\}$ can be represented as $f = \sum_{i=1}^r c_i m_i$, where $c_i \in K^*$ and $m_i \in T_R$. If σ is a term ordering and $f \in R$, we denote by $M_\sigma(f)$ the *highest monomial* of f with respect to σ and by $T_\sigma(f)$ the *highest term* of f with respect to σ, i.e. $M_\sigma(f)$ divided by its coefficient.

A polynomial g is said to be *reducible* modulo a set $\{f_1, \ldots, f_r\}$ iff $T_\sigma(g)$ is a multiple of $T_\sigma(f_i)$ for some i. A set of generators $F = \{f_1, \ldots, f_r\}$ of a polynomial ideal I is referred to as its *Gröbner basis*, if every nonzero element $g \in I$ is reducible modulo F. (Gröbner bases allow of numerous equivalent definitions, see [2]).

A Gröbner basis is said to be *reduced*, if none of its elements contains terms that are divisible by the highest terms of other elements of the basis. With respect to a given term ordering, every ideal has a unique reduced Gröbner basis, up to a permutation of its elements.

Definition. Let I be an ideal in R and let $\{f_1, \ldots, f_r\}$ be the reduced Gröbner basis of I with respect to some term ordering σ. The *Gröbner basis bound* is $\gamma = \gamma_\sigma(I) = \max\{\deg(f_j) : j = 1, \ldots, r\}$.

Now suppose that the ideal I is *homogeneous*, i.e. that it equals the direct sum of its homogeneous components, $I = \oplus_{k \geq 0} I^k = \oplus_{k \geq 0} I \cap R^k$, where R^k stands for the space of all homogeneous polynomials of degree k. The *Hilbert function* of I is the integer-valued function $h_I(t) = \dim R^t / I^t$. A classical result of commutative algebra reads that the function $h_I(t)$ becomes polynomial in t when t is big enough.

Definition. This polynomial is called *the Hilbert polynomial of the ideal* I and denoted by $H_I(t)$. The minimal number $\rho = \rho(I)$ such that $h_i(t) = H_I(t)$ for all $t \geq \rho$ is referred to as the *regularity bound* of the ideal I.

It is known that the Hilbert polynomial can be written as

$$H_I(t) = \sum_{j=0}^{d} a_j \binom{s+j}{j}$$

for suitable $a_j \in \mathbf{Z}$, where $d \leq n$ and $a_d \in \mathbf{N}$. There exists a simple combinatorial procedure to find the coefficients a_j from the reduced Gröbner basis of the ideal I with respect to an arbitrary term ordering σ, see, e.g., [10].

3. Symbols and their prolongations

Let K be a field of characteristic zero. Let V and W be finite-dimensional vector spaces over K, $n = \dim_K V$, $m = \dim_K W$.

Definition. A *symbol* of order k is a linear subspace $G \subset S^k V^* \otimes_K W$.

Elements of the symbol can be viewed as vector valued homogeneous polynomials in n variables, i.e. homogeneous forms on the space V taking values in W. A symbol can also be defined by its dual space, i.e. its annihilator

$$A = \operatorname{Ann} G = \{\theta : (\theta, \omega) = 0 \text{ for all } \omega \in G\} \subset S^k V \otimes_K W^*,$$

which will be also referred to as the *cosymbol* associated with G.

Example 3.1 *Let V be a two-dimensional space with basis e_1, e_2 and $\dim W = 1$. Let x_1, x_2 the corresponding dual basis of V^*. Then the spaces $G_2 = \langle x_2^2 \rangle$ and $A_2 = \langle e_1^2, e_1 e_2 \rangle$ are mutually dual. (Here and below, we denote linear spans with angular brackets.)*

Example 3.2 *Under the same assumptions, the symbol $G_2 = \langle x_1 x_2 \rangle$ corresponds to the cosymbol $A_2 = \langle e_1^2, e_2^2 \rangle$.*

The abstract notion of symbol models that of the symbol of a system of differential equations. To obtain the cosymbol of a specific system, one has to take the highest order homogeneous parts of the (linearized) equations at a certain point. Any symbol may be viewed as the symbol of a linear system of d.e. with constant coefficients. Thus, the previous examples come from the overdetermined systems $\{\partial^2 u/\partial x_1^2 = 0, \partial^2 u/\partial x_1 x_2 = 0\}$ and $\{\partial^2 u/\partial x_1^2 = 0, \partial^2 u/\partial x_2^2 = 0\}$, respectively.

Accordingly, the prolongation procedure for differential equations leads to the following procedure for symbols. To define the prolongation of a symbol G_k, one takes its dual space A_k, computes its prolongation A_{k+r} and goes back to the dual space G_{k+r}. By definition, the *r-th prolongation* A_{k+r} *of* A_k is generated by all the products of the elements of A_k by arbitrary homogeneous r-forms on V^*, i.e. it is the image of the natural mapping $S^r V \times A_k \longrightarrow S^r V \times S^k V \otimes_K W^* \longrightarrow S^{k+r} V \otimes_K W^*$.

Equivalently, the prolongation can be defined in terms of the symbol itself. If $x_1, \ldots, x_n$ is a basis of V^*, i.e. a system of linear coordinates in V, then one can define the derivative operators $\partial_i = \partial/\partial x_i : S^{k+1} V^* \otimes_K W \longrightarrow S^k V^* \otimes_K W$, regarding the elements of these tensor products as functions on V.

Definition. Let $G_k \subset S^k V^* \otimes_K W$ be a symbol. The *first prolongation of* G_k is

$$G_{k+1} = \{\omega \in S^{k+1} V^* \otimes_K W : \partial_i(\omega) \in G_k \text{ for all } i = 1, \ldots, n\}$$

and more generally the *r-th prolongation of* G_k is

$$G_{k+r} = \{\omega \in S^{k+r} V^* \otimes_K W : \partial_{i_1} \cdot \ldots \cdot \partial_{i_r}(\omega) \in G_k \text{ for all } i_j = 1, \ldots, n\}.$$

Example 3.3 *The first prolongation of the symbol considered in Example 3.1 is $G_3 = \langle x_2^3 \rangle$. For the symbol of Example 3.2 we have $G_3 = 0$.*

In the scalar case, i.e. $\dim_K W = 1$, the theory of symbols and their prolongations is nothing but the theory of homogeneous ideals in the polynomial ring $R = \oplus_{k\geq 0} S^k V^*$ on V all of whose generators have the same degree. More generally the theory of symbols and their prolongation is nothing but the theory of R-modules of finite presentation with all generators of the relations submodule having the same degree.

4. Cartan characters and involutivity

Involutivity is a certain property of a symbol G_k that ensures a regular behaviour of the sequence of its prologations $G_{k+1}, G_{k+2}, \ldots, G_{k+r} \ldots$

To give the precise definition of involutivity according to [9], we will use filtrations of the spaces of homogeneous forms induced by flags in the base space. Let $\Phi = \{0 = V_0 \subset V_1 \subset \ldots \subset V_n = V\}$ be a complete flag in the base space V of dimension n and $\Phi^* = \{V^* = (V^*)_0 \supset (V^*)_1 \supset \ldots \supset (V^*)_n = 0\}$ be the corresponding flag in the dual space. It is always possible to choose a basis $\{e_1, \ldots, e_n\}$ in V in such a way that $V_i = \langle e_1, \ldots, e_i\rangle$ and $(V^*)_i = \langle x_{i+1}, \ldots, x_n\rangle$, where $x_1, \ldots, x_n$ is the dual basis of V^*. Given a symbol $G_k \subset S^k V^* \otimes_K W$ and a flag Φ in V, we will use the notation $G_k^i = G_k \cap S^k(V^*)_i \otimes_K W$, $i = 0, \ldots, n$ for the induced filtration in G_k.

Let us consider for every $i = 1, \ldots, n$ the following sequence

$$0 \longrightarrow G_{k+1}^i \longrightarrow G_{k+1}^{i-1} \xrightarrow{\partial_i} G_k^{i-1},$$

where the two non-zero arrows are the natural injection and the differentiation over x_i, respectively. The sequence is obviously exact, which implies that $\dim_K G_{k+1}^{i-1} - \dim_K G_{k+1}^i \leq \dim_K G_k^{i-1}$ and hence $\dim_K G_{k+1} \leq \sum_i \dim_K G_k^i$.

Definition. A symbol $G_k \subset S^k V^* \otimes_K W$ is said to be *involutive* if there exists a flag Φ in the base space V such that $\dim_K G_{k+1} = \sum \dim_K G_k^i$.

Remark. If the above equality holds for one specific flag, then it is also true for almost every flag, i.e. for every flag in an open subset of the flag manifold. Such flags are referred to as *regular*.

Example 4.1 *The symbol considered in Example 3.1 above is involutive, because for any $V_1 \neq \langle x_2\rangle$ we have $\dim G_2^1 = 0$ and $\dim G_3 = 1$. On the other hand, the symbol G_2 of Example 3.2 is not involutive, since in this case $\dim G_3 = 0$ while for any choice of V_1 we have $\dim G_2^0 + \dim G_2^1 = 1$.*

Definition. Let $G_k \subset S^k V \otimes_K W$ be a symbol and Φ a flag in V. For any $i = 1, \ldots, n$ the *i-th Cartan character of G_k with respect to Φ* is $\alpha_k^i = \dim_K G_k^{i-1} - \dim_K G_k^i$.

The numbers α_k^i always satisfy inequalities $0 \leq \alpha_k^n \leq \alpha_k^{n-1} \leq \ldots \leq \alpha_k^1 \leq \alpha_k^0$ and, save for the trivial case $\dim V = 1$, one always has $\alpha_k^n = 0$. Furthermore for all flags in an open subset of the flag manifold the numbers α_k^1, $\alpha_k^1 + \alpha_k^2, \ldots, \alpha_k^1 + \alpha_k^2 + \ldots + \alpha_k^n$ are successively as large as possible, which is equivalent to say that the numbers α_k^n, $\alpha_k^n + \alpha_k^{n-1}, \ldots, \alpha_k^n + \alpha_k^{n-1} + \ldots + \alpha_k^1$ are successively as small as possible. Flags with this property will be likewise called regular.

Definition. The Cartan characters of a symbol are its characters computed with respect to a regular flag.

Example 4.2 *In the conditions of either of Examples 3.1 and 3.2 above, we have* $\alpha^1 = 1$, $\alpha^2 = 0$.

Inequalities

$$\dim_K G_{k+1}^{i-1} - \dim_K G_{k+1}^i \leq \dim_K G_k^i$$

and

$$\dim_K G_{k+1} \leq \sum \dim_K G_k^i$$

imply that $\alpha_{k+1}^i \leq \alpha_k^1 + \alpha_k^2 + \ldots + \alpha_k^i$ for all $i = 1, \ldots, n$ and $\dim_K G_{k+1} = \sum_{i=1}^n \alpha_{k+1}^i \leq \alpha_k^1 + 2\alpha_k^2 + \ldots + i\alpha_k^i + \ldots + (n-1)\alpha_k^{n-1} + n\alpha_k^n$. It follows that the involutivity of a symbol G_k can be reformulated in terms of its characters as follows. A symbol $G_k \subset S^k V \otimes_K W$ is involutive if and only if there exists a flag Φ in the base space V such that $\dim_K G_{k+1} = \alpha_k^1 + 2\alpha_k^2 + \ldots + n\alpha_k^n$.

It is well known ([9]) that if G_k is involutive, then so is its prolongation G_{k+r} for any $r \geq 0$ and

$$\dim_K G_{k+r} = \sum_{i=1}^n \alpha_k^i \binom{r+i}{i}.$$

In other words, if G_k is involutive, then the dimensions of $G_{k+1}, G_{k+2}, \ldots, G_{k+r}, \ldots$ behave as a polynomial in r.

If G_k is not involutive, then there exists an $r > 0$ such that G_{k+r} is involutive (Cartan-Kuranishi).

Definition. Let $G_k \subset S^k V \otimes_K W$ be a symbol. The *involutivity bound* ι for G_k is the smallest integer $k + r$, such that the r-th prolongation G_{k+r} is involutive.

There exists an estimation from above for the involutivity bound in terms of n, m, k, see [9]. The corresponding function $Q(n, m, k)$ is determined by double recursion, so that it has extremely fast growth, which apparently has nothing to do with the exact values for ρ. Thus, $Q(2, 1, 2) = 17$, but, as the previous examples show, any symbol $G \subset S^2\mathbf{R}^2$ of dimension 1 is either already involutive or becomes involutive after one prolongation.

Let Φ be a flag in V and $\{x_1, \ldots, x_n\}$ the corresponding basis of V^*. The variables $x_i, \ldots, x_i$ are called *multiplicative* for V_i and $x_{i+1}, \ldots, x_n$ are called *non-multiplicative* for V_i.

The following characterization of involutivity through the notions of multiplicative and non-multiplicative variables is useful for the applications.

Lemma 4.1 *A symbol* $G_k \subset S^k V^* \otimes W$ *is involutive if and only if for every* $i = 1, \ldots, n$ *the quotient space* G_{k+1}^{i-1}/G_{k+1}^i *has as a basis all the elements of the basis of* G_k^j *multiplied by the multiplicative variable* x_{i-1}*, where* $j = i - 1, \ldots, n$.

Proof. By the definition of involutivity

$$\dim_K G_{k+1}^{i-1} - \dim_K G_{k+1}^i = \dim_K G_k^{i-1} = \alpha_k^i + \ldots + \alpha_k^{n-1} + \alpha_k^n.$$

If $\{x_1, \ldots, x_n\}$ is a basis of V^*, then G_{k+1} is obtained by multiplying each element of a basis of G_k^i by the corresponding multiplicative variables $x_1, \ldots, x_i$. Now the lemma follows from the definition of G_{k+1}^i and from the definition of prolongation.

5. Relations between the three bounds

Lemma 5.1 *Let $G_k \subset S^k V^* \otimes_K W$ be a symbol. If $\dim_K W = 1$, then the regularity bound ρ of the homogeneous ideal corresponding to its dual A_k is less than or equal to $\iota - 1$.*

Proof. The Hilbert function $h(t) = \sum_{s=0}^{t} \dim_K G_s$ becomes a polynomial, when $t \geq \iota - 1$, by definition of involutivity bound. By the definition of regularity bound and by its minimality $\rho \leq \iota - 1$.

Lemma 5.2 *Let $G_k \subset S^k V^* \otimes_K W$ be a symbol. If $\dim_K W = 1$, then the Gröbner basis bound $\gamma_\sigma(I)$ of the homogeneous ideal I corresponding to its dual A_k with respect to a well defined degree preserving term ordering, which agrees with a regular flag on V, is less than or equal to the involutivity bound ι.*

Proof. Let $\{g_1, \ldots, g_t\}$ be a finite set of generators of the ideal generated by A_k. If each g_i is monomial, then $\{g_1, \ldots, g_t\}$ is a Gröbner basis of this ideal and $\gamma = k$. Now suppose that at least one g_i is not monomial. Choose a generic flag in V for G_k and let $\{x_1, \ldots, x_n\}$ be the corresponding basis of V. Choose a degree preserving plus lexicographic term ordering on T_R in such a way that $x_1 < x_2 < \ldots < x_n$. Without loss of generality we can suppose that $M_\sigma(g_i) \neq M_\sigma(g_j)$ for each pair i, j. Each S-polynomial $S(g_i, g_j) = L_i g_i - L_j g_j$, where $L_i = M_\sigma(g_i)/\gcd(M_\sigma(g_i), M_\sigma(g_j))$ and $L_j = M_\sigma(g_j)/\gcd(M_\sigma(g_i), M_\sigma(g_j))$ as in the Buchberger algorithm for Gröbner bases, has degree $k(i,j)$ which is less than or equal to $2k$ and its reduced polynomial $g_{i,j}$ with respect to $\{g_1, \ldots, g_t\}$ will appear in the $r(i,j)$-th prolongation of A_k, where $k(i,j) = k + r(i,j)$. Furthermore $M_\sigma(g_{i,j})$ is not a multiple of any $M_\sigma(g_s)$ and hence $k + r(i,j)$ is less than or equal to ι. If $\{g_i, g_{(i,j)} : i, j = 1, \ldots, t\}$ is a Gröbner basis, then the proof is completed. If not, we repeat the Buchberger procedure by finding the new S-polynomials and reducing them with respect to $\{g_i, g_{(i,j)} : i, j = 1, \ldots, t\}$. The leading monomials of the new polynomials obtained in this way will not be multiples of the leading monomials of any g_i and $g_{(i,j)}$ and their degrees will be less than or equal to ι. By repeating this argument, we can conclude that $\gamma \leq \iota$.

Example 5.1 *Let $A_n = \langle x_1^n \rangle \subset S^n V^* \otimes_K W$, where $\dim_K W = 1$. The Hilbert polynomial of the corresponding symbol is $H(t) = \sum_{j=0}^{n-1} \binom{t+n-1-j}{n-1}$. Then $\{x_1^n\}$ is a Gröbner basis of A_n and $\rho = 0 < k = \gamma = \iota = n$.*

This example shows that the difference between the involutivity and the regularity bound can be arbitrarily large.

Example 5.2 *Let $A_n = \langle x_1^n, \ldots, x_n^n \rangle \subset S^n V^* \otimes_K W$, where $\dim_K W = 1$. The Hilbert polynomial of the corresponding symbol is $H(t) = n^2$. Then $\{x_1^n, \ldots, x_n^n\}$ is a Gröbner basis of A_n and $k = \gamma = n < \rho = n^2 - n < \iota = n^2$.*

This example shows that the difference between the involutivity bound and the Gröbner basis bound can be arbitrarily large.

References

1. Bayer D. *The Division Algorithm and the Hilbert Scheme*. Ph.D. Thesis, Harvard (1982).
2. Buchberger B. *Some Properties of Gröbner Bases for Polynomial Ideals*. ACM SIGSAM Bull. Vol. 10 (1976), pp. 19–24.
3. Carrà-Ferro G. *Gröbner Bases and Hilbert Schemes I*. Computational aspects of commutative algebra, Academic Press London (1989), pp. 85–96.
4. Carrà-Ferro G.–Sit W. *On Term-Orderings and Rankings*, to appear in *Computational Algebra*, Marcel Dekker, New York 1993.
5. Cartan E. *Les systèmes différentiels extérieurs et leurs applications géométriques*. P., Gauthier-Villars, 1923.
6. Giusti M. *Some Effectivity Problems in Polynomial Ideal Theory*. Proc. Eurosam 84, Lect. Notes in Comp. Science Springer, vol. 174 (1984), pp. 159–171.
7. Kolchin E. R. *Differential Algebra and Algebraic Groups*, Academic Press, London (1973).
8. Möller H.–Mora T. *Upper and Lower Bounds for the Degree of Gröbner Bases*. Proc. of EUROSAM 84, Springer, Lect. Notes in Comp. Science, vol. 174 (1984), pp. 172–183.
9. Pommaret J.-F. *Nonlinear partial differential equations and Lie pseudogroups*. Gordon and Breach, N.Y. 1978.
10. Robbiano L. *Gröbner Bases: a Foundation for Commutative Algebra*. Preprint, 1989.

THE BIHAMILTONIAN APPROACH TO INTEGRABLE SYSTEMS*

PAOLO CASATI
Dottorato in Matematica, Università di Torino,
Via Principe Amedeo 8, I-10123 Torino, Italy

FRANCO MAGRI
Dipartimento di Matematica dell'Università di Milano,
Via C. Saldini 50, I-20133 Milano, Italy

and

MARCO PEDRONI
Dottorato in Matematica, Università di Milano,
Via C. Saldini 50, I-20133 Milano, Italy

Abstract. This paper seeks to suggest some unorthodox viewpoints on symmetries and reductions of Hamiltonian equations. Moreover, applicability of these viewpoints is supported by the study of a concrete example: Gelfand-Dickey's equations.

1. Introduction

The theory of integrable systems received a considerable impetus from the discovery (by the Princeton group, see [1]) of the inverse scattering method. It allows to solve the Korteweg-deVries equation [2]

$$u_t = \frac{1}{4}(u_{xxx} - 6uu_x), \tag{1}$$

that is therefore considered as the first example of (non-trivial) *infinite-dimensional integrable system*. Two properties of the KdV equation, already observed by the Princeton group (see [3] and [4]), are worth mentioning.

1. It is a *Hamiltonian* equation:

$$u_t = \frac{1}{4}(u_{xxx} - 6uu_x) = \{u, H_1\}_0, \tag{2}$$

whereby

$$\{F,G\}_0 = \int dF(dG)_x\, dx, \qquad H_1[u] = -\frac{1}{4}\int (u^3 + \frac{1}{2}u_x^2)\, dx. \tag{3}$$

2. The KdV equation has an infinite sequence of constants of motion, which are in involution w.r.t. the Poisson bracket (3). This implies the existence of an infinite sequence of vector fields that are symmetries for the KdV equation; the time evolutions along these vector fields give rise to equations called higher KdV equations. Their collection is called the KdV hierarchy.

* This work has been supported by the Italian M.U.R.S.T. and by the G.N.F.M. of the Italian C.N.R.

N. H. Ibragimov et al. (eds.),
Modern Group Analysis: Advanced Analytical and Computational Methods in Mathematical Physics, 101–110.

In [5] it was shown that the KdV equation admits another Hamiltonian representation:

$$u_t = \frac{1}{4}(u_{xxx} - 6uu_x) = \{u, H_0\}_1, \tag{4}$$

whereby

$$\{F,G\}_1 = \int dF\,((dG)_{xxx} - 4u(dG)_x - 2u_x dG)\,dx, \qquad H_0[u] = \frac{1}{4}\int u\,dx. \tag{5}$$

In other words, the KdV is a *bihamiltonian* equation. The existence of the integrals of motion can be easily deduced from this fact, as we shall see later.

After the KdV equation, a lot of nonlinear p.d.e. solvable by means of the inverse scattering method were found (for a complete list see [6]). A characteristic feature of these equations is that they exhibit solitons, that is solitary waves solutions (with additional interaction properties, see [6]); for this reason they are often called *soliton equations*. Among these, we are particularly interested in the Gelfand-Dickey's equations [7]. They are a system of n equations in n unknown functions, reducing to the KdV equation for $n = 1$; for $n = 2$ they are related to another important equation, namely the Boussinesq equation [8]:

$$u_{tt} - u_{xx} - u_{xxxx} + 3(u^2)_{xx} = 0. \tag{6}$$

A further step was performed by Sato [9], who introduced a system of infinite equations (called Sato's equations) in an infinite number of unknown functions. By means of a reduction process one can recover the Gelfand-Dickey's equations from Sato's ones.

The aim of this paper is to give the basic ideas of the bihamiltonian approach to integrable systems. In Sec. 2 we introduce the main objects of the bihamiltonian theory: Poisson and bihamiltonian manifolds, the Poisson pencil and its Casimirs functions, the bihamiltonian hierarchies. Finally we state a general reduction theorem for bihamiltonian manifolds. In Sec. 3 we show a remarkable class of bihamiltonian manifolds defined on a Lie group with a cocycle. In our opinion, this class is particularly useful for the geometric description of soliton equations. To support this claim, in the rest of Sec. 3 we shortly compare the definitions of Gelfand-Dickey's equations from Sato's viewpoint and from the bihamiltonian viewpoint, with special attention to the KdV case.

2. The bihamiltonian theory

The purpose of this section is to introduce the basic ideas of the bihamiltonian approach to integrable systems. The main point is to replace the family of first integrals in involution, usually considered in the theory of integrable systems, with a second Poisson bracket on the manifold. In our opinion, this variant is particularly useful for the study of soliton equations. In this section the language is purely geometrical. The manifolds are supposed to be smooth and finite-dimensional, and the maps are of class C^∞. However, we think that many results can be extended to infinite-dimensional manifolds. We point out that the example dealt with in the next section is infinite-dimensional.

2.1. Poisson manifolds, Casimir functions, and symplectic foliation

Poisson manifolds are the natural setting for the theory of hamiltonian systems. The term "Poisson manifold" was introduced by Lichnerowicz in [10], but its definition can be found already in [11]; the key properties go back to the works of Lie and Dirac. We refer to [12] for a more detailed exposition of the topics in this subsection.

Definition 2.1 *Let M be a C^∞-differentiable manifold. A* Poisson bracket *on M is an $\mathbf{R}$-bilinear map*

$$\{\cdot,\cdot\} : C^\infty(M) \times C^\infty(M) \to C^\infty(M) \tag{7}$$

with the following properties:

1. *Antisymmetry.*
2. *Jacobi identity: $\{\{f,g\},h\} + \{\{h,f\},g\} + \{\{g,h\},f\} = 0$.*
3. *Leibnitz rule: $\{fg,h\} = f\{g,h\} + \{f,h\}g$.*

A Poisson manifold *is a C^∞-differentiable manifold with a Poisson bracket. The vector field $X_h = \{\cdot,h\}$ is called the* Hamiltonian vector field with Hamiltonian h. *A* Casimir function *is a function h such that $X_h = 0$, that is a function which is in involution with any other function on M.*

It is well-known that a symplectic manifold has a natural Poisson bracket defined on it. This bracket has the following property: the only Casimir functions are the constant functions. In other words:

$$X_h = 0 \Longrightarrow h = \text{constant}. \tag{8}$$

On a generic Poisson manifold we can have non trivial Casimir functions.

Remark 2.2 *Let $(M,\{\cdot,\cdot\})$ be a Poisson manifolds; the application $h \mapsto X_h$, assigning to any function h its Hamiltonian vector field X_h, is a Lie-algebra homomorphism, that is*

$$X_{\{f,g\}} = [X_f, X_g]. \tag{9}$$

It follows that $\{f,g\} = 0 \Rightarrow [X_f,X_g] = 0$, that is, Hamiltonian vector fields associated to functions in involution commute.

Remark 2.3 (symplectic foliation) *Suppose that $\{h_1,\ldots,h_r\}$ is a maximal set of functionally independent Casimir functions on a Poisson manifold M. Then it can be shown that the submanifolds*

$$S_{\underline{c}} = \{\, m \in M \mid h_i(m) = c_i \ \forall i \,\} \tag{10}$$

are symplectic manifolds. In other words, every Poisson manifold is (locally) foliated by symplectic manifolds. The so-called symplectic leaves *$S_{\underline{c}}$ may also be defined as the maximal integral varieties of the distribution $\{X_f\}_{f\in C^\infty(M)}$ (this distribution being integrable on account of Remark 2.2).*

2.2. Bihamiltonian manifolds, Poisson pencil and its Casimir functions

As Poisson manifolds are the natural setting for hamiltonian systems, bihamiltonian manifolds are, in our opinion, the natural setting for *integrable* hamiltonian systems. In this section we shortly review the basic points of the theory of bihamiltonian manifolds, referring to [13] and [14] for further details.

Definition 2.4 *Let M be a manifold endowed with two Poisson brackets, $\{\cdot,\cdot\}_0$ and $\{\cdot,\cdot\}_1$. We say that M is a* bihamiltonian manifold *if the linear combination,*

$$\{f,g\}_\lambda := \{f,g\}_0 - \lambda\{f,g\}_1, \tag{11}$$

of these brackets satisfies the Jacobi identity for any value of the real or complex parameter, λ. This means that the cyclic compatibility condition,

$$\begin{aligned}&\{f,\{g,h\}_0\}_1 + \{h,\{f,g\}_0\}_1 + \{g,\{h,f\}_0\}_1 + \\ &\{f,\{g,h\}_1\}_0 + \{h,\{f,g\}_1\}_0 + \{g,\{h,f\}_1\}_0 = 0,\end{aligned} \tag{12}$$

is valid for any triple of functions, (f,g,h), on M. In this case, the bracket, $\{\cdot,\cdot\}_\lambda$, is called the Poisson pencil, *defined by $\{\cdot,\cdot\}_0$ and $\{\cdot,\cdot\}_1$ on M.*

Let us consider $\{\cdot,\cdot\}_\lambda$ as a deformation of $\{\cdot,\cdot\}_0$, and let us assume that there exist Casimir functions of the Poisson pencil which are deformations of the Casimir functions of $\{\cdot,\cdot\}_0$. This means that we assume the existence of one-parameter families of functions, $h(\lambda)$, such that

$$\{h(\lambda),f\}_\lambda = 0 \tag{13}$$

for any function f, and for any value of λ, which, for $\lambda \to \infty$, tend to a Casimir function, h_0, of $\{\cdot,\cdot\}_0$. Accordingly, we assume that $h(\lambda)$ admits a Laurent expansion in λ,

$$h(\lambda) = \sum_{k\geq 0} h_k\lambda^{-k}, \tag{14}$$

starting from h_0. Eq.(13) immediately implies the recurrence relation,

$$\{h_k,\cdot\}_1 = \{h_{k+1},\cdot\}_0, \tag{15}$$

on the coefficients h_k. If we set $X_f = \{\cdot,f\}_0$ and $Y_f = \{\cdot,f\}_1$, then Eq.(15) can be rewritten as

$$Y_{h_k} = X_{h_{k+1}}. \tag{16}$$

Now we shall prove that if $h(\lambda)$ and $l(\lambda)$ are two Casimir functions, then their coefficients commute w.r.t. both Poisson brackets. Indeed Eq.(15) implies

$$\{h_k,l_j\}_0 = \{h_{k-1},l_j\}_1 = \{h_{k-1},l_{j+1}\}_0 = \ldots = \{h_0,l_{j+k}\}_0 = 0, \tag{17}$$

and, similarly,

$$\{h_k,l_j\}_1 = \{h_{k+1},l_j\}_0 = 0. \tag{18}$$

Therefore, we conclude that the coefficients of the Laurent expansion of the Casimir's functions of the pencil are in involution with respect to both brackets generating the pencil. This remark is the basis of the bihamiltonian approach to integrable systems. It suggests considering the vector fields,

$$\Phi_k = \{h_k, \cdot\}_1 = \{h_{k+1}, \cdot\}_0, \tag{19}$$

as good candidates for descriptions of integrable systems. They commute as a consequence of (18):

$$[\Phi_j, \Phi_k] = 0. \tag{20}$$

Their integrability depends on the number and on the independence of the first integrals, h_k, which will not be considered here. We shall conclude these remarks by showing that Eq.(19) can also be written in the form,

$$\Phi_k = \{h^k(\lambda), \cdot\}_\lambda, \tag{21}$$

where

$$h^k(\lambda) := \sum_{j=0}^{k} h_j \lambda^{k-j}. \tag{22}$$

The Φ_k are, consequently, simultaneously Hamiltonian with respect to all the Poisson brackets of the pencil.

2.3. Bihamiltonian reduction and projected hierarchy

We now recall an interesting Poisson reduction canonically defined on every bihamiltonian manifold (see [13], Sec.1, for the proof). It follows from the observation that the Casimir functions of $\{\cdot,\cdot\}_0$ generate a Poisson subalgebra with respect to $\{\cdot,\cdot\}_1$, because of compatibility condition (12). Thus we can use the functions of the center, K_0, of $\{\cdot,\cdot\}_0$ in two distinct ways:

1. To define the symplectic leaves of $\{\cdot,\cdot\}_0$ as level sets of these functions.
2. To define an integrable distribution, D, on M, generated by the vector fields $Y_k = \{k, \cdot\}_1$ where $k \in K_0$.

Let us choose a specific symplectic leaf, S, of $\{\cdot,\cdot\}_0$, and let us denote by E the foliation induced on S by D. The leaves of E are the intersections of S with the leaves of D. We shall assume that E is sufficiently regular and that there exists a quotient space, $N = S/E$, and we denote the canonical immersion of S in M and the canonical projection of S onto N by $i : S \to M$ and $\pi : S \to N$ respectively. Then

Proposition 2.5 *The quotient space, $N = S/E$, is a bihamiltonian manifold. On N there exists a unique Poisson pencil, $\{\cdot,\cdot\}_N^\lambda$, such that*

$$\{f, g\}_N^\lambda \circ \pi = \{F, G\}_M^\lambda \circ i \tag{23}$$

for any pair of functions, F and G, which extend the functions f and g of N into M, and are constant on D. Technically, this means that F satisfies the conditions,

$$F \circ i = f \circ \pi \tag{24}$$

$$\{F, k\}_1 = 0, \tag{25}$$

for any function $k \in K_0$.

As a simple corollary of this proposition one easily proves

Proposition 2.6 *Let h_k the coefficients of the series (14). The bihamiltonian vector fields,*

$$\Phi_k = \{h_k, \cdot\}_1 = \{h_{k+1}, \cdot\}_0, \tag{26}$$

are tangent to the symplectic leaf S and are projectable on the quotient space $N = S/E$. The projected vector fields are bihamiltonian with respect to the reduced Poisson pencil on N, and consequently commute.

Thus we have concluded our presentation of the main points of the bihamiltonian theory. In the next section we shall suggest how to apply it in order to derive Gelfand-Dickey's equations.

3. Soliton equations and their bihamiltonian setting

In this section we shall describe Sato's approach to soliton equations, based on the introduction of a hierarchy of systems of infinite equations in an infinite number of unknown functions, called Sato's equations [9]. This approach is mainly algebraic, and is founded on the theory of formal pseudodifferential operators. After this description, we shall introduce the tools allowing to set soliton equations in the framework of the bihamiltonian theory. This part is simply sketched. A more comprehensive study of the relations between Sato's equations and bihamiltonian systems may be found in [13], [14] and [15].

3.1. Sato's and Gelfand-Dickey's equations

Let Q be a monic pseudodifferential operator of degree 1:

$$Q = \partial + q_0 + q_1\partial^{-1} + q_2\partial^{-2} + \dots \tag{27}$$

where the q_i are functions of x and ∂ is the derivative w.r.t. x. We can set (without loss of generality) $q_0 = 0$. Let us fix $k \in \mathbf{N}$; the k-th Sato's equation is [16]

$$\frac{\partial Q}{\partial t_k} = [(Q^k)_+, Q], \tag{28}$$

where A_+ is the purely differential part (zero-order term included) of the pseudodifferential operator A, and the powers Q^k can be computed by means of the commutation relations

$$[\partial^j, q] = \sum_{k=1}^{j} \binom{j}{k} \frac{\partial^k q}{\partial x^k}\partial^{j-k} \tag{29}$$

$$[\partial^{-j}, q] = \sum_{k=1}^{\infty} (-1)^k \binom{j+k-1}{k} \frac{\partial^k q}{\partial x^k}\partial^{-(j+k)}. \tag{30}$$

Recalling that $\deg[A,B] \leq \deg A + \deg B - 1$, one has that

$$\deg[(Q^k)_+, Q] = -1 = \deg\frac{\partial Q}{\partial t_k}. \tag{31}$$

Indeed $[(Q^k)_+, Q] = -[(Q^k)_-, Q]$, as a consequence of $[Q^k, Q] = 0$; it follows that $\deg[(Q^k)_+, Q] = \deg[(Q^k)_-, Q] \leq -1$. Hence Eq.(28) represents a system of infinite differential equations in the q_i. For example, for $k = 1$ Eq.(28) takes the form

$$\frac{\partial q_i}{\partial t_1} = q_{i_x} \quad \forall i, \tag{32}$$

while for $k = 2$ it becomes

$$\begin{cases} \dfrac{\partial q_1}{\partial t_2} = q_{1_{xx}} + 2q_{2_x} \\ \dfrac{\partial q_2}{\partial t_2} = q_{2_{xx}} + 2q_{3_x} + 2q_1 q_{1_x} \\ \dfrac{\partial q_3}{\partial t_2} = q_{3_{xx}} + 2q_{4_x} - 2q_1 q_{1_{xx}} + 4q_{1_x} q_2 \\ \cdots\cdots\cdots \end{cases} \tag{33}$$

As a particular case of the Sato hierarchy, one gets the n-th Gelfand-Dickey hierarchy [7], hereinafter denoted by GD(n). It suffices to observe that the constraint $Q^n = L$, with

$$L = \partial^n - \sum_{k=0}^{n-2} u_k \partial^k, \tag{34}$$

is consistent with Eq.(28). So the infinite system (28) reduces itself to a system of $n-1$ equations in the $n-1$ functions u_k; these equations coincides with the GD(n) eqs. For example, for $n = 2$ one has

$$Q^2 = L, \qquad L = \partial^2 - u, \tag{35}$$

so that the q_i can be expressed in terms of u:

$$\begin{cases} q_1 = -\frac{1}{2}u \\ q_2 = \frac{1}{4}u_x \\ q_3 = -\frac{1}{8}u_{xx} - \frac{1}{8}u^2 \\ \cdots\cdots\cdots \end{cases} \tag{36}$$

Hence the k-th Sato's equation (28) becomes a single equation in u; in particular,

$$\text{for } k = 1: \quad \frac{\partial u}{\partial t_1} = u_x; \tag{37}$$

$$\text{for } k = 2: \quad \frac{\partial u}{\partial t_2} = 0; \tag{38}$$

$$\text{for } k = 3: \quad \frac{\partial u}{\partial t_3} = \frac{1}{4}u_{xxx} - \frac{3}{2}uu_x. \tag{39}$$

Eq.(39) is the famous Korteweg-De Vries equation [2]. One can show that the hierarchy GD(3) is related to the Boussinesq equation [8].

3.2. Lie groups with cocycles and Poisson manifolds. The bihamiltonian setting of Gelfand-Dickey's equations

In order to give a bihamiltonian interpretation of the GD(n) eqs., we shall study the Poisson brackets that are naturally defined on the dual of a Lie algebra of a Lie group. We recall that a Lie group is a C^∞-differentiable manifold with a group structure which is compatible with the differentiable structure [12]. It is due to Lie [17] the remarkable observation that the dual $\mathcal{G}^*$ of a Lie algebra $\mathcal{G}$ has a natural Poisson bracket, given by

$$\{f,g\}(S) = \langle S, [df(S), dg(S)]\rangle. \tag{40}$$

This formula needs an explanation: f and g are real functions on $\mathcal{G}^*$; their differentials $df(S)$, $dg(S)$ are elements in $\mathcal{G}^{**} \simeq \mathcal{G}$; $\langle S, \cdot\rangle$ is simply the valuation of S on an element in $\mathcal{G}$.

As we want to deal with bihamiltonian manifolds, we have to find two different Poisson brackets on $\mathcal{G}^*$, or, in other words, a Poisson pencil on $\mathcal{G}^*$. This can be done by considering a Lie group G endowed with a cocycle, that is with an application

$$\sigma : G \to \mathcal{G}^* \tag{41}$$

such that

$$\sigma(g_1 g_2) = \sigma(g_1) + \mathrm{Ad}^*_{g_1}\sigma(g_2), \tag{42}$$

where Ad^* is the coadjoint action of G on $\mathcal{G}^*$ (see [12], App. 5). Let us set $\Sigma = d\sigma(e)$, where e is the identity of G; then $\Sigma : \mathcal{G} \to \mathcal{G}^*$ satisfies

$$\Sigma([v_1, v_2]) = \mathrm{ad}^*_{v_1}\Sigma v_2 - \mathrm{ad}^*_{v_2}\Sigma v_1 \tag{43}$$

on account of Eq.(42). If we suppose that Σ is antisymmetric, then it can be shown [12, p.210] that

$$\{f,g\}_\Sigma(S) = \langle \Sigma(df(S)), dg(S)\rangle + \langle S, [df(S), dg(S)]\rangle \tag{44}$$

is a Poisson bracket on the manifold $\mathcal{G}^*$. Hence every cocycle σ of G (such that Σ is antisymmetric) induces a Poisson bracket on $\mathcal{G}^*$. Now we observe that, $\forall A \in \mathcal{G}^*$, the application

$$\sigma_A : g \longmapsto \mathrm{Ad}^*_g A - A, \qquad g \in G, \tag{45}$$

is a cocycle of G (furthermore, it is a coboundary; see [12], App. 5), whose differential at the identity is

$$d\sigma_A(e) = \Sigma_A : v \longmapsto \mathrm{ad}^*_v A, \qquad v \in \mathcal{G}. \tag{46}$$

Consequently, if Σ is another cocycle of G, we can consider the pencil of cocycles $\Sigma_\lambda = \Sigma + \lambda\Sigma_A$, that gives rise to the corresponding Poisson pencil

$$\begin{aligned}\{f,g\}_\lambda(S) &:= \{f,g\}_{\Sigma_\lambda}(S) = \langle \Sigma_\lambda(df(S)), dg(S)\rangle + \langle S, [df(S), dg(S)]\rangle = \\ &= \langle \Sigma(df(S)), dg(S)\rangle + \langle S, [df(S), dg(S)]\rangle + \lambda\langle \Sigma_A(df(S)), dg(S)\rangle = \\ &= \langle \Sigma(df(S)), dg(S)\rangle + \langle S, [df(S), dg(S)]\rangle + \lambda\langle A, [df(S), dg(S)]\rangle.\end{aligned} \tag{47}$$

At this point we are ready to introduce the bihamiltonian structure of the GD(n) eqs. Let G be any simple group, which we shall think of as a matrix group, and let $G(x) = C^\infty(S^1, G)$ be the set of C^∞-functions from S^1 to G; we denote by x the coordinate on S^1. $G(x)$ inherits a Lie group structure from G. Moreover, its Lie algebra is $\mathcal{G}(x) = C^\infty(S^1, \mathcal{G})$, and the dual of $\mathcal{G}(x)$ is identified with $\mathcal{G}(x)$ by means of the separating bilinear form

$$\langle u, v\rangle = \int_{S^1} \mathrm{tr}(u(x)v(x))dx, \qquad u, v \in \mathcal{G}(x). \tag{48}$$

It is easy to check that the application

$$\sigma(g) = g_x g^{-1} \tag{49}$$

is a cocycle of $G(x)$. In this case Σ is simply the derivation w.r.t. x. Thus in order to provide a bihamiltonian setting for the GD(n) eqs. we have to:

1. Choose a Lie group G.
2. Choose an element A in $\mathcal{G}(x)$.
3. Exhibit a Casimir function $h(\lambda)$ of $\{\cdot,\cdot\}_\lambda$ which is a deformation of a Casimir function h_0 of $\{\cdot,\cdot\}_0$:

$$h(\lambda) = \sum_{k\geq 0} h_k\lambda^{-k}, \quad \{h(\lambda), \cdot\}_\lambda = 0. \tag{50}$$

This program has been completed for the case $n = 2$ (corresponding to the KdV hierarchy) in [13] and in [14]; in this case G is $SL(2)$, and A is a vector of minimal weight (as suggested by Drinfeld and Sokolov in [18]). Here we limit ourselves to give explicitly the bihamiltonian definition of the KdV hierarchy, alternative to the previous Sato's definition.

We have to recall that on every Poisson manifold it is canonically defined a *Poisson tensor* (sending 1-forms into vector fields) by

$$\langle df, Pdg\rangle = \{f, g\}. \tag{51}$$

For example, the Poisson tensor associated to the Poisson bracket (47) is

$$(P_\lambda)_S V = V_x + [V, S + \lambda A]. \tag{52}$$

Hence the differential $V(\lambda)$ of a Casimir function $H(\lambda)$ of $\{\cdot,\cdot\}_\lambda$ must satisfy the so-called stationary equation

$$V(\lambda)_x + [V(\lambda), S + \lambda A] = 0. \tag{53}$$

It is possible to show that:

1. There exists a unique solution of (53) of the form $V(\lambda) = \sum_{k\geq -1} V_k\lambda^{-k}$ such that $V(\lambda)^2 = \lambda I$. Such solution is an exact 1-form, that is, there exists a function $H(\lambda)$ such that $V(\lambda) = dH(\lambda)$.
2. The equations of the *matrix* KdV hierarchy may be written in bihamiltonian form,

$$\frac{\partial S}{\partial t_k} = V_{kx} + [V_k, S] = [A, V_{k+1}], \tag{54}$$

and consequently the vector fields (54) commute.

3. The usual scalar KdV hierarchy is obtained from the matrix KdV hierarchy by means of the reduction process described in Sec. 2.3.

In the above mentioned papers we also gave a bihamiltonian interpretation of several features of soliton equations, such as dressing transformations, τ-function, Lax and zero-curvature representations. In [19] it is shown how to perform the bihamiltonian reduction in the general case of a simple Lie algebra $\mathcal{G}$. The case $G = SL(n)$, corresponding to GD(n), will be treated in more detail in a subsequent paper.

References

1. C. S. Gardner, J. M. Green, M. D. Kruskal, R. M. Miura, *Korteweg-de Vries Equation and Generalizations. VI. Methods for Exact Solution*, Comm. Pure Appl. Math. **27** (1974), 97–133.
2. D. J. Korteweg, G. de Vries, Phil. Mag. **39** (1895), 422–443.
3. C. S. Gardner, *Korteweg-de Vries Equation and Generalizations. IV. The Korteweg-de Vries Equation as a Hamiltonian System*, J. Math. Phys. **12** (1971), 1548–1551.
4. R. M. Miura, C. S. Gardner, M. D. Kruskal, *Korteweg-de Vries Equation and Generalizations. II. Existence of Conservation Laws and Constants of Motion*, Jour. Math. Phys. **9**, No. 8 (1968), 93–98.
5. F. Magri, *A Simple Model of the Integrable Hamiltonian Equation*, J. Math. Phys. **19** (1978), 1156–1162.
6. F. Calogero, A. Degasperis, *Spectral Transform and Solitons*, North-Holland, Amsterdam, 1982.
7. I. M. Gelfand, L. A. Dickey, *Fractional Powers of Operators and Hamiltonian Systems*, Funct. Anal. Appl. **10** (1976), 259–273.
8. M. J. Boussinesq, J. Math. Pures Appl. **7** (1872), 55–108.
9. M. Sato, Y. Sato, *Nonlinear PDEs in Applied Sciences* (US-Japan Seminar, Tokyo), P. Lax and H. Fujita eds., North-Holland, Amsterdam, 1982.
10. A. Lichnerowicz, *Les Variétés de Poisson et leurs Algèbres de Lie associées*, J. Diff. Geom. **12** (1977), 253–300.
11. M. Flato, A. Lichnerowicz, D. Sternheimer, *Algèbres de Lie attachées à une Variété canonique*, J. Math. Pures Appl. **54** (1975), 445–480.
12. P. Libermann, C. M. Marle, *Symplectic Geometry and Analytical Mechanics*, Reidel, Dordrecht, 1987.
13. P. Casati, F. Magri, M. Pedroni, *Bihamiltonian Manifolds and the τ-function*, Mathematical Aspects of Classical Field Theory 1991 (M. J. Gotay, J. E. Mardsen, V. E. Moncrief, eds.), Contemporary Mathematics **132**, American Mathematical Society, Providence.
14. P. Casati, F. Magri, M. Pedroni, *Bihamiltonian Manifolds and Sato's Equations*, to appear in The Verdier Memorial Conference on Integrable Systems, Actes du Colloque International de Luminy, 1991 (O. Babelon, P. Cartier, Y. Kosmann-Schwarzbach, eds.), Progress in Mathematics, Birkhäuser, Basel.
15. G. Magnano, F. Magri, *Poisson-Nijenhuis Structures and Sato Hierarchy*, Reviews in Math. Phys. **3**, No. 4 (1991), 403–466.
16. J. Satsuma, D. Takahashi, T. Tokihiro, *An Elementary Introduction to Sato Theory*, in Recent Developments in Soliton Theory (J. Satsuma, S. Takeno, M. Wadati eds.), Progress of Theoretical Physics Supplement, No. 94 (1988), 210–241.
17. S. Lie, *Transformationsgruppen, I, II, III*, Teubner, Berlin (1890). Reprinted by Chelsea Publishing Company, New York.
18. V. G. Drinfeld, V. V. Sokolov, *Lie Algebras and Equations of Korteweg-de Vries Type*, J. Sov. Math. **30** (1985), 1975–2036.
19. P. Casati, M. Pedroni, *Drinfeld Sokolov Reduction on a Simple Lie Algebra from the Bihamiltonian Point of View*, Lett. Math. Phys. **25** (1992), 89–101.

CONSERVATION LAWS IN DISSIPATIVE SOLIDS

GIACOMO CAVIGLIA
Department of Mathematics, University,
Via Alberti 4, 16132 Genova

and

ANGELO MORRO
DIBE, University,
Viale Causa 13, 16145 Genova

1. Introduction

The existence of conservation laws for a system of partial differential equations is of great help in investigating the properties of the solutions. For example, such is the case in proving existence of solutions, study of singularities, analysis of integrability properties of the system. Of course the more conservation laws are known for a system, the more tools are available for such investigations. This is a natural motivation for the growing attention addressed to the subject (cf. [1, 2, 3, 4]).

Realistic models of continuous media should incorporate dissipation. In the mechanical context this is most often accomplished via the viscoelastic solid, which leads to integro-differential equations for the unknown displacement field. Analogous mathematical settings occur in other contexts (e.g., electromagnetism) whenever the constitutive equations model memory effects. The mathematical procedure for the derivation of conservation laws is often based on the use of Laplace or Fourier transformation with respect to time. In both cases the pertinent problem takes an equivalent static character and we are faced with systems of equations involving space derivatives only. Fourier transformation applies typically to stationary problems and eventually provides results which are of interest per se in the frequency domain, without inversion to the time domain. Incidentally, this case is often the counterpart of a physical context where the pertinent fields causing the phenomenon are time-harmonic. Laplace transformation applies naturally to initial-value problems. The determination of results, here conservation laws, in the time domain requires that appropriate inverse Laplace transforms are performed. This is an additional, non-trivial problem which motivates a greater detail in the description of the procedure via the Laplace transformation.

Lately Jiang [5, 6] has investigated the possibility of conservation laws in linear viscoelasticity thus addressing the attention to materials with memory. The context, however, is restricted to conservation laws with a vanishing density. Next we have developed a Noetherian approach to conservation laws, with a non-vanishing density, for the dynamics of viscoelastic solids [7]. In essence, the procedure is performed for

N. H. Ibragimov et al. (eds.),
Modern Group Analysis: Advanced Analytical and Computational Methods in Mathematical Physics, 111–123.

the corresponding version of the problem in terms of the Laplace transform of the unknown field. The dependence on the time t is thus represented by the occurrence of a parameter p. Starting from the observation that invariance of the action is meant to within a divergence, we derive general conditions for the sought density and flux. For definiteness, a number of explicit solutions are given which constitute conservation laws. Recently Zhang [8] has derived a two-states conservation law for the dynamics of elastic solids and discussed its application to crack analysis. To our mind this subject deserves further attention. Perhaps, more involved models might be of interest, both mathematically and practically.

Meanwhile we have considered a direct approach to conservation laws; in words, solutions for a desired conservation law are derived while taking into account that the pertinent unknown field has to satisfy the given system of equations. This has been performed in connection with viscoelasticity [9, 10] and heat conduction with memory [11].

The purpose of this paper is twofold. First, to show how to proceed in the search for conservation laws in viscoelasticity. Second, to provide specific results thus showing the effectiveness of the procedure. Both purposes are accomplished in connection with an initial-value problem which is investigated in the framework of the Laplace transform. The essential notation is specified as follows.

Let Ω be an open, bounded, regular region in three-dimensional space occupied by a homogeneous viscoelastic solid in its stress-free configuration. The motion of the solid is described by the function $\mathbf{y}(\mathbf{x},t)$ giving the position vector $\mathbf{y}$ at time t of the particle located at $\mathbf{x}$ in the stress-free configuration. Through the displacement vector field $\mathbf{u}$,

$$\mathbf{u}(\mathbf{x},t) = \mathbf{y}(\mathbf{x},t) - \mathbf{x}, \qquad \mathbf{x} \in \Omega,$$

we determine the infinitesimal strain tensor $\mathbf{E}(\mathbf{x},t) = \mathrm{sym}\nabla\mathbf{u}(\mathbf{x},t)$, sym denoting the symmetric part and ∇ the gradient operator. Letting the body be at rest up to $t = 0$ we write the constitutive equation for the Cauchy stress as

$$\mathbf{T}(\mathbf{x},t) = \mathbf{G}(t)\mathbf{E}(\mathbf{x},0) + \int_0^t \mathbf{G}(t-\tau)\dot{\mathbf{E}}(\mathbf{x},\tau)d\tau, \quad t > 0 \tag{1}$$

where the relaxation function $\mathbf{G}$ is taken to satisfy the usual symmetry relations. Since discontinuities may occur at $t = 0$, in (1) the integral is over $(0,t]$ and $\mathbf{E}(\mathbf{x},0)$ stands for $\mathbf{E}(\mathbf{x},0^+)$. The equation of motion

$$\rho\ddot{\mathbf{u}} = \nabla\cdot\mathbf{T} + \mathbf{F} \tag{2}$$

in the unknown field $\mathbf{u}(\mathbf{x},t)$ is associated with the initial conditions

$$\mathbf{u}(\mathbf{x},0) = \mathbf{d}(\mathbf{x}), \quad \dot{\mathbf{u}}(\mathbf{x},0) = \mathbf{v}(\mathbf{x}).$$

2. Direct approach to conservation laws

Consider a system of partial differential equations of the form

$$g_i(x_\alpha, \phi_j, \phi_{j\alpha}, \phi_{j\alpha\beta}) = 0, \tag{3}$$

in the unknown functions $\phi_j(x_\alpha)$, where latin (greek) indices vary from 1 to $n(m)$, $\phi_{j\alpha} = \partial\phi_j/\partial x_\alpha$, $\phi_{j\alpha\beta} = \partial^2\phi_j/\partial x_\alpha \partial x_\beta$. Let $\mathbf{J}$ be a vector function of components $J_\alpha(x_\beta, \phi_j, \phi_{j\beta})$. A conservation law for the system (3) is an equation of the form

$$D_\alpha J_\alpha \doteq 0 \tag{4}$$

in the unknowns J_α, where the symbol $\doteq$ means that equality holds on solutions to (3), D_α is the total derivative with respect to x_α, namely

$$D_\alpha = \frac{\partial}{\partial x_\alpha} + \phi_{j\alpha}\frac{\partial}{\partial \phi_j} + \phi_{j\beta\alpha}\frac{\partial}{\partial \phi_{j\beta}}.$$

Whenever (4) holds the vector $\mathbf{J}$ is said to be conserved or divergence-free. The direct approach consists in finding the general solution to (4) subject to the constraint (3).

For definiteness, henceforth we let $n = 4$ and $m = 3$; $x_i, i = 1,2,3$, stand for the Cartesian coordinates and $x_4 = t$ is the time. Similarly, we regard $I = J_4$ as the conserved density and J_i as the components of the corresponding flux. Then we write the conservation law (4) as

$$D_i J_i + D_t I \doteq 0. \tag{5}$$

To determine conservation laws, for the system governed by the equation of motion (2) and the constitutive equation (1), we now apply the Laplace transform operator $\mathcal{L} : \mathcal{L}\{h(\mathbf{x},t)\} = \tilde{h}(\mathbf{x},p)$ to obtain

$$\rho\tilde{\mathbf{u}}(\mathbf{x},p) - \frac{1}{p^2}\nabla\cdot[p\tilde{\mathbf{G}}(p)\nabla\tilde{\mathbf{u}}] + \frac{1}{p^2}\tilde{\mathbf{F}}(\mathbf{x},p) + \rho[\frac{1}{p^2}\mathbf{v}(\mathbf{x}) + \frac{1}{p}\mathbf{d}(\mathbf{x})].$$

In component form we can write

$$\rho\tilde{u}_i = \frac{1}{2p}\tilde{A}_{ijkl}\tilde{u}_{lkj} + \tilde{f}_i \tag{6}$$

where

$$\tilde{A}_{ijkl} = \tilde{G}_{ijkl} + \tilde{G}_{ikjl}, \qquad \tilde{f}_i = \frac{1}{p^2}\tilde{F}_i + \rho(\frac{v_i}{p^2} + \frac{d_i}{p});$$

the dependence on the parameter p is henceforth understood and not written. For later convenience we observe that, by definition, the tensor $\tilde{\mathbf{A}}$ satisfies the symmetry properties

$$\tilde{A}_{ijkl} = \tilde{A}_{ikjl} = \tilde{A}_{ljki} = \tilde{A}_{klij}.$$

Back to (5), the formal application of the Laplace transform yields

$$D_i\tilde{J}_i + p\tilde{I} - I_0 \doteq 0 \tag{7}$$

where I_0 is the restriction of I to $t = 0$. Letting

$$\tilde{J}_i = \hat{J}_i(\mathbf{x}, \tilde{\mathbf{u}}, \nabla\tilde{\mathbf{u}}), \qquad \tilde{I} = \hat{I}(\mathbf{x}, \tilde{\mathbf{u}}, \nabla\tilde{\mathbf{u}})$$

we can write (7) as

$$\frac{\partial \hat{J}_i}{\partial x_i} + \frac{\partial \hat{J}_i}{\partial \tilde{u}_j}\tilde{u}_{ji} + \frac{\partial \hat{J}_i}{\partial \tilde{u}_{jh}}\tilde{u}_{jhi} + p\hat{I} - I_0 \doteq 0\,. \tag{8}$$

Our purpose is to solve (8) for $\hat{J}_i$ and $\hat{I}$ and to determine their inverse Laplace transform so as to find conservation laws in the physical space; in applying this procedure we will identify I_0 with the expression, at $t = 0$, of the inverse transform of $\hat{I}$.

As a starting point observe that, under conditions of non-degeneracy [1], because of (6) there exist three real functions $B_k(\mathbf{x}, \tilde{\mathbf{u}}, \nabla\tilde{\mathbf{u}})$ such that (8) is equivalent to the validity of

$$\frac{\partial \hat{J}_i}{\partial x_i} + \frac{\partial \hat{J}_i}{\partial \tilde{u}_j}\tilde{u}_{ji} + \frac{\partial \hat{J}_i}{\partial \tilde{u}_{jh}}\tilde{u}_{jhi} + p\hat{I} - I_0 = B_k\Big(\frac{1}{p}\tilde{A}_{kihj}\tilde{u}_{jhi} + 2\tilde{f}_k - 2\rho\tilde{u}_k\Big) \tag{9}$$

for arbitrary values of $\tilde{u}_j$, $\tilde{u}_{jh}$, $\tilde{u}_{jhi}$. The analysis of (9) is straightforward [9]. The arbitrariness of $\tilde{u}_{jhi}$, along with the symmetry with respect to the indices h and i, implies that

$$\frac{\partial \hat{J}_i}{\partial \tilde{u}_{jh}} + \frac{\partial \hat{J}_h}{\partial \tilde{u}_{ji}} = \frac{2}{p}B_k\tilde{A}_{kihj}. \tag{10}$$

Regard (10) as a differential equation for the unknown functions $\hat{J}_i$. The integrability conditions for (10) are found by observing that suitable combinations of its partial derivatives with respect to the spatial gradient of $\tilde{\mathbf{u}}$ lead to

$$\frac{p}{2}\Big(\frac{\partial^2 \hat{J}_i}{\partial \tilde{u}_{jh}\partial \tilde{u}_{sq}} + \frac{\partial^2 \hat{J}_i}{\partial \tilde{u}_{jq}\partial \tilde{u}_{sh}}\Big) = \tilde{A}_{kihj}\frac{\partial B_k}{\partial \tilde{u}_{sq}} - \tilde{A}_{khqs}\frac{\partial B_k}{\partial \tilde{u}_{ji}} + \tilde{A}_{kiqj}\frac{\partial B_k}{\partial \tilde{u}_{sh}}.$$

The left-hand side is symmetric in the pairs j, s and h, q; on the contrary, the right-hand side is symmetric with respect to the pair h, q only. Hence we have to impose a compatibility condition, which is obtained by taking the skew part of both sides with respect to the indices j and s. This yields

$$\Big(\tilde{A}_{kihj}\frac{\partial B_k}{\partial \tilde{u}_{sq}} - \tilde{A}_{kihs}\frac{\partial B_k}{\partial \tilde{u}_{jq}}\Big)_{(ihq)} = 0, \tag{11}$$

the subscript (ihq) denoting symmetrization over i, h, q.

The homogeneous, linear system (11), in the unknowns $\partial B_k/\partial \tilde{u}_{sq}$, has to hold for any choice of the coefficients $\tilde{A}_{kihj}$ obeying the symmetry conditions induced by $\mathbf{G}$. Take the body to be isotropic and let

$$\tilde{A}_{kihj} = \gamma\delta_{kj}\delta_{ih} + \eta(\delta_{kh}\delta_{ij} + \delta_{ki}\delta_{hj}),$$

where γ and η are parameterized by p. Substitution into (11) leads to the necessary condition

$$\frac{\partial B_k}{\partial \tilde{u}_{sq}} = \delta_{ks}\alpha_q, \tag{12}$$

where α_q is a function of $\mathbf{x}, \tilde{\mathbf{u}}, \nabla\tilde{\mathbf{u}}$. This condition is also sufficient in the sense that, upon substitution of (12) into (11), the expression within round brackets vanishes identically.

Equation (12) yields the expression of B_k. By the integrability conditions of (12) we obtain that α_q cannot depend on $\nabla\tilde{\mathbf{u}}$, whence it follows that B_k may be represented as

$$B_k = \frac{1}{2}[\tilde{u}_{kq}(\alpha_q + \bar{\alpha}_q) + \beta_k + \bar{\beta}_k] \tag{13}$$

with $\alpha_q, \beta_k, \bar{\alpha}_q, \bar{\beta}_k$, as functions of $\mathbf{x}$, $\tilde{\mathbf{u}}$, and the parameter p; the quantities $\bar{\alpha}_q$ and $\bar{\beta}_k$ are considered for later convenience.

Upon substitution of (13) into (10) we conclude that the general solution to (10) can be written as

$$\hat{J}_i = \frac{1}{p}\Big[\frac{1}{2}(\tilde{u}_{kq}\alpha_q + \beta_k)\tilde{A}_{kihj}\tilde{u}_{jh} - \frac{1}{4}\alpha_i\tilde{A}_{jqsr}\tilde{u}_{jq}\tilde{u}_{rs} + (\tilde{u}_{kq}\bar{\alpha}_q + \bar{\beta}_k)\tilde{G}_{kihj}\tilde{u}_{jh} - \frac{1}{2}\bar{\alpha}_i\tilde{G}_{jqsr}\tilde{u}_{jq}\tilde{u}_{rs}\Big] + \nu_{ijh}\tilde{u}_{jh} + k_i\,, \tag{14}$$

where ν_{ijh} and k_i depend on $\mathbf{x}, \tilde{\mathbf{u}}$ while $\nu_{ijh} = -\nu_{hji}$. Comparison of (10), (13) and (14) yields

$$\begin{aligned}
&\frac{1}{p}\Big\{\frac{1}{2}\Big[\tilde{u}_{kq}\Big(\frac{\partial\alpha_q}{\partial x_i} + \frac{\partial\alpha_q}{\partial\tilde{u}_l}\tilde{u}_{li}\Big) + \frac{\partial\beta_k}{\partial x_i} + \frac{\partial\beta_k}{\partial\tilde{u}_l}\tilde{u}_{li}\Big]\tilde{A}_{kihj}\tilde{u}_{jh} - \frac{1}{4}\Big(\frac{\partial\alpha_i}{\partial x_i} + \frac{\partial\alpha_i}{\partial\tilde{u}_l}\tilde{u}_{li}\Big)\\
&\quad\times\tilde{A}_{jqsr}\tilde{u}_{jq}\tilde{u}_{rs} + \Big[\tilde{u}_{kq}\Big(\frac{\partial\bar{\alpha}_q}{\partial x_i} + \frac{\partial\bar{\alpha}_q}{\partial\tilde{u}_l}\tilde{u}_{li}\Big) + \frac{\partial\bar{\beta}_k}{\partial x_i} + \frac{\partial\bar{\beta}_k}{\partial\tilde{u}_l}\tilde{u}_{li}\Big]\tilde{G}_{kihj}\tilde{u}_{jh}\\
&\quad-\frac{1}{2}\Big(\frac{\partial\bar{\alpha}_i}{\partial x_i} + \frac{\partial\bar{\alpha}_i}{\partial\tilde{u}_l}\tilde{u}_{li}\Big)\tilde{G}_{jqsr}\tilde{u}_{jq}\tilde{u}_{rs}\Big\} + \frac{\partial\nu_{ijh}}{\partial x_i}\tilde{u}_{jh} + \frac{\partial\nu_{ijh}}{\partial\tilde{u}_l}\tilde{u}_{li}\tilde{u}_{jh} + \frac{\partial k_i}{\partial x_i} + \frac{\partial k_i}{\partial\tilde{u}_l}\tilde{u}_{li}\\
&+p\hat{I} - I_0 + [\tilde{u}_{kq}(\alpha_q + \bar{\alpha}_q) + \beta_k + \bar{\beta}_k](\rho\,\tilde{u}_k - \tilde{f}_k) = 0
\end{aligned} \tag{15}$$

where α_q, β_k, $\bar{\alpha}_q$, $\bar{\beta}_k$, ν_{ijh} and k_i are functions of $\mathbf{x}$ and $\tilde{\mathbf{u}}$.

The next step consists in finding the form of the functions α_q, β_k, $\bar{\alpha}_q$, $\bar{\beta}_k$, ν_{ijh}, k_i, and $\hat{I}$ which satisfy (15). The conservation law (15) is in fact a family of conservation laws. Actually, it seems that in view of the arbitrariness allowed in the determination of $\hat{I}$ almost any choice of $\hat{\mathbf{J}}$, gives rise to a conservation law. However, we have to take into account two basic facts. First, the expressions for the admissible functions should be simple enough to allow for the application of the inverse Laplace trasform; in practice, this is guaranteed by taking polynomial dependences on $\tilde{\mathbf{u}}$ and its spatial derivatives. Second, the arbitrariness in the choice of $\hat{I}$ is restricted by the requirement that the resulting conservation law is not trivial. In this connection it is worth observing that the first three lines in (15) cannot simply be removed by collecting them into the definition of $\hat{I}$; this would imply the appearance of a spatial divergence in the expression of $\hat{I}$, which ultimately should lead to a trivial conservation law in the physical space.

For the sake of definiteness and with a view to physical applications, we let $\hat{I}$ depend at most linearly on $\nabla\tilde{\mathbf{u}}$. Look at (15) as a third order expression in the partial derivatives $\tilde{u}_{kq}$ that must be satisfied for suitable choices of $\tilde{I}$. Then the vanishing of the contributions due to third-order and second-order terms implies that

$$\bar{\alpha}_k = 2ax_k + b_k, \qquad \alpha_k = cx_k + e_k\,, \tag{16}$$

$$\bar{\beta}_k = a\tilde{u}_k + \psi_k, \qquad \beta_k = \frac{1}{2}c\tilde{u}_k + \phi_k\,, \tag{17}$$

$$\nu_{ijh} = \frac{\partial z_{ih}}{\partial\tilde{u}_j}\,, \tag{18}$$

where a, b_k, c, e_k are only allowed to depend on the parameter p; $z_{ih} = z_{ih}(\tilde{\mathbf{x}}, \tilde{\mathbf{u}}, p)$, with $z_{ih} = -z_{hi}$, and $\phi_k = \phi_k(\mathbf{x}, p)$, $\psi_k = \psi_k(\mathbf{x}, p)$. As a consequence, (15) reduces to

$$\frac{1}{p}\Big(\frac{1}{2}\frac{\partial \phi_k}{\partial x_i}\tilde{A}_{kihj}\tilde{u}_{jh} + \frac{\partial \psi_k}{\partial x_i}\tilde{G}_{kihj}\tilde{u}_{jh}\Big) + \frac{\partial^2 z_{ih}}{\partial \tilde{u}_j \partial x_i}\tilde{u}_{jh} + \frac{\partial k_i}{\partial x_i} + \frac{\partial k_i}{\partial \tilde{u}_l}\tilde{u}_{li} + p\hat{I} - I_0 + \\ \big\{\tilde{u}_{kq}\big[(2a+c)x_q + b_q + e_q\big] + \big(a + \tfrac{1}{2}c\big)\tilde{u}_k + \phi_k + \psi_k\big\}\big(\rho\,\tilde{u}_k - \tilde{f}_k\big) = 0 \tag{19}$$

and $\hat{J}_i$ takes the form

$$\hat{J}_i = \frac{1}{p}\Big[\frac{1}{2}\big(\tilde{u}_{kq}\alpha_q + \beta_k\big)\tilde{A}_{kihj}\tilde{u}_{jh} + \big(\tilde{u}_{kq}\bar{\alpha}_q + \bar{\beta}_k\big)\tilde{G}_{kihj}\tilde{u}_{jh} \\ - \frac{1}{4}\alpha_i\tilde{A}_{jqsr}\tilde{u}_{jq}\tilde{u}_{rs} - \frac{1}{2}\bar{\alpha}_i\tilde{G}_{jqsr}\tilde{u}_{jq}\tilde{u}_{rs}\Big] + \frac{\partial z_{ih}}{\partial \tilde{u}_j}\tilde{u}_{jh} + k_i\,. \tag{20}$$

3. Explicit structure of some conservation laws

To obtain simple but physically significant examples we consider the particular case when a, c, b_q, e_q, ϕ_k and z_{ih} vanish. In view of (16)-(18), the expressions (19) and (20) reduce to

$$\frac{1}{p}\frac{\partial \psi_k}{\partial x_i}\tilde{G}_{kihj}\tilde{u}_{jh} + \frac{\partial k_i}{\partial x_i} + \frac{\partial k_i}{\partial \tilde{u}_h}\tilde{u}_{hi} + p\hat{I} - I_0 + \psi_k(\rho\tilde{u}_k - \tilde{f}_k) = 0, \tag{21}$$

$$\hat{J}_i = \frac{1}{p}\psi_k\tilde{G}_{kihj}\tilde{u}_{jh} + k_i. \tag{22}$$

Among the conservation laws are the balance laws of linear and angular momentum. With the purpose of obtaining these laws we choose $\hat{I}$ as

$$\hat{I} = -\frac{1}{p}\psi_k\rho\tilde{u}_k. \tag{23}$$

Now we show how the balance laws follow. By (23), eq. (21) reduces to

$$\frac{1}{p}\frac{\partial \psi_k}{\partial x_i}\tilde{G}_{kihj}\tilde{u}_{jh} + \frac{\partial k_h}{\partial x_h} + \frac{\partial k_h}{\partial \tilde{u}_j}\tilde{u}_{jh} - I_0 - \psi_k\tilde{f}_k = 0\,. \tag{24}$$

The arbitrariness of $\tilde{u}_{jh}$ gives

$$\frac{1}{p}\frac{\partial \psi_k}{\partial x_i}\tilde{G}_{kihj} + \frac{\partial k_h}{\partial \tilde{u}_j} = 0$$

whence

$$k_h = -\frac{1}{p}\frac{\partial \psi_k}{\partial x_i}\tilde{G}_{kihj}\tilde{u}_j + \hat{\sigma}_h \tag{25}$$

where $\hat{\sigma}_h$ is a function of $\mathbf{x}$ parameterized by p. Meanwhile (24) becomes

$$\frac{\partial k_h}{\partial x_h} - I_0 - \psi_k\tilde{f}_k = 0\,. \tag{26}$$

Letting w_k be the inverse Laplace transform of $\tilde{w}_k := \psi_k/p^2$, by (23) we have

$$I = -w_k * \rho\dot{u}_k - \rho w_k d_k \tag{27}$$

and

$$I_0 = -\rho w_k(0) d_k$$

where $w_k(0)$ denotes the restriction to $t = 0$ of w_k. Substitution of k_h from (25) into (26) yields

$$-\frac{1}{p}\frac{\partial^2 \psi_k}{\partial x_h \partial x_i}\tilde{G}_{kihj}\tilde{u}_j + \frac{\partial \hat{\sigma}_h}{\partial x_h} + \rho w_k(0) d_k - \psi_k \tilde{f}_k = 0$$

where only the first term depends on $\tilde{u}_j$. Hence

$$\frac{\partial^2 \psi_k}{\partial x_h \partial x_i} = 0. \tag{28}$$

The functions $\hat{\sigma}_h$ are then characterized through the divergence $\partial\hat{\sigma}_h/\partial x_h$ by

$$\frac{\partial \hat{\sigma}_h}{\partial x_h} = -\rho w_k(0) d_k + \psi_k \tilde{f}_k. \tag{29}$$

By (28) we can write

$$w_k = \lambda_k + (\chi_{kh} + \omega_{kh}) x_h \tag{30}$$

where λ_k, χ_{kh}, and ω_{kh} are arbitrary functions of t with $\chi_{kh} = \chi_{hk}$, $\omega_{kh} = -\omega_{hk}$. Meanwhile (29) can be written as

$$\frac{\partial \hat{\sigma}_h}{\partial x_h} = \rho[p\tilde{w}_k - w_k(0)] d_k + \tilde{w}_k \tilde{F}_k + \rho\tilde{w}_k v_k$$

whence we get the divergence of $\sigma_h = \mathcal{L}^{-1}(\hat{\sigma}_h)$ as

$$\frac{\partial \sigma_h}{\partial x_h} = \rho\dot{w}_k d_k + w_k * F_k + \rho w_k v_k \tag{31}$$

The inverse Laplace transform of (22) and account of (1) and (25) yield

$$J_i = w_k * T_{ki} + \chi_{kl} * (G_{klhj} * \dot{u}_j + G_{klhj} d_j) + \sigma_i. \tag{32}$$

The functions (27) and (32) provide the desired conservation law.

To get a more suggestive form of the balance equation we consider the global counterpart. By (5) and (32), integration over $\Omega \times (0, t]$ gives

$$V(\bar{I} - \bar{I}_0) = -\int_0^t \int_\Omega [\chi_{kl} * T_{kl} + \frac{\partial \sigma_i}{\partial x_i}]\, dx\, d\tau - \int_0^t \int_{\partial\Omega} w_k * T_{ki} n_i\, da\, d\tau$$

where V is the volume of Ω, a superposed bar denotes the spatial mean in Ω, $\mathbf{n}$ is the outward normal, dx is the volume element and da is the surface element. Account of (31) and some rearrangement yield

$$V\rho\overline{w_k * \dot{u}_k} - V\int_0^t \overline{\chi_{kl} * T_{kl}}\, d\tau = \int_0^t (\overline{w_k * F_k} + \rho\overline{w_k v_k}) d\tau + \int_0^t \int_{\partial\Omega} w_k * T_{ki} n_i\, da\, d\tau.$$

The arbitrariness of λ_k, χ_{kl}, ω_{kl} provides three separate conservation laws, namely

$$V(\lambda * \overline{\rho \mathbf{u}})(t) = V \int_0^t (\lambda * \overline{\mathbf{F}} + \lambda \overline{\rho \mathbf{v}}) d\tau + \int_0^t \lambda * \int_{\partial\Omega} \mathbf{Tn} \, da \, d\tau, \tag{33}$$

$$V\rho[\chi * \mathrm{sym}(\overline{\mathbf{x} \otimes \dot{\mathbf{u}}})](t) + V \int_0^t (\chi * \overline{\mathbf{T}}) dt' = V \int_0^t [\chi * \mathrm{sym}(\overline{\mathbf{x} \otimes \mathbf{F}}) +$$
$$\chi \rho \, \mathrm{sym}(\overline{\mathbf{x} \otimes \mathbf{v}})] d\tau + \int_0^t \chi * \int_{\partial\Omega} \mathrm{sym}(\mathbf{x} \otimes \mathbf{Tn}) \, da \, d\tau, \tag{34}$$

$$V\rho[\omega * (\overline{\mathbf{x} \times \dot{\mathbf{u}}})](t) = V \int_0^t (\omega * \overline{\mathbf{x} \times \mathbf{F}} + \omega\rho * \overline{\mathbf{x} \times \mathbf{v}}) d\tau + \int_0^t \omega * \int_{\partial\Omega} \mathbf{x} \times \mathbf{Tn} da \, d\tau, \tag{35}$$

where λ, χ, ω are arbitrary scalar-valued functions on $\mathbb{R}$; really, (33)-(35) hold with independent λ's, χ's, and ω's for the various components. If, rather, we regard λ, χ, ω as independent of t we have

$$\rho \bar{\dot{\mathbf{u}}}(t) = \int_0^t \overline{\mathbf{F}} \, d\tau + \rho \overline{\mathbf{v}} + \frac{1}{V} \int_0^t \int_{\partial\Omega} \mathbf{Tn} \, da \, d\tau, \tag{36}$$

$$\rho \, \mathrm{sym}(\overline{\mathbf{x} \otimes \dot{\mathbf{u}}})(t) + \int_0^t \overline{\mathbf{T}} d\tau =$$
$$\int_0^t \mathrm{sym}(\overline{\mathbf{x} \otimes \mathbf{F}}) d\tau + \rho \, \mathrm{sym}(\overline{\mathbf{x} \otimes \mathbf{v}}) + \frac{1}{V} \int_0^t \int_{\partial\Omega} \mathrm{sym}(\mathbf{x} \otimes \mathbf{Tn}) \, da \, d\tau, \tag{37}$$

$$\rho \overline{\mathbf{x} \times \dot{\mathbf{u}}}(t) = \int_0^t (\overline{\mathbf{x} \times \mathbf{F}}) d\tau + \rho \overline{\mathbf{x} \times \mathbf{v}} + \frac{1}{V} \int_0^t \int_{\partial\Omega} \mathbf{x} \times \mathbf{Tn} \, da \, d\tau. \tag{38}$$

Incidentally, (36) and (38) are the integral version of balance equations for linear and angular momentum. They give the means $\rho\bar{\dot{\mathbf{u}}}$ and $\rho\overline{\mathbf{x} \times \dot{\mathbf{u}}}$ in terms of the boundary and initial data, namely $\mathbf{Tn}$ on $\partial\Omega \times \mathbb{R}^+$ and $\mathbf{v}$ on Ω, and the known body force $\mathbf{F}$ on $\Omega \times \mathbb{R}^+$. Equation (37), instead, is unusual as a balance equation; it provides an estimate on $\rho \, \mathrm{sym}\,(\overline{\mathbf{x} \times \dot{\mathbf{u}}})$ or $\overline{\mathbf{T}}$ when information is given on one of the two quantities.

In connection with (37) we mention a result by Signorini [12] (cf. [13], §220) about stress means. Signorini's formula can be written as

$$V\overline{\mathbf{T}} = V \mathrm{sym}(\overline{\mathbf{x} \otimes (\mathbf{F} - \rho \ddot{\mathbf{u}})}) + \int_{\partial\Omega} \mathrm{sym}(\mathbf{x} \otimes \mathbf{Tn}) da$$

and this relation follows from (37) by simply differentiating with respect to t. Accordingly we can regard (34) as a generalization of Signorini's theory of stress means.

4. Further conservation laws

Means involving displacements. An uncommon conservation law follows by considering the case when the density I involves the displacement $\mathbf{u}$, rather than the velocity $\dot{\mathbf{u}}$. Formally, let $\tilde{w}_k = \psi_k/p$ and $w_k = \mathcal{L}^{-1}\{\psi_k/p\}$. By paralleling, step by step, the previous procedure we have

$$I = -w_k * \rho u_k \tag{39}$$

where, again, w_k is given by (30). Then from

$$\frac{\partial \hat{\sigma}_h}{\partial x_h} = \frac{1}{p}\tilde{w}_k \tilde{F}_k + \frac{1}{p}\rho \tilde{w}_k v_k$$

we have

$$\frac{\partial \sigma_h}{\partial x_h} = w_k * \mathcal{F}_k + \rho v_k \mathcal{W}_k \tag{40}$$

where $\mathcal{F}_k$ and $\mathcal{W}_k$ are the integrals, over the time interval, of F_k and w_k. Meanwhile

$$J_i = w_k * G_{kihj} * u_{jh} - \chi_{kh} * G_{khij} * u_j + \sigma_i.$$

By (6) and (39), (40), integration over $\Omega \times [0,t)$ yields

$$\rho\overline{w_k * u_k} = \int_0^t \overline{w_k * \mathcal{F}_k}\, d\tau + \int_0^t \rho\overline{v_k \mathcal{W}_k}\, d\tau + \frac{1}{V}\int_0^t \int_{\partial\Omega} (w_k * G_{kijh} * u_{jh} - \chi_{kh} * G_{khij} * u_j) n_i \, da\, d\tau$$

which in fact gives the three conservation laws

$$\rho(\lambda * \overline{\mathbf{u}})(t) = \int_0^t \lambda * \overline{\mathbf{F}}\, d\tau + \rho \mathbf{v} \int_0^t \int_0^\tau \lambda(s) ds\, d\tau + \frac{1}{V}\int_0^t \int_{\partial\Omega} \lambda * \mathbf{n}\,\mathbf{G} * \nabla\mathbf{u}\, da\, d\tau,$$

$$[\chi * \operatorname{sym}(\overline{\mathbf{x} \otimes \rho\mathbf{u}})](t) = \int_0^t [\chi * \operatorname{sym}(\overline{\mathbf{x} \otimes \mathbf{F}})]\, d\tau + \rho \operatorname{sym}(\overline{\mathbf{x} \otimes \mathbf{v}}) \int_0^t \int_0^\tau \chi(s) ds\, d\tau + \frac{1}{V}\int_0^t \int_{\partial\Omega} [\chi * \operatorname{sym}(\mathbf{x} \otimes \mathbf{n}\,\mathbf{G} * \nabla\mathbf{u}) - \chi * \mathbf{G} * (\mathbf{u} \otimes \mathbf{n})] da\, d\tau,$$

$$\rho(\omega * \overline{\mathbf{x} \times \mathbf{u}})(t) = \int_0^t (\omega * \overline{\mathbf{x} \times \mathbf{F}}) d\tau + \rho \overline{\mathbf{x} \times \mathbf{v}} \int_0^t \int_0^\tau \omega(s) ds\, d\tau + \frac{1}{V}\int_0^t \int_{\partial\Omega} \omega * \mathbf{x} \times (\mathbf{n}\,\mathbf{G} * \nabla\mathbf{u}) da\, d\tau,$$

whereby the knowledge of $\mathbf{u}$ and $\nabla\mathbf{u}$ on the boundary $\partial\Omega$ and of the body force $\mathbf{F}$ on $\Omega \times (0,t]$ leads at once to estimates on the means $\overline{\mathbf{u}}$ and $\overline{\mathbf{x} \otimes \mathbf{u}}$.

Two-states conservation laws. Every self-adjoint linear system does always possess a reciprocity relation [1, 14, 15, 4] which allows the determination of a conservation law canonically associated with every pair of solutions to the given equations. Conversely, it has been shown in linear elastostatics [16, 17] and in linear elastodynamics [18] that a reciprocal theorem follows from the knowledge of a complete set of conservation laws; it takes the explicit form of a Betti theorem whereby, given two elastic states, the work done by the external forces of the first over the displacements of the second is related to the work done by the external forces of the second over the displacements of the first. We observe that the system (6) is linear and self-adjoint. Thus an analogue of Betti's reciprocal theorem is likely to hold.

In this regard, consider a viscoelastic state $(\mathbf{u}^\star, \mathbf{E}^\star, \mathbf{T}^\star)$ and denote by a superposed $\star$ the pertinent quantities. Then look at the conservation law (21) and choose k_i and $\hat{I}$ as

$$k_i = -\frac{1}{p}\frac{\partial \psi_k}{\partial x_h}\tilde{G}_{khil}\tilde{u}_l\,, \tag{41}$$

$$\hat{I} = -\frac{1}{p}[\psi_k(\rho\tilde{u}_k - \tilde{f}_k) + \tilde{u}_k(\rho\psi_k - \tilde{f}_k^\star)]\,, \tag{42}$$

where $\tilde{f}_k^\star$ corresponds to $\tilde{\mathbf{F}}^\star, \tilde{\mathbf{v}}^\star, \tilde{\mathbf{d}}^\star$. Let $I_0 = 0$. By (41) and (42), eq. (21) becomes

$$-\frac{1}{p}\frac{\partial^2 \psi_k}{\partial x_h \partial x_i}\tilde{G}_{khil}\tilde{u}_l + \tilde{u}_l(\rho\psi_l - \tilde{f}_l^\star) = 0$$

which is satisfied whenever ψ_l coincides with the solution $\tilde{u}_l^\star$ to the counterpart of (6). Hence the explicit expressions for k_i and $\hat{I}$ become

$$k_i = -\frac{1}{p}\frac{\partial \tilde{u}_k^\star}{\partial x_h}\tilde{G}_{khil}\tilde{u}_l\,, \tag{43}$$

$$\hat{I} = -\frac{1}{p}\tilde{u}_k^\star(\rho\tilde{u}_k - \tilde{f}_k) + \frac{1}{p}\tilde{u}_k(\rho\tilde{u}_k^\star - \tilde{f}_k^\star). \tag{44}$$

The application of the inverse Laplace transform shows that (44) is consistent with the choice $I_0 = 0$ and provides the following result.

Let $(\mathbf{u}, \mathbf{E}, \mathbf{T})$ and $(\mathbf{u}^\star, \mathbf{E}^\star, \mathbf{T}^\star)$ be viscoelastic states on $\overline{\Omega} \times \mathbb{R}^+$ corresponding to the same relaxation function $\mathbf{G}$, to the body forces $\mathbf{F}$ and $\mathbf{F}^\star$, to the initial data $\mathbf{d}$, $\mathbf{v}$ and $\mathbf{d}^\star$, $\mathbf{v}^\star$, respectively. Then, for any $t \in \mathbb{R}^{++}$, the conservation law (5) holds with

$$J_i = \int_0^t (u_k^\star * G_{kihj} * u_{jh} - u_{hk}^\star * G_{khil} * u_l)(\tau)d\tau$$

$$I = \int_0^t \{-u_k * [g * F_k^\star + \rho(gv_k^\star + d_k^\star)] + u_k^\star * [g * F_k + \rho(gv_k + d_k)]\}(\tau)d\tau$$

where $\mathbf{u}^\star$ is the solution with body force $\mathbf{F}^\star$ and initial data $\mathbf{d}^\star$ and $\mathbf{v}^\star$.

Energy conservation law. So far we have considered choices of $\hat{I}$ which are linear in $\nabla\tilde{\mathbf{u}}$. Now we let $\hat{I}$ be non-linear and show that a new unexpected conservation law follows. Let α_q, β_k, $\bar{\alpha}_q$, ν_{ijh} and k_i vanish, and let $\bar{\beta}_k = -p^2(p\tilde{u}_k - d_k)$. By (14),

$$\hat{J}_i = -(p\tilde{u}_k - d_k)p\tilde{G}_{kihj}\tilde{u}_{jh}$$

whence

$$J_i = -\dot{u}_k * T_{ki}. \tag{45}$$

Meanwhile (45) becomes

$$(d_{ki} - p\tilde{u}_{ki})p\tilde{G}_{kihj}\tilde{u}_{jh} + p\hat{I} - I_0 - p^2(p\tilde{u}_k - d_k)(\rho\tilde{u}_k - \tilde{f}_k) = 0.$$

Then, letting $I_0 = 0$, we have

$$\begin{aligned}\hat{I} = &\tilde{u}_{ki}\tilde{T}_{ki} - \frac{1}{p}d_{ki}\tilde{T}_{ki} + (p\tilde{u}_k - d_k)\rho(p\tilde{u}_k - d_k) + (p\tilde{u}_k - d_k)\rho d_k \\ &-\tilde{u}_k\tilde{F}_k - \tilde{u}_k\rho v_k - \rho(p\tilde{u}_k - d_k)d_k + \frac{1}{p}d_k\tilde{F}_k + \frac{1}{p}\rho d_k v_k\end{aligned}$$

whence

$$I = u_{ki} * T_{ki} - d_{ki}\mathcal{T}_{ki} + \rho\dot{u}_i * \dot{u}_i - u_i * F_i - \rho(u_i - d_i)v_i + d_i\mathcal{F}_i \qquad (46)$$

where $\mathcal{T}_{ki}$ and $\mathcal{F}_i$ are the integrals, up to the present time t, of T_{ki} and F_i. Observe that, by (5), integration over $\Omega \times (0,t]$ yields

$$\bar{I}(t) - \bar{I}(0) = -\frac{1}{V}\int_0^t \int_{\partial\Omega} J_i n_i \, da d\tau.$$

Application to (45)-(46) gives

$$\overline{u_{ki} * T_{ki}} - \overline{d_{ki}\mathcal{T}_{ki}} + \overline{\rho\dot{u}_i * \dot{u}_i} - \overline{u_i * F_i} - \overline{\rho(u_i - d_i)v_i} + \overline{d_i\mathcal{F}_i} = \frac{1}{V}\int_0^t \int_{\partial\Omega} \dot{u}_k * T_{ki} \, da d\tau$$

which provides the left-hand side in terms of the known quantity $\dot{u}_k * T_{ki}$ at the boundary $\partial\Omega$. The result becomes more suggestive if we are looking at the homogeneous case when **d**, **v**, and **F** vanish, in that the knowledge of $\dot{u}_k * T_{ki}$ at $\partial\Omega$ yields the mean energy density $\overline{u_{ki} * T_{ki} + \rho\dot{u}_i * \dot{u}_i}$.

Since viscoelasticity is a model which accounts for dissipation one might think that conservation of energy is out of the question. Yet an energy-like conservation law has been derived which consists of the energy flux vector $\dot{u}_k * T_{ki}$ and the energy density $u_{ki} * T_{ki} + \rho\dot{u}_i * \dot{u}_i$, in close analogy with the power of the stress, per unit area, and the deformation and kinetic energy densities. Formally, though, the memory (and dissipation) inherent in the viscoelastic model results into the convolution product instead of the inner product in the corresponding elastic case.

5. Conservation laws via Noether's theorem

Owing to the application of Noether's theorem for the derivation of conservation laws it is worth investigating the possible connection between divergence-free fields (solutions to (5)) and Noetherian fields (conserved vectors that are generated via Noether's approach). This connection would have the advantage of providing divergence-free fields with the physical meaning usually associated with Noetherian fields in terms of the generators of infinitesimal invariance transformations.

The equations of motion in the form (6) admit a variational formulation in terms of the Lagrangian (density)

$$\hat{L}(\tilde{\mathbf{u}}, \nabla\tilde{\mathbf{u}}, p) = \frac{1}{2p}\tilde{G}_{ijkl}\tilde{u}_{ij}\tilde{u}_{kl} - \tilde{f}_i\tilde{u}_i + \frac{1}{2}\rho p\tilde{u}_i\tilde{u}_i.$$

Consider the transformation

$$x_i \to x_i + \epsilon \hat{h}_i(\mathbf{x}, \tilde{\mathbf{u}}, \nabla\tilde{\mathbf{u}}), \qquad \tilde{u}_i \to \tilde{u}_i + \epsilon\hat{\xi}_i(\mathbf{x}, \tilde{\mathbf{u}}, \nabla\tilde{\mathbf{u}}),$$

where ϵ is a small parameter. The invariance of the action functional under this transformation implies that

$$\hat{L}D_i\hat{h}_i + \frac{\partial\hat{L}}{\partial x_i}\hat{h}_i + \frac{\partial\hat{L}}{\partial\tilde{u}_i}\hat{\xi}_i + \frac{\partial\hat{L}}{\partial\tilde{u}_{ij}}(D_j\hat{\xi}_i - \tilde{u}_{ik}D_j\hat{h}_k) - D_i\hat{M}_i + p\hat{I} - I_0 \doteq 0.$$

Here $\hat{M}_i$ denotes an arbitrary set of functions of $\mathbf{x}, \tilde{\mathbf{u}}, \nabla\tilde{\mathbf{u}}$ arising from the fact that the invariance of the functional is meant up to a divergence term. Since the unknowns $\tilde{u}_i$ are taken to satisfy the Euler-Lagrange equations,

$$\frac{\partial\hat{L}}{\partial\tilde{u}_i} - D_j\frac{\partial\hat{L}}{\partial\tilde{u}_{ij}} = 0,$$

we can write the invariance condition as

$$D_i\hat{J}_i + p\hat{I} - I_0 \doteq 0$$

where

$$\hat{J}_i = \hat{L}\hat{h}_i + \frac{\partial\hat{L}}{\partial\tilde{u}_{ji}}(\hat{\xi}_j - \tilde{u}_{jk}\hat{h}_k) - \hat{M}_i.$$

This expression for $\hat{J}_i$ establishes the desired connection between the two approaches. Operatively, the search for the general solution is usually more involved than in the direct approach and then, often, appropriate restrictions on the form of the generators are introduced. For instance, in [7] we solve the problem by letting $\hat{h}_i = \hat{h}_i(\mathbf{x}, \tilde{\mathbf{u}})$, $\hat{\xi}_i = \hat{\xi}_i(\mathbf{x}, \tilde{\mathbf{u}})$ and $\hat{M}_i$ be linear in $\nabla\tilde{\mathbf{u}}$.

In conclusion, in searching for conservation laws, technical advantages weigh in favour of the direct approach. Meanwhile, the advantage of Noether's approach is that the meaning of the conserved quantities is usually more immediate.

Remark. Conservation laws associated with time-harmonic wave propagation in dissipative solids are investigated in [19], §6.6.

Acknowledgments

The research leading to this work has been partially supported by Italian Ministry MURST through the Research Project "Problemi di evoluzione nei fluidi e nei solidi".

References

1. Olver, P. J.: Applications of Lie Groups to Differential Equations. New York: Springer, 1986.
2. Ovsiannikov, L. V.: Group Analysis of Differential Equations. New York: Academic Press 1982.
3. Anderson, R. L., Ibragimov, N. H.: Lie-Bäcklund Transformations in Applications. Philadelphia: SIAM 1979.

4. Fushchich, V. I., Nikitin, A. G.: New and old symmetries of the Maxwell and Dirac equations. Sov. J. Part. Nucl. **14**, 1-22 (1983).
5. Jiang, Q.: Conservation laws in viscoelastostatics. Acta Mechanica **56**, 219-227 (1985).
6. Jiang, Q.: Conservation laws in linear viscoelastodynamics. J. Elasticity **16**, 213-219 (1986).
7. Caviglia, G., Morro, A.: A general approach to conservation laws in viscoelasticity. Acta Mechanica **66**, 191-204 (1988).
8. Zhang, Ch.: On some conservation laws in transient elastodynamics. Acta Mechanica **83**, 187-193 (1990).
9. Caviglia, G., Morro, A.: Conservation laws in viscoelasticity. Quart. Appl. Math. **48**, 503-516 (1990).
10. Caviglia, G., Morro, A.: On the generation of conservation laws in viscoelasticity. Acta Mechanica **81**, 91-95 (1990).
11. Caviglia, G., Morro, A.: Conservation laws in heat conduction with memory. Rend. Matem. **9**, 369-381 (1989).
12. Signorini, A.: Alcune proprietà di media nella elastostatica ordinaria. Rend. Accad. Naz. Lincei **15**, 151-156 (1932).
13. Truesdell, C., Toupin, R.: The classical field theories. In: Encyclopedia of Physics (Flügge, S., ed.), III/1. Berlin: Springer 1960.
14. Caviglia, G.: Symmetry transformations, isovectors, and conservation laws. J. Math. Phys. **27**, 972-978 (1986).
15. Caviglia, G.: Composite variational principles and the determination of conservation laws. J. Math. Phys. **29**, 812-816 (1988).
16. Olver, P. J.: Conservation laws in elasticity. I. General results. Arch. Rational Mech. Anal. **85**, 112-129 (1984).
17. Gurtin, M. E., Sternberg, E.: On the linear theory of viscoelasticity. Arch. Rational Mech. Anal. **11**, 291-356 (1963).
18. Olver, P. J.: Conservation laws in elasticity. III. Planar linear anisotropic elastostatics. Arch. Rational Mech. Anal. **102**, 167-181 (1988).
19. Caviglia, G., Morro, A.: Inhomogeneous Waves in Solids and Fluids. Singapore: WSPC 1992.

EXACT SOLUTIONS OF THE BOLTZMANN EQUATION

C. CERCIGNANI
Department of Mathematics
Politecnico di Milano
Piazza Leonardo da Vinci 32
20133 Milano, Italy

Abstract. After a brief introduction, the Boltzmann equation is introduced and the problem of finding exact solutions is discussed. A survey of specific methods and solutions is given. In particular, the solution of the space homogeneous equation is discussed in detail, with due attention to the early work of Truesdell and Muncaster on the moment equations, the unpublished contribution of R. Krupp and the subsequent results of Bobylev, Krook and Wu. In the case of space inhomogeneous solutions, after recalling Nikol'skii transformation that produces solutions of this kind from space homogeneous ones, the homoenergetic affine solutions of the moment equations due to Galkin and Muncaster are reviewed, together with an existence result of the author.

1. Introduction

The idea that gases are formed of elastic molecules rushing hither and thither at large speeds, colliding and rebounding according to the laws of elementary mechanics is, of course, an old idea. The physical development of the kinetic theory of gases was, however, accomplished in the nineteenth century [1, 2].

Although the rules generating the dynamics of these systems are easy to describe, the phenomena associated with this dynamics are not so simple. The typical evolution equation in kinetic theory is the Boltzmann equation which quickly became an important tool for investigating the properties of dilute gases. It has more recently also become a practical tool for the aerospace engineers, when they started to remark that flight in the upper atmosphere must face the problem of a decrease in the ambient density with increasing height, till one may well question the validity of the Euler and Navier-Stokes equations, which are usually introduced on the basis of a continuum model which does not take into account the molecular nature of a gas. It is to be remarked that the use of these equations can also be based on the kinetic theory of gases, which justifies them as asymptotically useful models when the mean free path is negligible.

According to kinetic theory the basic description of the evolution of a not-too-dense gas is in terms of a function of position $\mathbf{x}$, velocity $\mathbf{v}$, time t, $f = f(\mathbf{x}, \mathbf{v}, t)$, which gives the probability density of finding a molecule at position $\mathbf{x}$, with velocity $\mathbf{v}$, at time t. The usual quantities such as density ρ, bulk velocity $\mathbf{v}$, stresses p_{ij} (including pressure), temperature T, heat flux $\mathbf{q}$, are obtained as moments of the basic unknown f, through simple formulas that one can find in the textbooks [3–5].

The Boltzmann equation reads as follows [3–5]:

$$\frac{\partial f}{\partial t} + \mathbf{v} \cdot \frac{\partial f}{\partial \mathbf{x}} + \mathbf{X} \cdot \frac{\partial f}{\partial \mathbf{v}} = Q(f, f) \tag{1.1}$$

N. H. Ibragimov et al. (eds.),
Modern Group Analysis: Advanced Analytical and Computational Methods in Mathematical Physics, 125–136.

where

$$Q(f,f) = \int\int (f'f'_* - ff_*)B(\mathbf{v}-\mathbf{v}_*,\mathbf{n})d\mathbf{v}_*d\mathbf{n}. \tag{1.2}$$

Here $B(\mathbf{v}-\mathbf{v}_*,\mathbf{n})$ is a kernel associated with the details of the molecular interaction, f', f'_*, f_* are the same thing as f, except for the fact that the argument $\mathbf{v}$ is replaced by $\mathbf{v}'$, $\mathbf{v}'_*$, $\mathbf{v}_*$. The latter is an integration variable having the meaning of the velocity of a molecule colliding with the molecule of velocity $\mathbf{v}$, whose evolution we are following, while $\mathbf{v}'$ and $\mathbf{v}'_*$ are the velocities of two molecules entering a collision which will bring them into a pair of molecules with velocities $\mathbf{v}$ and $\mathbf{v}_*$; $\mathbf{n}$ is a unit vector defining the direction of approach of two colliding molecules. For details, see Refs [3–5]. The purpose of this paper is to review exact solutions of the Boltzmann equation; most of them will occur in the particular case when $B(\mathbf{v}-\mathbf{v}_*,\mathbf{n})$ depends only upon the angle θ between $\mathbf{v}-\mathbf{v}_*$ and $\mathbf{n}$. This case was first met by Maxwell when investigating molecules that interact according to inverse power laws of the distance between two molecules; if the repulsion force varies as the fifth inverse power of distance, it turns out that $B(\mathbf{v}-\mathbf{v}_*,\mathbf{n})$ is a well precise function of θ that can be expressed in terms of elliptic functions. That is why the molecules interacting as just described are called Maxwellian molecules. It has become customary, for various reasons, however, to consider the Boltzmann equation with the assumption that $B(\mathbf{v}-\mathbf{v}_*,\mathbf{n})$ depends only upon the angle θ, but is not exactly the function calculated by Maxwell; in this case one talks of Maxwellian pseudo-molecules and the integral of $B(\mathbf{v}-\mathbf{v}_*,\mathbf{n})$ on the unit sphere is assumed to be finite, although this is not the case for Maxwell molecules. What we shall say later about Maxwell molecules applies to Maxwell pseudo-molecules as well.

2. Maxwellian solutions

The simplest solution of the Boltzmann equation has the same form for any possible choice of $B(\mathbf{v}-\mathbf{v}_*,\mathbf{n})$ and was discussed by Maxwell, who introduced a sort of weak form of Boltzmann equation [1] before the latter was explicitly written [2]. This solution is called Maxwellian distribution, or simply Maxwellian, and depends by five parameters which can be conveniently organized as two scalars a, b and three components of a vector $\mathbf{u}$. A Maxwellian reads as follows

$$f(v) = a\exp(-b\,|\,\mathbf{v}-\mathbf{u}\,|^2) \tag{2.1}$$

If this expression is inserted in the right hand side of Eq. (1.1), the latter vanishes. Thus if $a, b, \mathbf{u}$ do not depend upon $\mathbf{x}$ and t and the external force $\mathbf{X}$ is absent, the function in Eq. (2.1) makes the left hand side vanish too and is thus an exact solution of the Boltzmann equation. If $\mathbf{X}$ does not vanish but depends on position only and is conservative with potential energy $V = V(\mathbf{x})$:

$$\mathbf{X} = -\frac{\partial V}{\partial \mathbf{x}} \tag{2.2}$$

then (2.1) still provides a solution if $a = a_0\exp(-2bV)$ with a_0 constant, $\mathbf{u} = 0$, $b =$ constant (Maxwell-Boltzmann distribution).

If $\mathbf{X} = 0$ it is easy to determine all the local Maxwellian solutions, *i. e.* solutions having the form (2.1) with $a, b, \mathbf{u}$ possibly depending on $\mathbf{x}$ and t. These were calculated by Boltzmann and correspond to expansions and contractions, possibly accompanied by solid body rotations. In particular if the number of particles per unit volume varies with time but is the same at each point at any given instant of time, the solid body rotation is absent and one obtains the following expressions for $a, b, \mathbf{u}$:

$$a = a_0, \qquad b = b_0(1 + t/t_0)^2, \qquad \mathbf{u} = (\mathbf{x} - \mathbf{x}_0)/(t_0 + t), \tag{2.3}$$

where the suffix zero denotes a constant. a_0 is necessarily positive, but t_0 might be negative; if it is, however, the solution ceases to exist at $t = -t_0 = |t_0|$ and describes an implosion, in which all the molecules end up in a single point, of position $\mathbf{x}_0$, in a finite time $|t_0|$. If t_0 is positive then the solution describe an expansion in which all the molecules, sooner or later, flow toward infinity.

It is also possible to discuss the case when $\mathbf{X}$ is not zero (see Ref. 3). In fact both f and $\log f$ satisfy the equation obtained by equating to zero the left hand side of Eq. (1.1). $\log f$ is more handy because the expression to be equated to zero is a third degree polynomial in the components of $\mathbf{v}$. In the case in which the body force is conservative these solutions where described by Boltzmann, but slightly more general forces can be considered. In the case of a general steady potential there is only the steady solution afforded by the Maxwell-Boltzmann distribution mentioned above. There are, however, exceptions: thus in a force field $\mathbf{X} = -k\mathbf{x}$ (harmonic oscillator) there is a solution in which the distribution function oscillates with a period which is one half of the natural period of the field.

3. Nikol'skii's transformation

The Maxwellian solutions discussed in the previous section are, so to speak, a happy accident, arising from the circumstance that both sides of the equation vanish. It is much more difficult to obtain solutions for which both sides are nonzero. This is particularly true for space dependent solutions; in fact the term with the space derivative in Eq. (1.1) is the only one that, even in the absence of body forces, mixes, so to speak, the velocity variables with the space-time variables.

It is therefore appropriate to look for transformations that produce space-dependent solutions from space-homogeneous ones, as was first suggested by Nikol'skii [6]. The following transformation suggests itself: if $h(t, \mathbf{v})$ is a homogeneous solution, then look for a solution of the form

$$f(t, \mathbf{x}, \mathbf{v}) = h(T(t), (1 + t/t_0)\mathbf{v} - (\mathbf{x} - \mathbf{x}_0)/t_0) \tag{3.1}$$

In fact for $\mathbf{X} = 0$, the left hand side has the property

$$\frac{\partial f}{\partial t} + \mathbf{v} \cdot \frac{\partial f}{\partial \mathbf{x}} = T'(t)\frac{\partial f}{\partial T} \tag{3.2}$$

and so under the transformation (3.1) the left hand side of the equation undergoes a similarity transformation.

One should now obtain a similarity property for the right hand side. It is clear that such a property will not apply unless the kernel $B(\mathbf{v} - \mathbf{v}_*, \mathbf{n})$ is homogeneous of some degree. Now, the general theory of the Boltzmann equation [3–5] informs us that this is the case for inverse power potentials. In fact if the repulsion force between the molecules varies as the inverse s-th power of the distance, then

$$B(\mathbf{v}-\mathbf{v}_*,\mathbf{n}) = |\mathbf{v}-\mathbf{v}_*|^g B(\theta) \tag{3.3}$$

where

$$g = \frac{s-5}{s-1} \tag{3.4}$$

and θ is the angle that was defined before. Please note that $s = 5$ corresponds to Maxwell molecules; $s \to \infty$ formally corresponds to hard sphere molecules.

We shall assume then that $B(\mathbf{v}-\mathbf{v}_*,\mathbf{n})$ has the following similarity property

$$B(\lambda\mathbf{v}-\lambda\mathbf{v}_*,\mathbf{n}) = \lambda^g B(\mathbf{v}-\mathbf{v}_*,\mathbf{n}) \tag{3.5}$$

where g is some real number ($g > 0$ corresponds to the so called hard potentials, $g < 0$ to soft potentials). Under this assumption we have that

$$Q(f(\lambda\mathbf{v}), f(\lambda\mathbf{v})) = \lambda^{-(g+3)}Q(f,f)(\lambda\mathbf{v}) \tag{3.6}$$

and hence Eq. (3.2) gives (if we remark that Q commutes with a translation in velocity space)

$$\frac{\partial f}{\partial t} + \mathbf{v}\cdot\frac{\partial f}{\partial \mathbf{x}} = T'(t)\frac{\partial f}{\partial T} = T'(t)(1+t/t_0)^{g+3}Q(f,f). \tag{3.7}$$

Then the Boltzmann equation is satisfied if the factor in front of $Q(f,f)$ is unity. This gives

$$T(t) = (g+2)^{-1}t_0[1-(1+t/t_0)^{-g-2}] \qquad (g \neq -2)$$
$$T(t) = t_0\log(1+t/t_0) \qquad (g=-2) \tag{3.8}$$

where the integration constant has been taken in such a way that $T(0) = 0$. Please note that $T(t)$ is positive for positive values of t.

Thus indeed Eq. (3.1) produces space inhomogeneous solutions from space homogeneous ones.

4. Space homogeneous solutions: moment methods

Having found that one can obtain space dependent solutions from homogeneous ones, the latter look more interesting. As a matter of fact the space homogeneous Boltzmann equation is more or less well understood. We know existence and uniqueness theorems in several function spaces and the overall behaviour of the solution for sufficiently general initial data: the distribution function changes in such a way as to tend to a Maxwellian distribution for $t \to \infty$. Of course the details of how this trend develops in time has its own interest, and this, together with the above remark, explains the continued interest in space homogeneous solutions [7, 8].

In the case of Maxwell molecules we can even assert that in a certain sense we know the solution explicitly for general initial data. In order to explain the meaning of this sentence, we introduce the notion of a moment of the distribution function. We first define the average $\overline{\phi}$ of a function ϕ of the molecular velocity $\mathbf{v}$ at a certain point $\mathbf{x}$ and time t by means of the following relation:

$$\rho\overline{\phi} = \int \phi(\mathbf{v})f(\mathbf{v})d\mathbf{v} \tag{4.1}$$

where ρ is the density of the gas

$$\rho = \int f(\mathbf{v})d\mathbf{v} \tag{4.2}$$

which is necessary to introduce in Eq. (4.1) in order to have a consistent meaning of the word average (a constant function must have the same constant as average value).

The first important average value is the bulk velocity $\mathbf{u}$ which is the average of $\mathbf{v}$; this permits to introduce the random velocity $\mathbf{c} = \mathbf{v} - \mathbf{u}$, for which obviously $\overline{\mathbf{c}} = 0$. The next important quantity is the second order moment ${}^2\mathbf{M}$ with components:

$$M_{ij} = \rho\overline{c_i c_j} \tag{4.3}$$

which has the physical meaning of stress tensor. In a similar way one can introduce higher order moments ${}^n\mathbf{M}$ with components

$$M_{k_1k_2\ldots k_n} = \rho\overline{c_{k_1}c_{k_2}\ldots c_{k_n}} \tag{4.4}$$

For Maxwell molecules, in the space homogeneous case one can prove that one can derive evolution equations for the moments in such a way that the equations for the components of the moment of a given order n are decoupled from the moments of order higher than n. In order to state this result in a more precise form it is expedient (following Ikenberry and Truesdell [9]) to introduce the spherical moments which are given by the averages

$$\mathbf{P}_{2r/s} = \rho\overline{|\mathbf{c}|^{2r}\mathbf{Y}_s} \tag{4.5}$$

where $\mathbf{Y}_s$ is the irreducible tensor polynomial of order s in the components of $\mathbf{c}$. These polynomials are uniquely defined as being given by components of tensors that are polynomials in the components of $\mathbf{c}$, have the highest degree term equal to the tensor with components $c_{k_1}c_{k_2\ldots}c_{k_s}$ and vanish when contracted on any pair of indices. Any polynomial in $\mathbf{c}$ can be expressed uniquely in terms of the polynomials $|\mathbf{c}|^{2r}\mathbf{Y}_s$; as a consequence the moment ${}^n\mathbf{M}$ can be expressed in terms of $\mathbf{P}_{2r/s}$ with $n = 2r + s$ (*e.g.* ${}^2\mathbf{M} = \mathbf{P}_{0/2} + (\mathbf{P}_{2/0}/3)\mathbf{I}$, where $\mathbf{I}$ is the fundamental tensor and we have omitted the sign of tensor product, because $\mathbf{P}_{2/0}$ is actually a scalar).

It is now possible to state the following

Theorem [9]. In a gas of Maxwellian molecules, if f possesses moments of orders $0, 1, \ldots, 2r + s$, then

$$\int |\mathbf{c}|^{2r}\mathbf{Y}_s Q(f,f)d\mathbf{v} = -c_{2r/s}\rho\mathbf{P}_{2r/s} + \mathbf{G}_{2r/s} \tag{4.6}$$

where $c_{2r/s}$ are constants that can be explicitly given (in terms of quadratures) and $\mathbf{G}_{2r/s}$ is a bilinear function of spherical moments with order higher than zero such that in each product the sum of the orders is $2r+s$ (hence the moments of order $2r+s$ or higher do not appear).

For Maxwell molecules it is then possible to write the moment equations (or equations of evolution of moments) in the following way:

$$\dot{\mathbf{P}}_{2r/s} + c_{2r/s}\rho\mathbf{P}_{2r/s} = \mathbf{G}_{2r/s} \tag{4.7}$$

Since these are linear equation at each order (ρ is constant and $\mathbf{G}_{2r/s}$ only depends on moments of lower order), they can be solved explicitly in a recursive way to yield the following

Theorem [9, 5]: For a homogeneous gas of Maxwellian molecules, the infinite system of equations for moments possesses a unique solution for each choice of the initial values. The solution exists globally in time and tends exponentially to the values appropriate to a Maxwell distribution.

5. Space homogeneous solutions: the distribution function

In the previous section we have shown that the moments of the distribution function can be determined in closed form, at least in principle, for a gas of Maxwell molecules (or pseudomolecules), in a space homogeneous situation. The problem of determining the distribution function itself remains to be discussed. To the best of the author's knowledge, this problem was first attacked by Roy S. Krupp [10], then a student at M. I. T., in a report "submitted to the Department of Aeronautics and Astronautics on 19 June 1967 in partial fulfillment of the requirements for the degree of Master of Science". His results were never published in a scientific journal and thus remained largely unknown, till they were partly rediscovered about ten years later by Krook and Wu [11], in U. S., and by Bobylev [12], in Soviet Union.

The method proposed by Krupp [10] (see also Bobylev [12]) was to introduce the Fourier transform $\phi(t,\mathbf{k}) = \rho\exp(-i\mathbf{k}\cdot\mathbf{v})$ of the distribution function $f(t,\mathbf{v})$ and obtain an evolution equation for ϕ

$$\frac{\partial\phi}{\partial t} + c\rho\phi = \int \phi(\frac{1}{2}(\mathbf{k}+\mathbf{n}k),t)\phi(\frac{1}{2}(\mathbf{k}-\mathbf{n}k),t)B(\mathbf{k}\cdot\mathbf{n}/k)d\mathbf{n} \tag{5.1}$$

where $k = |\mathbf{k}|$ is the magnitude of $\mathbf{k}, c$ is a constant (the integral of $B(\mathbf{k}\cdot\mathbf{n}/k)$ over the unit sphere described by $\mathbf{n}$) and ρ (the density) equals $\phi(t,0)$ and is constant because of mass conservation (put $\mathbf{k}=0$ in Eq. (5.1)).

The greatest simplification in Eq. (5.1) is that one has to deal with just a two-fold integral rather than a fivefold one.

The method proposed by Krupp to solve Eq. (5.1) has not been published by subsequent authors. Thus it seems worthwhile to mention it in detail, especially because it is based on a rather unusual idea. He considers only isotropic solutions of the Boltzmann equation, *i. e.* solutions depending on the magnitude of $\mathbf{v}$: hence the Fourier transform ϕ will depend on k^2 and t only. He also assumes that the Fourier transform is an entire analytic function of order one in k^2, so that it has the canonical representation

$$\phi(k^2,t) = \rho \exp[-k^2/(4b)] \prod_{m=0}^{\infty} \left[1 - \frac{k^2}{z_m(t)}\right] \exp[k^2/z_m(t)] \tag{5.2}$$

where b is a constant and $z_m = z_m(t)$ are the zeroes of ϕ.

In order to obtain the evolution of the zeroes in time, Krupp inserts the expansion (5.2) into Eq. (5.1), then divides by ϕ and takes the residue at each zero. The resulting system of infinitely many equations is rather complicated. There are two exceptions, however: 1) if ϕ has no zeroes, then the equation is identically satisfied by the factor in front of infinite product in (5.2) (this factor is nothing else than the Fourier transform of a Maxwellian with $\mathbf{u} = 0$); 2) if ϕ has just one zero, say z, then one obtains a very simple evolution equation for z, *i. e.*

$$\dot{z} = \lambda z \tag{5.3}$$

$$\lambda = \frac{1}{4} \int (1 - (\mathbf{n} \cdot \mathbf{k}/k)^2) B(\mathbf{n} \cdot \mathbf{k}/k) d\mathbf{n} \tag{5.4}$$

and hence

$$z(t) = z_0 e^{\lambda t}. \tag{5.5}$$

Eq. (5.2) immediately delivers ϕ

$$\phi(k^2,t) = \rho \exp\{-k^2[1/(4b) - 1/z(t)]\} \left[1 - \frac{k^2}{z(t)}\right]. \tag{5.6}$$

Taking the inverse Fourier transform yields the exact solution

$$f(v^2,t) =$$

$$\rho[\pi e(t)/b] \exp[-bv^2/e(t)][(3e(t) - 2)/2e(t)) + bv^2(1 - e(t))/(e(t))^2] \tag{5.7}$$

where, if t_0 is an arbitrary constant related to z_0 such that $\lambda t_0 > \log(5/2)$, we let

$$e(t) = 1 - \exp[-\lambda(t + t_0)] \tag{5.8}$$

The solution in (5.7) was re-obtained by Bobylev [12] and Krook and Wu [11] and is usually known as BKW mode, although another K should perhaps appear here. Let us briefly recall the method used by Bobylev, who started from the use of symmetries of the Fourier transformed Boltzmann equation. In fact if $\phi(\mathbf{k},t)$ is a solution, then $\phi(a\mathbf{k}, t + b)$ is another solution, with arbitrary a and b; let us look for a solution invariant with respect to a subgroup of this group for which $b = \log a/\mu$, where μ is a suitable constant, *i. e.* a solution such that

$$\phi(\mathbf{k},t) = \phi(a\mathbf{k}, t + \log a/\mu) \tag{5.9}$$

Since this is an identity for any a, we can also let a depend on t; in particular, if $a = e^{-\mu t}$, we obtain

$$\phi(\mathbf{k},t) = \phi(e^{-\mu t}\mathbf{k}, 0) = \psi(\mathbf{k}e^{-\mu t}). \tag{5.10}$$

This particular solution, if it exists, is such that

$$(\frac{\partial\phi}{\partial t})_{t=0} = -\mu\mathbf{k}\cdot\frac{\partial\psi}{\partial\mathbf{k}} \tag{5.11}$$

where now the argument of ψ in the right hand side is simply $\mathbf{k}$. Eq. (5.1) then simply becomes

$$-\mu\mathbf{k}\cdot\frac{\partial\psi}{\partial\mathbf{k}} + c\rho\psi = \int\psi(\frac{1}{2}(\mathbf{k}+\mathbf{n}k))\psi(\frac{1}{2}(\mathbf{k}-\mathbf{n}k))B(\mathbf{k}\cdot\mathbf{n}/k)d\mathbf{n}. \tag{5.12}$$

Let us look for a solution of this nonlinear equation where μ plays the role of a disposable parameter (eigenvalue). We easily find a solution for $\mu = 0$

$$\psi_0 = \rho\exp(ak^2) \tag{5.13}$$

where a is an arbitrary constant. This solution is nothing else than the Fourier transform of a Maxwellian if $a < 0$; if $a > 0$ it is nonphysical because it cannot be a Fourier transform. Still it makes perfectly sense as a solution of Eq. (5.12). In order to extract some useful thing from it let us remark that if $\mu \neq 0$ the first term in (5.12) when calculated for ψ_0 would produce the same function multiplied by k^2 (times a constant); thus one is led to thinking that $\psi_1 = k^2\psi_0$ might be a solution with $\mu \neq 0$. At first this seems to work because the right hand side gives a constant times $k^4\psi_0$ and in the left hand side we obtain a term of the same kind and two terms of the form $k^2\psi_0$, so that one may hope in some cancellation; we have, however, a difficulty because ρ must be the value of ψ at $k = 0$ which is zero with the chosen *ansatz*. This suggests however, another possibility, namely $\psi_0(1+\alpha k^2)$ where α is another constant and this time we win, because we have enough constants to cancel all the coefficients. α turns out to be equal to $-a$ and μ is $\lambda/2$, where λ is given by (5.4). Thus we obtain the similarity solution of Eq. (5.1)

$$\phi(k^2,t) = \rho(1 - ak^2e^{-\lambda t})\exp(ak^2e^{-\lambda t}) \tag{5.14}$$

This solution is, however, physically acceptable only in a generalized sense because it can be shown to have no energy. In addition, if we let t go to infinity, ϕ tends to ρ whose inverse transform is $\rho\delta(\mathbf{v})$ where ρ is a Dirac delta function, describing again a gas of motionless particles.

In order to obtain a more interesting solution we make use of another symmetry of the Boltzmann equation; if we multiply a solution ϕ of Eq. (5.1) by $\exp[-k^2/(4b)]$ where b is an arbitrary constant, we obtain another solution. This is sometimes called the Bobylev symmetry although it was discovered by Morgenstern [13] in the 1950's (in terms of f rather than of ϕ). Thus we immediately obtain another solution

$$\phi(k^2,t) = \rho(1 - ak^2e^{-\lambda t})\exp[-k^2/(4b) + ak^2e^{-\lambda t}]. \tag{5.15}$$

Here we recognize Krupp's solution (5.6) with a zero $z = a^{-1}e^{-\lambda t}$.

One may look for other solutions of Eq. (5.12), but so far those which have been found do no seem to lead to distribution functions which are everywhere positive.

6. Space inhomogenous solutions

As remarked in Sect. 3, one can obtain a space inhomogeneous solution from a homogeneous one, by means of Nikol'skii's transformation (3.1). Thus, by applying this transformation to the BKW mode, one obtains a solution first obtained by Muncaster [14]. The solutions delivered by Nikol'skii's transformation are a rather wide class, but are characterized by the fact that they have the same bulk velocity, *i. e.*

$$\mathbf{u}(t,\mathbf{x}) = (\mathbf{x}-\mathbf{x}_0)/(t+t_0). \tag{6.1}$$

They describe a spherical expansion for $t_0 > 0$; if $t_0 < 0$ they describe a spherical compression, which, however, ceases to exist for $t = -t_0$, when all the gas is concentrated at $\mathbf{x} = \mathbf{x}_0$.

It is interesting to evaluate the moments ${}^n\mathbf{M}$ with components given by Eq. (4.4). A simple calculation gives

$${}^n\mathbf{M}(t) = {}^n\mathbf{M}_h(T(t))(1+t/t_0)^{-n+3} \tag{6.2}$$

where ${}^n\mathbf{M}_h$ denotes the corresponding moment of the homogeneous solution h. Since these are given by the relaxation theorem discussed in Sect. 4, we can obtain the moments ${}^n\mathbf{M}(t)$ in explicit way by recurrence. This result was obtained by Galkin [15].

We remark that the moments (6.2) are space-homogeneous; this applies in particular to the trace of ${}^2\mathbf{M}(t)$, which is proportional to the thermal energy per unit volume of the gas. For this reason Galkin's solution is called a homo-energetic dilatation [5].

Homo-energetic dilatations are particular cases of more general solutions called homoenergetic affine flows. The book by Truesdell and Muncaster [5] gives a unified discussion of homoenergetic affine flows. The defining properties are the following:

a) The body force (per unit mass) $\mathbf{X}$ acting on the molecules is constant:

$$\mathbf{X} = \text{const.} \tag{6.3}$$

b) The moments formed with the peculiar velocity $\mathbf{c}$ are space-homogeneous.
c) The bulk velocity $\mathbf{u}$ is an affine function of position $\mathbf{x}$:

$$\mathbf{u} = \mathbf{K}(t)\mathbf{x} + \mathbf{u}_o(t). \tag{6.4}$$

An analysis of the balance equations based on a), b) and c) immediately leads to the following restrictions on $\mathbf{K}$ and $\mathbf{u}_o$:

$$\begin{cases} \dot{\mathbf{K}} + \mathbf{K}^2 = 0, \\ \dot{\mathbf{K}}\mathbf{v}_0 + \mathbf{K}\mathbf{v}_0 = \mathbf{X}. \end{cases} \tag{6.5}$$

The general solution of this system is:

$$\begin{cases} \mathbf{K}(t) = [\mathbf{I} + t\mathbf{K}(0)]^{-1}\mathbf{K}(0) \\ \mathbf{u}_0(t) = [\mathbf{I} + t\mathbf{K}(0)]^{-1}[\mathbf{u}_0(0) + t\mathbf{X} + \frac{1}{2}t\mathbf{K}(0)\mathbf{X}], \end{cases} \tag{6.6}$$

where $\mathbf{I}$ is the 3×3 identity matrix. This solution exists globally for $t > 0$ if the eigenvalues of $\mathbf{K}(0)$ are nonnegative; otherwise the solutions ceases to exist for $t = t_o$, where $-t_o^{-1}$ is the largest, in absolute value, among the negative eigenvalues of $\mathbf{K}(0)$.

In particular, if

$$[\mathbf{K}(0)]^2 = 0, \tag{6.7}$$

then $[\mathbf{I} + t\mathbf{K}(0)]^{-1} = \mathbf{I} - t\mathbf{K}(0)$ and therefore $\mathbf{K}(t)$ is independent of time. $\mathbf{u}$ is then steady if and only if

$$\mathbf{K}(0)\mathbf{X} = \mathbf{0}, \tag{6.8}$$

and if $\mathbf{u}_o(0)$ is chosen in such a way that

$$\mathbf{K}(0)\mathbf{u}_o(0) = \mathbf{X}. \tag{6.9}$$

In particular, this is always possible if $\mathbf{X} = \mathbf{0}$.

Eq. (6.7) is satisfied if and only if a coordinate system exists for which the matrix representation of $\mathbf{K}(0)$ is given by

$$((K_{ij})) = \begin{pmatrix} 0 & 0 & 0 \\ K & 0 & 0 \\ 0 & 0 & 0 \end{pmatrix}. \tag{6.10}$$

For a simple proof of this, see [16]. When (6.7) applies one talks, for obvious reasons, of a homoenergetic shear flow.

As shown by Galkin [15, 17–19] and Truesdell [20], the second order moment equations for a Maxwell gas, associated with a homoenergetic affine flow, are decoupled from those of higher order and can be solved explicitly. This result generalizes the result given above for homoenergetic dilatations.

The most interesting moment equations are those for p_{11}, p_{12}, $p = (p_{11} + p_{22} + p_{33})/3$:

$$\begin{aligned} \tau\dot{p} + \frac{2}{3}\mathcal{T}p_{12} &= 0 \\ \tau\dot{p}_{12} + p_{12} + \mathcal{T}p_{11} &= 0 \\ \tau\dot{p}_{11} - p + p_{11} &= 0 \end{aligned} \tag{6.11}$$

Here τ is a relaxation time, given by:

$$\tau = 1/(3\lambda) = \mu/p \tag{6.12}$$

where λ was defined in Eq. (5.4) and μ is the viscosity of the gas when the Boltzmann equation agrees with the Navier-Stokes equations, while $\mathcal{T}$ is the nondimensional number

$$\mathcal{T} = \tau K = \frac{\mu}{p}K \tag{6.13}$$

This is a system of three linear, homogeneous, first order differential equations that possesses a set of linearly independent solutions of the form:

$$p = p^0 \exp(\chi t/\tau); \qquad p_{12} = p_{12}^0 \exp(\chi t/\tau); \qquad p_{11} = p_{11}^0 \exp(\chi t/\tau) \tag{6.14}$$

where χ is one of the roots of the third degree equation

$$\chi(\chi+1)^2 = \frac{2}{3}\mathcal{T}^2 \tag{6.15}$$

This equation can be solved explicitly by means of the formulas by Scipione del Ferro and Cardano. It is enough here to remark that it certainly has a real root R; the other two roots can be easily expressed in terms of R and turn out to be complex with real part $-1 - R/2$ and imaginary part $\pm[R(1+\frac{3R}{4})]^{1/2}$. This immediately shows that R must be positive. Thus, in general, the solution of system (6.11) will possess a part which grows exponentially in time and another part that oscillates with an exponentially decaying amplitude. The asymptotically dominating part agrees with the Navier-Stokes equations with an error $O(\mathcal{T}^2)$. The exponential growth of the dominating part can be easily explained by the circumstance that the solution under consideration forbids any heat diffusion; thus the work done by the tangential stress heats up the gas and increases the pressure, which in turn increases the stress components p_{11} and p_{12} because viscosity increases with pressure.

Condition b) above holds for the solutions obtained by Truesdell [20] and Galkin [15, 17–19]. For analyses relating directly to the distribution function f, this condition is transformed [16] into

b′) the variable $\mathbf{x}$ appears in f only through the bulk velocity

$$f = f(\mathbf{c}, t). \tag{6.16}$$

where $\mathbf{c} = \mathbf{v} - \mathbf{u}$ is the random velocity.

In Ref [16] the present author attacked the problem of showing that Eqs. (2.5) are also sufficient for a solution of the Boltzmann equation for homoenergetic affine flows to exist. In particular he showed that $f(\mathbf{c}, t)$ must satisfy the following equation

$$\frac{\partial f}{\partial t} - \frac{\partial f}{\partial c} \cdot \mathbf{K}c = Q(f, f), \tag{6.17}$$

and proved an existence theorem for this equation, not only for Maxwell's molecules but for general kernels, as explained by the following.

Existence Theorem: There exists a solution f of Eq. (1.1), where the kernel $B(\mathbf{v} - \mathbf{v}_*, \mathbf{n})$ of the collision term $Q(f, f)$ does not grow more than quadratically in $\mathbf{v}, \mathbf{v}_*$ and the initial mass density, energy density, and H-functional ($= \int f \log f d\mathbf{v}$) are finite at time 0. These functionals remain bounded for $0 \le t \le T$. The time T is arbitrary provided $\mathbf{K}(0)$ has no negative eigenvalues. If $\mathbf{K}(0)$ possesses negative eigenvalues and t_0^{-1} is their largest absolute value, then T must not be larger than t_0.

One can also prove [16] the following

Uniqueness Theorem: Let f be the solution delivered by the Existence Theorem. If ${}^4\mathbf{M}(0)$ exists, so does ${}^4\mathbf{M}(t)$ in $[0, T]$, and then the solution of Eq. (6.10) is unique.

Here we want to deal briefly with the question whether it is possible to generalize Nikol′skii transformation to arbitrary homoenergetic affine flows. In other words, whether one can reduce Eq. (6.17) to a homogeneous Boltzmann equation, by a suitable transformation of variables. This transformation, if it exists, must be necessarily anisotropic in velocity space; this immediately rules out non-Maxwell molecules because the relative speed $|\mathbf{v} - \mathbf{v}_*|$ does not undergo a similarity transformation if $\mathbf{v}$ and $\mathbf{v}_*$ undergo the (same) anisotropic transformation. But, even if we restrict ourselves to Maxwell molecules, the method does not work because the transformation from the unprimed to primed variables does not commute with an anisotropic transformation in velocity space. Thus it seems impossible to generalize Nikol'skii transformation to homoenergetic affine flows different form a pure dilatation and we must be satisfied with the solution of the moment equations rather than obtaining the distribution function in a closed form.

References

1. Maxwell, J. C. (1867) "On the Dynamical Theory of Gases", *Philosophical Transactions of the Royal Society of London* **157**, 49–88.
2. Boltzmann, L. (1872) "Weitere Studien über das Wärmegleichgewicht unter Gasmolekülen", *Sitzungsberichte Akad. Wiss.*, Vienna, part II, **66**, 275–370.
3. Cercignani, C. (1988) *The Boltzmann equation and its applications*, Springer, New York.
4. Cercignani, C. (1990) *Mathematical Methods in Kinetic Theory*, Plenum Press, New York.
5. Truesdell, C., and Muncaster, R. G. (1980) *Fundamentals of Maxwell's Kinetic Theory of a Simple Monatomic Gas*, Academic Press, New York.
6. Nikol'skii, A. A. (1964) "The three-dimensional expansion-contraction of rarefied gas with power interaction functions", *Soviet Physics-Doklady* **8**, 639–641.
7. Ernst, M. H. (1983) "Exact Solutions of the Nonlinear Boltzmann Equation and Related Kinetic Equations" in *Nonequilibrium Phenomena I. The Boltzmann Equation*, Lebowitz, J. L., and Montroll, E. W., eds., North Holland, Amsterdam.
8. Bobylev, A. V. (1988), "The theory of the nonlinear spatially uniform Boltzmann equation for Maxwell molecules", *Sov. Sci. Rev. C., Math. Phys.* **7**, 111–233.
9. Ikenberry, E., and Truesdell, C., (1956) "On the pressures and the flux of energy in a gas according to Maxwell's kinetic theory, I", *Journal of Rational Mechanics and Analysis* **5**, 1–54.
10. Krupp, R. S. (1967) "A nonequilibrium solution of the Fourier transformed Boltzmann equation", submitted to the Department of Aeronautics and Astronautics in partial fulfillment of the requirements for the degree of Master of Science, M.I.T., Cambridge, Mass.
11. Krook, M., and Wu, T. T. (1976) "Formation of Maxwellian tails", *Physical Review Letters* **36**, 1107–1109; (1977) "Exact solutions of the Boltzmann equation", *The Physics of Fluids* **20**, 1589–1595.
12. Bobylev, A. V. (1976) "Some properties of Boltzmann's equation for Maxwell molecules", *Sov. Phys. Dokl.* **20**, 820, 822; **21**, 632.
13. Morgenstern, D. (1955) "Analytical studies related to Maxwell-Boltzmann equation", *J. Rational Mech. Anal.*, **4**, 533–565.
14. Muncaster, R. G. (1979) "On generating exact solutions of the Maxwell-Boltzmann equation", *Archive for Rational Mechanics and Analysis* **70**, 79–90.
15. Galkin, V.S. (1966) "Exact solutions of the kinetic moment equations of a mixture of monatomic gases", *Fluid Dynamics* **1**, 29–34.
16. Cercignani, C. (1989) "Existence of homoenergetic affine flows for the Boltzmann equation", *Archive for Rational Mechanics and Analysis* **105**, 377–387.
17. Galkin, V. S. (1956) "On a solution of the kinetic equation", *PMM (in Russian)* **20**, 445–446.
18. Galkin, V. S. (1958) "On a class of solutions of Grad's moment equations", *PMM* **22**, 532–536.
19. Galkin, V. S. (1964) "One-dimensional unsteady solutions of the equations for the kinetic moments of a monatomic gas", *PMM* **28**,226–229.
20. Truesdell,C., (1956) "On the pressures and the flux of energy in a gas according to Maxwell's kinetic theory, II", *Journal of Rational Mechanics and Analysis* **5**, 55–128.

EINSTEIN EQUATIONS WITH 1 PARAMETER SPACELIKE ISOMETRY GROUP

YVONNE CHOQUET-BRUHAT
Université Paris VI

Introduction

In a recent series of interesting articles V. Moncrief has obtained reduced hamiltonians for vacuum Einstein, or Einstein-Maxwell-Higgs equations. We shall in this paper show how one can obtain by direct methods, without making use of a hamiltonian formalism, the splitting of vacuum Einstein equations into a system elliptic on each spacelike slice on the one hand, and on the other hand a hyperbolic system which is an harmonic map equation from a pseudoriemannian manifold, whose metric depends on the solution of the elliptic system, into a fixed 2-dimensional symmetric riemannian space. We discuss, following with some variants ideas of Cameron and Moncrief, the general solution of the system, depending on the topology of space.

1. Definitions

The space time is a manifold of the type $V_4 \equiv M \times \mathbb{R}$ with M a smooth 3-dimensional manifold. It is endowed with a pseudo riemannian metric $\underline{g}$ of hyperbolic signature such that the submanifolds $M_t \equiv M \times \{t\}$ are space like and the lines $\{x\} \times \mathbb{R}$ are time like. We suppose that $\underline{g}$ admits a $U(1)$, space like, isometry group, more precisely we suppose that

1) M is a principal fiber bundle with group and typical fiber $U(1) \cong T^1 \cong S^1$. We denote by Σ its basis, a 2-dimensional smooth manifold which we suppose oriented, by π the projection $M \to \Sigma$. Then $M \times \mathbb{R}$ is a principal fibre bundle with group and typical fiber $U(1)$ while its basis is $\Sigma \times \mathbb{R}$ and the projection $\Pi : M \times \mathbb{R} \to \Sigma \times \mathbb{R}$ is $(x,t) \mapsto (\pi(x),t)$. If xe^{ia} denotes the right action of $e^{ia} \in U(1)$ on $x \in M$, then its action on $(x,t) \in M \times \mathbb{R}$ is (xe^{ia},t).
2) The metric $\underline{g}$ is invariant under the action of $U(1)$ on $M \times \mathbb{R}$.

The above hypothesis are those made in the classical Kaluza Klein theory. It is well known that the pseudo riemannian manifold $(M \times \mathbb{R}, \underline{g})$ satisfies the invariance hypothesis if and only if it is of the form

$$\underline{g} \equiv \pi^* g + (\pi^* e^{2\gamma})\underline{\theta}^2 \tag{1}$$

where g is a pseudo riemannian metric and $e^{2\gamma}$ a scalar function on $\Sigma \times \mathbb{R}$ while $\underline{\theta}$ is a $U(1)$ connection 1-form on $M \times \mathbb{R}$. The hypothesis on the signature of $\underline{g}$ and on time and space (which led to the writing as $e^{2\gamma}$ of the scalar function) are satisfied

N. H. Ibragimov et al. (eds.),
Modern Group Analysis: Advanced Analytical and Computational Methods in Mathematical Physics, 137–145.

if g is of lorentzian signature on $\Sigma \times \mathbb{R}$, with appropriate hypothesis on the nature of Σ_t and $\{\xi\} \times \mathbb{R}$.

Let $(x^\alpha) = (x^1, x^2, t)$ and x^3 be coordinates adapted to a local trivialization over $U_\Sigma \times \mathbb{R}$, U_Σ domain of a chart in Σ where coordinates are x^1, x^2. In such coordinates

$$\pi^* g \equiv g_{\alpha\beta}\, dx^\alpha dx^\beta \text{ defines a metric on } \Sigma \times \mathbb{R}\ \underline{\theta} \equiv \theta^3 \equiv dx^3 + A_\alpha dx^\alpha,$$

$A_\alpha\, dx^\alpha$ is a 1-form on $U_\Sigma \times \mathbb{R}$, representing the connection but does not extend to a 1-form on $\Sigma \times \mathbb{R}$ if the bundle $M \to \Sigma$ is not trivial. Under a change of adapted coordinates $x^\alpha = f^\alpha(x'^\beta)$, $x^3 = f(x'^\beta, x'^3)$ we have

$$A'_{\alpha'} = \frac{\partial x^\alpha}{\partial x^{\alpha'}} A_\alpha + \frac{\partial f}{\partial x^{\alpha'}}.$$

2. Kaluza Klein formulas

General formulas expressing the Ricci tensor of $\underline{g}$ in terms of the Ricci tensor of g, the scalar function γ and A are known for an arbitrary Lie group (cf. CB-DM 2). They read here in the adapted frame dx^α, θ^3

$$\begin{aligned}
\underline{R}_{\alpha\beta} &\equiv R_{\alpha\beta} - \partial_\alpha\gamma\partial_\beta\gamma - \nabla_\alpha\partial_\beta\gamma - \frac{1}{2}e^{2\gamma}F_{\alpha\lambda}F^\lambda_\beta \\
\underline{R}_{\alpha 3} &\equiv -\frac{1}{2}e^{-\gamma}\nabla_\lambda(e^{3\gamma}F^\lambda_\alpha) \\
\underline{R}_{33} &\equiv -e^{2\gamma}(g^{\alpha\beta}\nabla_\alpha\partial_\beta\gamma + g^{\alpha\beta}\partial_\alpha\gamma\partial_\beta\gamma - \frac{1}{4}e^{2\gamma}F_{\alpha\beta}F^{\alpha\beta}).
\end{aligned}$$

∇ is the covariant derivative in the metric g (note $\partial_\alpha\gamma = \partial\gamma/\partial x^\alpha$ because $\partial\gamma/\partial x^3 = 0$), F is a 2-form on $\Sigma \times \mathbb{R}$, representant of the curvature $\underline{F}$ of the $U(1)$ connection $\underline{\theta}$:

$$\underline{F} = d\underline{\theta} = \pi^* F.$$

The expression of F in the frame dx^α is

$$F_{\alpha\beta} \equiv \nabla_\alpha A_\beta - \nabla_\beta A_\alpha.$$

3. Equations $\underline{R}_{33} = 0$ and $\underline{R}_{\alpha 3} = 0$

The equation

$$\underline{R}_{33} = 0 \qquad \text{(Ia)}$$

is a non linear wave equation in the metric g for the scalar function γ.

The equations $\underline{R}_{\alpha 3} = 0$ express the vanishing of a 1-form on $\Sigma \times R$, codifferential of the 2-form $e^{3\gamma}F$. Equivalently it is the vanishing of the differential of the metric g adjoint of this form, i.e.,

$$dE = 0, \quad \text{with } E \equiv e^{3\gamma}\,{}^*F, \qquad ({}^*F)_\alpha = \frac{1}{2}\eta_{\alpha\beta\lambda}F^{\beta\lambda}, \qquad F_{\alpha\beta} = e^{-3\gamma}\eta_{\alpha\beta\lambda}E^\lambda.$$

η the volume 3-form of $\Sigma \times R$ in the metric g. Denote by $\bar{E}_t$ the 1-form induced on Σ_t by E, by $E_t^{\$}$ the scalar function on Σ_t obtained by projecting E on $\mathbb{R}$. The equation $dE = 0$ is equivalent to

$$\bar{E}_t = \bar{E}_{t_0} + \int_{t_0}^{t} \bar{d}E_\tau^{\$}\, d\tau, \qquad \bar{d} \text{ exterior differential on } \Sigma.$$

We deduce from this relation, when t_0 is fixed

$$E = p^*\bar{E}_{t_0} + d\alpha, \qquad \alpha \equiv \int_{t_0}^{t} E_\tau^{\$}\, d\tau, \qquad p : \Sigma \times \mathbb{R} \to \Sigma \text{ the trivial projection}$$

$\bar{E}_{t_0}$ is a 1-form on Σ such that $\bar{d}\bar{E}_{t_0} = 0$. By the Hodge theory it can be written, if it is square integrable in some given riemannian metric m on Σ

$$\bar{E}_{t_0} = \bar{d}\bar{\beta} + \bar{H},$$

with $\bar{\beta}$ a function on Σ and $\bar{H}$ a harmonic 1-form on (Σ, m). The general solution of $dE = 0$ has thus be found to be

$$E = d\omega + H, \qquad H = p^*\bar{H}$$

$\omega = p^*\bar{\beta} + \alpha$ a scalar function on $\Sigma \times \mathbb{R}$ called the twist potential.

If Σ is compact there is exactly one harmonic form (independently of the choice of m) in each cohomology class, there are b_1 (the first Betti number of Σ) linearly independent harmonic 1-forms. If the 2-dimensional manifold Σ is moreover orientable then $b_1 = 2G$, where G is the genus of Σ (maximum number of cuts along imbedded circles one can inflict without destroying the connectedness, $G(S^2) = 0$, $G(T^2) = 1$).

When E, hence F, is known we can find A locally only if

$$dF = 0 \qquad \text{equivalently} \qquad \nabla.(^*F) = 0.$$

That is the twist potential must satisfy the equation

$$\nabla_\alpha(e^{-3\gamma} g^{\alpha\beta}\partial_\beta\omega) + \nabla_\alpha(e^{-3\gamma}H^\alpha) = 0. \tag{Ib}$$

When H is given the equations (Ia), (Ib) are a system of non linear wave equations for γ and ω, hyperbolic if g is of lorentzian signature.

To deduce from F globally on $\Sigma \times I$ the connection 1-form $\underline{\theta}$ it is necessary that F be the representant on $\Sigma \times I$ of a curvature 2-form $\underline{F}$ on the $U(1)$ bundle $M \times I$. It has been proved by Cameron (for a compact Σ) that these bundles are characterized by their Chern number, an integer n which obeys the formula for the representant F of the curvature of a connection

$$n = \frac{1}{2\pi}\int_{\Sigma_t} F \qquad \text{(for a trivial bundle } n = 0)$$

The integral does not depend on t if F is closed and expresses then a necessary and sufficient for the existence of $\underline{\theta}$ on the bundle.

4. Conformal formulas

The previous paragraphs have transformed the equations $\underline{R}_{\alpha 3} = 0$ and $\underline{R}_{33} = 0$ into a system of nonlinear wave equations when the metric g is known. The remaining equations $\underline{R}_{\alpha\beta} = 0$ are a system of Einstein equations on the 3-manifold $\Sigma \times I$ with sources. It has proved convenient through a long history of construction of solutions to introduce a metric conformal to g and to rewrite the equations using this metric. One sets

$$g = e^{-2\gamma}\tilde{g}.$$

We recall that the Christoffel symbols of g and $\tilde{g}$ are then linked by

$$\Gamma^{\lambda}_{\alpha\beta} = \tilde{\Gamma}^{\lambda}_{\alpha\beta} - \delta^{\lambda}_{\alpha}\partial_{\beta}\gamma - \delta^{\lambda}_{\beta}\partial_{\alpha}\gamma + g^{\lambda\mu}g_{\alpha\beta}\partial_{\mu}\gamma$$

which gives

$$\nabla_{\alpha}\partial_{\beta}\gamma = \tilde{\nabla}_{\alpha}\partial_{\beta}\gamma + 2\partial_{\alpha}\gamma\partial_{\beta}\gamma - \tilde{g}_{\alpha\beta}\tilde{\Delta}_{1}\gamma, \quad \text{with} \quad \tilde{\Delta}_{1}\gamma \equiv \tilde{g}^{\lambda\mu}\partial_{\lambda}\gamma\partial_{\mu}\gamma$$

$$\Delta\gamma = e^{2\gamma}(\tilde{\Delta}\gamma - \tilde{\Delta}_{1}\gamma)$$

since $\Sigma \times I$ is 3-dimensional.

The equation $\underline{R}_{33} = 0$ reads then with F replaced by its value in terms of ω

$$e^{-4\gamma}\underline{R}_{33} \equiv \tilde{\Delta}\gamma + \frac{1}{2}e^{-4\gamma}\tilde{g}^{\alpha\beta}(\partial_{\alpha}\omega + H_{\alpha})(\partial_{\beta}\omega + H_{\beta}) = 0. \tag{Ia}$$

while the equation for the twist potential ω gives

$$\tilde{\nabla}^{\alpha}\partial_{\alpha}\omega - 4\tilde{g}^{\alpha\beta}\partial_{\alpha}\gamma\partial_{\beta}\omega = -e^{4\gamma}\tilde{\nabla}^{\alpha}(e^{-4\gamma}H_{\alpha}). \tag{Ib}$$

Well known formulas (cf. CB-DM 1) give, on a 3-dimensional manifold

$$R_{\alpha\beta} \equiv \tilde{R}_{\alpha\beta} + \tilde{\nabla}_{\alpha}\partial_{\beta}\gamma + \tilde{g}_{\alpha\beta}(\tilde{\Delta}\gamma - \tilde{\Delta}_{1}\gamma) - \partial_{\alpha}\gamma\partial_{\beta}\gamma$$

from which we deduce that $\underline{R}_{\alpha\beta} = 0$ together with the equations (Ia) is equivalent to this equation and

$$\tilde{R}_{\alpha\beta} = e^{-4\gamma}(\partial_{\alpha}\omega + H_{\alpha})(\partial_{\beta}\omega + H_{\beta}) + 4\partial_{\alpha}\gamma\partial_{\beta}\gamma. \tag{II}$$

Inspection of the equations leads to the following theorem

Theorem (Moncrief) When the harmonic form H is zero the equations (I) are the harmonic map equations from the manifold $(\Sigma \times I, \tilde{g})$ into the hyperbolic 2-space $\mathcal{H} = (\mathbb{R}^2, 4d\gamma d\gamma + e^{-4\gamma}\, d\omega d\omega)$ and the equations (II) are the Einstein equations on $\Sigma \times I$ with source the stress energy tensor of this harmonic map.

5. Cauchy problem

The unknown are ω, γ and $\tilde{g}$ on a manifold of the type $\Sigma \times I$, I an interval of $\mathbb{R}$ containing 0. The Cauchy data are to be given on a 2-manifold Σ which will be imbedded as $\Sigma_0 \equiv \Sigma \times \{0\}$ in $\Sigma \times I$. The intrinsic Cauchy data are (in addition to a 1-form $\bar{H}$ harmonic in some given metric m on Σ) on the one hand four scalar functions ω_0, γ_0 and ω_0', γ_0' which will be respectively the values induced on Σ_0 by ω, γ and the values of the scalar product of the gradients of ω, γ with the unit normal ν to Σ_0 in $(\Sigma \times I, \tilde{g})$, on the other hand a metric $\bar{g}_0$ on Σ and a symmetric covariant 2-tensor k_0 which will be the first and second fundamental forms of Σ_0 as submanifold of $(\Sigma \times I, \tilde{g})$. We shall determine H on $\Sigma \times I$ by the pull back of $\bar{H}$ under the mapping p determined from $\Sigma \times I$ onto Σ_0 by the the orthogonal trajectories of the family of the Σ_t in $(\Sigma \times I, \tilde{g})$.

If we can determine scalar functions ω, γ and a metric $\tilde{g}$ on $\Sigma \times I$ taking these initial data and satisfying (I), (II) it will give a solution of the original vacuum Einstein equations on the bundle V_4 if and only if the initial data satisfy the integral condition (n characterizes the bundle structure)

$$\frac{1}{2\pi}\int_\Sigma \bar{F} = n.$$

$\bar{F}$, 2-form induced on $\Sigma \times \{0\}$ by F, depends indeed only on the initial data: Let x^a, $a = 1, 2$, be local coordinates in Σ and (x^a, t) coordinates in $\Sigma \times I$

$$\bar{F} \equiv \frac{1}{2}\bar{F}_{ab}\, dx^a\, dx^b,$$
$$\bar{F}_{ab} \equiv \{e^{-4\gamma}\tilde{\eta}_{ab0}\tilde{g}^{0\mu}(\partial_\mu \omega + H_\mu)\}_{t=0} \equiv e^{-4\bar{\gamma}}(\det \bar{g})^{1/2}\omega'$$

by the hypothesis made on H.

6. $2+1$ decomposition

The usual procedures to solve 4-dimensional Einstein equations can be applied to solve the Einstein equations with sources (II) on $\Sigma \times I$, though in this 3-dimensional case we may expect special properties since the Ricci tensor determines the full Riemann tensor. We choose a zero shift without restricting the generality and set

$$\tilde{g} = -N^2\, dt^2 + \bar{g}_t, \qquad \bar{g}_t \text{ a properly riemannian metric on } \Sigma_t,$$
$$k_t = -\frac{1}{2N}\partial_t \bar{g}_t;$$

to avoid cumbersome writing we suppress the index t in the following formulas in local coordinates. Indices are raised with $\bar{g}_t$:

$$\bar{g}_t = \bar{g}_{ab}\, dx^a\, dx^b, \qquad k_{ab} = -\frac{1}{2N}\partial_t \bar{g}_{ab}$$

We have the following identities for the Ricci tensor

$$\tilde{R}_{ab} \equiv \bar{R}_{ab} - N^{-1}\partial_t k_{ab} - 2k_{ac}k^c_b + \tau k_{ab} - N^{-1}\bar{\nabla}_a \partial_b N, \qquad \tau \equiv k^a_a$$
$$\tilde{R}_{0a} \equiv N(-\bar{\nabla}_b k^b_a + \partial_a \tau)$$
$$\tilde{R}_{00} \equiv N\bar{\nabla}^a \partial_a N + N\partial_t \tau - N^2 k^a_b k^b_a$$

from which results for the Einstein tensor

$$2N^2\tilde{S}^{00} \equiv \bar{R} - k^a_b k^b_a + \tau^2.$$

The equations (II) imply on each Σ_t the equations

$$-\bar{\nabla}_b k^b_a + \partial_a\tau = J_a \equiv N^{-1}\left\{\frac{1}{2}(\partial_a\omega + H_a)\partial_t\omega + 2\partial_a\gamma\partial_t\gamma\right) \tag{IIa}$$

$$\bar{R} - k^a_b k^b_a + \tau^2 = Q \equiv \frac{3}{2}N^{-2}\{e^{-4\gamma}(\partial_t\omega)^2 + 4(\partial_t\gamma)^2\} + \bar{g}^{ab}\{e^{-4\gamma}(\partial_a\omega + H_a)(\partial_b\omega + H_b) + 4\partial_a\gamma\partial_b\gamma\} \tag{IIb}$$

In particular on Σ_0 the equations (IIa), (IIb) depend only on the Cauchy data, they are *constraints* to be satisfied by these data. To solve the constraints, or more generally to solve (IIa), (IIb) on each slice Σ_t, we consider the conformal method (cf. CB-Y): one sets on Σ_t

$$\bar{g}_{ab} = e^{2\lambda}b_{ab}, \qquad \text{with } b_t \text{ some given metric on } \Sigma_t,$$
$$k^b_a = e^{-2\lambda}h^b_a + \frac{1}{2}\delta^b_a\tau, \qquad h \text{ is then traceless.}$$

The equations (IIa) become, with $D_{(b)}$ the covariant derivative, $\Delta_{(b)}$ and $R(b)$ the laplacian and the riemann curvature in the metric b

$$D_{(b)c}h^c_a - \frac{1}{2}e^{2\lambda}\partial_a\tau = -e^{2\lambda}J_a \equiv L_a \tag{IIa'}$$

$$2\Delta_b\lambda - R(b) + e^{-2\lambda}(h^a_b h^b_a + A) - \frac{1}{2}e^{2\lambda}\tau^2 + B = 0 \tag{IIb'}$$

where

$$A \equiv N^{-2}e^{4\lambda}\left\{\frac{1}{2}e^{-4\gamma}(\partial_t\omega + H_0)^2 + 2(\partial_t\gamma)^2\right\},$$
$$B \equiv b^{ab}\left\{\frac{1}{2}e^{-4\gamma}(\partial_a\omega + H_a)(\partial_b\omega + H_b) + 2\partial_a\gamma\partial_b\gamma\right\}$$

Supposing ω_t, γ_t, ω'_t, γ'_t known, as well as the metric b_t and τ (also depending on t), the system (IIa'), (IIb') is a non linear elliptic system for λ and h on Σ_t, but even if τ is constant on Σ_t it decouples into a linear elliptic system for h and a non linear elliptic equation for λ, once h is known, only if we consider that $e^{2\lambda}J_a$, A and B are known. The quantity B depends only on b_t, ω_t and γ_t. It has been shown by Moncrief (at least when $H = 0$) that $N^{-1}e^{2\lambda}\partial_t\omega$ and $N^{-1}e^{2\lambda}\partial_t\gamma$ are the conjugate variables to ω and γ in an hamiltonian formalism. If these quantities are known on Σ_t (as well as b_t, ω_t, γ_t) the same is true of $e^{2\lambda}J_a$ and A. When b, A, B, h and τ are known on Σ_t the non linear elliptic equation (IIb') has always a positive solution on Σ, except in the case $A + h^a_b h^b_a = 0$ and $\tau = 0$ where the following additional hypothesis is required (χ Euler-Poicaré number of Σ)

$$\int_\Sigma (B - 2R(b))\, dv(b) \equiv \int_\Sigma B\, dv(b) - 4\pi\chi = 0$$

The linear elliptic system for h reads when $\partial_a \tau = 0$

$$D_{(b)c} h^c_a = L_a, \qquad h^a_a = 0.$$

It admits a solution if and only if L is L_2-orthogonal to the solution of the homogeneous adjoint system which acts on vector fields and reads

$$D_{(b)a} u_b + D_{(b)b} u_a - b_{ab} D_{(b)c} u^c = 0$$

Its solutions are the conformal Killing fields of b. The system for h has solutions if

$$\int_\Sigma J_a u^a\, dv(b) = 0, \quad \text{for all } u \text{ conformal Killing field of } b$$

The condition is the same for different choices of b in the same conformal class.

7. Mixed elliptic and hyperbolic system

The system (IIa'), (IIb') holds on each Σ_t. When b and λ are known on each Σ_t, $t \in I$, then $\tilde{g}$ is known on $\Sigma \times I$ if we also know N. If τ is given on $\Sigma \times I$ and h known then, to have Ricci$(\underline{g}) = 0$, N must satisfy the linear elliptic equation on each Σ_t

$$\underline{R}_{00} N^{-1} - N^2 e^{-2\gamma} \underline{R}_{33} \equiv \bar{\Delta} N - N(k^a_b k^b_a + q) + \partial_t \tau = 0,$$

$$q \equiv 2(\partial_t \gamma)^2 + \frac{1}{2} e^{-4\gamma} (\partial_t \omega + H_0)^2$$

i.e.,

$$\Delta_b N - N(e^{-2\lambda} h^a_b h^b_a + \frac{1}{2} e^{2\lambda} \tau^2 + e^{2\lambda} q) + e^{2\lambda} \partial_t \tau\} = 0. \tag{IIc}$$

This equation has one positive solution if Σ is compact, $\partial_t \tau > 0$ and the coefficient of N (which is ≥ 0) is not identically zero.

On the other hand we must have by the definitions

$$k_t = \partial_t \bar{g}_t \equiv e^{2\lambda} (\partial_t b + 2b \partial_t \lambda).$$

equivalently, the covariant h being defined as $h_{ab} = b_{ac} h^c_b = \tilde{g}_{ac} k^c_b$,

$$h_t = e^{2\lambda} \left\{ \partial_t b_t - \frac{1}{2} (b^\$_t . \partial_t b_t) b_t \right\}, \qquad \tau = 4 \partial_t \lambda + b^\$_t . \partial_t b_t .$$

Suppose b_t of the form

$$b_t = \varphi^*_t c$$

with c some given metric on Σ, and φ_t^* the pull back on $\Sigma_t \cong \Sigma$ by a diffeomorphism $\varphi_t : \Sigma_t \to \Sigma$ by $x \mapsto y = \varphi_t(x)$. We denote by ξ_t the vector field

$$\xi_t : \Sigma \to T\Sigma, \quad \text{by} \quad y \mapsto (\partial_t \varphi_t) \circ (\varphi_t^{-1})(y)$$

We find by straightforward computation

$$(\varphi_t^{-1})^*(\partial_t b_t)_{ab} \equiv \frac{\partial \varphi^c}{\partial x^a}\frac{\partial \varphi^d}{\partial x^b}\partial_t b_{cd} = D_a\xi_b + D_b\xi_a,$$

where D is the covariant derivative in the metric c. the equation (IIIb) gives

$$D_a\xi_b + D_b\xi_a - 2D_c\xi^c = f_{ab} \tag{III'}$$

where f is the image by φ_t of h_t and $\partial_t\varphi_t = \xi_t \circ \varphi_t$.

The system (IIIb′) is elliptic in ξ. The adjoint homogeneous system has for general solution the sum of a pure trace and a traceless transverse, on (Σ, c), symmetric 2-tensor. Since f_{ab} is traceless the system (IIIb′) has solutions on Σ_t only if

$$(i2) \int_{\Sigma_t} f_{ab}u^{ab}\, dv_c = 0 \quad \text{for all tranverse traceless tensors } u. \tag{i2}$$

Equations (II), (IIIa) can be rewritten as elliptic equations on (Σ, c).

Conclusion

1) When ω and γ are known on $\Sigma \times I$ together with their conjugate variables, τ and b_t the system (IIa), (IIb) is for each t an elliptic system for λ and h.
2) When λ and h are given, (IIIa), (IIIb'), (IIIc) is an elliptic system for N and ξ for each t, together with an ordinary differential equation for φ_t.

A solution of the systems obtained in 1) and 2) determines a metric $\tilde{g}$ on $\Sigma \times I$.

3) When $\tilde{g}$ is known on $\Sigma \times I$ the equations (I) are a harmonic map equation $(\Sigma \times I, \tilde{g}) \to \mathcal{H}$.

The splitting studied above leads naturally to an iteration scheme which will in principle determine all unknown from the Cauchy data ω_0, γ_0, ω_0', γ_0' satisfying the integral condition (i1), if the obtained quantities satisfy (i2) which is not clear if there are transverse traceless 2-tensors on Σ, i.e., if the genus of Σ is $G \geq 1$.

Supposing the metric $\underline{g}$ has been determined by the mixed elliptic-hyperbolic system; we still have to show that we have solved the full system of Einstein equations, i.e., that the equations $\underline{R}_{A3} = 0$, $\underline{R}_{0\alpha} = 0$, $\underline{S}_{00} = 0$ imply $\underline{R}_{ab} = 0$. They imply $\underline{R} = 0$ and the Bianchi identities give

$$\underline{\nabla}_A \underline{R}^{AB} \equiv 0$$

A straightforward computation shows that together with the previous equations they imply on each Σ_t

$$\bar{\nabla}_a(\underline{R}^{ab}e^{-2\gamma}N) = 0 \; :$$

$\underline{R}^{ab}e^{-2\gamma}N$ is transverse and traceless in the metric $\bar{g}_t$, $\underline{R}^{ab}$ is zero if there are no such tensor fields on Σ.

8. Case $\Sigma = S^2$

If Σ is diffeomorphic to S^2 (genus $G = 0$) the Betti number b_1 is zero, there is no harmonic 1-form, except zero.

On the sphere S^2 any riemannian metric $\bar{g}$ is conformally equivalent to the canonical metric s, i.e., there exists a positive function $e^{2\lambda}$ on S^2 and a diffeomorphism $\varphi : S^2 \to S^2$ such that

$$\bar{g} \equiv e^{2\lambda}\varphi^* s.$$

It is therefore no restriction to set

$$\bar{g}_t = e^{2\lambda_t}\varphi_t^* s.$$

It is known that there is no transverse traceless 2-tensor on S^2 except zero. The integral condition (i2) is always trivially satisfied. On the other hand a solution of the mixed elliptic-hyperbolic system satisfies the full Einstein equations. The case of Σ diffeomorphic to S^2 is therefore a case where we can use the previously constructed mixed system.

9. Manifolds of genus greater than 1

If the 2-dimensional oriented manifold Σ has a genus $G \geq 1$ the space of classes of conformally equivalent riemannian metrics on Σ, with diffeomorphisms φ homotopic to the identity is called the Teichmüller space $\mathcal{T}$ of Σ.

In the case $G = 1$ (Σ diffeomorphic to the torus T^2) every metric is conformal to a flat metric ($R = 0$) but there is a 2-parameters family of conformally inequivalent flat metrics: the Teichmüller space is $\mathbb{R}^2$.

In the case $G > 1$ the Teichmüller space can be endowed with the structure of the manifold $\mathbb{R}^{6G-6}$. In each equivalence class there is a metric of scalar curvature -1, unique up to isometries. We can therefore choose the metric b_0 to have scalar curvature -1 and there is $\mathbb{R}^{6G-6}$ such possible choices up to isometries, but there is no reason that b_t stays in the same equivalence class: in fact the vector space of traceless transverse 2-tensors in a metric on Σ is isomorphic to the tangent space to $\mathcal{T}$, of dimension $6G - 6$. Evolution equations must be used to determine b_t.

References

1. V. Moncrief "Reduction of Einstein Equations for vacuum Space Times with $U(1)$ Spacelike isometry groups" Annals of Physics, 167 n°1, 1986, pp. 118–142.
2. J. Cameron "The reduction of vacuum Einstein equations with space like $U(1)$ isometry group" Doctoral thesis Yale University 1991.
3. Y. Choquet-Bruhat and C. DeWitt-Morette "Analysis manifolds and Physics" Part I: Basics, North Holland 1982, Part II: 92 applications, North Holland 1989.
4. Y. Choquet-Bruhat and J. York "The Cauchy Problem" in General Relativity and Gravitation, A. Held ed. Plenum 1980.

LIE POINT SYMMETRIES AND DYNAMICAL SYSTEMS*

G. CICOGNA
Dip. di Fisica dell'Università
p.a Torricelli 2
56126 PISA, Italy

Abstract. The notion and the peculiar properties of Lie point symmetries in the context of first-order time evolution ODE are examined, also in comparison with different situations (e.g. PDE, Hamiltonian problems). We obtain in particular that Lie symmetries provide an extension to the nonlinear case of a useful "reduction lemma", and some applications concerning the symmetry properties inherited by the dynamics restricted to the center manifold. These results may be important e.g. in the classical Hopf problem of finding periodic bifurcating solutions. Some examples are also considered.

1. Introduction

We want to present some considerations on the role played by general (nonlinear) Lie point (LP) symmetries in dynamical systems (DS), i.e. in first-order ODE time-evolution problems:

$$\dot{u} = f(t, u) \qquad u = u(t) \in R^n \tag{1}$$

where the dot means d/dt and standard smoothness properties for the functions f are tacitly assumed, t is the time variable, and where f may also possibly depend on some real "control" parameters λ, in order that bifurcation problems can be included in the present scheme.

The main point is actually that LP symmetries in DS possess some strongly singular features, and their role appears to be completely different with respect to the properties they exhibit in any other context, as we shall discuss. To illustrate this point, it is useful to briefly consider, for sake of comparison, the case of a time-evolution problem expressed instead by a PDE, namely the Fokker-Planck equation, which is in fact a typical example from this point of view.

2. LP symmetries in PDE's: the Fokker-Planck equation

Let

$$\frac{\partial f}{\partial t} = -\frac{\partial}{\partial x}(a(x)\, f) + \frac{1}{2}\frac{\partial^2 f}{\partial x^2} \tag{2}$$

be a generic 1-dimensional Fokker-Planck equation. If one looks for the LP symmetries

$$\mathbf{v} = \xi(t, x, f)\frac{\partial}{\partial x} + \tau(t, x, f)\frac{\partial}{\partial t} + \varphi(t, x, f)\frac{\partial}{\partial f} \tag{3}$$

---

* *Based in part on joint work with G. Gaeta, Math. Inst. Univ. of Utrecht, NL*

N. H. Ibragimov et al. (eds.),
Modern Group Analysis: Advanced Analytical and Computational Methods in Mathematical Physics, 147–153.

admitted by this equation, one finds that [1, 2, 3]:

1) if $a(x)$ satisfies a Riccati equation of the form

$$a^2 + \frac{da}{dx} = \mu x^2 + \mu' x + \mu_0 \tag{4}$$

(where μ, μ', μ_0 are arbitrary constants), then there are exactly 4 nontrivial LP symmetries (their expressions and further details can be found in be above mentioned references)

2) if $a(x)$ satisfies another (less restrictive) differential equation, there are just 2 nontrivial LP symmetries

3) *otherwise*, i.e. in the "generic" case, only trivial symmetries are permitted for the PDE (2).

Incidentally, it can be a curious and remarkable fact (in favour of symmetries) that in almost all concrete applications, the most frequently used Fokker-Planck equations are just the symmetrical ones 1) or 2) !

This example shows a typical case, in the sense that only under very special conditions some LP symmetries are admitted by the PDE. An essentially similar result is true even for the case of second-order ODE's (for some examples, see [4, 5, 6, 7]). The situation is drastically different for first-order DS, where a infinite number of LP symmetries is "in principle" present (but this has nothing to do with the infinite number of conservation laws as in Bäcklund transformations etc.), and some strong difficulties cannot be avoided, in general, as we shall discuss in the next Section.

3. LP symmetries in DS, general properties

Given the DS (1), the determining equations for the LP symmetry

$$\gamma = \sum_{i=1}^{n} \varphi_i(t,u)\frac{\partial}{\partial u_i} + \tau(t,u)\frac{\partial}{\partial t} \tag{5}$$

are given by [8, 9, 5, 6, 10]

$$\tau \dot{f}_j + f_j \dot{\tau} + f_j f_i \partial_i \tau - \dot{\varphi}_j + \varphi_i \partial_i f_j - f_i \partial_i \varphi_j = 0 \tag{6}$$

(where $\partial_i = \partial/\partial u_i$). For concreteness, we shall restrict our attention to *autonomous* problems, and to time-independent LP symmetries, i.e. we assume $f = f(u)$ and $\tau = 0$, $\varphi = \varphi(u)$. In this case, the determining equations (6) acquire the nice form of a Lie commutator

$$[\gamma,\ \gamma_f] = 0 \tag{7}$$

where

$$\gamma_f = \sum f_i \frac{\partial}{\partial u_i} \equiv f\partial_u \tag{8}$$

is the generator of the dynamical flow (which is a - trivial - LP symmetry, of course).

In principle, there is some method [8, 10] in order to find n independent LP symmetries $\gamma_1, \ldots, \gamma_n$; the most general time-independent LP symmetry for the given DS may be written as

$$\gamma = \sum h_i \gamma_i \tag{9}$$

where $h_i(u)$ are first integrals of the DS (or functions thereof): then the set of these symmetries is not only a (∞–dimensional) algebra, but rather a (n–dimensional) *module* over the set of first integrals. In any case, therefore, one has the problem of finding the first integrals of the given DS, a problem which may be in practice very difficult, and which has in general only local solutions. In addition, *even if* everything is regular and well defined, it may happen that a LP symmetry is very singular, as the foregoing example (exceedingly simple !) may show. Consider in R^2 the DS, with $u \equiv (x, y)$

$$\begin{aligned} \dot{x} &= y \\ \dot{y} &= xy \end{aligned} \tag{10}$$

It is easily seen that all solutions, apart from $y = 0$, satisfy the condition $y - x^2/2 = \text{const}$, but the two solutions lying along the lines $y = 0$ and $y = x^2/2$ are "exceptional", in the sense that all nearby solutions exhibit a completely different behaviour in their time evolution. This implies that any nontrivial LP symmetry must be singular along these two solutions. Indeed, despite the simplicity of the DS (10), we find the following quite complicated and strongly singular (as expected) LP symmetry

$$\begin{aligned} \gamma = &\left(x + \frac{y}{\sqrt{|x^2 - 2y|}} \ln \left| \frac{x - \sqrt{|x^2 - 2y|}}{x + \sqrt{|x^2 - 2y|}} \right| \right) \frac{\partial}{\partial x} + \\ &\left(2y + \frac{xy}{\sqrt{|x^2 - 2y|}} \ln \left| \frac{x - \sqrt{|x^2 - 2y|}}{x + \sqrt{|x^2 - 2y|}} \right| \right) \frac{\partial}{\partial y} \end{aligned} \tag{11}$$

A similar result about the singularities of the LP symmetries is clearly also true in the presence of isolated periodic orbits, homo- (hetero-) -clinic orbits, separatrices [11, 12], etc.

The above discussion shows some of the difficulties one usually meets when searching for the set of the LP symmetries of a DS. Anyway, let us remark that the results presented in the foregoing sections do *not* require the knowledge of *all* LP symmetries of the given DS, but it would be sufficient to know just one of them.

4. A "reduction lemma" and applications to center manifolds

The foregoing results can be interpreted as suitable extensions of well known properties of *linear* symmetries to the case of general LP symmetries.

Theorem 1 *Let $\gamma = \varphi \partial_u$ be a LP symmetry for a DS. If*

$$U = \{u \in R^n | \varphi(u) = 0\} \neq \emptyset \tag{12}$$

then U is an invariant manifold under the dynamical flow.

This is a "reduction lemma" ([13]): in fact, it implies that

$$f|_U : U \to U \tag{13}$$

and therefore that one can restrict the dynamical problem to the flow on the manifold U. This result can be viewed as the nonlinear counterpart of the (geometrical part of the) well known "Equivariant bifurcation lemma" [14], so-called because it was introduced in the context of bifurcation theory, in the presence of linear symmetry properties: precisely, the manifold U plays here the same role played by the "fixed subspace" under the isotropy subgroups in the linear case.

An example of the application of Theorem 1 is the following. Putting $u \equiv (x, z)$ with $x \in R^r$, $z \in R^{n-r}$, the following DS (where X, Z, ζ are arbitrary regular functions of the indicated arguments)

$$\begin{aligned} \dot{x}_i &= X_i(x) \\ \dot{z}_j &= Z_j(x)(z_j - \zeta_j(x)) + X(x) \cdot \partial_x \zeta_j(x) \end{aligned} \tag{14}$$

admits the LP symmetry generated by

$$\gamma = \sum_j (z_j - \zeta_j(x)) \frac{\partial}{\partial z_j} \tag{15}$$

and the flow-invariant manifold U is just described by the equations

$$z_j = \zeta_j(x) \tag{16}$$

The second result we want to propose deals with the important notion of center manifold [15] of a DS. We have the following [16]:

Theorem 2 *If the DS admits a (locally) uniquely defined center manifold W, then W is an invariant manifold under all LP symmetries γ of the problem, and the dynamical problem restricted to W inherits in a natural way an "induced" LP symmetry*

$$\widehat{\gamma} \equiv \gamma|_W \tag{17}$$

Let us notice that, apart from some technical complications, the result can be suitably extended also to the case that the center manifold is *not* unique, which is actually the usual case.

There are some nontrivial consequences of this result. First, it implies that the center manifold must be expressed by means of functions which depend only on the quantities $\kappa_\alpha(u)$ which are left invariant by the LP symmetries, i.e. such that [11]

$$\gamma \cdot \kappa_\alpha = 0 \tag{18}$$

(this may be useful for explicitly finding the center manifold itself). Next, it is well known that the first step in the study of a DS is precisely the restriction to the center manifold: indeed the "relevant" dynamics occurs just along the center manifold, which is not, in general, a linear subspace. The above result ensures that LP symmetries are inherited from the original problem to the reduced one, showing in some sense that nonlinear LP symmetries are the "natural tool" to examine symmetry properties of DS.

Finally, it can also happen that this inherited symmetry $\widehat{\gamma}$, obtained through the reduction of the dynamics to the center manifold, turns out to be a *linear* one, even if the symmetry of original problem was *not* linear. This may be interesting because clearly, in this case, standard (i.e. linear) equivariant theory is not able to give any (symmetry) constraints on the form of the reduced equations. A specially relevant case occurs when one deals with a Hopf bifurcation problem and the reduced symmetry is a rotation symmetry, due to its relationship with the appearance of periodic bifurcating solutions. The following example (chosen among a very large family of examples [16]) can illustrate this point.

Consider for $u \in R^3$, $u \equiv (x, y, z)$ the bifurcation problem

$$\begin{aligned} \dot{x} &= xf(\lambda, r^2, w) - yg(\lambda, r^2, w) \\ \dot{y} &= yf(\lambda, r^2, w) + xg(\lambda, r^2, w) \\ \dot{z} &= -e^y h(\lambda, r^2, w) + yzf(\lambda, r^2, w) + xzg(\lambda, r^2, w) \end{aligned} \tag{19}$$

where

$$r^2 = x^2 + y^2, \quad w = ze^{-y} \tag{20}$$

and f, g, h are arbitrary smooth functions. A LP symmetry for this DS is

$$\gamma = x\frac{\partial}{\partial y} - y\frac{\partial}{\partial x} + xz\frac{\partial}{\partial z} \tag{21}$$

Choosing e.g. $f = \lambda - r^2$, $h = -w + r^2 - 2r^2 f$, leaving g arbitrary, one obtains a standard Hopf bifurcation problem. At the critical point $\lambda = 0$, the center manifold is given by

$$z = e^y r^2 \tag{22}$$

(notice its invariance under the LP symmetry γ, in agreement with (18); the same manifold (22) is a flow-invariant manifold even for $\lambda \neq 0$). Once reduced to the center manifold, the problem admits a linear SO_2 symmetry, generated by

$$\widehat{\gamma} = x\frac{\partial}{\partial y} - y\frac{\partial}{\partial x} \tag{23}$$

and a SO_2−symmetric periodic solution bifurcates at $\lambda > 0$, along the lines $r^2 = \lambda$, $z = e^y r^2$, with frequency $\omega = g(\lambda, \lambda, \lambda)$.

5. LP symmetries and coordinate transformations

In this section we consider some relationships existing between LP symmetries and different types of coordinate transformations which can be introduced in the study of a DS.

First, there is a close and nontrivial relationship between LP symmetries and the classical Poincaré-Birkhoff procedure [17] for reducing the DS to normal form. We cannot give here details on this point, which would require a quite lengthy and technical discussion (see [18]). Rather, we want to devote some words to the specially interesting case of Hamiltonian DS, i.e. systems of the form (with $u \equiv (q,p)$, $u \in R^{2n}$)

$$\begin{aligned} \dot{q} &= \frac{\partial H}{\partial p} \\ \dot{p} &= -\frac{\partial H}{\partial q} \end{aligned} \tag{24}$$

Together with LP symmetries admitted by these systems, one can consider the infinitesimal coordinate transformations

$$q \to q' = q + \varepsilon\vartheta(q,p) \qquad ; \qquad p \to p' = p + \varepsilon\psi(q,p) \tag{25}$$

which have the property of leaving invariant the Hamiltonian, and - more in particular - the *canonical* infinitesimal coordinate transformations leaving invariant the Hamiltonian. All these transformations are ultimately related to invariance properties of the problem: it is therefore a natural question to ask what type of relationship exists between them. The answer is summarized in the following statement [7].

Theorem 3 *There may be LP symmetries admitted by* (24) *which are not canonical transformations, or which are also canonical transformations but do not leave invariant the Hamiltonian (actually, the Hamiltonian may be changed only by the addition of some "Casimir function"* [9]*). Similarly, there may be infinitesimal coordinate transformations leaving invariant the Hamiltonian, but which are not LP symmetries. If a coordinate transformation leaving invariant the Hamiltonian is canonical, then it is also a LP symmetry admitted by the problem, and each constant of motion $G = G(q,p)$ generates not only a canonical transformation which leaves invariant the Hamiltonian, but also a LP symmetry, called Hamiltonian Lie symmetry* [9], *and given by*

$$\gamma = \gamma^{(G)} = \frac{\partial G}{\partial p_i}\frac{\partial}{\partial q_i} - \frac{\partial G}{\partial q_i}\frac{\partial}{\partial p_i} \tag{26}$$

References

1. G. Cicogna and D. Vitali, J. Phys. A **22**, L453 (1989); and **23** L85 (1990)
2. P. Rudra, J. Phys. A **23** (1990) 1663
3. G. Saccomandi, this volume
4. T. Sen, Phys. Lett. A **122** (1987) 327
5. G.W. Bluman and S. Kumei, "Symmetries and differential equations" (Springer, Berlin, 1989)
6. H. Stephani, "Differential equations. Their solution using symmetries" (Cambridge Univ. Press, 1989)
7. G. Cicogna and G. Gaeta, Nuovo Cim. B, **107**, 1085 (1992)
8. L.V. Ovsjannikov, "Group properties of differential equations" (Novosibirsk, 1962); and "Group analysis of differential equations" (Academic Press, New York, 1982)
9. P.J. Olver, "Applications of Lie groups to differential equations" (Springer, Berlin, 1986)
10. G. Cicogna and G. Gaeta, Ann. Inst. H. Poincaré **56** (1992) 375
11. G.W. Bluman, SIAM J. Appl. Math. **50** (1990) 1706
12. G. Gaeta, Int. J. Theor. Phys. (1993), to appear
13. G. Cicogna, J. Phys.A **23** (1990) L1339; G. Gaeta, Acta Appl. Math. **28** (1992) 43
14. G. Cicogna, Lett. Nuovo Cim. **31** (1981) 600; and Boll. Un. Mat. Ital. **3**-A (1984) 131; A. Vanderbauwhede, "Local bifurcation and symmetry" (Pitman, Boston, 1982)

15. D. Ruelle, "Elements of differentiable dynamics and bifurcation theory" (Academic Press, London 1989); A. Vanderbauwhede, "Center manifolds, normal forms, and elementary bifurcations", Dynamics Reported, **2**, 89 (1989)
16. G. Cicogna and G. Gaeta, Phys. Lett. A **172** (1993) 361
17. V.I. Arnold, "Geometrical methods in the theory of ordinary differential equations" (Springer, New York, 1983); C. Elphick, E. Tirapegui, M.E. Brachet, P. Coullet and G. Iooss, Physica D **29** (1987) 95
18. G. Cicogna and G. Gaeta, J. Phys. A **23** (1990) L799; and **25** (1992) 1535.

SYMMETRIES OF THE NONLINEAR HEAT EQUATION

PETER A. CLARKSON and ELIZABETH L. MANSFIELD
Department of Mathematics, University of Exeter, Exeter, EX4 4QE, UK and Program in Applied Mathematics, University of Colorado, Boulder, CO 80309-0526, U.S.A.

Abstract. Classical and nonclassical symmetries of the nonlinear heat equation $u_t = u_{xx} + f(u)$ are considered. The method of Differential Gröbner bases is used to find the conditions on $f(u)$ under which symmetries other than the trivial translational symmetries exist. A catalogue of symmetry reductions is given including some new symmetry reductions for the linear heat equation.

1. Introduction

The classical method for finding symmetry reductions of PDES is the Lie group method of infinitesimal transformations [3, 4, 16, 25, 28, 34]. To apply the classical method to the nonlinear heat equation

$$u_t = u_{xx} + f(u), \tag{1.1}$$

where x and t are the independent variables, $f(u)$ is an arbitrary sufficiently differentiable function and subscripts denote partial derivatives, we consider the one-parameter Lie group of infinitesimal transformations in (x, t, u) given by

$$\begin{aligned} \tilde{x} &= x + \varepsilon\xi(x,t,u) + O(\varepsilon^2), \\ \tilde{t} &= t + \varepsilon\tau(x,t,u) + O(\varepsilon^2), \\ \tilde{u} &= u + \varepsilon\phi(x,t,u) + O(\varepsilon^2), \end{aligned} \tag{1.2}$$

where ε is the group parameter. Requiring that (1.1) is invariant under this transformation yields an overdetermined, linear system of equations for the infinitesimals $\xi(x,t,u)$, $\tau(x,t,u)$ and $\phi(x,t,u)$. The associated Lie algebra of infinitesimal symmetries is the set of vector fields of the form

$$\mathbf{v} = \xi(x,t,u)\partial_x + \tau(x,t,u)\partial_t + \phi(x,t,u)\partial_u. \tag{1.3}$$

There have been several generalizations of the classical Lie group method for symmetry reductions. Ovsiannikov [28] developed the method of partially invariant solutions. Bluman and Cole [2], in their study of symmetry reductions of the linear heat equation, proposed the so-called nonclassical method of group-invariant solutions, also known as the "method of conditional symmetries" [11, 12, 13] (see also [14, 27, 29, 30, 38]) and the "method of partial symmetries of the first type" [37]. In this method, the original PDE (1.1) is augmented with the invariant surface condition

$$\psi \equiv \xi(x,t,u)u_x + \tau(x,t,u)u_t - \phi(x,t,u) = 0, \tag{1.4}$$

N. H. Ibragimov et al. (eds.),
Modern Group Analysis: Advanced Analytical and Computational Methods in Mathematical Physics, 155–171.

which is associated with the vector field (1.3). By requiring that both (1.1) and (1.4) are invariant under the transformation (1.2) one obtains an overdetermined, nonlinear system of equations for the infinitesimals $\xi(x,t,u)$, $\tau(x,t,u)$ and $\phi(x,t,u)$, which appear in both the transformation (2) and the supplementary condition (1.4). The set of solutions is larger than for the classical method since the number of determining equations is smaller. However, it should be emphasized that the associated vector fields do not form a vector space, still less a Lie algebra, since the invariant surface condition (1.4) depends upon the particular reduction. Subsequently, these methods were further generalized by Olver and Rosenau [26, 27] to include "weak symmetries" and, even more generally, "side conditions" or "differential constraints", and they concluded that "the unifying theme behind finding special solutions of PDES is not, as is commonly supposed, group theory, but rather the more analytic subject of overdetermined systems of PDES".

In this paper we consider both classical and nonclassical symmetries of the nonlinear heat equation (1.1), and find conditions on $f(u)$ under which symmetries, other than trivial translational ones, may exist. The method used to find the necessary conditions on $f(u)$, and from there to find the symmetries, is that of differential Gröbner bases (DGBS) [19, 21]. This method yields the "triangulation" of a system of PDES from which solutions can be obtained in a systematic way.

The basic algorithm used to generate a DGB for a system of PDES is the Kolchin-Ritt algorithm, which forms an essential part of the StandardForm algorithm [31, 32] and the diffGbasis algorithm [20]. It appears to have been formally written down first by Carrà-Ferro [5]. The Kolchin-Ritt algorithm has been used for some time, intuitively if not formally, to yield integrability conditions for linear and orthonomic systems. With hindsight, one can see that the integrability conditions generated by the Janet-Riquier algorithm, which completes a system of orthonomic PDES to involutive form (cf. for example [17, 18, 36] and references therein), are produced by exactly the calculations performed in the Kolchin-Ritt algorithm. The production of integrability conditions require a total degree ordering on the derivative terms. On the other hand, the elimination ideals, or triangulation, of a system of PDES require a purely lexicographical ordering.

Recently, the limits of the Kolchin-Ritt algorithm in calculating a DGB of a nonlinear system were found, and sufficient conditions for obtaining a DGB for a nonlinear system of PDES were formulated [19]. The Kolchin-Ritt algorithm generates a DGB only "up to" (in a sense to be defined) a certain set of differential coefficients, and these coefficients must not lie in the ideal. Proving that these coefficients do not lie in the ideal has no algorithmic solution as yet, and there are reasons as to why it may not have one at all. The condition is also one of the sufficient conditions in [21] for a DGB of a nonlinear system. Thus far, attempts to formulate algorithms that circumnavigate the difficulty lead to processes that do not terminate in finite time [5, 24]. Fortunately in many cases one can prove the condition directly, and there is a range of techniques that can be used when the condition fails. Despite the limitation, the Kolchin-Ritt algorithm is the fundamental tool in the theory, and improvements to efficiency and practicality of this algorithm is the most important problem to be faced from an applied point of view.

The next section contains an outline of the theory of differential Gröbner bases.

After establishing the notation and formulae required, we show how to calculate the triangulation of a system of PDES using this theory. A variation of the Kolchin-Ritt algorithm is given that appears to be much faster than the one presented in [5, 21], at least for the examples calculated in this paper. We also give a "skeletal" form of the Kolchin-Ritt algorithm which we found useful in extracting information to reduce the complexity of the calculations.

In the third section of this paper, classical symmetries of the general nonlinear heat equation are catalogued for all those nonlinear $f(u)$ where a non-trivial symmetry is possible. Similarly, nonclassical symmetries are given in the fourth section, including some nonclassical symmetries for the linear heat equation.

A number of symbolic manipulation programmes have been developed, e.g. in MACSYMA [6], MAPLE [31, 32, 33], MATHEMATICA [1], MUMATH [15] and REDUCE [35] that calculate the determining equations for classical Lie symmetries of PDES. A survey of the different packages presently available and a discussion of their strengths and applications is given in [6]. The determining equations for symmetries solved in this paper were calculated using the MACSYMA programme SYMGRP.MAX [6]. The triangulations of these systems of PDES were calculated using the MAPLE package diffgrob2 [20].

2. Differential Gröbner Bases

After introducing some notation, we discuss the triangulation of a system of PDES and give some algorithms using which a triangulation can be determined.

2.1. Some notation

The determining equations for symmetries of PDES can be regarded as polynomials in the infinitesimals ξ, τ and ϕ, their derivatives, and the variables $\{x, t, u\}$, with complex coefficients. We establish our notation and formulae required for the statement of the algorithms in generic co-ordinates for ease of reference.

Let f be a differential polynomial; that is, a polynomial with complex coefficients in the independent variables $\{x_1, \ldots, x_n\}$, the functions $\{u_1, \ldots, u_m\}$ and the derivative terms $\{\mathrm{D}^\alpha u_j\}$ where

$$\mathrm{D}^\alpha u_j = \frac{\partial^{|\alpha|} u_j}{\partial x_1^{\alpha_1} \ldots \partial x_n^{\alpha_n}}.$$

The set of all such differential polynomials (d.p.'s) is denoted $R_{n,m}$ or $\mathbb{C}_{\text{diff}}(x_1, \ldots, x_n, u_1, \ldots, u_m)$.

The *highest derivative* term occurring in a d.p., f, is denoted $HDT(f)$.

The *highest power* of the $HDT(f)$ occurring in f is denoted $Hp(f)$.

The *highest coefficient*, $H\text{coeff}(f)$, is the d.p., $\text{coeff}(f, HDT(f)^{Hp(f)})$.

The *head* of f is $\text{Head}(f) = H\text{coeff}(f) HDT(f)^{Hp(f)}$.

The *separant* of f, the highest coefficient of $\mathrm{D}^\alpha f$ for any non-zero multi-index α, is denoted $Sep(f)$.

A *pseudo-reduction* of a d.p. f by $G = \{g_1, g_2, \ldots, g_k\} \subset R_{n,m}$ effects elimination from f of any terms that are of the form $\mathrm{D}^\alpha HDT(g_i)$, for some $i \in \{1, \ldots, k\}$ and $\alpha \in \mathbf{N}^n$.

Let a derivative term DT occur in f to some power p. Suppose there exists an $\alpha \in \mathbb{N}^n$ such that $D^\alpha HDT(g) = DT$ for some $g \in G$. If $\alpha = 0$ assume further that $p \geq Hp(g)$. A pseudo-reduction to f' of f by g is given by the formulae

$$f' = \begin{cases} \dfrac{H\text{coeff}(D^\alpha g)f - \text{coeff}(f,\, DT^p)DT^{p-1}D^\alpha g}{Z}, & \text{if } \alpha \neq 0 \\ \dfrac{H\text{coeff}(g)f - \text{coeff}(f,\, DT^p)HDT(g)^{(p-Hp(g))}g}{Z}, & \text{if } \alpha = 0 \end{cases}$$

where Z is the greatest common divisor of $H\text{coeff}(D^\alpha g)$ and $\text{coeff}(f,\, DT^p)$.

The *pseudo-normal* form of f with respect to a set G is obtained when no further pseudo-reduction with respect to any member of G is possible, and is denoted $\text{normal}^p(f, G)$.

The *diffSpolynomial* of two d.p.'s is obtained first by cross-differentiating and then by cross-multiplying by the highest coefficients, and subtracting.

Let f_1 and f_2 be two d.p.'s with the same highest unknown. Let α_1 and α_2 be the smallest multi-indices possible such that $D^{\alpha_1}HDT(f_1) = D^{\alpha_2}HDT(f_2)$.

Let $Z = \gcd(H\text{coeff}(D^{\alpha_1}f_1),\, H\text{coeff}(D^{\alpha_2}f_2))$. If both $\alpha_1,\, \alpha_2 \neq 0$, define

$$\text{diff}S\text{poly}(f_1, f_2) = \frac{H\text{coeff}(D^{\alpha_1}f_1)D^{\alpha_2}f_2 - H\text{coeff}(D^{\alpha_2}f_2)D^{\alpha_1}f_1}{Z}.$$

If $\alpha_1 = 0$ and $\alpha_2 \neq 0$, then

$$\text{diff}S\text{poly}(f_1, f_2) = \frac{H\text{coeff}(f_1)HDT(f_1)^{(Hp(f_1)-1)}D^{\alpha_2}f_2 - H\text{coeff}(D^{\alpha_2}f_2)f_1}{Z}$$

and similarly if $\alpha_1 \neq 0$ and $\alpha_2 = 0$.

If $\alpha_1 = \alpha_2 = 0$ so that $HDT(f_1) = HDT(f_2)$, or if f_1 and f_2 have different highest unknowns, then the differential S polynomial is defined to be

$$\text{diff}S\text{poly}(f_1,\, f_2) = \frac{\text{Head}(f_2)f_1 - \text{Head}(f_1)f_2}{\gcd(\text{Head}(f_1),\, \text{Head}(f_2))}.$$

For $G \subset R_{n,m}$ let $S(G)$ be the multiplicative set in $R_{n,m}$ generated by the set of factors of all the highest coefficients and separants of the g in G. We assume that $\{1, -1\} \in S(G)$.

2.2. The triangulation of a system of PDE

Given a system of PDES, $\Sigma \subset R_{n,m}$, we define the ideal generated by Σ to be all those equations that can be obtained from the elements of Σ by differentiating and adding, and multiplying by arbitrary elements of $R_{n,m}$. In other words,

$$I(\Sigma) = \Big\{\sum_{\alpha,i} g_{\alpha,i}D^\alpha f_i : f_i \in \Sigma,\, g_{\alpha,i} \in R_{n,m},\, \alpha \in \mathbb{N}^n\Big\}.$$

If the system is linear, we allow multiplication by polynomials in the variables $x_1, \ldots, x_n$ over $\mathbb{C}$ only.

A *differential Gröbner basis* of $I(\Sigma)$ is defined to be a set of generators G of $I(\Sigma)$ such that every element of $I(\Sigma)$ pseudo-reduces to zero with respect to G.

Differential Gröbner Bases (DGBS) of ideals of differential polynomials have many useful properties and can be used to solve, in theory at least, a great many problems. These include being able to find all elimination ideals, integrability conditions and all compatibility conditions of a system of nonlinear PDES [21].

A DGB of a system of PDES depends on the ordering of derivative terms. The standard lexicographic ordering based on $u_1 < u_2 < \ldots < u_m$ and $x_1 < x_2 < \ldots < x_n$ is given by $D^\alpha u_j > D^\beta u_k$ if $u_j > u_k$, else $j = k$ and $\alpha_n > \beta_n$, else $\alpha_n = \beta_n, \ldots, \alpha_{n-j} = \beta_{n-j}$, $\alpha_{n-j-1} > \beta_{n-j-1}$ for some j such that $0 \le j \le n-2$.

Given a DGB for a system Σ of nonlinear PDES, we have the following theorem [21]:

Theorem: Elimination Ideals. *If Σ is a* DGB *for $I(\Sigma)$ in a lexicographic ordering with $u_1 < u_2 < \ldots < u_m$ and $x_1 < x_2 < \ldots < x_n$ then $\Sigma \cap \mathbb{C}_{\mathrm{diff}}(x_1, \ldots, x_p, u_1, \ldots, u_q)$ generates $I(\Sigma) \cap \mathbb{C}_{\mathrm{diff}}(x_1, \ldots, x_p, u_1, \ldots, u_q)$.* □

In other words, every condition involving only the first q unknowns with derivatives with respect to the first p variables, obtainable from the original set of PDES by differentiating, adding, multiplying by like d.p.'s and so on, can be "read off" from a DGB of the system in a lexicographic order.

The sets $I(\Sigma) \cap \mathbb{C}_{\mathrm{diff}}(x_1, \ldots, x_p, u_1, \ldots, u_q)$ are called the elimination ideals of $I(\Sigma)$, while the DGB listed in decreasing order is called the "triangulation" of the system.

2.3. The algorithms

The basic algorithm used to calculate a DGB is the Kolchin-Ritt algorithm [5, 19, 21, 31, 32]. In this algorithm, pairs of d.p.'s are cross-differentiated and cross-multiplied to make the leading derivative terms in each d.p. cancel. The result is then pseudo-reduced with respect to all other d.p.'s to obtain a new d.p. This is done systematically until the result of cross-differentiating all pairs of d.p.'s given or obtained pseudo-reduce to zero.

In practice, there are three difficulties in calculating a DGB. The first is the problem of the build-up of differential coefficients obtained by repeated cross-multiplication and pseudo-reduction. If the coefficient of a highest derivative term of any d.p. contains a derivative term, then Kolchin-Ritt no longer necessarily suffices to calculate a DGB. Sufficient conditions may be found in [21], but the problem is not in general "solved", since one of the sufficient conditions given in [21] is not algorithmic. It is well known that the complexity of the Buchberger algorithm for computing the Gröbner basis of a polynomial ideal is doubly exponential [22]. Since the Kolchin-Ritt algorithm is modelled on the Buchberger algorithm, we have that the complexity of the Kolchin-Ritt algorithm is at least that of Buchberger's. However the main problem is that of "expression swell" in which the length of the expressions calculated exceed the memory limits of the available computer.

In spite of these problems, a DGB can often be obtained by a clever choice of the order in which pairs are cross-differentiated. We present here a version of the Kolchin-Ritt algorithm which for some systems is much faster than that given in [5, 21]. This version of the Kolchin-Ritt algorithm sorts the system after each new condition is found so that we are essentially finding all the lowest conditions (relative to the order) first. This seems to work well with those systems that are already partially sorted with respect to some lexicographical ordering on the dependent and independent variables, such as those in this paper.

The output statement of the algorithm shows the limits of the Kolchin-Ritt algorithm when one begins with, or obtains en route, an equation where the coefficient of the highest derivative contains differential terms. (The definition of $S(G)$ for any $G \subset R_{n,m}$ can be found in §2.1.) The proof of the output statement can be found in [21]. It shows that when the highest coefficient of a d.p. vanishes, for example when considering additional constraints or conditions, then one must re-do the calculation taking the additional conditions into account from the beginning. In particular, when an equation obtained factorises, and one wishes to consider a factor that does not include the highest derivative term for that equation, then one must re-start the algorithm including the desired factor at the outset.

We make some remarks about factors. It should be mentioned that if a product of functions is zero, say $f_1 f_2 = 0$, then in general one can only conclude that one of f_1, f_2 must be zero if the f_i are analytic. However, if one obtains, for example, a cubic in u_x, then the continuity of u ensures that u_x must be one of the roots of the polynomial.

Algorithm Kolchin-Ritt (sorting)

INPUT: Σ a system of d.p.'s, a termordering
OUTPUT: $\Sigma' \supset \Sigma$ such that
$S(\Sigma') \cap I(\Sigma') = \emptyset \Longrightarrow \forall f \in I(\Sigma')$
$\exists\ s \in S(\Sigma')$ such that sf pseudo-reduces to 0 with respect to Σ'.

```
pairsdone = {}
i = 1, j = 2
sort Σ into increasing order
Σ' = Σ
while i ≠ |Σ'| do
        let f_k be the k^th element of Σ' for 0 ≤ k ≤ |Σ'|
        if not {f_i, f_j} ∈ pairsdone then
            pairsdone=pairsdone ∪ {{f_i, f_j}}
            h =diffSpoly(f_i, f_j)
            h =normal^p(h, Σ')
            if h ≠ 0, Σ' = Σ' ∪ {h}
                    sort Σ' into increasing order
                    i = 1, j = 2
            else if j = |Σ'| then j = i + 2, i = i + 1
            else j = j + 1
end
```

We give now a related algorithm, the "direct search" algorithm. The "direct search" algorithm recursively looks for conditions that are lower in the order than those given. Its output is a "skeleton" of that of Kolchin-Ritt, but often provides useful information that allows the triangulation of the system to proceed faster.

Algorithm Direct Search

INPUT: F a set of d.p.'s, a termordering
OUTPUT: $G \supset F$

$G = F$
sort F into increasing order
let f_k be the kth element of F for $0 \leq k \leq |F|$
$i = |F| - 1$
$h = f_{|F|}$
while $i \neq 0$ do
  while $h \neq 0$ do
    $G = G \cup \{h\}$
    $k = h$
    $h = \text{normal}^p(\text{diff}S\text{poly}(h, f_i), G)$
  $h = \min\{k, f_i\}$
  $i = i - 1$
end

3. Classical Symmetries of $u_t = u_{xx} + f(u)$

The determining equations for the classical symmetries of the nonlinear heat equation (1.1) are

$$\begin{aligned}
f_1 &: \quad \tau_u = 0 \\
f_2 &: \quad \tau_x = 0 \\
f_3 &: \quad \tau_t - 2\xi_x = 0 \\
f_4 &: \quad \xi_u = 0 \\
f_5 &: \quad \phi_{uu} = 0 \\
f_6 &: \quad 2\phi_{xu} + \xi_t - \xi_{xx} = 0 \\
f_7 &: \quad \phi_t - \phi_{xx} + \phi_u f - \phi f_u - 2f\xi_x = 0.
\end{aligned}$$

These equations were calculated using a MACSYMA programme SYMGRP.MAX [6]. The SPDE package in REDUCE [35] appears unable to handle equations such as (1.1) which contain arbitrary functions.

Assume the lexicographic ordering given by $\phi > \tau > \xi > f$ and $x > t > u$. From f_2 and f_3 we immediately obtain

$$f_8 : \quad \xi_{xx} = 0.$$

Calculating diff$S$poly(f_6, f_7) we obtain

$$f_9 : \quad -2\phi_{ut} + 2\phi f_{uu} + 4f_u\xi_x - \xi_{xt} = 0.$$

Performing the "direct search" algorithm on f_5 and f_9 and reducing the results with respect to $\{\xi_u, \xi_{xx}, \phi_{uu}\}$ yields

$$k_1 : \quad \phi_u f_{uu} + \phi f_{uuu} + 2f_{uu}\xi_x = 0$$

$$k_2: \quad \left(-f_{uu}f_{uuuu}+2f_{uuu}^2\right)\phi+2f_{uu}f_{uuu}\xi_x=0$$
$$k_3: \quad -\xi_x f_{uu}^2\left(f_{uuuu}f_{uuu}^2-2f_{uu}f_{uuuu}^2+f_{uu}f_{uuu}f_{uuuuu}\right)=0.$$

Performing a "direct search" algorithm on f_6 and f_9 and reducing the results with respect to $\{\xi_u, \xi_{xx}, \phi_{uu}\}$ yields

$$j_1: \quad \phi_{xu}f_{uu}+\phi_x f_{uuu}=0$$
$$j_2: \quad \xi_{tt}+2\phi_x f_{uu}=0$$
$$j_3: \quad f_{uuu}\xi_{tt}+f_{uu}^2\xi_t=0.$$

Since $\xi_u = 0$ we must have, differentiating j_3 with respect to u and cross-multiplying, that

$$j_4: \quad \left(2f_{uuu}^2-f_{uu}f_{uuuu}\right)f_{uu}\xi_t\xi_{tt}=0.$$

We now consider the possibilities for $f(u)$; although it is possible that other "direct searches" will find other conditions, we can simplify the calculation of other conditions by putting $\xi_x = 0$ for example.

We solve those factors in k_3 and j_4 involving derivatives of $f(u)$. Setting $H = f_{uu}$ in k_3 we obtain (assuming that $H \neq 0$ and $\xi_x \neq 0$)

$$H_{uu}H_u^2-2HH_{uu}^2+HH_uH_{uuu}=0.$$

Dividing by HH_uH_{uu} we have

$$\frac{H_u}{H}-2\frac{H_{uu}}{H_u}+\frac{H_{uuu}}{H_{uu}}=0.$$

This integrates easily; the possibilities from k_3 are

(i) $f(u)=au^3+bu^2+cu+d$
(ii) $f(u)=(au+b)^{n+2}+cu+d$, for $n\geq 2$,
(iii) $f(u)=ae^{bu}+cu+d$
(iv) $f(u)=[\ln(au+b)]/a+cu+d$
(v) $f(u)=\left[(au+b)\ln(au+b)-(au+b)\right]/a^2+cu+d$, or
(vi) $\xi_x=0$

where a, b, c and d are arbitrary constants.

Next we look at the equation j_4. In this case we obtain

(i) $f(u)=au^2+bu+c$
(ii) $f(u)=\left[(au+b)\ln(au+b)-(au+b)\right]/a^2+cu+d$, or
(iii) $\xi_t=0$.

We shall not consider $f(u)$ linear for the classical case as the symmetries in that case are already well known [2]. Hence for those solutions $f(u)$ not of the form $u\ln(u)+bu+c$ or u^2+bu+c, with b and c constants, we must have that $\xi_t=0$.

If $\xi_x=0$, from the conditions k_1 and k_2 one can conclude that if $f(u)$ is not of the form $u\ln(u)+bu+c$ or u^2+bu+c, then only the trivial solution, $\xi=\xi_0$, $\tau=\tau_0$,

$\phi = 0$ with ξ_0 and τ_0 constants, is possible. This solution exists for all choices of $f(u)$ and corresponds to the travelling wave reduction

$$u(x,t) = w(z), \qquad z = x - ct,$$

where c is an arbitrary constant.

The cases where $f(u)$ is of the form $u\ln(u)+bu+c$ or u^2+bu+c require further attention since then the equation k_2 does not provide a condition for ϕ. For the other possibilities for $f(u)$, we must re-calculate the triangulation to obtain further conditions on ξ. For the remainder of this section, κ_1, κ_2, κ_3 and κ_4 are constants.

Case 3.1: $f(u) = u^2 + bu + c$. Here the triangulation of the system is (in addition to f_1, f_2, f_3, f_4):

$$4\phi + 8u\xi_x + 4b\xi_x + 3\xi_{xt} = 0, \qquad \xi_t = 0, \qquad \xi_x\left(b^2 - 4c\right) = 0.$$

Thus unless $f(u)$ is a perfect square, the only solution to the system is the trivial one. If $f(u) = u^2$ then

$$\xi = \kappa_1 x + \kappa_2, \qquad \tau = 2\kappa_1 t + \kappa_3, \qquad \phi = -2\kappa_1 u \tag{3.1}$$

and so for $\kappa_1 \neq 0$ (set $\kappa_1 = 1$) we obtain the symmetry reduction

$$u(x,t) = w(z)/(t + \tfrac{1}{2}\kappa_3), \qquad z = (x+\kappa_2)/(t+\kappa_3/2)^{1/2}. \tag{3.2}$$

Case 3.2: $f(u) = u\ln(u) + bu + c$. In the case $c = 0$, the triangulation of the system (in addition to f_1, f_2, f_3, f_4) is given by

$$u\phi_u - \phi = 0, \quad \phi_t - \phi = 0, \quad 2\phi_x + u\xi_t = 0, \quad \xi_x = 0, \quad \xi_{tt} - \xi_t = 0$$

which yields

$$\xi = \kappa_1 e^t + \kappa_2, \qquad \tau = \kappa_3, \qquad \phi = u\left(-\tfrac{1}{2}\kappa_1 x + \kappa_4\right)e^t \tag{3.3}$$

and so for $\kappa_3 \neq 0$ (set $\kappa_3 = 1$) we obtain the symmetry reduction

$$u(x,t) = w(z)\exp\left\{-\tfrac{1}{2}\kappa_1 x e^t + \tfrac{1}{4}\kappa_1^2 e^{2t} + \left(\tfrac{1}{2}\kappa_1\kappa_2 + \kappa_4\right)e^t\right\}, \quad z = x - \kappa_1 e^t - \kappa_2 t. \tag{3.4}$$

If $\kappa_3 = 0$ we obtain the symmetry reduction

$$u(x,t) = w(t)\exp\left\{\frac{-\frac{1}{4}\kappa_1 x^2 + \kappa_4 x}{\kappa_1 + \kappa_2 e^{-t}}\right\}. \tag{3.5}$$

The solutions for the other cases of $f(u)$ are calculated the same way and are listed below.

Case 3.3: $f(u) = u^n + bu + c$, $n \neq 0,1,2,3$. Unless $b = c = 0$ we obtain only the trivial solution. Otherwise

$$\xi = \kappa_1 x + \kappa_2, \qquad \tau = 2\kappa_1 t + \kappa_3, \qquad \phi = -2\kappa_1 u/(n-1) \tag{3.6}$$

and so for $\kappa_1 \neq 0$ (set $\kappa_1 = 1$) we obtain the symmetry reduction

$$u(x,t) = w(z)/(t+\kappa_3/2)^{1/(n-1)}, \qquad z = (x+\kappa_2)/(t+\kappa_3/2)^{1/2}. \tag{3.7}$$

Case 3.4: $f(u) = \ln(u) + bu + c$. Here we obtain only the trivial solution.
Case 3.5: $f(u) = e^{au} + bu + c$, $a \neq 0$. Unless $b = c = 0$ there is only the trivial solution. If $f(u) = e^u$ then

$$\xi = \kappa_1 x + \kappa_2, \qquad \tau = 2\kappa_1 t + \kappa_3, \qquad \phi = -2\kappa_1 \tag{3.8}$$

and so for $\kappa_1 \neq 0$ (set $\kappa_1 = 1$) we obtain the symmetry reduction

$$u(x,t) = w(z) - \ln(t + \kappa_3/2), \qquad z = \frac{x + \kappa_2}{(t + \kappa_3/2)^{1/2}}. \tag{3.9}$$

Case 3.6: $f(u) = u^3 + au^2 + bu + c$. Unless $ab - 9c = 0$ and $a^2 - 3b = 0$ there is only the trivial solution. The two algebraic conditions imply that $f(u)$ is a perfect cube. Setting $f(u) = u^3$ yields

$$\xi = \kappa_1 x + \kappa_2, \qquad \tau = 2\kappa_1 t + \kappa_3, \qquad \phi = -\kappa_1 u \tag{3.10}$$

and so for $\kappa_1 \neq 0$ (set $\kappa_1 = 1$) we obtain the symmetry reduction

$$u(x,t) = w(z)/(t + \kappa_3/2)^{1/2}, \qquad z = (x + \kappa_2)/(t + \kappa_3/2)^{1/2}. \tag{3.11}$$

4. Nonclassical Symmetries of $u_t = u_{xx} + f(u)$

There are two sorts of nonclassical symmetries, those where the infinitesimal τ is non-zero, and those where it is zero. In the first case, we can assume without loss of generality that $\tau \equiv 1$, while in the second case we can assume that $\xi \equiv 1$.

4.1. Nonclassical symmetries, $\tau \equiv 1$

The determining equations for the nonclassical symmetries, given $\tau \equiv 1$ are

$$\begin{aligned}
f_1 &: \quad \xi_{uu} = 0\\
f_2 &: \quad \phi_{uu} - 2\xi_{xu} + 2\xi\xi_u = 0\\
f_3 &: \quad 2\phi_{xu} - 2\phi\xi_u + 3\xi_u f + \xi_t - \xi_{xx} + 2\xi\xi_x = 0\\
f_4 &: \quad \phi_t - \phi_{xx} + \phi_u f - f_u\phi + 2\phi\xi_x - 2\xi_x f = 0.
\end{aligned}$$

These were calculated using the MACSYMA programme SYMGRP.MAX [6].

The result of the "direct search" algorithm on $\{f_1, f_2, f_3\}$ in the ordering $x > t > u$ is the equation $\xi_u f_{uuuu} = 0$. Hence, unless $\xi_u = 0$, $f(u)$ must be cubic in u. In the case $\xi_u = 0$, another "direct search" on the system led to the same condition k_3 for $f(u)$ obtained in the classical case. If $\xi_x = \xi_u = 0$ then we obtain the condition $\phi\left(2f_{uuu}^2 - f_{uu}f_{uuuu}\right) = 0$. As in the previous Section, we consider the various possibilities for $f(u)$ separately.
Case 4.1.1: $\tau \equiv 1$, $f(u) = u^3 + bu^2 + cu + d$. The direct search algorithm yields the equation $\xi_u\left(2\xi_u^2 + 9\right) = 0$.
Subcase 4.1.1i $2\xi_u^2 + 9 = 0$. Here a triangulation of the system yields the infinitesimals

$$\xi = \tfrac{3}{2}\mathrm{i}\sqrt{2}\,(u + b/3), \qquad \phi = \tfrac{3}{2}(u^3 + bu^2 + cu + d), \tag{4.1}$$

and the associated invariant surface condition is

$$\tfrac{3}{2}\mathrm{i}\sqrt{2}\,(u+b/3)u_x + u_t - \tfrac{3}{2}(u^3+bu^2+cu+d) = 0. \tag{4.2}$$

Using this to eliminate u_t in the PDE

$$u_t = u_{xx} + u^3 + bu^2 + cu + d, \tag{4.3}$$

yields

$$u_{xx} + \tfrac{3}{2}\mathrm{i}\sqrt{2}\,(u+b/3)u_x - \tfrac{1}{2}(u^3+bu^2+cu+d) = 0,$$

which can be linearized by the transformation $u(x,t) = -\sqrt{2}\,\mathrm{i}\,\phi_x(x,t)/\phi(x,t)$ yielding

$$2\sqrt{2}\,\phi_{xxx} + 2\mathrm{i}b\phi_{xx} - \sqrt{2}\,c\phi_x - \mathrm{i}d\phi = 0. \tag{4.4}$$

By solving this and then requiring that u also satisfies (4.2) we obtain the following exact solution of (4.3)

$$u(x,t) = \frac{\kappa_1 m_1\phi_1(x,t) + \kappa_2 m_2\phi_2(x,t) + \kappa_3 m_3\phi_3(x,t)}{\kappa_1\phi_1(x,t) + \kappa_2\phi_2(x,t) + \kappa_3\phi_3(x,t)},$$

where

$$\begin{aligned}
\phi_1(x,t) &= \exp\left\{\tfrac{1}{2}\sqrt{2}\,\mathrm{i}\,m_1 x - \left(\tfrac{1}{2}m_1^2 + m_2m_3\right)t\right\},\\
\phi_2(x,t) &= \exp\left\{\tfrac{1}{2}\sqrt{2}\,\mathrm{i}\,m_2 x - \left(\tfrac{1}{2}m_2^2 + m_3m_1\right)t\right\},\\
\phi_3(x,t) &= \exp\left\{\tfrac{1}{2}\sqrt{2}\,\mathrm{i}\,m_3 x - \left(\tfrac{1}{2}m_3^2 + m_1m_2\right)t\right\},
\end{aligned}$$

m_1, m_2 and m_3 are the distinct roots of the cubic $u^3+bu^2+cu+d=0$ and κ_1, κ_2 and κ_3 are arbitrary constants.

Setting $\xi_u = 0$ we obtain the triangulation

$$\begin{aligned}
&\phi + \xi_x(u+b/3) = 0\\
&\xi_t - 3\xi_{xx} + 2\xi\xi_x = 0\\
&3\xi_{xxx} - 3\xi\xi_{xx} - \xi_x\left(b^2-3c\right) = 0\\
&\xi_{xx}\left(-3\xi_{xx} + 2\xi\xi_x\right) = 0\\
&\xi\xi_x^2\left(-b^2+3c\right)\left(3c-b^2-\tau\right)\left(9c-3b^2-2\xi^2+6\xi_x\right) = 0\\
&\xi_x\left(-27d-2b^3+9bc\right) = 0.
\end{aligned}$$

Unless the algebraic condition $-27d-2b^3+9bc=0$ on the coefficients of $f(u)$ is satisfied, the only solution is the trivial one ($\xi=\xi_0$ a constant, $\phi=0$). So assume that $\xi_x \neq 0$ and that $-27d-2b^3+9bc=0$. There are thus several cases to consider. (Note that if both $b^2-3c=0$ and $-27d-2b^3+9bc=0$ then $f(u)=(u+b/3)^3$.)
Subcase 4.1.1ii $\xi_u=0$, $\xi_{xx}=0$, $f(u)=(u+b/3)^3$. In this case we have from the triangulation equations that $\xi_t + 2\xi\xi_x = 0$ and we thus have that

$$\xi = (x+\kappa_1)/(2t+\kappa_2), \qquad \phi = -(u+b/3)/(2t+\kappa_2). \tag{4.5}$$

These are equivalent to 3.1 and thus also yield the (classical) scaling reduction 3.2.
Subcase 4.1.1iii $\xi_u = 0$, $-3\xi_{xx} + 2\xi\xi_x = 0$, $f(u) = (u + b/3)^3$. In this case we have $\xi_t = 0$ and the Kolchin-Ritt algorithm on the equations $\{-3\xi_{xx} + 2\xi\xi_x, -3\xi_{xxx} + 3\xi\xi_{xx}\}$ leads to the condition $2\xi_x\left(3\xi_x - \xi^2\right) = 0$. Thus the non-trivial solution in this case is

$$\xi = -3/(x + \kappa_1), \qquad \phi = -3(u + b/3)/(x + \kappa_1)^2 \tag{4.6}$$

and so we obtain the nonclassical symmetry reduction

$$u(x,t) = (x + \kappa_1)w(z) - \tfrac{1}{3}b, \qquad z = \tfrac{1}{2}x^2 + \kappa_1 x + 3t. \tag{4.7}$$

Subcase 4.1.1iv $\xi_u = 0$, $27d + 2b^3 - 9bc = 0$, $b^2 - 3c \neq 0$. In this case we have $\xi_t = 0$ and $9c - 3b^2 - 2\xi^2 + 6\xi_x = 0$, leading to the solution

$$\xi = 3\mu\tan(\mu x + \kappa_1) \qquad \phi = -\mu(3u + b)\sec^2(\mu x + \kappa_1), \tag{4.8}$$

where $\mu^2 = \frac{1}{6}(b^2 - 3c)$, and so we obtain the nonclassical symmetry reduction

$$u(x,t) = \cot(\mu x + \kappa_1)w(z) - \tfrac{1}{3}b, \qquad z = \cos(\mu x + \kappa_1)\exp(3\mu^2 t). \tag{4.9}$$

We remark that the roots of the cubic $u^3 + bu^2 + cu + (9c - 2b^2)b/27 = 0$ are $-\frac{1}{3}b$, $-\frac{1}{3}b + \sqrt{2}\,\mu$ and $-\frac{1}{3}b - \sqrt{2}\,\mu$, i.e. they are collinear and the distances from the outer roots to the central one are equal.
Case 4.1.2: $\tau \equiv 1$, $f(u) = au + b$. A preliminary triangulation yields the condition $\xi_u^3 = 0$. Reducing the equations with respect to $\xi_u = 0$ we obtain $\phi_{uu} = 0$. Inserting $\phi = A(x,t)u + B(x,t)$ into the equations, we obtain the system

$$\begin{aligned} g_1 : &\quad 2A_x + \xi_t - \xi_{xx} + 2\xi\xi_x = 0 \\ g_2 : &\quad A_t - A_{xx} + 2\,(A - a)\,\xi_x = 0 \\ g_3 : &\quad B_t - B_{xx} + 2\,(B - b)\,\xi_x - aB + bA = 0. \end{aligned}$$

Note that we no longer have an over-determined system. For an arbitrary function $u(x,t)$ let $\Phi(u;\xi) = u_t - u_{xx} + 2u\xi_x$, and set $\Phi(\xi;\xi) = \eta$. Let $A > \xi$. Then cross-differentiating g_1 and g_2 and reducing, we obtain

$$4\xi_{xx}(A - a) = \Phi(\eta;\xi).$$

We first take care of the special case, $\xi_{xx} = 0$.
Subcase 4.1.2i $\tau \equiv 1$, $\xi_{xx} = 0$, $f(u) = au + b$. Using the ordering $\xi > A$ we obtain from g_1 and $\xi_{xx} = 0$ that $A_{xxx} = 0$. Substituting $\xi = \alpha(t)x + \beta(t)$ and $A = \delta(t)x^2 + \epsilon(t)x + \gamma(t) + a$ into g_1 and g_2 we obtain the following system:

$$\begin{aligned} &4\delta + \alpha' + 2\alpha^2 = 0 \\ &2\epsilon + \beta' + 2\alpha\beta = 0 \\ &\delta' + 2\delta\alpha = 0 \\ &\epsilon' + 2\epsilon\alpha = 0 \\ &\gamma' - 2\delta + 2\gamma\alpha = 0. \end{aligned}$$

Eliminating δ from the first and third of these equations yields the equation $\alpha'' + 6\alpha\alpha' + 4\alpha^3 = 0$, which can be linearized using the transformation $\alpha = \zeta'/(2\zeta)$. Successively solving for α, δ, ϵ, β and γ we arrive at the solution

$$\xi = \frac{(2t+\kappa_4)\,x + \frac{1}{2}\kappa_1 t + \kappa_2}{2(t^2+\kappa_4 t+\kappa_5)}, \qquad A = -\frac{\frac{1}{2}x^2 + \frac{1}{4}\kappa_1 x + t + \kappa_3}{2(t^2+\kappa_4 t+\kappa_5)} + a \tag{4.10}$$

where κ_1, κ_2, κ_3, κ_4 and κ_5 are arbitrary constants. Solving g_3 for B with ξ and A as given in (4.10) yields

$$B = \frac{2x^2\left[\kappa_6 e^{at} - b\right] + x\left[\kappa_7 e^{at} - b\kappa_1\right] + 4a(\kappa_8 + t\kappa_6)\,e^{at} - 4b(t+\kappa_3)}{8a(t^2+\kappa_4 t+\kappa_5)},$$

where κ_6, κ_7 and κ_8 are arbitrary constants

Subcase 4.1.2ii $\tau \equiv 1$, $\xi_{xx} \neq 0$, $f(u) = au + b$. Pseudo-reducing g_1, g_2 with respect to $\xi_{xx}(A - a) = \frac{1}{4}\Phi(\eta;\xi)$, we obtain two equations k_1 and k_2 in ξ whose highest derivative terms in a *total degree* ordering are ξ_{xxxxxx} and ξ_{xxxxx}. (A total degree ordering is determined first by the degree of the terms and then by a lexicographic ordering.) Reducing k_1 with respect to k_2 in this ordering yields a condition k_3 for ξ whose highest derivative term is ξ_{xxxxt}. Now $\{k_2, k_3\}$ forms a DGB for $I(\{k_1, k_2\})$ with $S(\{k_2, k_3\}) = \{\pm(\xi_{xx})^n : n \in \mathbb{N}\}$. (We have stipulated that $\xi_{xx} \neq 0$ in this case.) Since we have a DGB for the conditions satisfied by ξ, our next step is to apply the Initial Data algorithm [31]. A DGB in a total degree ordering is the required input for this algorithm. Thus we see that a formal solution to the equations for ξ depends on four arbitrary functions and one arbitrary constant. In fact we can assign $\xi(0,t)$, $\xi_x(0,t)$, $\xi_{xx}(0,t)$, $\xi_{xxx}(0,t)$ and $\xi_{xxxx}(0,0)$ and the other coefficients for a formal expansion for ξ will be determined.

Assuming that $\xi_{xx}(0,t)$ is a constant, and that the other arbitrary functions and constant are zero, one can derive a solution to the equations for ξ and A (i.e. g_1 and g_2):

$$\xi = 3x^2/(3c - x^3), \qquad A = a + 3x/(3c - x^3) \tag{4.11}$$

where c is a constant, simply by summing the formal Taylor series obtained from k_2 and k_3. Solving g_3 for B with ξ and A as given in (4.11) yields $B = b + 3bcx/[a(3 - cx^3)]$. Therefore the infinitesimals in this case are

$$\xi = \frac{3x^2}{3c - x^3}, \qquad \phi = (au+b) + \frac{3x(au+b)}{a(3c - x^3)} \tag{4.12}$$

and so we obtain the symmetry reduction

$$u(x,t) = xw(z)\,e^{at} - b/a, \qquad z = \tfrac{1}{6}x^2 + t + c/x. \tag{4.13}$$

Case 4.1.3: Nonclassical, $\tau \equiv 1$, $f(u)$ neither cubic nor linear. In this case we have k_3 or $2f_{uuu}^2 - f_{uu}f_{uuuu} = 0$. The various possibilities for $f(u)$ have already been listed in §3. For all these possibilities, we obtain the same symmetries as for the classical case, with the classical ξ/τ yielding the nonclassical ξ and the classical ϕ/τ yielding the nonclassical ϕ. However, for some of the cases, this is harder to prove than for the classical determining equations.

4.2. NONCLASSICAL SYMMETRIES, $\tau \equiv 0$, $\xi \equiv 1$

We come now to those nonclassical symmetries where $\tau \equiv 0$, and without loss of generality we can take $\xi \equiv 1$. In this case we have the system consisting of the equation (1.1), the invariant surface condition (1.4) and one additional determining equation for ϕ:

$$\begin{aligned} f_1 &: \quad u_t - u_{xx} - f(u) = 0 \\ f_2 &: \quad u_x - \phi(x,t,u(x,t)) = 0 \\ f_3 &: \quad \phi_t - \phi_u\phi_x - \phi^2\phi_{uu} - \phi\phi_{xu} + f\phi_u - f_u\phi = 0. \end{aligned}$$

The equation f_3 was calculated using the MACSYMA program SYMGRP.MAX [6].

Cross-differentiating f_1 and f_2 and reducing with respect to f_3 yields an additional condition

$$f_4 : \quad \phi_{xx} + \phi\phi_{xu} - \phi_u\phi_x = 0.$$

Now

$$\frac{\mathrm{d}\phi}{\mathrm{d}x} = \phi_x + \phi_u\phi \qquad \text{and} \qquad \frac{\mathrm{d}\phi_x}{\mathrm{d}x} = \phi_{xx} + \phi_{ux}\phi,$$

where $\mathrm{d}/\mathrm{d}x$ denotes total derivative with respect to x, so that f_4 can be written

$$\frac{\mathrm{d}}{\mathrm{d}x}\ln\left(\frac{\phi_x}{\phi}\right) = -\frac{\phi_x}{\phi},$$

implying that $\phi = \theta(u,t)\left[x + k(t)\right]$. Setting this result into f_3 and equating coefficients of powers of x to zero we obtain

$$\theta_{uu} = 0, \qquad k'(t) = 0, \qquad \theta_t - 2\theta_u\theta + f\theta_u - f_u\theta = 0.$$

Hence $\theta(u,t) = \alpha(t)u + \beta(t)$. The third condition can be written as

$$\theta_t - 2\theta\theta_u - \theta^2(f/\theta)_u = 0.$$

which can be integrated to yield

$$f(u) = -\frac{1}{\alpha^2}(\alpha\beta' - \alpha'\beta) + \left(\frac{\alpha'}{\alpha^2} - 2\right)(\alpha u + \beta)\ln(\alpha u + \beta) + c\,(\alpha u + \beta)$$

if $\alpha \neq 0$, or $\beta = de^{bt}$ and $f(u) = bu + c$ if $\alpha = 0$, where b, c and d are constants.

Case 4.2.1: $\alpha \neq 0$. The function f is independent of t and hence $\alpha = a$, $\beta = b$, where a and b are constants, giving

$$f(u) = -2\,(au + b)\ln(au + b) + c\,(au + b)\,. \tag{4.14}$$

Finally we solve f_1 (i.e. (1.1)) and f_2 (i.e. (1.4)) for u, where we know $\phi = (au + b)(x + k)$. From f_2 we obtain

$$u = \mu(t)\exp\left\{\tfrac{1}{2}a(x + k)^2\right\} - b/a.$$

Putting this into f_1 yields equation for $\mu(t)$

$$\frac{\mathrm{d}\mu}{\mathrm{d}t} = a(c + 1)\mu(t) + 2a\mu(t)\ln[a\mu(t)].$$

Hence for $f(u)$ given by (4.14) we obtain the solution

$$u(x,t) = a^{-1}\exp\left\{me^{-2at} + \tfrac{1}{2}(1+c) + \tfrac{1}{2}a(x+k)^2\right\} - b/a \tag{4.15}$$

where k and m are arbitrary constants.

Case 4.2.2: $\alpha = 0$. We have $f(u) = bu + c$, and $\phi = de^{bt}(x+k)$, where b, c, d and k are constants. Now $u_x = de^{bt}(x+k)$ which implies

$$u(x,t) = \tfrac{1}{2}de^{bt}(x+k)^2 + \mu(t). \tag{4.16}$$

Substituting this into (1.1) yields an equation for $\mu(t)$:

$$\frac{d\mu}{dt} - b\mu(t) = de^{bt} + c.$$

Hence we obtain finally the solution

$$u(x,t) = \left[\tfrac{1}{2}d(x+k)^2 + dt + m\right]e^{bt} - c/b,$$

where m is an arbitrary constant.

5. Discussion

In this paper we have demonstrated that the method of DGBS has enabled all symmetries of the nonlinear heat equation (1.1) to be found; the list it provides of possible analytic functions $f(u)$ for which symmetries may exist is definitive. The use of DGBS has made the analysis of overdetermined systems of PDES more tractable and whilst the diffgrob2 package needs to be used interactively at present, nevertheless it is effective in solving the overdetermined systems of determining equations for classical and nonclassical symmetries of (1.1).

It is not clear how the direct method, developed by Clarkson and Kruskal [8] for finding symmetry reductions of PDES may be applied to equations such as (1.1) which contain arbitrary functions. Nucci and Clarkson [23] (see also [9]) demonstrated that there exist symmetry reductions of the Fitzhugh-Nagumo (FN) equation, (1.1) with $f(u) = u(1-u)(u-a)$, where a is a constant, which are obtainable using the nonclassical method due to Bluman and Cole [2], but not using the direct method due to Clarkson and Kruskal [8]. Hence the nonclassical method is more general than the direct method, at least as it was originally formulated. Clarkson and Hood [7], following Olver and Rosenau [26, 27], have shown that every reduction obtained by the direct method is necessarily obtainable using the nonclassical method (see also [39]). A recent extension of the direct method by Estevez [10] does find all the symmetry reduction solutions of the FN equation found by the nonclassical method. These results pose the following important open question: "for which PDES does the nonclassical method yield more symmetry reductions than the direct method?" Furthermore, it remains an open question to determine *a priori* which PDES possess symmetry reductions that are not obtainable using the classical Lie group approach.

Acknowledgements

We would like to thank the Program in Applied Mathematics, University of Colorado at Boulder, for their hospitality, where this work was completed. The support of SERC (grant GR/H39420) is gratefully acknowledged. PAC is also grateful for support through a Nuffield Foundation Science Fellowship and NATO grant CRG 910729.

References

1. G. Baumann, "Lie Symmetries of Differential Equations", Universität Ulm, preprint (1992).
2. G.W. Bluman and J.D. Cole, *J. Math. Mech.*, **18** (1969) 1025–1042.
3. G.W. Bluman and J.D. Cole, "*Similarity Methods for Differential Equations*," *Appl. Math. Sci.*, **13**, Springer-Verlag, Berlin (1974).
4. G.W. Bluman and S. Kumei, "*Symmetries and Differential Equations*," *Appl. Math. Sci.*, **81**, Springer-Verlag, Berlin (1989).
5. G. Carrá-Ferro, *Proc. AAECC5,* Menorca, Spain, (1987) 129–140.
6. B. Champagne, W. Hereman and P. Winternitz, *Comp. Phys. Comm.*, **66** (1991) 319–340.
7. P.A. Clarkson and S. Hood, *Europ. J. Appl. Math.*, **3** (1992) 381–414.
8. P.A. Clarkson and M.D. Kruskal, *J. Math. Phys.*, **30** (1989) 2201–2213.
9. P.A. Clarkson and E.L. Mansfield, in "*Applications of Analytic and Geometric Methods to Nonlinear Differential Equations*," [Ed. P.A. Clarkson], Kluwer, Dordrecht (1993) to appear.
10. P.G. Estevez, *Phys. Lett.*, **171A** (1992) 259–261.
11. W.I. Fushchych, *Ukrain. Mat. Zh.*, **43** (1991) 1456–1470.
12. W.I. Fushchych and A.G. Nikitin, "*Symmetries of Maxwell's Equations*," D. Reidel, Dordrecht (1987).
13. W.I. Fushchych and R.Z. Zhdanov, *Physics Reports*, **172** (1989) 123–174.
14. G. Gaeta, *J. Phys. A: Math. Gen.*, **23** (1990) 3643–3645.
15. A. Head, "Lie: a MUMATH program for the calculation of the Lie algebra of differential equations", preprint (1990).
16. J.M. Hill, "*Differential Equations and Group Methods for Scientists and Engineers*," CRC Press, Boca Raton (1992).
17. M. Janet, *J. Math. Pure Appl.*, **3** (1920) 65–151.
18. M. Janet, "*Leçons sur les Systèmes d'Équations*," Gauthier-Villars (1929).
19. E.L. Mansfield, "*Differential Gröbner Bases*", Ph.D. Thesis, University of Sydney, Australia (1992).
20. E.L. Mansfield, "diffgrob (Release 2) User's Manual".
21. E.L. Mansfield and E.D. Fackerell, "Differential Gröbner Bases", Macquarie University, preprint **92/108** (1992).
22. E. Mayr and A. Meyer, *Adv. Math.*, **46** (1982) 305–329.
23. M.C. Nucci and P.A. Clarkson, *Phys. Lett.*, **164A** (1992) 49–56.
24. F. Olivier, "Canonical Bases: Relations with Standard Bases, Finiteness Conditions and Application to Tame Automorphisms", École Polytechnique Research Report **LIX/RR/90/14** (1990).
25. P.J. Olver, "*Applications of Lie Groups to Differential Equations*," *Graduate Texts Math.*, **107**, Springer-Verlag, New York (1986).
26. P.J. Olver and P. Rosenau, *Phys. Lett.*, **114A** (1986) 107–112.
27. P.J. Olver and P. Rosenau, *SIAM J. Appl. Math.*, **47** (1987) 263–275.
28. L.V. Ovsiannikov, "*Group Analysis of Differential Equations*," [Trans. W.F. Ames], Academic, New York (1982).
29. E. Pucci, *J. Phys. A: Math. Gen.*, **25** (1992) 2631–2640.
30. E. Pucci and G. Saccomandi, *J. Math. Anal. Appl.*, **163** (1992) 588–598.
31. G.J. Reid, *J. Phys. A: Math. Gen.*, **23** (1990) L853–L859.
32. G.J. Reid, *Europ. J. Appl. Math.*, **2** (1991) 293–318.
33. G.J. Reid, *Europ. J. Appl. Math.*, **2** (1991) 319–340.

34. C.A. Rogers and W.F. Ames, "*Nonlinear Boundary Value Problems in Science and Engineering*," Academic Press, Boston (1992).
35. F. Schwarz, *Computing*, **34** (1985) 91–106; *SIAM Rev.*, **30** (1988) 450.
36. O. Stormark, "Formal and local solvability of PDEs", preprint **TRITA-MAT-1988-11** (1988).
37. E.M. Vorob'ev, *Acta Appl. Math.*, **24** (1991) 1–24.
38. P. Winternitz, "Conditional symmetries and conditional integrability for nonlinear systems", preprint **CRM-1709**, Montreal (1990).
39. S. Zidowitz, "Nichtklassische Symmetrierduktionen hydrodynamischer Plasmagleichungen", Master's thesis, Institut für Mathematische Physik, Technische Universität Carolo-Wilhelmina zu Branschweig, Germany (1992).

SYMMETRIES OF TIME DEPENDENT HAMILTONIAN SYSTEMS

A. DEWISME and S. BOUQUET
Centre d'Études de Vaujours,
Commissariat à l'Energie Atomique,
BP n° 7, 77181 Courtry, France

and

P.G.L. LEACH
Department of Mathematics and Applied Mathematics,
University of Natal, Durban 4001, South Africa

Abstract. The method of Lie symmetries is applied to the problem defined by the time dependent Hamiltonian $H = \frac{p^2}{2} + V(q,t)$. This shows the relation between generalized symmetries and first integrals. The geometrical and variational symmetries are found. A special case of the linear oscillator with time-dependent frequency gives an example of application of the methods developped in the text.

1. Introduction

We consider the time-dependent Hamiltonian

$$H = \frac{p^2}{2} + V(q,t) \tag{1}$$

from which we derive the system

$$\begin{cases} \dot{q} = p \\ \dot{p} = f(q,t) \end{cases} \tag{2}$$

where

$$f = \frac{\partial V}{\partial q} \tag{3}$$

or the equivalent equation

$$\ddot{q} = f(q,t) \tag{4}$$

There are two important ways to find first integrals of this equation. The first one is the direct approach [1-3]: a form of the first integral I must be chosen and put into the equation

$$D_t(I) = I_t + pI_q + f(q,t)I_p = 0 \tag{5}$$

This method gives result but it forgets the origin of the problem. On the contrary, the second method is the use of mechanics [4].

It is then interesting to test the Lie symmetries on this problem. The full study is contained in [5-6] and we want to sum up the results in this proceeding. The first

N. H. Ibragimov et al. (eds.),
Modern Group Analysis: Advanced Analytical and Computational Methods in Mathematical Physics, 173–179.

two sections deal with the relations between the first integrals and the (generalized) symmetries. The sections 4 et 5 studies the geometrical and variational symmetries. An integrable case of the linear oscillator with time dependent frequency concludes this text.

2. An extension of the Liouville theorem

H.R. Lewis, P.G.L. Leach, S. Bouquet and M.R. Feix have proved in [4] that if (2) has a first integral then a second one can be derived explicitly and the system is then totally integrable. The tool used is generating functions but a simpler way to obtain the second first integral is presented here.

If a first integral $I(q,p,t)$ is known then (1) is reduced to a first order equation

$$I = I(q,p,t) \qquad \Longleftrightarrow \qquad p = G(q,I,t) \tag{6}$$

One can solve it by looking for an integrating factor, i.e. a function α which satisfies

$$\alpha_t + \alpha_q G + \alpha G_q = 0 \tag{7}$$

Such a function is given by the first integral I itself $\alpha = G_I$ because of the following equality:

$$G_{It} + G_{Iq} G + G G_q = \frac{1}{I_p} \frac{\partial f}{\partial p} = 0 \tag{8}$$

The new first integral is then

$$J = \int_0^q G_I(q', I, t)\, \mathrm{d}q' - \int_0^t (G\, G_I)(0, I, t')\, \mathrm{d}t' \tag{9}$$

The existence of this first integral is already known from Liouville theorem but the drawback of this theorem is not to provide explicit formulae as the one given here.

In the autonomous case, H is defined by

$$I = H = \frac{p^2}{2} + V(q) \tag{10}$$

and J is then given by the well-known formula

$$J = \int_0^q \frac{\mathrm{d}q'}{\sqrt{2\,(I - V(q'))}} - t \tag{11}$$

There is a connection with symmetries: the integrating factor $\alpha = G_I$ gives a symmetry of the first order equation (8) and of the system (2)

$$X = I_p \frac{\partial}{\partial q} - I_q \frac{\partial}{\partial p} \tag{12}$$

One can wonder if it is the general form of a symmetry of (2). This is the aim of the next section.

3. From symmetries to first integrals

Let's consider a symmetry of (2). We can put it in its evolutionary form

$$X = \xi(q,p,t)\frac{\partial}{\partial q} - \eta(q,p,t)\frac{\partial}{\partial p} \tag{13}$$

if X is of the form (12) then it satisfies

$$\frac{\partial \xi}{\partial q} + \frac{\partial \eta}{\partial q} = 0 \tag{14}$$

If this condition is verified, there exists a function $A(q,p,t)$ such that

$$\begin{cases} A_p = \xi \\ A_q = -\eta \end{cases} \tag{15}$$

With the help of the determining equations, one obtains that $D_t(A)$ does not depend on q or p:

$$D_t(A) = B(t) \tag{16}$$

We get a first integral

$$I(q,p,t) = A(q,p,t) - \int B(t)\,\mathrm{d}t \tag{17}$$

and the initial symmetry is of the form (12).

This is a nice result because for a general equation of second order a symmetry allow only to get a reduced equation of first order. In the Hamiltonian case and if the condition (14) is verified a first integral is obtained directly and a second one follow by the way explained in section 2. The system is consequently completely integrable.

4. The Lie symmetries

It is well-known that the symmetries of (2) of the form

$$X = \xi(q,t)\,\frac{\partial}{\partial t} + \eta(q,t)\,\frac{\partial}{\partial q} \tag{18}$$

verify

$$\xi = a(t) \tag{19}$$

$$\eta = \left(\frac{1}{2}a'(t) + \beta\right) q + b(t) \tag{20}$$

(if f is not linear in q, which is the case of the linear oscillator). We use new notations to express f

$$a = \rho^2 \qquad T = \int \frac{\mathrm{d}t}{\rho^2} \qquad Q = \frac{q}{\rho} - e^{\beta T}\int e^{-\beta T}\frac{b}{\rho^3}\,\mathrm{d}T \tag{21}$$

and obtain

$$f = e^{\beta T} \rho^{-3} \left\{ -\phi(e^{-\beta T} Q) + \int \left(\rho \ddot{b} - \ddot{\rho} b \right) e^{-\beta T} \, \mathrm{d}T \right\} + \frac{\ddot{\rho}}{\rho} q \tag{22}$$

where ϕ is an arbitrary function. With the variables Q and T the initial equation (4) reduces to

$$\ddot{Q} = -e^{\beta T} \phi(e^{-\beta T} Q) \tag{23}$$

where the dot means derivation with respect to T. The symmetry is then

$$X = \frac{\partial}{\partial T} + \beta Q \frac{\partial}{\partial Q} \tag{24}$$

and (23) is easily reduced to a first order equation

$$y \frac{\mathrm{d}y}{\mathrm{d}x} + 2\beta y + \phi(x) + \beta^2 x = 0 \tag{25}$$

by using variables

$$x = Q e^{-\beta T} \qquad\qquad y = \dot{Q} e^{-\beta T} - \beta x \tag{26}$$

This equation is not integrable in the general case but there are three exceptions. The first one is ϕ linear in q, the case of the linear oscillator we treat in the last section and the second one is $\beta = 0$ which is the subject of the next section.

The third one appears when the equations (4) or (23) have two independent symmetries. After some calculations, (23) has a symmetry

$$X_1 = \xi_1(T) \frac{\partial}{\partial T} + \eta_1(Q, T) \frac{\partial}{\partial Q} \tag{27}$$

distinct from (24) if and only if ϕ has the following form:

$$\psi = -B \left(Q e^{-\beta T} + K \right)^{M+1} + K\beta^2 - \left(\frac{\beta M}{M+4} \right)^2 \left(Q e^{-\beta T} + K \right) \tag{28}$$

where K and M are arbitrary constants. This case can be reduced to the autonomous case by defining

$$\tau = -\frac{M+4}{2M\beta} exp \left(-\frac{2M\beta}{M+4} T \right) \qquad\qquad \lambda = exp \left(-\frac{4\beta}{M+4} T \right) \left(Q e^{-\beta T} + K \right) \tag{29}$$

and the equation is then

$$\ddot{\lambda} = B \lambda^{M+1} \tag{30}$$

which is readily integrable. After this isolated case, we treat a much more general one.

5. The variational symmetries

The case $\beta = 0$ appear more naturally by using the variational point of view: the system (2) is the system of Euler-Lagrange equations associated to the Lagrangian

$$L = \frac{1}{2}\dot{q}^2 - V(q,t) \tag{31}$$

It is then interesting to know which Lie symmetries of (2) are variational symmetries too. After some calculations one finds that they are defined by

$$\beta = 0 \tag{32}$$

The Noether theorem gives then a first integral

$$I = \frac{1}{2}\left[\rho(p-\dot{\alpha}) - \dot{\rho}(q-\alpha)\right]^2 + \chi\left(\frac{q-\alpha}{\rho}\right) \tag{33}$$

where

$$b = \rho^3 \left(\frac{\alpha}{\rho}\right)' \qquad \phi = \chi' \tag{34}$$

This first integral was already found by an other method in [7].

The second first integral can then be calculated by the method explained in section 2

$$J = K\left(\frac{q-\alpha}{\rho}, I\right) - \int \frac{dt}{\rho^2} \tag{35}$$

where

$$K(u,v) = \int \frac{du}{\sqrt{2\left(v - \chi(u)\right)}} \tag{36}$$

These expressions are the first integrals of an autonomous Hamiltonian derived from H by using appropriate variables

$$T = \int \frac{d}{\rho^2} \qquad Q = \frac{q-\alpha}{\rho} \tag{37}$$

The new Hamiltonian is the I

$$H_1 = I = \frac{1}{2}\dot{Q}^2 + \chi(Q) \tag{38}$$

We could have found the first integrals by two others methods. The first one is the integration of the reduced equation (8) and the second one is the results of section 3 because it is easy to verify that the condition (14) is satisfied. It is interesting to notice that these two methods and the Noether theorem give the same result but none can be generalized to a more general case.

6. The case of the linear oscillator

The equation (4) is

$$\ddot{q} + \omega^2(t)q = 0 \tag{39}$$

Following [8] we look for symmetries of the following type

$$X = u(t)\,\frac{\partial}{\partial t} + \frac{1}{2}\,u'(t)q\,\frac{\partial}{\partial q} \tag{40}$$

The determining equations reduce to

$$u''' + 4\,\omega^2 u' + 4\,\omega\omega' u = 0 \tag{41}$$

This equation is not integrable for all ω but if we suppose that u is a polynomial of degree 2

$$u = at^2 + bt + c \tag{42}$$

then the first term of (41) vanishes which makes the equation integrable

$$\omega = \frac{K}{at^2 + bt + c} \tag{43}$$

The evolutionary form of the symmetry is

$$X' = \left[\frac{1}{2}\,u' - up\right]\frac{\partial}{\partial q} + \left[-\frac{1}{2}\,u'p + \left(a + \frac{K^2}{u}q\right)\right]\frac{\partial}{\partial p} \tag{44}$$

and it verifies the condition (14) of section 3

$$\frac{\partial}{\partial q}\left[\frac{1}{2}\,u' - up\right] + \frac{\partial}{\partial p}\left[-\frac{1}{2}\,u'p + \left(a + \frac{K^2}{u}q\right)\right] = \frac{1}{2}\,u' - \frac{1}{2}\,u' = 0 \tag{45}$$

We obtain a first integral:

$$I = -u\left(\frac{p^2}{2} + \omega^2\frac{p^2}{2}\right) + u'\frac{pq}{2} - a\frac{q^2}{2} \tag{46}$$

which is quite close from the Hamiltonian. The method of section 2 gives another first integral

$$J = \frac{1}{\sqrt{4K^2 - \delta}}\ \text{arc sin}\left(\frac{\epsilon q}{2}\sqrt{\frac{-4K^2 - \delta}{4Iu}}\right) - \frac{1}{\sqrt{-\delta}}\ \text{arc tg}\left(\frac{u'}{\sqrt{-\delta}}\right) \tag{47}$$

where

$$\epsilon = \pm 1 \qquad \delta = b^2 - 4ac \tag{48}$$

These complex formulas appear clearer if they are compared with those of the autonomous case

$$I = H = \frac{p^2}{2} + \omega^2\frac{p^2}{2} \tag{49}$$

and

$$J = \text{arc sin}\left(\frac{q}{q_0}\right) - \omega t \tag{50}$$

In the two cases, I must be considered as an amplitude and J as a phase.

7. Conclusion

The results of sections 2 and 3 implies that the number of first integrals and of symmetries are not arbitrary. The system (2) has zero first integral or is integrable. In this case, there are an infinity of first integrals and of (generalized) symmetries. In the case of non-integrability there exists zero symmetry or one which does not verify the condition (14).

The study of the geometrical symmetries gives the first integral known since [7] but it is interesting, it explains why this first integral was found and why it cannot be generalized: the existence of the first integral is essentially connected with the existence of a geometrical symmetry.

The main problem is now to find explicit generalized symmetries and it would be interesting to know if it is easier than the search of first integrals.

References

1. Lewis H.R. and Leach P.G.L. (1982) J. Math. Phys. **23** 237
2. Goedert J. and Lewis H.R. (1987)J. Math. Phys. **28** 728
3. Feix M.R., Bouquet S. and Lewis H.R. (1987) Physica D **28** 80
4. Lewis H.R., Leach P.G.L., Bouquet S. and Feix M.R. (1992) J. Math. Phys. **33** 591
5. Leach P.G.L., Bouquet S. and A. Dewisme 'Symmetries of Hamiltonian one-dimensional systems' submitted to J. Non-Lin. Mechs.
6. Dewisme A. and Bouquet S. to appear in J. Math. Phys. **34** (march 1993)
7. Leach P.G.L., Lewis H.R. and Sarlet W. (1984) J. Math. Phys. **25** 486
8. B.K. Shivanoggi L. Muilenberg (1991) Phys Lett A **154** 1

QUASILINEAR HYPERBOLIC SYSTEMS: REDUCTION TO AUTONOMOUS FORM AND WAVE PROPAGATION

ANDREA DONATO and FRANCESCO OLIVERI
Department of Mathematics, University of Messina,
Contrada Papardo, Salita Sperone 31
98166 Sant'Agata, Messina, Italy.

Abstract. We consider quasilinear hyperbolic systems with coefficients depending explicitly on space and time variables. Assuming that these systems are left invariant by two Lie groups of transformations having commuting infinitesimal operators, it is shown how to reduce them to autonomous form. A physical example admitting two groups of transformations is considered; the application of the procedure allows us to characterize an exact particular solution. Finally, some considerations about the linearization of the model are pointed out.

1. Introduction

In order to study the propagation of nonlinear waves (either discontinuity or shock waves) in continuous media, one is led to consider quasilinear hyperbolic systems of first order. In one space dimension one has:

$$\frac{\partial \mathbf{u}}{\partial t} + A(\mathbf{u}, x, t)\frac{\partial \mathbf{u}}{\partial x} = \mathbf{B}(\mathbf{u}, x, t) \tag{1}$$

where $\mathbf{u}$ is the unknown column vector having N components while A is a $N \times N$ square matrix and $\mathbf{B}$ a column vector with N components, both known functions of their arguments.

The dependence of A on $\mathbf{u}$ is related to the nonlinearity of the problem while its dependence on x is related, for instance, to propagation problems into non-uniform regions. The dependence of $\mathbf{B}$ on $\mathbf{u}$ and x is often obtained by considering axi-symmetric problems.

Some problems of physical interest described by systems like (1) are concerned with the point explosion problem in fluids [1] also in the presence of a magnetic field [2] with cylindrical or spherical symmetry, the viscoelasticity materials [3], the fluid filled elastic tubes and nonlinear elastic rods with variable cross-section [4], the rate-type materials [5] and so on.

Let $\mathbf{u}(x,t)$ be a piecewise C^1-continuous solution with discontinuities in the first order derivatives occurring across a line $\phi(x,t) = 0$. If we denote with π the jump

$$\pi = \left(\frac{\partial \mathbf{u}}{\partial \phi}\right)_{\phi=0^+} - \left(\frac{\partial \mathbf{u}}{\partial \phi}\right)_{\phi=0^-} \neq 0 \tag{2}$$

the compatibility conditions for the propagation of weak discontinuities are

$$(\phi_t I + \phi_x A(\mathbf{u}_0(x,t), x, t))\pi = 0 \tag{3}$$

N. H. Ibragimov et al. (eds.),
Modern Group Analysis: Advanced Analytical and Computational Methods in Mathematical Physics, 181–189.

where I is the identity $N \times N$ matrix and $\mathbf{u}_0(x,t)$ represents a solution of the system (1) characterizing the unperturbed state in which the discontinuity wave propagates.

In general it is very difficult to study the propagation of weak discontinuities in a state characterized by $\mathbf{u}_0(x,t) \neq$ const. because it is not easy to find a solution to a nonlinear problem and because the evolution law of weak discontinuities is a little bit complicated. Of course there are the same problems when one studies the propagation of shock waves.

On the contrary, when the governing system is in autonomous form, i.e.

$$\frac{\partial \mathbf{w}}{\partial \tau} + \tilde{A}(\mathbf{w}) \frac{\partial \mathbf{w}}{\partial \eta} = \tilde{\mathbf{B}}(\mathbf{w}) \tag{4}$$

there may exist constant solutions $\mathbf{w} = \mathbf{w}_0$ such that $\tilde{\mathbf{B}}(\mathbf{w}_0) = 0$ or also simple wave solutions when $\tilde{\mathbf{B}}(\mathbf{w}) \equiv 0$, i.e., when the system is also homogeneous; consequently the corresponding study of the propagation of discontinuity or shock waves in constant states generally can be carried out easier.

So the idea is to check the conditions allowing to perform a suitable change of variables

$$\begin{aligned} \tau &= \tau(\mathbf{u}, x, t) \\ \eta &= \eta(\mathbf{u}, x, t) \\ \mathbf{w} &= \mathbf{w}(\mathbf{u}, x, t) \end{aligned} \tag{5}$$

such that the system (1) transforms to another one having the form (4). In this way, constant solutions, characterized by $\tilde{\mathbf{B}}(\mathbf{w}_0) = 0$, of the transformed system are non-constant solutions of the original system and the propagation, for instance, of weak discontinuities in a constant state of (4) correspond to the propagation of weak discontinuities in a non-constant state of (1).

The plan of the paper is the following. In section 2 the fundamental theorem [6] allowing for the reduction of a general nonautonomous system to autonomous form is sketched. In section 3 it is provided an application to the system of equations suitable to describe thin walled elastic tubes filled with an incompressible fluid through the use of two suitable Lie groups of transformations [7,8,9,10]. Moreover, some considerations related to the linearization of the transformed autonomous system are pointed out.

2. Reduction to autonomous form

Systems like (1) are very often left invariant by one-parameter Lie groups of transformations [7,8,9,10] allowing to look for similarity solutions. In this context [6], we consider a general nonlinear system of partial differential equations of first order

$$Q_R(x, t, \mathbf{u}, \frac{\partial \mathbf{u}}{\partial x}, \frac{\partial \mathbf{u}}{\partial t}) = 0, \qquad R = 1, \ldots M, \qquad \mathbf{u} \in \mathbb{R}^N \tag{6}$$

and assume that it is left invariant by the following two one-parameter Lie groups of transformations having as infinitesimal operators

$$\zeta_1 = X_1(x, t, u_A) \frac{\partial}{\partial x} + T_1(x, t, u_A) \frac{\partial}{\partial t} + \sum_{A=1}^{N} U_{1A} \frac{\partial}{\partial u_A} \tag{7}$$

$$\zeta_2 = X_2(x,t,u_A)\frac{\partial}{\partial x} + T_2(x,t,u_A)\frac{\partial}{\partial t} + \sum_{A=1}^{N} U_{2A}\frac{\partial}{\partial u_A} \tag{8}$$

such that

$$[\zeta_1,\zeta_2] = \zeta_1\zeta_2 - \zeta_2\zeta_1 = 0. \tag{9}$$

Of course the invariance of the system (6) with respect to the groups characterized by (7)-(8) requires some compatibility conditions, involving the coefficients of the system as well as the infinitesimal generators of the groups, to hold. In [6] it has been shown that, under suitable conditions, the system (6), admitting the commuting operators (7)-(8), can be constructively transformed to autonomous form. Let us recall briefly the procedure.

First of all, we introduce "canonical variables" (with respect to the Lie group whose infinitesimal operator is ζ_1) τ, ξ, $\mathbf{v}$ characterized by

$$\zeta_1\tau = 1, \qquad \zeta_1\xi = 0, \qquad \zeta_1\mathbf{v} = 0 \tag{10}$$

whereupon the infinitesimal operator ζ_1, written in terms of the new variables, becomes

$$\hat{\zeta}_1 = \frac{\partial}{\partial \tau} \tag{11}$$

i.e., simply characterizes a translation in the variable τ.

Since the system (6) is left invariant by the Lie group whose infinitesimal operator is ζ_1, in the canonical variables it must assume the form:

$$\hat{Q}_R(\xi,\mathbf{v},\frac{\partial \mathbf{v}}{\partial \xi},\frac{\partial \mathbf{v}}{\partial \tau}) = 0. \tag{12}$$

Furthermore, it becomes possible to write the operator ζ_2 in terms of the new variables as follows

$$\hat{\zeta}_2 = (\zeta_2\tau)\frac{\partial}{\partial \tau} + (\zeta_2\xi)\frac{\partial}{\partial \xi} + \sum_{A=1}^{N}(\zeta_2 v_A)\frac{\partial}{\partial v_A} = \Theta\frac{\partial}{\partial \tau} + \Xi\frac{\partial}{\partial \xi} + \sum_{A=1}^{N}\Lambda_A\frac{\partial}{\partial v_A}. \tag{13}$$

By using the circumstance that the operators ζ_1 and ζ_2 commute, we have:

$$\begin{aligned} \zeta_1(\zeta_2\tau) &= \zeta_2(\zeta_1\tau) = 0 \\ \zeta_1(\zeta_2\xi) &= \zeta_2(\zeta_1\xi) = 0 \\ \zeta_1(\zeta_2\mathbf{v}) &= \zeta_2(\zeta_1\mathbf{v}) = 0; \end{aligned} \tag{14}$$

consequently, Θ, Ξ and Λ_A $(A = 1,\ldots,N)$ are left invariant by the Lie group corresponding to ζ_1, whereupon they may depend at most on the invariants of ζ_1, namely ξ and $\mathbf{v}$.

If $\Theta(\xi,\mathbf{v}) \neq 0$ we introduce

$$\theta = \tau + \Phi(\xi,\mathbf{v}) \tag{15}$$

with Φ satisfying the condition

$$\Theta + \Xi\frac{\partial \Phi}{\partial \xi} + \sum_{A=1}^{N}\Lambda_A\frac{\partial \Phi}{\partial v_A} = 0. \tag{16}$$

On the contrary, if $\Theta(\xi, \mathbf{v}) = 0$ then we may choose $\Phi = 0$ and $\theta = \tau$. In both cases we are able to write $\hat{\zeta}_2$ in the form:

$$\hat{\zeta}_2 = \Xi \frac{\partial}{\partial \xi} + \sum_{A=1}^{N} \Lambda_A \frac{\partial}{\partial v_A} \tag{17}$$

and we may introduce new canonical variables η, $\mathbf{w}$ (with respect to the Lie group having $\hat{\zeta}_2$ as infinitesimal operator) defined by:

$$\hat{\zeta}_2 \eta = 1, \qquad \hat{\zeta}_2 \mathbf{w} = 0 \tag{18}$$

by means of which the system (12) writes in autonomous form:

$$\tilde{Q}_R(\mathbf{w}, \frac{\partial \mathbf{w}}{\partial \eta}, \frac{\partial \mathbf{w}}{\partial \theta}) = 0. \tag{19}$$

This procedure has been successfully applied [6] to reduce systems like (1) to autonomous form in the case when they admit either the dilatation or the spiral group. Moreover, an example involving the projective group has been considered in [11] in connection with the system governing the axi-symmetric motion of a fluid in the presence of a magnetic field. In the next section we consider an example, connected with a system of equations suitable to model a thin walled elastic tube filled with an incompressible fluid, where more general and unusual groups of transformations are involved.

3. Fluid filled elastic tubes: unusual groups

Flows in large blood vessels can be described by considering thin walled elastic tubes filled with an incompressible fluid. The governing equations in one space dimensions [12] are in the form

$$\frac{\partial \mathbf{u}}{\partial t} + A(\mathbf{u}, x) \frac{\partial \mathbf{u}}{\partial x} = \mathbf{B}(\mathbf{u}, x, t) \tag{20}$$

with:

$$\mathbf{u} = \begin{bmatrix} p \\ v \end{bmatrix} \qquad A = \begin{bmatrix} v & S/S_p \\ k & v \end{bmatrix} \qquad \mathbf{B} = \begin{bmatrix} -(vS_x + \Psi)/S_p \\ f - kP_x \end{bmatrix} \tag{21}$$

where v is the velocity of the fluid in the x-direction, $k = 1/\rho$ with ρ the constant density of the fluid, S the cross-sectional area that, neglecting the viscosity, is of the form $S = S(p, x)$, p and $P(x,t)$ are the transmural pressure and the external pressure applied to the tube respectively, $\Psi(x, t, p, v)$ is the volume outflow function per unit length and $f(x, t, p, v)$ is the viscous retarding force; moreover, here and in the sequel the subscripts stand for the derivatives with respect to the indicated variables.

The full group analysis of this system has been carried out in [13]. Among the infinitesimal operators of the Lie group characterized in [13] it is possible to select the following two:

$$\zeta_1 = \frac{\gamma(t)}{\omega'(x)} \frac{\partial}{\partial x} + \frac{\gamma(t)}{\omega'(x)} [a(x)(p - p_0) + b(x)] \frac{\partial}{\partial p} + \frac{\dot{\gamma}(t) + a(x)\gamma(t)v/2}{\omega'(x)} \frac{\partial}{\partial v} \tag{22}$$

$$\zeta_2 = \gamma(t)\frac{\partial}{\partial t} + \frac{\dot\gamma(t)\omega(x)}{\omega'(x)}\frac{\partial}{\partial x} + \frac{\dot\gamma(t)\omega(x)}{\omega'(x)}[a(x)(p-p_0)+b(x)]\frac{\partial}{\partial p}$$
$$+ (\ddot\gamma(t) + \frac{a(x)\dot\gamma(t)v}{2})\frac{\omega(x)}{\omega'(x)}\frac{\partial}{\partial v} \qquad (23)$$

where p_0 is an arbitrary constant, $\gamma(t)$, $a(x)$ and $b(x)$ are arbitrary functions of their indicated arguments, $\omega(x) = \int \exp(-\int \frac{a(x)}{2}dx)dx$, and $' = \frac{d}{dx}$, $\dot{} = \frac{d}{dt}$.

It is rather simple to show that

$$[\zeta_1, \zeta_2] = 0. \qquad (24)$$

Of course the request of invariance of the system (20) with respect to the groups of transformations possessing the operators ζ_1 and ζ_2 requires some compatibility conditions to be satisfied by the coefficients of the system allowing for the determination of some classes of constitutive laws [13]. By imposing the invariance with respect to the Lie group corresponding to the infinitesimal operator ζ_1, the following conditions must be verified:

$$g_x + [a(x)(p-p_0)+b(x)]g_p = a(x)g \qquad (25)$$
$$\begin{aligned} 4\gamma(t)F_x &+ 4\gamma(t)[a(x)(p-p_0)+b(x)]F_p + [4\dot\gamma(t) + 2a(x)\gamma(t)v]F_v \\ &= 4a(x)\gamma(t)F - 2\gamma(t)v[a'(x)g + 2a'(x)(p-p_0) + 2b'(x)] \\ &- [2\dot\gamma(t) + a(x)\gamma(t)v][a(x)g + 2a(x)(p-p_0) + 2b(x)] \end{aligned} \qquad (26)$$
$$\begin{aligned} 4\gamma(t)H_x &+ 4\gamma(t)[a(x)(p-p_0)+b(x)]H_p + [4\dot\gamma(t) + 2a(x)\gamma(t)v]H_v \\ &= 2a(x)\gamma(t)H - 4k\gamma(t)[a'(x)(p-p_0) + b'(x)] \\ &- \gamma(t)v^2(2a'(x) + a^2(x)) - 4(\ddot\gamma(t) + a(x)\dot\gamma(t)v) \\ &- 2ka(x)\gamma(t)[a(x)(p-p_0)+b(x)] \end{aligned} \qquad (27)$$

where we set

$$g(p,x) = \frac{S(p,x)}{S_p(p,x)} \qquad (28)$$
$$F(x,t,p,v) = \frac{vS_x + \psi}{S_p} \qquad (29)$$
$$H(x,t,p,v) = -f(x,t,p,v) + kP_x(x,t). \qquad (30)$$

Once that the canonical variables associated with ζ_1 are introduced in the form

$$\begin{aligned} t &= t, \qquad \xi = \frac{\omega(x)}{\gamma(t)} \\ \Pi &= \omega'^2(x)(p-p_0) - \int b(x)\omega'^2(x)dx \\ V &= \omega'(x)v - \frac{\dot\gamma(t)}{\gamma(t)}\omega(x) \end{aligned} \qquad (31)$$

from the compatibility conditions (25)-(27) one obtains:

$$g(p,x) = \omega'^{-2}(x)g_0(\Pi) \qquad (32)$$

$$
\begin{aligned}
F(x,t,p,v) &= \omega'^{-2}(x)\Big(-\frac{1}{2}\Big(\frac{\dot\gamma(t)\omega(x)}{\gamma(t)\omega'(x)}+\frac{V}{\omega'(x)}\Big)\big(a(x)(g_0(\Pi)+2\Pi) \\
&+ 2a(x)\int b(x)\omega'^2(x)dx + 2b(x)\omega'^2(x)\big) + \hat F(t,\Pi,V)\Big) \qquad (33)\\
H(x,t,p,v) &= \omega'^{-1}(x)\Big(-ka(x)\omega'^{-1}(x)\big(\Pi+\int b(x)\omega'^2(x)dx\big) + kb(x)\omega'(x) \\
&+ \frac{\ddot\gamma(t)}{\gamma(t)}\omega(x) + \frac{a(x)\dot\gamma^2(t)\omega^2(x)}{2\gamma^2(t)\omega'(x)} \\
&+ \frac{a(x)\dot\gamma(t)\omega(x)}{\gamma(t)\omega'(x)}V + \frac{a(x)V^2}{2\omega'(x)} + \hat H(t,\Pi,V)\Big) \qquad (34)
\end{aligned}
$$

where $\hat F(t,\Pi,V)$ and $\hat H(t,\Pi,V)$ represent arbitrary function s of their arguments.
In the new variables (31) the system (20) results written in the form:

$$
\begin{aligned}
\Pi_t + \frac{1}{\gamma(t)}V\Pi_\xi + \frac{g_0(\Pi)}{\gamma(t)}V_\xi + \frac{\dot\gamma(t)}{\gamma(t)}g_0(\Pi) + \hat F(t,\Pi,V) &= 0 \qquad (35)\\
V_t + \frac{1}{\gamma(t)}VV_\xi + \frac{k}{\gamma(t)}\Pi_\xi + \frac{\dot\gamma(t)}{\gamma(t)}V + \hat H(t,\Pi,V) &= 0. \qquad (36)
\end{aligned}
$$

Moreover, in terms of the canonical variables (31) the infinitesimal operator ζ_2 reduces to

$$
\hat\zeta_2 = \gamma(t)\frac{\partial}{\partial t}; \qquad (37)
$$

thus, under the assumption that the system (35)-(36) admits the infinitesimal operator (37), it follows that $\hat F(t,\Pi,V)$ and $\hat H(t,\Pi,V)$ must be such that

$$
\begin{aligned}
\gamma(t)\frac{\partial\hat F}{\partial t} + \dot\gamma(t)\hat F + \ddot\gamma(t)g_0(\Pi) &= 0 \qquad (38)\\
\gamma(t)\frac{\partial\hat H}{\partial t} + \dot\gamma(t)\hat H + \ddot\gamma(t)V &= 0. \qquad (39)
\end{aligned}
$$

Consequently we have

$$
\begin{aligned}
\hat F(t,\Pi,V) &= -\frac{\dot\gamma(t)}{\gamma(t)}g_0(\Pi) + \frac{1}{\gamma(t)}\tilde F(\Pi,V) \qquad (40)\\
\hat H(t,\Pi,V) &= -\frac{\dot\gamma(t)}{\gamma(t)}V + \frac{1}{\gamma(t)}\tilde H(\Pi,V) \qquad (41)
\end{aligned}
$$

where $\tilde F(\Pi,V)$ and $\tilde H(\Pi,V)$ are arbitrary functions of their arguments.
By introducing the new variable

$$
\tau = \int\frac{1}{\gamma(t)}dt \qquad (42)
$$

the infinitesimal operator (37) reduces to

$$
\tilde\zeta_2 = \frac{\partial}{\partial\tau} \qquad (43)
$$

and the system (35)-(36) finally assumes the following autonomous form:

$$\begin{aligned} \Pi_\tau + V\Pi_\xi + g_0(\Pi)V_\xi + \tilde{F}(\Pi, V) &= 0 \qquad (44)\\ V_\tau + VV_\xi + k\Pi_\xi + \tilde{H}(\Pi, V) &= 0. \qquad (45) \end{aligned}$$

First of all we want to stress that to a constant solution $\Pi = \Pi_0$, $V = V_0$ satisfying the conditions

$$\tilde{F}(\Pi_0, V_0) = 0, \qquad \tilde{H}(\Pi_0, V_0) = 0 \qquad (46)$$

will immediately correspond a non-constant solution of the original system (20) given by:

$$\begin{aligned} p(x) &= p_0 + \omega'^{-2}(x)(\Pi_0 + \int b(x)\omega'^2(x)dx) \qquad (47)\\ v(x,t) &= \frac{\dot{\gamma}(t)\omega(x)}{\gamma(t)\omega'(x)} + \frac{V_0}{\omega'(x)}. \qquad (48) \end{aligned}$$

Other solutions to the system (20) can be obtained by looking for particular solutions of the autonomous system (44)-(45) (for instance in the form of travelling waves).

It is worthwhile to remark that the results found above can be applied to a particular model already described in literature. In fact, the system of equations (20) with the constitutive law

$$S(p, x) = \Omega_0(x)\exp(\frac{p - p_0}{\rho(c(p, x)c(p_0, x))}) \qquad (49)$$

where

$$c(p, x) = (\alpha_0 + \alpha_1 p)(1 + nx) \qquad (50)$$

α_0, α_1 and n being constants of suitable dimensions and $\Omega_0(x) = S(p_0, x)$ is a given function representing the area at $p = p_0$, has been considered in [14] to characterize a model of an aortic pathway. The constitutive law (49)-(50) belongs to the more general class considered above allowing the governing system to admit the infinitesimal operators ζ_1 and ζ_2. To ascertain this it is enough to take

$$\begin{aligned} a(x) &= \frac{-2n}{1 + nx}\\ b(x) &= a(x)(p_0 + \frac{\alpha_0}{\alpha_1}) \qquad (51)\\ g_0(\Pi) &= \rho\alpha_1^2\Pi^2. \end{aligned}$$

Consequently, by making use of the canonical variables defined by (31) and (42), we may characterize a non-constant solution in the form

$$\begin{aligned} p(x) &= -\frac{\alpha_0}{\alpha_1} + \frac{\Pi_0}{(1 + nx)^2} \qquad (52)\\ u(x,t) &= \frac{U_0}{1 + nx} + \frac{x + \frac{1}{2}nx^2}{1 + nx}\frac{\dot{\gamma}(t)}{\gamma(t)}. \qquad (53) \end{aligned}$$

The reduction to autonomous form can be considered, in the case of the 2×2 systems, as a necessary first step in order to apply a possible linearization procedure. To this purpose, let us note that if we require in the system (44)-(45)

$$\tilde{F}(\Pi, V) = \tilde{H}(\Pi, V) = 0 \tag{54}$$

for every Π and V, namely we restrict the constitutive laws which must be in the form (33)-(34 with

$$\tilde{F} = -\frac{\dot{\gamma}(t)}{\gamma(t)} g_0(\Pi), \qquad \tilde{H} = -\frac{\dot{\gamma}(t)}{\gamma(t)} V, \tag{55}$$

it results that the autonomous system is homogeneous and, by means of the well known hodograph transformation, can be linearized.

Another case leading to a linearization may be obtained by requiring

$$g_0(\Pi) = -2\Pi, \qquad \tilde{F} = 0, \qquad \tilde{H} = \frac{V}{k_0} \tag{56}$$

with k_0 positive constant.

In fact, with this assumption, the system (44)-(45) belongs to the linearizable class of first order quasilinear systems considered in [15] (see also [4,5] for other similar examples).

Under the conditions (56), the change of variables

$$\Pi = -\frac{1}{2z^2}, \qquad V = \frac{w}{z}, \qquad \eta = \xi + k_0 \frac{w}{z}, \qquad \hat{\tau} = \tau - k_0 \ln|z| \tag{57}$$

transforms the system (44)-(45) to the form:

$$z_{\hat{\tau}} + w_\eta = 0, \qquad w_{\hat{\tau}} + \frac{k}{z^2} z_\eta = -\frac{w}{k_0} \tag{58}$$

which in turn is equivalent to the nonlinear second order equation

$$(z_{\hat{\tau}} + \frac{z}{k_0})_{\hat{\tau}} + (\frac{k}{z})_{\eta\eta} = 0. \tag{59}$$

Finally, the latter equation, by means of the Bäcklund transformation [16]

$$\begin{aligned} d\eta^* &= k_1(k_0 z_{\hat{\tau}} + z)d\eta + \frac{k k_0 k_1}{z^2} z_\eta d\hat{\tau} \\ \tau^* &= \hat{\tau} + k_0 \ln|k_1 z| \\ z^* &= \frac{k k_0 k_1}{z} \end{aligned} \tag{60}$$

k_1 being another arbitrary constant, reduces to the linear telegraph equation

$$k_0 \frac{\partial^2 z^*}{\partial \tau^{*2}} + \frac{\partial z^*}{\partial \tau^*} - k_0 {k_1}^2 k \frac{\partial^2 z^*}{\partial \eta^{*2}} = 0 \tag{61}$$

It may be of some interest to observe that the condition $g_0(\Pi) = -2\Pi$ given in (56) is satisfied when the system (44)-(45) is completely exceptional, i.e., all the possible admissible discontinuity waves never evolve into shocks.

Finally, we remark that the reduction to autonomous form and the consequent linearization of the governing system considered in this section is obtained with constitutive laws which are more general than those considered in [4].

4. Conclusions

Axi-symmetric wave propagation problems in non-uniform regions are usually described by first order quasilinear systems which are nonautonomous, i.e., the coefficients depend explicitly on space and time variables. Very often it occurs that the governing systems result invariant with respect to infinitesimal Lie groups of transformations. Assuming that there exist at least two Lie groups with commuting infinitesimal operators, it is possible to introduce suitable canonical variables allowing to write the system of equations in autonomous form. In this way some special non-constant solutions, characterizing suitable states where one can consider for instance the propagation of weak discontinuities, may be easily found. The procedure is applied to a specific model of physical interest admitting two unusual infinitesimal operators. Constitutive laws allowing to linearize the model are settled by using a procedure applicable to autonomous systems: so the reduction to the autonomous form results to be essential.

Acknowledgement

This work was partially supported by C.N.R. under contract n. 90.04202.MZ78.

References

1. A. Donato, W. F. Ames, On the evolution of weak discontinuities in a state characterized by similarity solutions, Int. J. Non-linear Mech. **23**, 167–174, 1988.
2. N. Manganaro, F. Oliveri, Group analysis approach in magnetohydrodynamics: weak discontinuity propagation in a non constant state, Meccanica **24**, 71-78, 1989.
3. D. Fusco, A. Palumbo, Similarity solutions and model constitutive laws to viscoelasticity with applications to nonlinear wave propagation, ZAMP **40**, 78–92, 1989.
4. N. Manganaro, Linearization of nonautonomous models describing fluid filled elastic tubes and nonlinear elastic rods with variable cross-section, to appear in Acta Mechanica, 1993.
5. N. Manganaro, G. Valenti, Group analysis and linearization procedure for a nonautonomous model describing rate-type materials, to appear in J. Math. Phys., 1993.
6. A. Donato, F. Oliveri. When nonautonomous equations are equivalent to autonomous ones, to appear in Applicable Analysis, 1993.
7. L. V. Ovsiannikov, Group analysis of differential equations, Nauka, Moscow, 1978 (English Edition translated by W. F. Ames, Academic Press, New York, 1982).
8. W. F. Ames, Nonlinear partial differential equations in engineering, vol. 2, Academic Press, New York, 1976.
9. P. J. Olver, Applications of Lie groups to differential equations, Springer, New York, 1986.
10. G. W. Bluman, S. Kumei, Symmetries and differential equations, Springer, New York, 1989.
11. A. Donato, F. Oliveri, Reduction to autonomous form by group analysis and exact solutions of axi-symmetric MHD equations, to appear in "Similarity, symmetry and solutions of nonlinear boundary value problems", J. Math. Computer Mod., 1993.
12. T. J. Pedley, The fluid mechanics of large bloof vessels, Cambridge University Press, Cambridge, 1980.
13. D. Fusco, Group analysis and constitutive laws for fluid filled elastic tubes, Int. J. Non-linear Mech. **19**, 565–574, 1984.
14. M. Anliker, R. L. Rockwell, E. Ogden, Non-linear analysis of flow pulses and shock waves in arteries. I & II, ZAMP **22**, 217–246 & 563–581, 1971.
15. D. Fusco, N. Manganaro, A class of linearizable models and generation of material response functions to nonlinear hyperbolic heat conduction, J. Math. Phys. **32**, 3043–3046, 1991.
16. C. Rogers, T. Ruggeri, A reciprocal Bäcklund transformation: application to a nonlinear hyperbolic model in heat conduction, Lettere al Nuovo Cimento **2**, 289–296, 1985.

4 Conclusions

[illegible]

FINITE DIFFERENCE MODELS ENTIRELY INHERITING SYMMETRY OF ORIGINAL DIFFERENTIAL EQUATIONS

V. A. DORODNITSYN
Keldysh Institute of Applied Mathematics
Russian Academy of Sciences
Miusskaya Pl. 4,
125047, Moscow, Russia

Abstract. The present paper is concerned with the continuous groups of transformations in a space of discrete variables. The criteria of invariance of difference equations, difference grids and difference functionals are obtained. The finite-difference analogy of Noether's construction is developed.

The main idea of the work is to construct finite difference equations together with difference grids admitting the group which is isomorphic to the group of the original differential model. This leads to existence of exact invariant finite difference solutions, conservation laws for invariant variation problems and conservation of other useful properties of initial differential equations.

1. Transformation groups in the spaces of continuous and discrete variables

The nonlocal nature of mesh variables and operators demands the consideration of infinite dimensional spaces.

We denote by Z the space of sequences of continuous variables x, u, u_1, u_2, ... (where $x = (x^1, x^2, \dots, x^n)$ – independent variables, $u = (u^1, u^2, \dots, u^m)$ – dependent variables, u_1 ($u_i^k = \dfrac{\partial u^k}{\partial x^i}$, $i = 1, \dots, n$; $k = 1, \dots, m$) – partial derivatives of the first order, u_s are the derivatives of s order from u^k.

By z we mention an arbitrary finite number of coordinates of a vector x, u, u_1, u_2, ... and z^i is its i-th coordinate.

Let A be the space of the locally analytical functions Fz with a finite number of variables z. We denote by D_i the total differentiation in i-th direction:

$$D_i = \frac{\partial}{\partial x^i} + u_i^k \frac{\partial}{\partial u^k} + u_{ji}^k \frac{\partial}{\partial u_j^k} + \dots . \tag{1}$$

Let's regard the sequences of formal power series of one parameter a: $f^i(z,a) = \sum_{k=0}^{\infty} A_k^i(z) a^k$, $i = 1, 2, \dots$, where $A_k^i(z) \in A$, when $A_0^i \equiv z^i$.

We denote by $\widetilde{Z}$ the space of sequences of formal power series f^i:

$$(f^1(z,a), f^2(z,a), \dots, f^s(z,a), \dots),$$

and consider in $\widetilde{Z}$ transformations:

$$z^{i*} = f^i(z,a), \qquad i = 1, 2, \dots, \tag{2}$$

N. H. Ibragimov et al. (eds.),
Modern Group Analysis: Advanced Analytical and Computational Methods in Mathematical Physics, 191–201.

which transform the sequence z^i into the sequence z^{i*}. Transformations (2) form the group if [3, 5]:

$$z^{i*} = f^i(z,a) = e^{aX}(z^i) \equiv \sum_{s=0}^{\infty} \frac{a^s}{s!} X^{(s)}(z^i), \qquad i = 1, 2, \dots \tag{3}$$

where X is infinitesimal operator (generator) of the group:

$$X = \xi^i(z)\frac{\partial}{\partial x^i}, \quad i = 1, 2, \dots, \qquad \xi^i(z) = \left.\frac{\partial f^i(z,a)}{\partial a}\right|_{a=0}, \tag{4}$$

We fix an arbitrary value of a group parameter $a = h > 0$ and form the pair of operators with the help of tangent field (1). Such operators will be called *operators of discrete shift to the right* and *to the left* correspondingly:

$$\underset{+h}{S}_i = e^{+h^i D_i} \equiv \sum_{s=1}^{\infty} \frac{(h^i)^{s-1}}{s!} D_i^s, \qquad \underset{-h}{S}_i = e^{-h^i D_i} \equiv \sum_{s=1}^{\infty} \frac{-h^i)^{s-1}}{s!} D_i^s.$$

With the help of $\underset{+h}{S}$, $\underset{-h}{S}$ we construct the pair of operators of discrete (finite-difference) *differentiation to the right* and *to the left*:

$$\underset{+h}{D}_i = \frac{\underset{+h}{S}_i - 1}{h^i_+}, \qquad \underset{-h}{D}_i = \frac{1 - \underset{-h}{S}_i}{h^i_-}, \tag{5}$$

where $\{h^i_+\}$ is the set of right steps, $\{h^i_-\}$ is the set of left steps.

We introduce the formal power series of special type:

$$\underset{h}{u_1} = \underset{h}{u^k_i} = \frac{1}{h^i_+}(\underset{+h}{S}_i - 1)(u^k), \qquad i = 1, 2, \dots, n \tag{6}$$

$$\underset{h}{u_{\bar{1}}} = \underset{h}{u^k_{\bar{1}}} = \frac{1}{h^i_-}(1 - \underset{-h}{S})(u^k); \quad \underset{h}{u_2} = \underset{h}{u^k_{ij}} = \frac{1}{h^i_+}(\underset{+h}{S}_i = 1)(\underset{h}{u^k_j}), \quad \dots$$

where $\underset{h}{u_s}$ will be called the *finite-difference* (discrete, mesh) *derivative* of order s. In the odd case the formal series $\underset{h}{u_{2k+1}}$ we shall call the *right* difference derivative, $\underset{h}{u_{\overline{2k+1}}}$ – the *left* difference derivatives.

The set of points $x = x(\alpha) = \underset{+h}{S}^{\alpha}(x)$, $\alpha = \pm 1, \pm 2, \dots$ we denote by $\underset{h}{\omega}$ and it will be called the regular (uniform) difference grid. Hereafter, we employ the notation $\underset{h}{Z}$ for $(x, u, \underset{h}{u_1}, \underset{h}{u_2}, \underset{h}{u_3}, \dots, h)$ and it will be called *mesh* (grid) *space*.

We define the action of discrete shift $\underset{\pm h}{S}$ on functions from A: $\underset{\pm h}{S}(F(z)) = F(\underset{\pm h}{S}(z))$, and so this makes it possible to find the difference derivatives for $F \in A$ with a special finite-difference Leibniz rule ([5]):

$$\begin{aligned} \underset{+h}{D}(FG) &= \underset{+h}{D}(F)G + F\underset{+h}{D}(G) + h\underset{+h}{D}(F)\underset{+h}{D}(G), \\ \underset{-h}{D}(FG) &= \underset{-h}{D}(F)G + F\underset{-h}{D}(G) - h\underset{-h}{D}(F)\underset{-h}{D}(G). \end{aligned} \tag{7}$$

Let us consider a formal one-parameter group G_1 of transformations in $\widetilde{Z}$.

In order to apply the formal group G_1 to the finite-difference objects we need to select such transformations, *under which the sense of the finite-difference derivatives is conserved.*

First finite-difference derivative $\underset{h}{u_1}$ is defined by:

$$\underset{h}{u_1} h_+ = \underset{+h}{S}(u) - u \tag{8}$$

The "conservation of the sense" $\underset{h}{u_1}$ means that $\underset{h}{u_1^*}$ is defined by the same relation (8) in the space of new variables.

Let the following operator correspond to the group G_1:

$$X = \xi\frac{\partial}{\partial x} + \eta\frac{\partial}{\partial u} + \sum_{s\geq 1}\zeta_s\frac{\partial}{\partial u_s} + \sum_{l\geq 1}\underset{h}{\zeta_l}\frac{\partial}{\partial \underset{h}{u_l}} + (\xi(\underset{+h}{S}(z)) - \zeta(z))\frac{\partial}{\partial h_+} \tag{9}$$

where ξ, η, ζ_s – arbitrary functions from A and $\underset{h}{\zeta_l}$ are certain power series of h with arbitrary coefficients from A.

Theorem 1 ([5, 6]). Let G_1 be the formal one parameter group with the operator (9). Let the equation (8) be the invariant manifold of G_1 in every point of $\widetilde{Z}$. Then the following series of the relations for the coordinates of the operator (9) is true:

$$\zeta_s = D^s(\eta - \xi u_1) + \xi u_{s+1}, \qquad s = 1, 2, \ldots,$$

i.e., the formulas of Lie-Bäcklund groups ([2, 3]).

The invariance of (8) ensures the conservation of the sense of a first difference derivative under the transformations of G_1. It can be shown that the invariance of (8) is sufficient for the conservation of the sense of the second, third etc. derivatives [5, 6, 8].

The finite-difference equation can be written on the mesh $\underset{h}{\omega}$ in the form:

$$F(z) = 0, \qquad F \in \underset{h}{A} \tag{10}$$

where $\underset{h}{A}$ is the space of analytical functions of a finite number of variables from $\underset{h}{Z}$. This equation is written on the finite number of points (difference pattern, stencil) of difference mesh $\underset{h}{\omega}$ which is uniform or not.

Let the mesh be defined by the equation:

$$\Omega(z, h_+) = 0, \qquad \Omega \in \underset{h}{A} \tag{11}$$

In contrast to a continuous case the equation (11) should be included in the invariance condition in the case of the group transformations in $\underset{h}{Z}$, because the arbitrary grid $\underset{h}{\omega}$ need not to admit the given group.

Theorem 2 ([8]). Let a one-parameter group G_1 be given with the help of the operator (9). For the difference equation (10) admits the group G_1 on the mesh (11) it is necessary and sufficient that the following conditions be fulfilled:

$$XF(z)\big|_{(10,11)} = 0, \qquad X\Omega(z,h_+)\big|_{(10,11)}. \tag{12}$$

Uniform difference mesh $\underset{h}{\omega}$ is mainly widespread method of discretization of the space of independent variables. The transformations of a formal one-parameter group alternating x can change a mesh, i.e., it may become nonregular and this can influence on finite-difference equations on $\underset{h}{\omega}$. Thus we select the *class of admitted transformations conserving the mesh regularity*.

It is clear that when the independent variable is the invariant of G_1, then such class of transformations does not change the difference mesh and also the uniform one. But this condition is not necessary.

Theorem 2 ([6, 8]). For a mesh $\underset{h}{\omega}$ be regular under the transformations of a group G_1, i.e., $h_+^* = h_-^*$, it is necessary and sufficient that in each point of $\underset{h}{Z}$:

$$\underset{+h}{D}\underset{-h}{D}(\xi(z))\big|_{F(z)=0,h^+=h^-} = 0 \tag{13}$$

is true.

The most widespread symmetries satisfy to this criterion – translations, dilations, rotations, Lorentz transformations, etc. .

2. The method of finite-difference invariants

The conditions (12) can be used for invariance check of difference equations together with a grid. The following method ([6, 8]) gives the possibility to *construct* difference models according to a given differential model and its known symmetry.

Let we have the system of differential equations of the order k:

$$F_\alpha(z) = 0, \qquad \alpha = 1,\dots,m, \tag{14}$$

admitting a known group G_r.

We construct the system of difference equations

$$F_\alpha(z) = 0, \qquad \alpha = 1,\dots,m, \tag{15}$$

on a difference mesh:

$$\Omega_\beta(h_i^+, z) = 0, \qquad \beta = 1,\dots,n, \tag{16}$$

admitting the same group G_r in $\underset{h}{Z}$.

Because the system (14) admits the known group G_r, the full set of functional independent differential invariants of order k is known:

$$(J^1(z), J^2(z),\dots,J^\tau(z)) \qquad J^\alpha \in A.$$

Then system (14) can be written in the invariant form ([1]):

$$\Phi_\alpha(J^1(z), J^2(z), \dots, J^\tau(z)) \qquad \alpha = 1, \dots, m. \tag{17}$$

Group G_r in $\underset{h}{Z}$ has $(\tau + n)$ functionally independent difference invariants:

$$I^1(z), I^2(z), \dots, I^{\tau+n}(z).$$

The next step is constructing the set of (τ) invariants with the necessary approximation, i.e., for each invariant I^α in continuous space $\widetilde{Z}$ is true:

$$I^\alpha(z) J^\alpha(z) + O(h_i^k) \qquad \alpha = 1, \dots, m, \quad \tau = 1, \dots, n. \tag{18}$$

Finally we have:

$$\begin{gathered} \Phi_\alpha(I^1(z), I^2(z), \dots, I^\tau(z)) \\ \omega_\beta(I^1(z), I^2(z), \dots, I^{\tau+n}(z)) \end{gathered} \tag{19}$$

where ω_β is invariant representation of difference mesh (16). The choice of ω_β is defined by convenience only.

Because functions $\Phi_\alpha(I^1, I^2, \dots, I^\tau)$ are locally analytical of their arguments, then we get from (18) that difference equations $\Phi_\alpha = 0$ approximates the corresponding differential with the k-th order.

Example 1. Let us consider the equation of heat transfer with the relaxation of the heat flow:

$$\tau_0 u_{tt} + u_t = k_0 u_x u_{xx}, \tag{20}$$

where k_0, τ_0 are positive constants.

This equation admits the following operators [9]:

$$\begin{gathered} X_1 = \frac{\partial}{\partial t}; \quad X_2 = \frac{\partial}{\partial x}; \quad X_3 = \frac{\partial}{\partial u}; \\ X_4 = e^{-\frac{1}{\tau_0}} \frac{\partial}{\partial u}; \quad X_5 = x\frac{\partial}{\partial x} + 3u\frac{\partial}{\partial u}. \end{gathered} \tag{21}$$

All five operators above satisfy the conditions of invariant uniformity of a grid: $\underset{-h}{D}\underset{+h}{D}(\xi^i) = 0$, $\underset{-\tau}{D}\underset{+\tau}{D}(\xi^i) = 0$, $i = 1, 2, \dots, 5$, so we can use the quadrilateral mesh $\underset{\tau}{\omega} \times \underset{h}{\omega}$ regular in each direction.

The full set of differential invariants for (21) consists of two invariants:

$$J_1 = \frac{\tau_0 u_{tt} + u_t}{u_x^{3/2}}; \qquad J_2 = \frac{u_{xx}}{u_x^{1/2}}; \tag{22}$$

With help of invariants (22) we can write the equation (20) in the invariant form: $J_1 = k_0 J_2$.

Using the prolongation formulas [5,6,8] for operators (21) in $\underset{h}{Z}$ one can obtain the isomorphic representation of the group G_1.

Then in the subspace $(x, t, u, \underset{\tau}{u_t}, \underset{\tau}{u_{\bar{t}}}, \underset{h}{u_x}, \underset{h}{u_{\bar{x}}}, \underset{\tau}{u_{t\bar{t}}}, \underset{h}{u_{x\bar{x}}}, \tau, h)$, we may choose only two invariants which approximate (22) with the order $O(\tau^2 + h^2)$. It can be shown that the following forms of the second order:

$$
\begin{gathered}
I_1 = \left(\tfrac{1}{2}\underset{h}{u_x} + \tfrac{1}{2}\underset{h}{u_{\bar{x}}}\right)^{-3/2} \left\{\tau_0 \underset{\tau}{u_{t\bar{t}}} + \tfrac{\tau_0}{2\tau}\left[\underset{\tau}{u_t}\left(e^{\frac{\tau}{\tau_0}} - 1\right) + \underset{\tau}{u_{\bar{t}}}\left(1 - e^{-\frac{\tau}{\tau_0}}\right)\right]\right\}, \\
I_2 = \underset{h}{u_{x\bar{x}}}\left(\tfrac{1}{2}\underset{h}{u_x} + \tfrac{1}{2}\underset{h}{u_{\bar{x}}}\right)^{-1/2},
\end{gathered}
\tag{23}
$$

are invariants for all operators (21) and approximate J_1, J_2 with the order $O(\tau^2+h^2)$.

Substituting them in invariant representation $J_1 = k_0 J_2$ one can obtain: $I_1 = k_0 I_2$ or in the equivalent form

$$
\begin{gathered}
\tau_0 \underset{\tau}{u_{t\bar{t}}} + \frac{1}{2}\underset{\tau}{u_{\bar{t}}}\left\{\frac{\tau_0}{\tau}\left(1 - e^{-\frac{\tau}{\tau_0}}\right)\right\} + \frac{1}{2}\underset{\tau}{u_t}\left\{\frac{\tau_0}{\tau}\left(e^{\frac{\tau}{\tau_0}} - 1\right)\right\} = \\
k_0 \underset{h}{u_{x\bar{x}}}\left(\frac{1}{2}\underset{h}{u_x} + \frac{1}{2}\underset{h}{u_{\bar{x}}}\right).
\end{gathered}
\tag{24}
$$

The finite-difference equation (24) admits the all set of operators (21) and approximates the differential equation (20) with the order $(\tau^2 + h^2)$ on a regular mesh $\underset{\tau}{\omega} \times \underset{h}{\omega}$.

Example 2. The heat equation with a special type of nonlinear heat conductivity:

$$
u_t = (u^{-4/3}u_x)_x, \tag{25}
$$

possesses the widest symmetry group [10]:

$$
\begin{gathered}
X_1 = \frac{\partial}{\partial t}; \quad X_2 = \frac{\partial}{\partial x}; \quad X_3 = 2t\frac{\partial}{\partial t} + x\frac{\partial}{\partial x}; \\
X_4 = 2x\frac{\partial}{\partial x} - 3u\frac{\partial}{\partial u}; \quad X_5 = x^2\frac{\partial}{\partial x} - 3xu\frac{\partial}{\partial u}.
\end{gathered}
\tag{26}
$$

The equation (25) can be written in the following invariant form:

$$
\frac{(u^{-1/3})_{xx}}{u_t} = -\frac{1}{3}. \tag{27}
$$

The operator X_5 destroys the regularity of a uniform grid in x-direction and conserves the regularity in t-direction.

Starting from the full set of finite-difference invariants of operators (26), prolonged into grid space $\underset{h}{Z}$, one can construct the following difference equation:

$$
\underset{\tau}{u_t} = -\frac{3}{h}\left(\frac{\hat{u}^{-1/3} - u^{-1/3}}{h^+} - \frac{u^{-1/3} - \check{u}^{-1/3}}{h^-}\right), \tag{28}
$$

where $h = \frac{2h^+h^-}{h^+ + h^-}$.

The equation (28) admits the whole set of operators (2) on the following invariant nonregular grid:

$$h^+ = \varepsilon_0 \frac{\sqrt{\tau}}{\sqrt[3]{u\hat{u}}}; \qquad h^- = \varepsilon_0 \frac{\sqrt{\tau}}{\sqrt[3]{u\check{u}}}; \tag{29}$$

where $\varepsilon_0 = \text{const}$, $\tau = \text{const}$ is a regular time step of a grid. Equations (28),(29) approximate the initial one (25) with the order $O(\varepsilon_0^2)$ and yield the explicit finite-difference scheme of a divergent type.

Equations (28),(29) are equivalent to the following mapping:

$$\underset{+\tau}{S}(u) = u + \frac{3u^{1/3}}{2\varepsilon_0^2}(\hat{u}^{1/3} + \check{u}^{1/3})(\hat{u}^{1/3} - 2u^{1/3} + \check{u}^{1/3}), \tag{30}$$

where $\underset{+\tau}{S}$ is the shift operator in t-direction.

3. The finite-difference analogy of Noether's construction

It is well known ([1, 2, 3]) that the Euler operator in the "continuous" space $\widetilde{Z}$ in *I-D* case can be written in the following form:

$$\frac{\delta}{\delta u} = \frac{\partial}{\partial u} + \sum_{s=1}^{\infty}(-1)^s D^s(\frac{\partial}{\partial u_s}. \tag{31}$$

Let us represent the Euler's operator in $\underset{h}{Z}$ on a regular mesh. We suppose that it acts upon functions $\mathcal{L}(x, u, \underset{h}{u_1}) \in \underset{h}{A}$, which are determined on a mesh $\underset{h}{\omega}$. It can be shown [6, 7, 8] that the operator (31) can be represented in $\underset{h}{Z}$ in such a form ([6, 7, 8])

$$\frac{\delta}{\delta u} = \frac{\partial}{\partial u} \underset{-h}{D}\left(\frac{\partial}{\partial \underset{h}{u_1}}\right), \tag{32}$$

where $\underset{-h}{D} = \sum_{s\geq 1} \frac{(-h)^{s-1}}{s!} D^s -$ is the operator of the left difference differentiation.

In general many dimensional case $\mathcal{L} = \mathcal{L}(x^i, u, \underset{h}{u_1}, \underset{h}{u_{\bar{1}}}, \underset{h}{u_2}, \underset{h}{u_{\bar{2}}}, \ldots)$, Euler's equation can be written in the following form [7, 8]:

$$\frac{\partial \mathcal{L}}{\partial u} = \frac{h_i^-}{h_i^+} \underset{-h}{D_i}\left(\frac{\partial \mathcal{L}}{\partial \underset{h}{u_i}}\right) - \frac{h_i^+}{h_i^-} \underset{+h}{D_i}\left(\frac{\partial \mathcal{L}}{\partial \underset{h}{u_{\bar{i}}}}\right), \qquad i = 1, 2, \ldots . \tag{33}$$

Let a finite-difference functional be given on a one-dimensional mesh $\underset{h}{\omega}$ in the following form:

$$L = \sum_{\underset{h}{\Omega}} \mathcal{L}(x, u, \underset{h}{u_1}) h^+. \tag{34}$$

Let a one-parameter group G_1 be given in the $\underset{h}{Z}$ with the operator (9). The mesh functional L will be invariant with respect to a group G_1 on a *regular* grid iff

[6, 7, 8]:

$$\xi\frac{\partial\mathcal{L}}{\partial x}+\eta\frac{\partial\mathcal{L}}{\partial u}+\left[\underset{-h}{D}(\eta)-\underset{h}{u_1}\underset{+h}{D}(\xi)\right]\frac{\partial\mathcal{L}}{\partial \underset{h}{u_1}}+\mathcal{L}\underset{+h}{D}(\xi)=0,\qquad \underset{-h}{D}\underset{+h}{D}(\xi)=0.$$

This necessary and sufficient conditions of invariance of the difference functional can easy be generalized for nonregular invariant grids and for many dimensional case [7, 8].

In contrast the continuous case [4], the invariance of a difference functional does not lead to the invariance of difference Euler's equation.

For simplicity we shall consider the one-dimensional case of a regular mesh.

Theorem 4 ([7, 8]). The Euler's equation (32) of invariant functional (34) is invariant on the uniform mesh $\underset{h}{\omega}$ iff:

$$\xi_u\left(\frac{\partial\mathcal{L}}{\partial x}+\underset{h}{u_{\bar{1}}}\frac{\partial\mathcal{L}}{\partial u}+\underset{h}{u_{1\bar{1}}}\frac{\partial\mathcal{L}}{\partial u_1}-\underset{-h}{D}\mathcal{L}\right)=0,\qquad \underset{-h+h}{D\,D}(\xi)=0. \tag{35}$$

It can be shown ([7, 8]) that a constant value of discrete functional under an arbitrary group transformation is reached on the solutions of the equation (for regular grids):

$$\xi\left(\frac{\partial\mathcal{L}}{\partial x}+\underset{-h}{D}\left(\underset{h}{u_1}\frac{\partial\mathcal{L}}{\partial \underset{h}{u_1}}-\mathcal{L}\right)\right)+\eta\left(\frac{\partial\mathcal{L}}{\partial u}-\underset{-h}{D}\left(\frac{\partial\mathcal{L}}{\partial \underset{h}{u_1}}\right)\right)=0 \tag{36}$$

The equation (36) was called ([7, 8]) the quasi-extremal equation and its each solutions – *quasi-extremals*.

Notice that the coefficients $\xi(x,u)$ and $\eta(x,u)$ are in the quasi-extreme equation (36). Therefore the equations for quasi-extremals for one and the same invariant functional for different groups will have different form.

So, generally speaking, in the difference variational problems the invariance of the mesh functionals does not mean the invariance of the Euler's equations. The set of solutions where a difference functional has it's stationary value depends on the group orbit. If $\xi\neq 0$, i.e., the group transformations change independent variable (and a difference mesh), then the stationarity of the action of a group is reached on the quasi-extremals, not on the extremals.

The mesh functional (34), which satisfies on invariant-uniform mesh to the condition

$$(\eta-\xi\underset{h}{u_i})\frac{\partial\mathcal{L}}{\partial u}+\underset{+h}{D}(\eta-\xi\underset{h}{u_{\bar{1}}})\frac{\partial\mathcal{L}}{\partial \underset{h}{u_1}}=\underset{+h}{D}(B(z)), \tag{37}$$

with any $B(z)\in \underset{h}{A}$, will be called *quasi-invariant*.

If a variational functional (34) satisfies to (37) with some function $B(z)\in \underset{h}{A}$ on the invariant uniform mesh $\underset{h}{\omega}$, then the Euler equation admits the operator (9) ([7, 8]).

The following operator identity is true ([7, 8]) for the one-dimensional uniform difference meshes:

$$\xi\frac{\partial\mathcal{L}}{\partial x}+\eta\frac{\partial\mathcal{L}}{\partial u}+\left[\underset{+h}{D}(\eta)\underset{h}{u_1}\underset{+h}{D}(\xi)\right]\frac{\partial\mathcal{L}}{\partial\underset{h}{u_1}}+\mathcal{L}\underset{+h}{D}(\xi)\equiv$$
$$\equiv\xi\left(\frac{\partial\mathcal{L}}{\partial x}+\underset{h}{u_{\bar{1}}}\frac{\partial\mathcal{L}}{\partial u}+\underset{h}{u_{1\bar{1}}}\frac{\partial\mathcal{L}}{\partial\underset{h}{u_1}}-\underset{-h}{D}(\mathcal{L})\right)+ \tag{38}$$
$$+(\eta-\xi\underset{h}{u_{\bar{1}}})\left[\frac{\partial\mathcal{L}}{\partial u}\underset{-h}{D}\left(\frac{\partial\mathcal{L}}{\partial\underset{h}{u_1}}\right)\right]+\underset{+h}{D}\left(\xi\check{\mathcal{L}}+(\eta-\xi\underset{h}{u_{\bar{1}}})\left(\frac{\partial\check{\mathcal{L}}}{\partial\underset{h}{u_1}}\right)\right),$$

where $\check{f}=\underset{-h}{S}(f)$.

For many dimensional case there are similar identities [7, 8].

Identity (38) can be used for constructing conservation laws for quasi-extremals of invariant difference functionals.

Example. On the regular grid $\underset{h}{\omega}$ we consider difference functional with lagrangian $\mathcal{L}=-u^{-2}-\frac{1}{2}\underset{h}{u_{\bar{x}}^2}$ which is invariant with respect to operators $X_1=\frac{\partial}{\partial x}$ and $X_2=2x\frac{\partial}{\partial x}+u\frac{\partial}{\partial u}$. Quasi-extremals corresponding to X_1 and X_2 defined by

$$\underset{h}{u_{x\bar{x}}}=\frac{\underset{h}{u_{\bar{x}}}}{\underset{h}{u_x}+\underset{h}{u_{\bar{x}}}}\frac{u+\check{u}}{u^2+\check{u}^2},\qquad\text{where }\check{u}=\underset{-h}{S}(u) \tag{39}$$

$$\underset{h}{u_{x\bar{x}}}=\frac{1}{u^3}+\frac{x}{u}\left(\underset{h}{u_{x\bar{x}}}(\underset{h}{u_x}+\underset{h}{u_{\bar{x}}})-\underset{h}{u_{\bar{x}}}\frac{u+\check{u}}{u^2\check{u}^2}\right) \tag{40}$$

The identity (38) yields the following conservation laws:

$$\underset{+h}{D}(\check{u}^{-2}+\underset{h}{u_{\bar{x}}^2})\big|_{(39)}=0,\qquad\underset{+h}{D}(x\check{u}^{-2}+x\underset{h}{u_{\bar{x}}^2}-u\underset{h}{u_{\bar{x}}})\big|_{(40)}=0.$$

4. The difference analogy of Noether's theorem

For many-dimensional case the following identity is true ([7, 8]):

$$\bar{\eta}^k\frac{\partial\mathcal{L}}{\partial u^k}+\underset{+h}{D_i}(\bar{\eta}^k)\frac{\partial\mathcal{L}}{\partial\underset{h^i}{u_i^k}}+\underset{-h}{D_i}(\bar{\eta}^k)\frac{\partial\mathcal{L}}{\partial\underset{h^i}{u_{\bar{i}}^k}}+\underset{+h}{D_i}(B^i)\equiv$$
$$\equiv\bar{\eta}^k\left(\frac{\partial\mathcal{L}}{\partial u^k}-\frac{h_-^i}{h_+^i}\underset{-h}{D_i}\left(\frac{\partial\mathcal{L}}{\partial\underset{h^i}{u_i^k}}\right)-\frac{h_+^i}{h_-^i}\underset{+h}{D_i}\left(\frac{\partial\mathcal{L}}{\partial\underset{h^i}{u_{\bar{i}}^k}}\right)\right) \tag{41}$$
$$+\underset{+h}{D_i}\left(B^i+\bar{\eta}^k\underset{-h}{S_i}\left(\frac{\partial\mathcal{L}}{\partial\underset{h^i}{u_i^k}}\right)+\underset{-h}{S_i}(\bar{\eta}^k)\frac{\partial\mathcal{L}}{\partial\underset{h^i}{u_{\bar{i}}^k}}\right)$$

where $\bar{\eta}^k=(\eta^k-\xi^\alpha u_{\bar{\alpha}}^k)$, $i=1,2,\ldots$.

Identity (41) makes it possible to state the difference analogy of Noether's theorem.

Theorem 5. Let the difference equation (or system)

$$F(z) = 0, \tag{42}$$

be given on a difference mesh $\underset{h}{\omega}$

$$\Omega(z, h) = 0 \tag{43}$$

Let (42) be the Euler equations of the difference functional with Lagrangian $\mathcal{L}$ on the mesh (43) and let (42), (43) define the invariant manifold with respect to the group G_1 with operators (9). Then the quasi-invariance of the functional (34) on the solutions of the Euler equation (42) on the mesh (43) with some functions $B^i(z)$, $i = 1, 2, \ldots$, is the necessary and sufficient condition for the existence of conservation law on the mesh (43) with the vector:

$$A_i = \xi^i \underset{-h}{S}{}_i(\mathcal{L}) + (\eta^k - \xi^\alpha u^k_{\bar{\alpha}}) \underset{-h}{S} \left(\frac{\partial \mathcal{L}}{\partial \underset{h}{u_i}} \right), \qquad i = 1, 2 \ldots .$$

Example. The finite-difference wave equation:

$$\underset{h}{u}_{1\bar{1}} - \underset{h}{u}_{2\bar{2}} = 0, \tag{44}$$

admits the same group that in continuous case, in particular:

$$X_1 = \frac{\partial}{\partial t}; \quad X_2 = \frac{\partial}{\partial x}; \quad X_3 = t\frac{\partial}{\partial x} + x\frac{\partial}{\partial t}.$$

Equation (44) is the extremal equation for Lagrangian:

$$L = \frac{1}{2}\underset{h}{u}_2^2 - \frac{1}{2}\underset{h}{u}_1^2.$$

The following coordinates of vector B and functions $\bar{\eta}$ provides the quasi-invariance conditions for difference functional with above L and conservation laws corresponding to X_1, X_2, X_3:

$$\alpha = 1\colon\ \bar{\eta}_1 = -\underset{h}{u}_1 - \underset{h}{u}_{\bar{1}}, \qquad B^1_1 = 0, \qquad B^2_1 = \underset{h}{u}_2 \underset{-h}{S}{}_1(\underset{h}{u}_2) - \underset{h}{u}_1 \underset{h}{u}_{\bar{1}},$$

$$\underset{+h}{D}{}_1(\underset{h}{u}_{\bar{1}}^2 + \underset{h}{u}_2 \underset{-h}{S}{}_{(}\underset{h}{u}_2)) + \underset{+h}{D}{}_2(-\underset{h}{u}_{\bar{2}}(\underset{h}{u}_1 + \underset{h}{u}_{\bar{1}}))\big|_{(44)} = 0;$$

$$\alpha = 2\colon\ \bar{\eta}_2 = -\underset{h}{u}_2 - \underset{h}{u}_{\bar{2}}, \qquad B^1_2 = \underset{h}{u}_2 \underset{h}{u}_{\bar{2}} - \underset{h}{u}_1 \underset{-h}{S}{}_2(\underset{h}{u}_1) \qquad B^2_2 = 0,$$

$$\underset{+h}{D}{}_1(\underset{h}{u}_1(\underset{h}{u}_2 + \underset{h}{u}_2)) + \underset{+h}{D}{}_2(\underset{h}{u}_{\bar{2}}^2 + \underset{h}{u}_1 \underset{-h}{S}{}_2(\underset{h}{u}_1))\big|_{(44)} = 0;$$

$$\alpha = 3\colon\ \bar{\eta}_3 = -t(\underset{h}{u}_2 + \underset{h}{u}_{\bar{2}}) - x(\underset{h}{u}_1 + \underset{h}{u}_{\bar{1}}),$$

$$B_3^1 = t\underset{h}{u}_2\underset{h}{u}_{\bar{2}} - (t+\tau)\underset{h}{u}_1 \underset{-h}{S}_2(\underset{h}{u}_1) - \underset{h}{u}_1 \underset{-h}{S}_2(u),$$

$$B_3^2 = -x\underset{h}{u}_1\underset{h}{u}_{\bar{1}} + (x+h)\underset{h}{u}_2 \underset{-h}{S}_1(\underset{h}{u}_2) + \underset{h}{u}_2 \underset{-h}{S}_1(u),$$

$$\underset{+h}{D}_1(x\underset{h}{u}_{\bar{1}}^2 + (x+h)\underset{h}{u}_2 \underset{-h}{S}_1(\underset{h}{u}_2) + t\underset{h}{u}_{\bar{1}}(\underset{h}{u}_2 + \underset{h}{u}_{\bar{2}}) + \underset{-h}{S}_1(u)\underset{h}{u}_2) +$$

$$+\underset{+h}{D}_2(-t\underset{h}{u}_{\bar{2}}^2 - (t+\tau)\underset{h}{u}_1 \underset{-h}{S}_2(\underset{h}{u}_1) - x\underset{h}{u}_{\bar{2}}(\underset{h}{u}_1\underset{h}{u}_{\bar{1}}) - \underset{-h}{S}_2(u)\underset{h}{u}_1)\Big|_{(44)} = 0.$$

References

1. Ovsyannikov, L.V., Group Analysis of Differential Equations, Academic Press, New York, 1982.
2. Ibragimov, N.H., Transformation Groups Applied to Mathematical Physics, Reidel, Dordrecht, 1985.
3. Olver, P.J., Application of Lie Groups to Differential Equations, Springer-Verlag, New York, 1986.
4. Noether, E., Invariante Variationsprobleme, Nachr. Konig. Gesell. Wissen. Göttingen, Math.-Phys. Kl., 1918, pp. 235–257.
5. Dorodnitsyn, V.A., Taylor's group and transformations conserving finite-difference derivatives, Keldysh Inst. Appl. Mathem., Preprint N. 67, 1987, pp. 1–31, (in Russian).
6. Dorodnitsyn, V.A., Transformation groups in mesh spaces, Jour. of Sov. Mathematics, vol. 55, N. 1, pp. 1490–1517, (June 1991), Plenum Publ. Corporation.
7. Dorodnitsyn, V.A., Finite-difference analogy of Noether's theorem, Doklady AN, 1993, vol. 328, N. 6, pp. 676–679 (in Russian).
8. Dorodnitsyn, V.A., Symmetry of Finite-Difference Equations, CRC Handbook of Lie Group Analysis of Differential Equations, Volume I: Symmetries, Exact Solutions and Conservation Laws, – CRC Press, 1993, (to appear).
9. Svirshchevskii, S.R., Group properties of the heat transfer model accounting heat flux relaxation, Keldysh Inst. of Appl. Math., Preprint N. 105, 1988, (In Russian).
10. Ovsyannikov, L.V., Group properties of a nonlinear heat equation, Doklady AN SSSR, 1959, 125, N. 3, pp. 492–495, (in Russian).

BOUNDARY CONDITION INVARIANCE

M. J. ENGLEFIELD
Department of Mathematics
Monash University
Clayton, Victoria 3168
AUSTRALIA

1. Introduction

Suppose that a partial differential equation is reduced to an ordinary differential equation by using new variables which are invariants of some subalgebra of the symmetry algebra of the equation. The number of possible reductions depends on the dimension d of the symmetry algebra $\mathcal{S}$. A group classification problem considers a family of equations of the same form, differing in the choice of some functions in the form, and determines how d depends on this choice of functions. For example, the diffusion-convection equation

$$\frac{\partial u}{\partial t} = \frac{\partial}{\partial x}\left[F(u)\frac{\partial u}{\partial x}\right] + G(u)\frac{\partial u}{\partial x} \tag{1}$$

admits, for arbitrary F and G, the 2-dimensional algebra generating translations in x and t; but there is a 3-dimensional symmetry algebra for the cases [2, 4]

$$F(u) = au^p, \qquad G(u) = cu^b \tag{2}$$

and

$$F(u) = ae^{pu}, \qquad G(u) = ce^{bu}. \tag{3}$$

If equation (1) is used to describe some physical problem, then the method of group-invariant solutions is more powerful if F and G can be chosen according to (2) or (3), since more different reductions of (1) to an ordinary differential equation will be available.

This paper discusses the extension of this idea to boundary conditions. Just as (2) and (3) give the cases of (1) that allow an extra symmetry, boundary conditions that admit some subalgebra of $\mathcal{S}$ may be determined and listed. The actual physical boundary conditions may then be modelled so that some symmetry is admitted. Reduction using this symmetry allows the model boundary conditions to be applied to the resulting ordinary differential equation. This extension of the group classification problem was used by Coggleshall and Axford [1] when considering group-invariant solutions of the radiation hydrodynamic equations.

N. H. Ibragimov et al. (eds.),
Modern Group Analysis: Advanced Analytical and Computational Methods in Mathematical Physics, 203–208.

2. Invariance conditions

If a group-invariant solution $u = S(x,t)$ is determined by a boundary condition on the curve $x = f(t)$ then this solution represents a part of the surface $u = S(x,t)$ bounded by the space (supporting) curve

$$x = f(t), \qquad u = S(f(t),t) = M(t). \tag{4}$$

This condition can be applied to the ordinary differential equation obtained by reducing a partial differential equation for u provided the curve (4) is an invariant curve of the subalgebra used for the reduction. If the generator is

$$V = \xi[x,t,u]\frac{\partial}{\partial x} + \tau[x,t,u]\frac{\partial}{\partial t} + \eta[x,t,u]\frac{\partial}{\partial u} \tag{5}$$

invariance of (4) requires $V\{x - f(t)\} = V\{u - M(t)\} = 0$ on the curve, i.e.

$$\xi[f(t),t,M(t)] = \tau[f(t),t,M(t)]f'(t) \tag{6}$$
$$\eta[f(t),t,M(t)] = \tau[f(t),t,M(t)]M'(t) \tag{7}$$

Equations (6) and (7) are linear boundary conditions on determining equations for ξ, τ and η; they are also differential equations for the functions f and M for which (4) is an invariant boundary condition; and they express the fact that the vector with components (ξ,τ,η) is tangent to the invariant curve (4). Thus (6) and (7) also result from substituting (4) into

$$\frac{dx}{\xi} = \frac{dt}{\tau} = \frac{du}{\eta} \tag{8}$$

The above work does not treat x and t on an equal footing, and does not include a boundary $t = t_0$. To avoid this it may be convenient to use a parametric representation

$$x = X(\theta), \qquad t = T(\theta), \qquad u = U(\theta) \tag{9}$$

of (4); substituting into (8) gives

$$X'(\theta)/\xi(\theta) = T'(\theta)/\tau(\theta) = U'(\theta)/\eta(\theta) \tag{10}$$

where $\xi(\theta) = \xi[X,T,U]$ etc. Equation (10) may be integrated to obtain $X(\theta)$, $T(\theta)$ and $U(\theta)$.

Since equations (8) are those solved for the invariant functions that define the new variables which allow a reduction of the equation, the required conditions (4) may also be written down after determining the invariant functions. In particular suppose $\partial\xi/\partial u = \partial\tau/\partial u = 0$, and the invariant functions are $\zeta(x,t)$ and $R(x,t,u)$ satisfying $V\zeta = VR = 0$. Then the possible invariant boundaries are $\zeta(x,t) = \zeta_B$, a reducing substitution is $R(x,t,u) = W(\zeta)$, and the invariant conditions have the form $R(x,t,u) = \text{constant} = W(\zeta_B)$.

As an example, consider (1) and (3) with $p = b$. Then the 3-dimensional symmetry algebra of the equation has the generators (5) with

$$\xi = c_1, \qquad \tau = b(c_2 - \gamma t), \qquad \eta = \gamma \tag{11}$$

Solving (6) gives the possible invariant boundaries and conditions

$$x = f(t) = K + c_1 t / b c_2, \qquad u = M(t) = K' \tag{12}$$

corresponding to $\gamma = 0$, and

$$x = K - \frac{c_1}{b} \log |t - c_2|, \qquad u = K' - \frac{1}{b} \log |t - c_2| \tag{13}$$

corresponding to $\gamma = 1$. The parametric form (9) of (13) is

$$x = K_1 + c_1 \theta, \qquad t = c_2 + K_2 e^{-b\theta}, \qquad u = K_3 + \theta. \tag{14}$$

The constants in the model boundaries given in (13) and (14) may now be chosen to give the best realization of the actual physical problem. The c_i determine the subalgebra to be used for reduction, whereas the same reduction applies for all values of the constants denoted by K_i. The different values of K_i give all the orbits of the subalgebra, and these orbits are possible invariant (supporting) curves.

Next consider the use of invariant functions, taking the case $\gamma = 1$. Solving (8) gives the invariants

$$\zeta = (t - c_2) e^{bx/c_1}, \qquad R = u - x/c_1 \tag{15}$$

if $c_1 \neq 0$, and otherwise

$$\zeta = x, \qquad R = u + \frac{1}{b} \log |t - c_2| \tag{16}$$

The substitution $R = W(\zeta)$ reduces (1) to an ordinary differential equation for $W(\zeta)$. The invariant boundaries are the curves $\zeta = \zeta_B$, and the invariant conditions on these curves have the form $R = K$. The boundary condition $W(\zeta_B) = K$ is applied to the ordinary differential equation.

3. Conditions on derivatives

Equations (4)–(10) may be extended to derivatives by considering a prolongation space. Various extensions of (6) and (7) have been given elsewhere [3]. Here the example (15) will be used to illustrate the derivation of invariant conditions from the criterion that they must be equivalent to $W'(\zeta_B) = $ (constant).

The reducing substitution $u = x/c_1 + W(\zeta)$ corresponding to (15) gives

$$u_x = 1/c_1 + W'(\zeta)(b\zeta/c_1), \qquad u_t = W'(\zeta) e^{bx/c_1}. \tag{17}$$

Hence invariant derivative conditions on the boundary $\zeta = \zeta_B$ have the form $u_x = K_4$ or $u_t = K_5 e^{bx/c_1}$; using ζ the latter could also be written $u_t = K_6/(t - c_2)$.

4. Tangential and Normal Derivatives

Consider again an invariant solution representing a surface $u = S(x, t)$ bounded by an invariant curve to which the vector with components (ξ, τ, η) is tangent. This vector shows that the slope of the surface in the (tangential) direction of the boundary curve is

$$u_T = \eta / (\xi^2 + \tau^2)^{1/2}. \tag{18}$$

This also follows from the invariant surface condition

$$\xi u_x + \tau u_t = \eta \tag{19}$$

because if $\mathbf{e}_T$ is a unit vector in the tangent direction to the boundary curve in the x–t plane, then

$$u_T = \mathbf{e}_T \cdot \nabla u = (\xi u_x + \tau u_t)/(\xi^2 + \tau^2)^{1/2}. \tag{20}$$

The result (18) expresses the fact that u_T is determined by the values of u on the boundary. The independent condition is on the normal derivative (cf. (20)

$$u_N = \mathbf{e}_N \cdot \nabla u = \pm(\tau u_x - \xi u_t)/(\xi^2 + \tau^2)^{1/2} \tag{21}$$

The allowed forms of invariant normal derivatives may now be determined as before using the reducing substitution. In the example considered above, the values (17) may be inserted with (11) into (21), showing that a suitable constant value assigned to $W'(\zeta_B)$ will satisfy a condition of the form

$$u_N = g(t) + K/g(t) \tag{22}$$

where

$$g(t) = (b/c_1)(t - c_2)[c_1^2 + b^2(t - c_2)^2]^{-1/2} \tag{23}$$

Here $(t - c_2)$ may be replaced by $\zeta_B e^{-bx/c_1}$ if more convenient.

In general, because $\zeta(x,t)$ satisfies (8), ξ and τ in (20) and (21) may be replaced by $-\zeta_t$ and ζ_x; the normal vector is $\nabla\zeta$.

The equation for normal derivatives corresponding to (7) may also be derived. Apply the first prolongation of (5) to

$$u_N - N(x,t) = 0$$

with u_N given by (21), and eliminate u_x and u_t using (19) and (21), to get

$$\begin{aligned}(\xi N_x + \tau N_t)(\xi^2 + \tau^2) &+ N[\tau^2\xi_x + \xi^2\tau_t - \xi\tau(\tau_x + \xi_t) - (\xi^2 + \tau^2)\eta_u] \\ &= \eta(\xi^2 + \tau^2)^{-1/2}[(\xi^2 - \tau^2)(\tau_x + \xi_t) + 2\xi\tau(\tau_t - \xi_x)] \\ &\quad + (\xi^2 + \tau^2)^{1/2}(\tau\eta_x - \xi\eta_t)\end{aligned} \tag{24}$$

For a given generator, (24) determines $N(x,t)$, the invariant normal derivative. The equation may be simplified, using (9), to a 1st order linear equation for $N\{\theta\}$, noting that $\xi N_x + \tau N_t = z(\theta)dN/d\theta$ where the factor $z(\theta)$ may be determined by inspection from $\xi\{\theta\} = z(\theta)X'(\theta)$ or $\tau\{\theta\} = z(\theta)T'(\theta)$. The arbitrary constant in the integral for $N\{\theta\}$ is related to the value to be assigned to $W'(\zeta_B)$; these constants are denoted by K in the appendix.

In the example used above, comparing (11) and (14) gives $z(\theta) = 1$, and after setting $(\xi^2 + \tau^2)^{-1/2} = c_1 g(t)/b(t - c_2)$, equation (24) reduces to

$$\frac{b^2(t - c_2)^3}{c_1^2 g^2(t)}\frac{dN}{dt} + c_1^2 N(t) = -2c_1^2 g(t) \tag{25}$$

where $g(t)$ is defined in (23). The general solution of (25) is given by (22).

The appendix gives the invariant normal derivatives for all cases of the diffusion-convection equation (1). In each result, the part independent of K (corresponding to some particular solution of the equation for $N\{\theta\}$) varies with the definition of ζ and R. The actual substitutions are therefore also given.

5. Conclusion

This paper has discussed group-invariant boundary conditions of various kinds, and shown how the form of these conditions may be determined for an equation once its symmetry algebra is known. At least for algebras of low dimension it seems feasible to list the possible invariant conditions, the associated subalgebras, and the corresponding reducing transformations. The work of Coggleshall and Axford [1] is an example of this procedure.

For a physical application of , say, an equation of the form (1), one may choose a model that allows (2) or (3) with suitable values of the constants. The physical boundaries may then be compared with the list of invariant boundaries, thereby choosing the constants c_i which determine the subalgebra to be used for reduction. For example this step would determine c_1, c_2, K_1 and K_2 in (14), or equivalently c_1, c_2 and a value of ζ in (15). Finally the physical boundary conditions may be compared to the available model boundary conditions to get suitable values of the remaining constants, for example K_3 in (14), or a value of r in (15), and the value of K in (22).

Appendix: other invariant normal derivatives

Consider (1) with either (2) or (3). In each case below, the form given for u_N is valid on a boundary $\zeta = $ (constant).

(i) For the generic case with $p \neq b$ and $p \neq 2b$, the substitutions to be made are $\zeta = (x-c_1)^{p-2b}(t-c_2)^{b-p}$, $u = W(\zeta) + r\log|x-c_1|$ for the exponential case (3), where $r = 1/(p-b)$, and $u = W(\zeta)(x-c_1)^r$ for the power case (2). Define

$$h(x,t) = (p-2b)(t-c_2)[(p-b)^2(x-c_1)^2 + (p-2b)^2(t-c_2)^2]^{-1/2}.$$

Then the invariant normal derivative has the form

$$\pm\, u_N = (x-c_1)^{q-1}[K_3 h(x,t) + K/h(x,t)] \tag{26}$$

where for the exponential case (3) $q = 0$ and $K_3 = r$, while for the power case (2) $q = 1/(p-b)$ and K_3 is determined by the boundary condition

$$u = (p-b)K_3(x-c_1)^q. \tag{27}$$

(ii) For the case $p = 2b$ the substitutions are $\zeta = (x-c_1)e^{-bt/c_2}$, $u = W(\zeta) + (t/c_2)$ for the exponential case (3), and $u = W(\zeta)e^{t/c_2}$ for the power case (2). Define

$$m(x) = b(x-c_1)[c_2^2 + b^2(x-c_1)^2]^{-1/2}. \tag{28}$$

Then, on the boundary $\zeta = \zeta_B$ the invariant normal derivative has the form

$$\pm\, u_N = [K_3 m(x) + K/m(x)]e^{qt} \tag{29}$$

where $K_3 = 1/c_2$ for the exponential case (3) and is determined for the power case (2) by the boundary condition $u = c_2 K_3 e^{t/c_2}$; $q = 0$ for the exponential case and $q = 1/c_2$ for the power case.

(iii) For the power case (2) with $p = b$, and $g(t)$ given by (23),

$$\pm\, u_N = e^{x/c_1}[K_3 g(t) + K/g(t)] \tag{30}$$

on the same boundary (13), where K_3 is determined by the boundary condition $u = K_3 e^{x/c_1}$. The substitution used is $u = W(\zeta)e^{x/c_1}$ with ζ given in (15).

(iv) For (1) with arbitrary F and G, $u_N =$ (constant) is invariant on any straight line boundary.

References

1. S. V. Coggleshall and R. V. Axford, *Phys. Fluid* **29** (1986) 2398.
2. M. J. Englefield and T. L. Katoanga, to appear in *Proc. XIX Int. Coll. Group Theor. Meth. Phys.*, eds. J. Mateas-Guilarte, M.A. del Olmo and M. Santander (to be published by CIEMAT, Spain).
3. M. J. Englefield, Proc. 1992 AMS-SIAM Summer Seminar (to appear).
4. A. Oron and P. Rosenau, *Phys. Lett.* **A118** (1986) 172.

NONLINEAR DIFFERENTIAL EQUATIONS, LIE SYMMETRIES, AND THE PAINLEVÉ TEST

N. EULER and W.-H STEEB
Department of Applied Mathematics and Nonlinear Studies, Rand Afrikaans University, P. O. Box 524, Auckland Park 2006, South Africa.A

Abstract. We study nonlinear ordinary and partial differential equations that depend explicitly on the independent variable. The study consists of calculating the Lie symmetries, performing the Painlevé test, and, in the case of ordinary differential equations, a study of the invertible point transformations. In particular we are interested in the link between the Painlevé test, Lie symmetries, and point transformations.

The Painlevé test is an important tool (see for example [3]) as a 'test' for integrability. Within this approach one can often obtaining exact solution, Lie-Bäcklund transformations, Lax pairs, and recursion operators. The Painlevé test also plays an important role in the study of the chaotic behaviour of a nonlinear differential equation. In this contribution we study the Painlevé test and its connection to point transformations and Lie symmetries. We consider nonlinear differential equations which depend explicitly on the independent variables. In particular we study a anharmonic oscillator and an externally driven anharmonic oscillator, the Sine-Gordon equation, and the Mikhailov equation. A REDUCE program is given to demonstrate the calculation of the Painlevé test with computer algebra.

1. Ordinary differential equations.

Definition: A system of ordinary differential equations in the complex domain has the Painlevé property when every solution has no movable singularities except poles. We can now state the following

Theorem: A necessary condition that an nth-order system of the form

$$\frac{d\mathbf{w}}{dz} = \mathbf{g}(\mathbf{w}),$$

where g is rational in $\mathbf{w}$, has the Painlevé property is that there is a Laurent expansion,

$$w_k = (z - z_1)^m \sum_{j=0}^{\infty} a_{kj}(z - z_1)^j,$$

(m is a negative integer) with $n-1$ arbitrary expansion coefficients (besides the pole position which is arbitrary).

Application of this theorem is known as the Painlevé test for ordinary differential systems or equations.

N. H. Ibragimov et al. (eds.), Modern Group Analysis: Advanced Analytical and Computational Methods in Mathematical Physics, 209–215.

Let us now study the explicitly time dependent anharmonic oscillator

$$\ddot{x} + f_1(t)\dot{x} + f_2(t)x + f_3(t)x^n = 0, \tag{1}$$

where f_1, f_2 and f_3 are arbitrary smooth functions and $\dot{x} = dx/dt$. We consider $n \neq -3, -1, 1$. We are interested in the conditions on f_1, f_2 and f_3 for
i) The existence of point transformations to the integrable equation

$$\frac{d^2X}{dT^2} + X^n = 0. \tag{2}$$

ii) The existance of non-trivial Lie symmetry vector fields.
iii) The Painlevé test.

With the invertible point transformation

$$X(T(t)) = f(t)x(t), \qquad T(t) = g(t), \tag{3}$$

for (2), we obtain $\ddot{x} + A_1\dot{x} + A_0 = 0$ where $A_1 = (2\dot{g}\dot{f} - f\ddot{g})/(f\dot{g})$ and $A_0 = x(\dot{g}\dot{f} - f\ddot{g})/(f\dot{g}) + \dot{g}^2 f^{n-1}x^n$, i.e.,

$$f(t) = Cf_3^{1/(n+3)} \exp\left(\int \frac{2f_1(t)}{n+3}dt\right), \quad g(t) = \int \frac{f_3^{1/2}(t)}{f^{(n-1)/2}(t)}dt.$$

The condition on the arbitrary functions are thus given by

$$f_2(t) = \frac{2(n+1)f_1^2}{(n+3)^2} - \frac{(n-1)f_1\dot{f}_3}{(n+3)^2 f_3} - \frac{(n+4)\dot{f}_3^{\,2}}{(n+3)^2 f_3^2} + \frac{2\dot{f}_1}{(n+3)} + \frac{\ddot{f}_3}{(n+3)f_3}.$$

The first integral for (2) can be obtained:

$$I(t,x,\dot{x}) = \frac{(f\dot{x} + \dot{f}x)^2}{2\dot{g}^2} + \frac{f^{n+1}x^{n+1}}{n+1}.$$

Let us consider $f_1 = c_1$, $f_2 = c_2$ and $f_3 = c_3$, where c_1, c_2 and c_3 are arbitrary constants. The above condition reduces to

$$\boxed{c_2 = \frac{2(n+1)}{(n+3)^2}c_1^2.} \tag{4}$$

We now investigate the Lie symmetry vector fields ([1]) for (2). Consider the following ansatz for the infinitesimal transformation

$$t' = t + \varepsilon h_1(t), \quad x' = x + \varepsilon h_2(t)x.$$

We find

$$h_1(t) = b_1 \exp\left(\frac{c_1 t(n-1)}{n+3}\right) + b_2, \quad h_2(t) = -\frac{2b_1c_1}{(n+3)} \exp\left(\frac{(n-1)c_1 t}{n+3}\right)$$

where c_1 and c_2 must satisfy condition (4). The generators are

$$G_1 = \frac{\partial}{\partial t}$$
$$G_2 = -\frac{2c_1}{n+3} \exp\left(\frac{c_1 t(n-1)}{n+3}\right) x \frac{\partial}{\partial x} + \exp\left(\frac{c_1 t(n-1)}{n+3}\right) \frac{\partial}{\partial t}$$

with $[G_1, G_2] = c_1(n-1)G_1/(n+3)$. In performing the Painlevé test for (2) we must consider the expansion

$$x(t) = \sum_{j=0}^{\infty} a_j (t-t_1)^{j+\alpha} \tag{5}$$

where $\alpha = 2/(1-n)$. We study three different cases for n:
$n = 2$: It follows that $\alpha = -2$ with the resonance $r = 6$. The condition that a_6 is arbitrary is

$$c_2 = \frac{6c_1^2}{25}. \qquad \text{(condition (4) with } n = 2)$$

$n = 3$: It follows that $\alpha = -1$ with the resonance $r = 4$. With the expansion

$$x(t) = \sum_{j=0}^{\infty} x_j(t) \phi(t)^{j-1}$$

we find that a_4 is arbitrary (with f_1, f_2 and f_3 arbitrary functions of t) if the following condition is satisfied:

$$\begin{aligned} 0 = {} & 9f_3^{(4)} f_3^3 - 54 f_3^{(3)} f_3' f_3^2 + 18 f_3^{(3)} f_3^3 f_1 - 36(f_3'')^2 f_3^2 \\ & +192 f_3''(f_3')^2 f_3 - 78 f_3'' f_3' f_3^2 f_1 + 36 f_3'' f_3^3 f_2 + 3 f_3'' f_3^3 f_1^2 \\ & -112(f_3')^4 + 64(f_3')^3 f_3 f_1 + 6(f_3')^2 f_1' f_3^2 - 72(f_3')^2 f_3^2 f_2 \\ & +90 f_3' f_2' f_3^3 - 27 f_3' f_1'' f_3^3 - 57 f_3' f_1' f_3^3 f_1 + 72 f_3' f_3^3 f_2 f_1 \\ & -14 f_3' f_3^3 f_1^3 - 54 f_2'' f_3^4 - 90 f_2' f_3^4 f_1 + 18 f_1^{(3)} f_3^4 + 54 f_1'' f_3^4 f_1 \\ & +36(f_1')^2 f_3^4 - 36 f_1' f_3^4 f_2 + 60 f_1' f_3^4 f_1^2 - 36 f_3^4 f_2 f_1^2 + 8 f_3^4 f_1^4. \end{aligned}$$

For $f_1 = c_1$, $f_2 = c_2$ and $f_3 = c_3$ the above condition reduces to

$$c_2 = \frac{2c_1^2}{9} \qquad \text{(condition (4) with } n = 3).$$

REMARK: *The above differential equation gives the condition for the anharmonic oscillator to be transformed to the second Painlevé transcendent ([2])*

$$\frac{d^2U}{dT^2} = 2U^3 + TU + a.$$

$n = 5$: It follows that $\alpha = -1/2$ with the resonance $r = 3$ (weak Painlevè test). The condition that a_3 is arbitrary is

$$c_2 = \frac{3c_1^2}{16}. \qquad \text{(condition (4) with } n = 5).$$

We conclude that the relation (4) gives the condition for the anharmonic oscillator to be transformed to the integrable equation, to have non-trivial Lie symmetry vector fields, and to pass the Painlevé test.

We now consider the externally driven anharmonic oscillator

$$\ddot{x} + c_1\dot{x} + c_2 x + x^3 = f(t),$$

where f is an arbitrary smooth function. The equation admits the Lie symmetry vector field

$$G_2 = -\frac{c_1}{3}\exp\left(\frac{c_1 t}{3}\right) x\frac{\partial}{\partial x} + \exp\left(\frac{c_1 t}{3}\right)\frac{\partial}{\partial t}$$

where

$$c_2 = \frac{2c_1^2}{9} \qquad \text{(condition (4) with } n = 3)$$

and $df/dt + c_1 f = 0$, i.e., $f(t) = C\exp(-c_1 t)$. We now perform the Painlevé test. With the expansion

$$x(t) = \sum_{j=0}^{\infty} x_j(t)\phi(t)^{j-1}$$

we find the following condition at the resonance $r = 4$:

$$-27\sqrt{-2}\frac{df}{dt} - 27\sqrt{-2}c_1 f - 18c_2c_1^2 + 4c_1^4 = 0$$

so that $c_2 = 2c_1^2/9$ and $df/dt + c_1 f = 0$. We conclude that the equation

$$\ddot{x} + c_1\dot{x} + \frac{2c_1^2}{9}x + x^3 = C\exp(-c_1 t)$$

admits G_2 and passes the Painlevé test. This equation has the first integral

$$I(t,x,\dot{x}) = \exp\left(\frac{4}{3}c_1 t\right)\left(\left(\dot{x} + \frac{c_1 x}{3}\right)^2 + \frac{1}{2}x^4 - 2Cx\exp(-c_1 t)\right).$$

2. Partial differential equations

In order to extend the Painlevè test to partial differential equations Weiss *et al* ([5]) made the following

Proposition: Seek solutions of a nonlinear partial differential equation in the form

$$u = \phi^\alpha \sum_{j=0}^{\infty} u_j \phi^j$$

where ϕ is a holomorphic function whose vanishing defines a noncharacteristic hypersurface. The Painlevé test is passed if α is a negative integer, the recursion relation is consistent and the series expansion contains the correct number of arbitrary functions. One can then expect that the partial differential equation is integrable (perhaps after some transformation).

Let us consider the explicitly space- and time-dependent Sine-Gordon equation

$$u_{tt} - u_{xx} + f(x,t)\sin u = 0$$

where f is an arbitrary smooth function and subscripts denotes partial derivatives. With the transformation $v = \exp(iu)$ the equation takes on the form

$$vv_{tt} - v_t^2 - vv_{xx} + v_x^2 + \frac{1}{2}f(x,t)(v^3 - v) = 0. \tag{6}$$

It is well known that for $f = 1$ the Sine-Gordon equation is integrable: it passes the Painlevé test and it can be solved by the inverse scattering transform. We are seeking a conditions on f such that (6) passes the Painlevé test. Equation (6) admits an expansion of the form

$$v = \sum_{j=0}^{\infty} v_j \phi^{j-2}$$

The resonances are $-1, 2$ and

$$v_0 = 4(\phi_x^2 - \phi_t^2)f^{-1}, \quad v_1 = 4(\phi_{tt} - \phi_{xx})f^{-1}.$$

At the resonance we obtain

$$(-f_{tt}f + f_t^2 + f_{xx}f - f_x^2)(\phi_t^2 - \phi_x^2)^2 = 0.$$

We are thus left with the following condition on f:

$$\boxed{f_t^2 - f_{tt}f = f_x^2 - f_{xx}f} \tag{7}$$

with general solution

$$f(x,t) = f_1(x+t)f_2(x-t).$$

Note: The equation

$$u_{tt} - u_{xx} + f_1(x+t)f_2(x-t)\sin u = 0 \tag{8}$$

can be transformed to $\bar{u}_{\xi\eta} = (1/4)\sin\bar{u}$ by the transformation

$$\xi(x,t) = \int^{x+t} f_1(s)\,ds, \quad \eta(x,t) = \int^{x-t} f_2(s)\,ds, \quad \bar{u}(\xi(x,t),\eta(x,t)) = u(x,t).$$

We have thus found an integrable case by applying the Painlevé test.

We now consider the explicitly space- and time-dependent Mikhailov equation

$$u_{xx} - u_{tt} = c_1 g_1(x,t)e^u + c_2 g_2(x,t)e^{-2u}$$

where g_1 and g_2 are arbitrary smooth functions. It is well known that for $g_1 = g_2 = 1$ the equation is integrable by a 3×3 matrix scattering problem. With the transformation $v = \exp(u)$, the Mikhailov equation transforms to

$$u(u_{xx} - u_{tt}) + u_t^2 - u_x^2 = c_1 g_1(x,t) u^3 + c_2 g_2(x,t). \tag{9}$$

The equation admits two branches $-2, 1$ with the resonances $-1, 2$. The combined conditions on g_1 and g_2, for both branches, is again given by (7). Thus, the equation

$$u_{xx} - u_{tt} + f_1(x+t) f_2(x-t)(c_1 e^u + c_2 e^{-2u}) = 0 \tag{10}$$

passes the Painlevé test.

Let us study some Lie symmetries for the above Sine-Gordon and Mikhailov equations. Consider

$$Z_1 = x\frac{\partial}{\partial t} + t\frac{\partial}{\partial x}$$
$$Z_2 = (x^2 + t^2)\frac{\partial}{\partial x} + 2xt\frac{\partial}{\partial t}$$

for the above equations. The condition on $f_1(s_1)$, $f_2(s_2)$, where $s_1 = x + t$ and $s_2 = x - t$, is as follows:
For Z_1 we obtain

$$f_2 s_1 \frac{df_1}{ds_1} - f_1 s_2 \frac{df_2}{ds_2} = 0$$

with solutions $f_1(s_1) = s_1^2$, $f_2(s_2) = s_2^2$ and $f_1(s_1) = s_1^{-2}$, $f_2(s_2) = s_2^{-2}$.
For Z_2 we obtain

$$s_1 \frac{df_1}{ds_1} + 2f_1 = 0, \quad s_2 \frac{df_2}{ds_2} + 2f_2 = 0$$

with solution $f_1(s_1) = s_1^{-2}$, $f_2(s_2) = s_2^{-2}$. Thus the equations

$$u_{tt} - u_{xx} + (x^2 - t^2)^{-2} \sin u = 0$$
$$u_{tt} - u_{xx} + (x^2 - t^2)^{-2}(c_1 e^u + c_2 e^{-2u}) = 0$$

pass the Painlevé test and admit Z_1 and Z_2.

Finally we give a REDUCE 3.4 program ([4]) to perform the Painlevé test for the anharmonic oscillator (1) with $n = 3$.

```
operator a;
depend phi,x; depend a(0),x; depend a(1),x; depend a(2),x;
depend a(3),x; depend a(4),x; depend f1,x; depend f2,x; depend f3,x;
for n:=0 step 1 until 4 sum (phi**(n-1))*a(n);
u := ws;
b := df(u,x,2)+f1*df(u,x)+f2*u+f3*u^3;
c := num(b);
d0 := coeffn(c,phi,0); d1 := coeffn(c,phi,1);
```

```
d2 := coeffn(c,phi,2); d3 := coeffn(c,phi,3);
d4 := coeffn(c,phi,4);
second(solve(d0=0,a(0)));
a(0):=part(ws,2); q1:=solve(d1=0,a(1));
part(q1,1); a(1):=part(ws,2);
q2:=solve(d2=0,a(2)); part(q2,1); a(2):=part(ws,2);
q3:=solve(d3=0,a(3)); part(q3,1); a(3):=part(ws,2);
q4:=solve(d4=0,a(4));
d4;
```

The output gives the condition on f_1, f_2, f_3 such that $a(4)$ is arbitrary.

References

1. Euler N. and Steeb W.-H., *Continuous Symmetries, Lie algebras and Differential Equations*, B.I. Wissenschaftsverlag, Mannheim, 1992.
2. Euler N., Steeb W.-H., Duarte L.G.S. and Moreira I.C. Int. J. Theor. Phys. **30**, 8, 1991.
3. Steeb W.-H. and Euler N., *Nonlinear Evolution Equation and Painlevé test*, World Scientific, Singapore, 1988.
4. Steeb W.-H. and Lewien D., *Algorithms and Computations with REDUCE*, B.I. Wissenschaftsverlag, Mannheim, 1992.
5. Weiss J., Tabor M. and Carneval G, J. Math. Phys. **24**, 522, 1983.

NON-ITERATIVE TRANSFORMATION METHODS EQUIVALENCE

RICCARDO FAZIO
Department of Mathematics, University of Messina,
Contrada Papardo, Salita Sperone 31
98166 Sant'Agata, Messina, Italy.

Abstract. The non-iterative numerical solution of nonlinear boundary value problems is a subject of great interest. The present note is concerned with the theory of non-iterative transformation methods. These methods are defined within group invariance theory. Here we prove the equivalence between the two non-iterative transformation methods defined by the stretching group and the spiral group respectively. Then we point out the prominent role of stretching transformations in solving non-iteratively boundary value problems governed by ordinary differential equations.

1. Introduction

The non-iterative numerical solution of nonlinear boundary value problems (BVPs) is a subject of current research. In this context transformation methods (TMs) are founded on the group invariance theory. These methods are initial value methods because they transform BVPs to initial value problems.

The first application of a non-iterative TM, without any consideration of group theory, was given by Töpfer in [25]. Acrivos, Shah and Petersen [1] first and Klamkin [14] later extended Töpfer's method respectively to a more general problem and to a class of problems. By following the work of Klamkin, Na [17, 18] showed the relation between the invariance properties (with respect to a linear group of transformation: the stretching group) of the considered problem and the applicability of a non-iterative TM. Moreover, Na considered the invariance with respect to a nonlinear group of transformations: the spiral group.

Belford [7], Ames and Adams [4] proposed non-iterative TMs for eigenvalue problems. A term paper was proposed by Klamkin [15].

The invariance of one and of two or more physical parameters, if they are involved in the mathematical model, were respectively proposed by Na [19] and by Scott, Rinschler and Na [23]. A term book was written by Na [20]. Fazio and Evans [12] introduced non-iterative TMs for free BVPs.

The paper of Na [17] has had a real impact, because of the utilization of the spiral group. As a result it has been thought that TMs ought to be generalized by introducing groups different from the stretching or spiral ones, see Belford [7], Ames and Ibragimov [6], Ames and Adams [5], Na [20] or Seshadry and Na [24, pp. 157-168]. However we shall prove in this note that the utilization of stretching or of spiral group are equivalent. In fact, we can extend the applicability of non-iterative TMs by some variable transformations linking different invariant groups, see Fazio [8, 9]. Through a variable transformation, we can sometimes state a given problem in an alternative form which, though equivalent, may benefit of established

N. H. Ibragimov et al. (eds.),
Modern Group Analysis: Advanced Analytical and Computational Methods in Mathematical Physics, 217–221.

resolution (non-iterative) methods. A remark is in order here. Non-iterative TMs are applicable only to particular classes of BVPs so that they have been considered as ad hoc methods, see Meyer [16, pp. 35-36], Na [20, p. 137] or Sachdev [22, p. 218].

In the next section we consider two-point BVPs and we prove that two apparently different TMs, defined by the spiral group and the stretching group, respectively, are indeed equivalent. It is worth mentioning that the quoted example due to Na [17] — see Na and Tang [21], Klamkin [15], Ames [2, p. 140], Na [20, pp. 155-158], Seshadri and Na [24, pp. 157-168] or Ames [3] — belongs to the characterized class of problems.

Finally, in the last section, we discuss the implications of the present work. There we indicate again how the present approach is by no means the only way to extend non-iterative TMs (see Na [19], Scott, Rinschler and Na [23] and Fazio [10, 11]).

2. On the transformation via spiral group

Let us introduce the class of two-point BVPs

$$\begin{aligned} \frac{d^2z}{dt^2} &= z^{1-2\delta}\Theta\left(tz^{-\delta}, \frac{dz}{dt}z^{\delta-1}\right) \qquad t \in (0,b) \\ \frac{dz}{dt}(0) &= \alpha z(0)^{1-\delta} \\ z(b)^{\varsigma}\Phi\left(\frac{dz}{dt}(b)z(b)^{\delta-1}\right) &= \beta \end{aligned} \tag{1}$$

where α and δ are arbitrary constants, $\Theta(\cdot,\cdot)$ and $\Phi(\cdot)$ are arbitrary functions of their arguments, $b > 0, \varsigma \neq 0$ and $\beta \neq 0$. Here the governing differential equation and the boundary condition at $t = 0$ are invariant with respect to the following stretching group of transformations

$$t^* = \lambda^{\delta} t \quad z^* = \lambda z \quad \lambda \in R^+ - \{0\} \quad . \tag{2}$$

The non-iterative numerical solution of (1) can be obtained by the following steps:

- we fix a value of $z^*(0)$, this defines a value of $\frac{dz^*}{dt^*}(0)$ according to the boundary condition;
- next we integrate numerically, with initial data $z^*(0)$ and $\frac{dz^*}{dt^*}(0)$, forwards in $[0, b^*]$ where b^* is defined by

$$b^* = b\left\{\left[\frac{z^*(b^*)^{\varsigma}\,\Phi\left(\frac{dz^*}{dt^*}(b^*)z^*(b^*)^{\delta-1}\right)}{\beta}\right]^{1/\varsigma}\right\}^{\delta} \quad ; \tag{3}$$

- finally, use of group properties provides

$$\lambda = \left[\frac{z^*(b^*)^{\varsigma}\Phi\left(\frac{dz^*}{dt^*}(b^*)z^*(b^*)^{\delta-1}\right)}{\beta}\right]^{1/\varsigma}$$

$$z(0) = \lambda^{-1} z^*(0) \quad ; \quad \frac{dz}{dt}(0) = \lambda^{\delta-1} \frac{dz^*}{dt^*}(0)$$
$$z(b) = \lambda^{-1} z^*(b^*) \quad ; \quad \frac{dz}{dt}(b) = \lambda^{\delta-1} \frac{dz^*}{dt^*}(b^*) \quad .$$

We note that b^* is defined implicitly, in (3), so that it can be considered as a root of an unknown function.

Via the simple variable transformation

$$t = t \quad ; \quad v = \ln z \tag{4}$$

the stretching group (2) is transformed to the following spiral group

$$t^* = \lambda^\delta t \quad ; \quad v^* = v + \mu \quad ; \quad \mu = \ln \lambda$$

where μ is the group parameter. (1), after (4), transforms to

$$\begin{gathered} \frac{d^2 v}{dt^2} = e^{-2\delta v} \Theta \left(t e^{-\delta v}, \frac{dv}{dt} e^{\delta v} \right) - \left(\frac{dv}{dt} \right)^2 \\ \frac{dv}{dt}(0) = \alpha e^{-\delta v(0)} \\ e^{\zeta\, v(b)}\, \Phi \left(\frac{dv}{dt}(b) e^{\delta\, v(b)} \right) = \beta \quad . \end{gathered} \tag{5}$$

Therefore, z(t) $\geq$ 0 for every t $\in (0, b)$. It is interesting to note that the example due to Na [17] (there $p = -1$, see Na and Tang [21] for $p = 0$ and $p = 1$)

$$\begin{gathered} \frac{d^2 v}{dt^2} + \frac{p+1}{t} \frac{dv}{dt} + q e^v = 0 \\ \frac{dv}{dt}(0) = 0 \\ v(1) = 0 \end{gathered} \tag{6}$$

is a particular case of (5): actually $\delta = -1/2$, $\Theta(\cdot, \cdot) = -\frac{p+1}{t} \frac{dv}{dt} e^{-v} - q + \left(\frac{dv}{dt} \right)^2 e^{-v}$, $\Phi(\cdot) = 1$, $\alpha = 0$, $b = 1$, $\zeta = 1$ and $\beta = 1$. Now, by inverting (4), problem (6) takes the form

$$\begin{gathered} \frac{d^2 z}{dt^2} + \frac{p+1}{t} \frac{dz}{dt} + q z^2 - \left(\frac{dz}{dt} \right)^2 z^{-1} = 0 \\ \frac{dz}{dt}(0) = 0 \\ z(1) = 1 \quad . \end{gathered} \tag{7}$$

Therefore, along the lines of the approach outlined above a non-iterative numerical solution for (6) can be obtained by solving (7).

We can say that the applicability of a non-iterative TM to the class of problems (5) is equivalent to the invariance of (1) with respect to the stretching group. The key point here is that invariant properties are conserved under the transformation (4).

A non-iterative TM can be introduced again if in (1), instead of the boundary condition at zero, we have $z(0) = 0$. However, under (4) from $z(0) = 0$ we get $v(0) = -\infty$ and this is not a suitable condition from a numerical viewpoint.

3. Discussion

The previous section indicates how two apparently different non-iterative TMs are equivalent under a simple variable transformation. The underlying idea of the mentioned transformation was to solve the transformed problem instead of the original one, see Fazio [8, 9]. Moreover, we specified a widely range of applicability for the stretching group.

In closing, we can indicate some further implications for this work. First of all, non-iterative TMs are ad hoc methods. Their applicability depend upon the invariance properties of the governing differential equation and upon the related boundary conditions. To introduce a variable transformation linking two different invariant groups is a possible way to extend the applicability of non-iterative TMs. However, it is a simple matter to exhibit differential equations not admitting any group of transformations, see for instance Hill [13, pp. 81-82]. Consequently, it is easy to realize that non-iterative TMs can not be extended to every BVP. On the other hand, a further extension is possible if one or more physical parameters are involved in the mathematical model, see Na [18] and Scott, Rinschler and Na [23], cf. also Na [20, Chapters 8 and 9]. Moreover, BVPs governed by the most general second order differential equation in normal form can be solved iteratively by introducing an extended stretching group via a numerical parameter, see Fazio [10, 11].

Acknowledgement

This work was partially supported by C.N.R. through G.N.F.M. and by M.U.R.S.T. through "Fondi 40% e 60%"

References

1. A. Acrivos, M. J. Shah and E. E. Petersen, Momentum and heat transfer in laminar boundary-layer flows of non-newtonian fluids past external surfaces, AIChE J., **6** (1960) 312-317.
2. W. F. Ames, Nonlinear partial differential equations in engineering, Vol. II, Academic Press, New York, 1972.
3. W. F. Ames, Applications of group theory in computation, A survey, in Numerical and Applied Mathematics Vol. 1.1, W. F. Ames ed., J. C. Baltzer AG, Basel, 1989.
4. W. F. Ames and E. Adams, Exact shooting and eigenparameter problems, Nonlinear Anal.; Theo. Meth. Applic., **1** (1976) 75-82.
5. W. F. Ames and E. Adams, Non-linear boundary and eigenvalue problems for the Emden-Fowler equations by group methods, Int. J. Non-linear Mech., **14** (1979) 35-42.
6. W. F. Ames and N. H. Ibragimov, Utilization of group properties in computation, in Group Theoretical Methods in Mechanics, Proceedings of the Joint IUTAM/IMV Symposium at Novosibirsk, 1978. N. H. Ibragimov & L. V. Ovsiannikov eds., pp. 9-23.

7. G. G. Belford, An initial value problem approach to the solution of eigenvalue problems, SIAM J. Numer. Anal., **6** (1969) 99-103.
8. R. Fazio, A noniterative transformation method applied to two-point boundary value problems, Appl. Math. Comput., **39** (1990) 79-87.
9. R. Fazio, Normal variables transformation method applied to free boundary value problems, Int. J. Comput. Math., **37** (1990) 189-199.
10. R. Fazio, The iterative transformation method: free or moving boundary value problems, in Computational and Applied Mathematics, Vol. I Algorithms and Theory, IMACS 13th World Congress, Dublin, Ireland, 22-26 July 1991, C. Brezinski & U. Kulish eds., North-Holland, Amsterdam, 1992, pp. 147-152.
11. R. Fazio, Similarity analysis for moving boundary parabolic problems, in Computational and Applied Mathematics, Vol. II Differential Equations, IMACS 13th World Congress, Dublin, Ireland, 22-26 July 1991, W. F. Ames & P. J. van der Houwen eds., North-Holland, Amsterdam, 1992, pp. 153-162.
12. R. Fazio and D. J. Evans, Similarity and numerical analysis for free boundary value problems, Int. J. Comput. Math., **31** (1990) 215-220, **39** (1991) 249.
13. J. M. Hill, Solution of differential equations by means of one-parameter groups, Research Notes in Math., **63**, Pitman, Boston, 1982.
14. M. S. Klamkin, On the transformation of a class of boundary value problems into initial value problems for ordinary differential equations, SIAM Rev., **4** (1962) 43-47.
15. M. S. Klamkin, Transformation of boundary value problems into initial value problems, J. Math. Anal. Appl., **32** (1970) 308-330.
16. G. H. Meyer, Initial Value Methods for Boundary Value Problems; Theory and Application of Invariant Imbedding, Academic Press, New York, 1973.
17. T. Y. Na, Transforming boundary conditions to initial conditions for ordinary differential equations, SIAM Rev., **9** (1967) 204-210.
18. T. Y. Na, Further extension on transforming from boundary value to initial value problems, SIAM Rev., **10** (1968) 85-87.
19. T. Y. Na, An initial value method for the solution of a class of nonlinear equations in fluid mechanics, J. Basic Eng. Tarns. ASME, **92** (1970) 91-99.
20. T. Y. Na, Computational Methods in Engineering Boundary Value Problems, Academic Press, New York, 1979.
21. T. Y. Na and S. C. Tang, A method for the solution of heat conduction with nonlinear heat generation, Z. Angew. Math. Mech., **49**$\frac{1}{2}$ (1969) 45-52.
22. P. L. Sachdev, Nonlinear ordinary differential equations and their applications, Marcel Dekker, New York, 1991.
23. T. C. Scott, G. L. Rinschler and T. Y. Na, Further extensions of an initial value method applied to certain nonlinear equations in fluid mechanics, J. Basic Eng. Trans. ASME, **94** (1972) 250-251.
24. R. Seshadri and T. Y. Na, Group Invariance in Engineering Boundary Value Problems, Springer-Verlag, New York, 1985.
25. K. Töpfer, Bemerkung zu dem Aufsatz von H. Blasius: Grenzschichten in Flussigkeiten mit kleiner Reibung, Z. Math. Phys., **60** (1912) 397-398.

REDUCTION PROCEDURES FOR A CLASS OF RATE-TYPE MATERIALS

D. FUSCO and N. MANGANARO
Department of Mathematics, University of Messina,
Contrada Papardo, Salita Sperone 31
98166 Sant'Agata, Messina, Italy.

Abstract. There are shown some reduction techniques which allow for determining exact solutions to a model describing rate-type materials. Special forms of the material response functions involved therein are also considered.

1. Introduction and general remarks

Let us consider the following system of first order partial differential equations

$$U_t + A(x,t,U)U_x = B(x,t,U) \tag{1}$$

where

$$U = \begin{bmatrix} u \\ v \end{bmatrix} \quad ; \quad A = \begin{bmatrix} a_{11} & a_{12} \\ a_{21} & a_{22} \end{bmatrix} \quad ; \quad B = \begin{bmatrix} b_1 \\ b_2 \end{bmatrix}$$

x and t are space and time coordinates, respectively. Here and in the following a subscript means for derivative with respect to the indicated variable.

We assume the system (1) to be strictly hyperbolic. That is tantamount to require the matrix A to admit two real distinct eigenvalues λ and μ (characteristic wave speeds) to which there correspond two left eigenvectors $l^{(\lambda)}, l^{(\mu)}$ as well as two right eigenvectors $d^{(\lambda)}, d^{(\mu)}$ spanning the Euclidean space E^2.

Recently a great deal of attention has been devoted to reduce (1) to some canonical forms which allow for the existence of Riemann-like invariant solutions or which can be linearized by means of suitable variable transformations.

Within the theoretical framework mentioned above, here we will show some results concerning the following system of equations

$$u_t + v_x = 0 \tag{2}$$

$$v_t + h(v,t)u_x = f(v,t) \tag{3}$$

In the case where u is the Lagrangian velocity and v represents the stress, system of equations (2), (3) is widely used to model one-dimensional deformation processes in nonlinear viscoelastic materials [1].

In (2), (3) $h(v,t)$ and $f(v,t)$ are material functions which measure, respectively, the istantaneous and nonistantaneous response of the material to an increment of the stress. Here we assume the condition

$$h(v,t) > 0 \tag{4}$$

N. H. Ibragimov et al. (eds.),
Modern Group Analysis: Advanced Analytical and Computational Methods in Mathematical Physics, 223–229.

to be fulfilled so that the model at hand is consistent with hyperbolic wave propagation. The wave speeds are given by

$$\lambda = -\mu = \{h(v,t)\}^{1/2} \tag{5}$$

to which there correspond the left eigenvectors of A

$$l^{(\lambda)} = [1, \quad \lambda^{-1}] \quad ; \quad l^{(\mu)} = [1, \quad \mu^{-1}]. \tag{6}$$

2. Riemann-like exact solutions along characteristics

In the case where system (2), (3) assumes an autonomous form (i.e. $h = h(v), f = f(v)$), owing to the assumption of strictly hyperbolicity (4), on introduction of the Riemann field variables

$$r(U) = \int l^{(\lambda)} \cdot dU = u + \gamma(v); \quad s(U) = \int l^{(\mu)} \cdot dU = u - \gamma(v) \tag{7}$$

with $\gamma(v) = \int \{h(v)\}^{-1/2} dv$ the model (2), (3) after left multiplying by $l^{(\lambda)}$ and $l^{(\mu)}$, takes the form

$$r_t + \lambda(r,s)r_x = \beta_1 \tag{8}$$

$$s_t + \mu(r,s)s_x = \beta_2 \tag{9}$$

where $\beta_1 = l^{(\lambda)} \cdot B = \beta, \quad \beta_2 = l^{(\mu)} \cdot B = -\beta$ and $\beta = \{h(v)\}^{-1/2} f(v)$.

Contrary to what happens in the corresponding homogeneous case (i.e.$f = 0$), equations (8), (9) show that r and s are no longer invariant along the characteristic curves associated to the wave speeds λ and μ, respectively.

Recent papers [2-4] have been devoted to study the consistency of systems like (8), (9) with a pair of equations of the form

$$F_t + \lambda(u,v)F_x = 0 \tag{10}$$

$$G_t + \mu(u,v)G_x = 0 \tag{11}$$

where the functions $F(u,v)$ and $G(u,v)$ are to be determined under the condition

$$\frac{\partial(F,G)}{\partial(u,v)} \neq 0. \tag{12}$$

The equations (10), (11) are to be considered as additional equations (constraints) which must be satisfied by the field variables u and v. Consequently, once the functional form of $F(u,v)$ and $G(u,v)$ has been determined, we are led to particular solutions of (2), (3) (or of (8), (9)) which allow for the existence of Riemann-like invariant quantities along the associated characteristic curves. By requiring $F(u,v)$ and $G(u,v)$ to specialize to forms of relevant interest in wave propagation it is possible to provide a vehicle either to determining exact solutions to governing systems like (2), (3) or to selecting in specific problems classes of model constitutive laws for the involved material response functions.

Taking (12) into account and assuming $\partial(F,G)/\partial(x,t) \neq 0$ (or equivalently $\partial(u,v)/\partial(x,t) \neq 0$) by means of the variable transformation $x = x(F,G), t = t(F,G)$ the full set of equations (8)-(11) reduces to the form

$$\beta_1 = \frac{r_G}{t_G}; \quad \beta_2 = \frac{s_F}{t_F} \tag{13}$$

$$x_G = \lambda t_G; \quad x_F = \mu t_F \tag{14}$$

The functions $F(u,v)$ and $G(u,v)$ as well as the solution $r(x,t), s(x,t)$ of (8), (9) (or $u(x,t)$ and $v(x,t)$ of (2), (3)) which allows the constraint equations (10), (11) to hold can be determined, in principle, by solving the set of coupled equations (13), (14).

According to the underlying ideas of the reduction procedures which have been widely used in Continuum Mechanics to solve initial and/or boundary value problems of physical interest [5-6], here we aim to develop an "ad hoc" technique in order to provide ansatzs to the problem at hand. Actually, we will be led to prescribe F and G to adopt some forms which, in turn, will permit us to specialize the material response functions $h(v)$ and $f(v)$ in order that the full set of conditions (13) and (14) can be integrated in a closed form.

Cross differentiation in (14) produces the second order equation

$$t_{FG} + \frac{\lambda_F}{\lambda - \mu} t_G - \frac{\mu_G}{\lambda - \mu} t_F = 0 \tag{15}$$

Along the lines of the analysis carried on in [3-5] the approach which will be developed here is based upon the following steps:

i) Reduction of equation (15) to a "canonical form" allowing for an explicit integration;

ii) Insertion of the general solution $t(F,G)$ of (15) into (13) and further requirement to the resulting conditions to be fulfilled.

Since $\lambda = -\mu$, the condition [7-9]

$$\frac{\partial}{\partial G}\left(\frac{\lambda_F}{\lambda - \mu}\right) = \frac{\partial}{\partial F}\left(\frac{\mu_G}{\mu - \lambda}\right) \tag{16}$$

is satisfied whence by means of the transformation

$$t^* = \lambda^{1/2} t \tag{17}$$

(15) specializes to the self-adjoint equation

$$t^*_{FG} - \left(\frac{\partial^2 \lambda^{1/2}}{\partial F \partial G} / \lambda^{1/2}\right) t^* = 0 \tag{18}$$

Solutions to equation (18) can be obtained in the form of a Bergmann-like series [6], [10]. However such a method of approach is not suitable to our purpose of achieving the integration in a closed form of the full set of equations (13), (14) which, ultimately, would lead to characterizing exact solutions to the governing system (2), (3).

In [9] there has been carried on an analysis in order to reduce equations like (18) to some "canonical forms" allowing for an explicit integration. Here we shall restrict our attention to the case where (18) takes on the form

$$t^*_{FG} = 0 \tag{19}$$

that is tantamount to require F and G to fulfill the following relationship

$$P(F) + Q(G) = \lambda^{1/2} \tag{20}$$

where P and Q are arbitrary functions. Bearing in mind the equations (10), (11) it is straightforward to ascertain that, without any loss of generality, during our further analysis we can set $P = F$ and $Q = G$. It is to be noticed that, according to (20), the condition (12) requires $dh/dv \neq 0$.

After (20), via the variable transformation (17), the solution of (15) is given by

$$t = \frac{M(F) + N(G)}{F + G} \tag{21}$$

where $M(F)$ and $N(G)$ are arbitrary functions of integration.

Next, by inserting (21) into (13) and taking (7) into account, after some algebra [4], it is easy to see that the functions $M(F)$ and $N(G)$ specialize, respectively, to

$$M(F) = \frac{k}{6}F^3 - \frac{N_0}{2}F^2 + N_1 F + M_2 \tag{22}$$

$$N(G) = \frac{k}{6}G^3 + \frac{N_0}{2}G^2 + N_1 G + N_2 \tag{23}$$

and the functions $h(v)$, $f(v)$ to obey the condition

$$8\frac{d}{dv}\left(\frac{h^{1/4}}{dh/dv}\right) - \frac{d}{dv}\left(\frac{f}{h}\right)\left\{\frac{k}{6}h^{3/4} + \nu\right\} = 0; \quad \nu = -2(M_2 + N_2) \tag{24}$$

Finally, taking into account (21), (22) and (23), from the pair of equations (13) and (14) we obtain:

$$x = \frac{k}{12}(G - F)(G + F)^3 + \frac{N_0}{6}(G + F)^3 + \frac{\nu}{2}(G - F) \tag{25}$$

$$t = \frac{k}{6}(F^2 - FG + G^2) + \frac{N_0}{2}(G - F) - \frac{\nu}{2(F + G)} \tag{26}$$

$$F = -\frac{8}{k\Omega(v)}u + \left\{h^{1/4} - \psi(v)\right\} \tag{27}$$

$$G = \frac{8}{k\Omega(v)}u + \psi(v) \tag{28}$$

where

$$\frac{d\Omega}{dv} = fh^{-5/4}\frac{dh}{dv}; \quad \psi(v) = \Omega^{-1}(v)\left\{\omega(v) + \psi_0\right\} \tag{29}$$

$$\frac{d\omega}{dv} = -\frac{h^{-3/4}}{k}\frac{dh}{dv}\left\{\frac{f}{h}\left(N_0 - \frac{k}{2}h^{3/4}\right) - \frac{k}{8}\Omega(v)\right\} \tag{30}$$

Relations (25) to (28) define, in the present case, the required solutions to (2), (3) which make the concerned functions F and G invariant along the characteristic curves $C^{(\lambda)}$ and $C^{(\mu)}$, respectively.

3. Linearization procedure

Here our main concern is to reduce the nonautonomous and nonhomogeneous model (2), (3) to linear form. To this end first we transform the governing nonautonomous equations to an autonomous form which can be linearized further by using suitable variable transformations.

Recently in [11] and [12] there has been shown that the canonical variables associated to a given Lie-group of transformations represent a useful tool to reduce nonautonomous systems to autonomous form.

In the present case it is possible to show [13] that the system (2), (3) admits the following one-parameter (ν) infinitesimal Lie group of transformations:

$$u^* = u + \nu(\alpha u + \beta x) + O(\epsilon^2) \tag{31}$$

$$\frac{v^* = v + \nu[(\alpha - dT}{dt)v + \delta(t)] + O(\epsilon^2)} \tag{32}$$

$$x^* = x \quad ; \quad t^* = t + \nu T(t) + O(\epsilon^2) \tag{33}$$

where $T(t)$ and $\delta(t)$ are unspecified functions, while α and β are arbitrary constants. Furthermore the material response functions $h(t,v)$ and $f(t,v)$ must adopt the form:

$$h(t,v) = \frac{\phi_0(\bar{v})}{T^2} \tag{34}$$

$$f(t,v) = \frac{e^{\alpha F(t)}}{T^2}\left\{\int \pi(t,\bar{v})T^2 e^{-\alpha F(t)}dt + \psi_0(\bar{v})\right\} \tag{35}$$

where

$$\bar{v} = Te^{-\alpha F(t)}v - \int \delta(t)e^{-\alpha F(t)}dt \tag{36}$$

$$\pi(t,\bar{v}) = -\frac{\beta h}{T}\frac{d\delta}{dt} - \frac{1}{T^2}\frac{d^2T}{dt^2}e^{\alpha F(t)}\left\{\int \delta(t)e^{-\alpha F(t)}dt + \bar{v}\right\} \tag{37}$$

and $F(t) = \int dt/T(t)$. Then, following [12] (see also [13] for details), we introduce the canonical variables associated to (31)-(33):

$$\tau = F(t) \tag{38}$$

$$\bar{u} = \xi(t,x,u) = e^{-\alpha F(t)}\left(u + \frac{\beta}{\alpha}x\right) \tag{39}$$

$$\bar{v} = \eta(t,v) = Te^{-\alpha F(t)}v - \int \delta(t)e^{-\alpha F(t)}dt \tag{40}$$

where ξ and η are the similarity variables connected to the group (31)-(33) which are determined by the characteristic equations:

$$\frac{dt}{T(t)} = \frac{du}{\alpha u + \beta x} = \frac{dv}{(\alpha - T'(t))v + \delta(t)} \tag{41}$$

Thus, taking into account of (34) and (35), by means of the variable transformation (38)-(40), the model (2), (3) reduces to the autonomous form:

$$\bar{u}_\tau - \bar{v}_x = -\alpha\bar{u} \tag{42}$$

$$\bar{v}_\tau - \phi_0(\bar{v})\bar{u}_x = \psi_0(\bar{v}) - \alpha\bar{v} \tag{43}$$

There are different possibilities to linearize a 2×2 quasilinear nonhomogeneous model like (42), (43) either by using Bäcklund transformations [6], [14] or using hodograph-like transformations [15-18]. Here we shall be concerned with three different cases.

i) We set $\alpha = \beta = 0$ and

$$\phi_0(\bar{v}) = \frac{(1+ke_0\bar{v})^2}{d_0}; \quad \psi_0(\bar{v}) = -\frac{e_0(1+ke_0\bar{v})\bar{v}}{d_0} \tag{44}$$

where k, e_0, d_0 are constants. The model (42), (43) supplemented by (44) has been studied in [15]. In particular there has been proved that under the variable transformation

$$\bar{x} = x - k\bar{u}; \quad \bar{\tau} = \tau - \frac{d_0}{e_0}\ln|1+ke_0\bar{v}| \tag{45}$$

the equations (42), (43) assume the linear form

$$\bar{u}_{\bar{\tau}} - \bar{v}_{\bar{x}} = 0 \tag{46}$$

$$\bar{v}_{\bar{\tau}} - \frac{1}{d_0}\bar{v}_{\bar{x}} = -\frac{e_0}{d_0}\bar{v}. \tag{47}$$

Furthermore all the results obtained in [15] concerning nonlinear wave propagation problems can be extended to the present case.

ii) Choosing

$$\psi_0(\bar{v}) = \alpha\bar{v} \tag{48}$$

and eliminating $\bar{u}$ between the pair of equations (42), (43) we get

$$\frac{\partial}{\partial\tau}\left(\frac{\bar{v}_\tau}{\phi_0 2}\right) + \alpha\frac{\bar{v}_\tau}{\phi_0} - \bar{v}_{xx} = 0 \tag{49}$$

Provided that

$$\phi_0(\bar{v}) = \frac{q_0 a^2}{\alpha}(\bar{v}_0 + v)^2; \quad q_0, a, \bar{v}_0 = \text{constant} \tag{50}$$

the nonlinear telegraph equation (49) reduces to the linear telegraph equation

$$p_0\frac{\partial^2\omega}{\partial\bar{\tau}^2} + q_0\frac{\partial\omega}{\partial\bar{\tau}} - \frac{\partial^2\omega}{\partial\bar{x}^2} = 0; \quad p_0 = \frac{q_0}{\alpha} \tag{51}$$

by means of the Bäcklund transformation [6], [14]

$$d\bar{x} = \left\{a\left(\frac{G_\tau}{\alpha} + G\right) + b\right\}dx + \frac{a}{\alpha}\bar{v}_x d\tau \tag{52}$$

$$G = -\frac{\alpha}{a^2 q_0(\bar{v}+\bar{v}_0)}\bar{\tau} = \tau + \frac{1}{\alpha}\ln|aG+b|; \quad \omega = \frac{1}{q_0(aG+b)} \tag{53}$$

with b constant.

iii) Finally, setting $\bar{x} = -x$ and $\alpha = \beta = \psi_0(\bar{v}) = 0$, then the model (42), (43) reduces formally to the homogeneous system describing nonlinear dielectric media in which the magnetic flux B and the magnetic field H are linearly related (i.e. $B = \mu_0 H, \mu_0 =$ constant). In this case the reduction to linear form is achieved by using the hodograph transformation [7], [19], so that the results there obtained can be easily extended also to the present case.

Acknowledgements

This work was supported by C.N.R. through G.N.F.M.

References

1. N. Cristescu, Dynamic Plasticity, North-Holland, Amsterdam (1967).
2. D. Fusco, N. Manganaro, Exact solutions to flows in fluid filled elastic tubes, in Differential Equations with Applications to Mathematical Physics, (W. F. Ames, J. V. Herod, E. M. Harrell eds.), Academic Press. To appear.
3. D. Fusco, N. Manganaro, Generation of exact solutions to a class of quasilinear hyperbolic models via reduction techniques, in print on Mathematical and Computer Modelling.
4. D. Fusco, N. Manganaro, Riemann invariants-like solutions for a class of rate-type materials. To appear.
5. C. Rogers, W. F. Shadwick, Bäcklund Transformations and Their Applications, Academic, New York (1982).
6. C. Rogers, W. F. Ames, Nonlinear Boundary Value Problems in Science and Engineering, Academic, New York (1080).
7. A. Donato, D. Fusco, Some application of the Riemann method to electromagnetic wave propagation in nonlinear media, ZAMM 60, 539 (1980).
8. C. Currò, D. Fusco, On a Class of quasilinear hyperbolic reducible systems allowing for special wave interactions, Z. angew. Math. Phys. 38, 580-594 (1987).
9. C. Currò, D. Fusco, Reduction to linear canonical forms and generation of conservation laws for a class of quasilinear hyperbolic systems, Int. J. Non-Linear Mech. 23, 25-35, (1988).
10. S. Bergman, Integral Operators in the Theory of Partial Differential Equations, Springer-Verlag, Berlin, New York (1986).
11. A. Donato, F. Oliveri, When nonautonomous equations are equivalent to autonomous ones, to appear.
12. N. Manganaro, Linearization of nonautonomous models describing fluid filled elastic tubes and nonlinear elastic rods with variable cross-section, in print on Acta Mechanica.
13. N. Manganaro, G. Valenti, Group analysis and linearization procedure for a nonautonomous model describing rate-type materials, J. Math. Phys. 34, (3), 1993.
14. C. Rogers, T. Ruggeri, A reciprocal Bäcklund transformation: application to a nonlinear hyperbolic model in heat conduction, Lettere al Nuovo Cimento 2 (44), 289 (1985).
15. B. Seymour, E. Varley, Exact solutions for a large amplitude waves in dispersive and dissipative systems, Studies Appl. Math. 72, 241 (1985).
16. D. Fusco, N. Manganaro, Linearization of a hyperbolic model for non-linear heat conduction through hodograph-like and Bäcklund transformations, Int. J. Non-Linear Mech. 24 (2), 99 (1989).
17. D. Fusco, N. Manganaro, Prominent features of a variables transformation for a class of quasilinear hyperbolic systems of first order, Nonlinear Wave Motion, A. Jeffrey ed., Longman, 71 (1989).
18. D. Fusco, N. Manganaro, A class of linearizable models and generation of material response functions to nonlinear hyperbolic heat conduction, J. Math. Phys. 32 (11), 3043 (1991).
19. A. Jeffrey, Quasilinear Hyperbolic Systems and Waves, Research Notes in Mathematics, Pitman, London (1976).

CONDITIONAL SYMMETRIES OF THE EQUATIONS OF MATHEMATICAL PHYSICS

WILHELM FUSHCHYCH
Institute of Mathematics
Ukrainian Academy of Sciences
Repin St. 4
252004 Kiev 4
Ukraine

Abstract. We briefly present the results of research in conditional symmetries of equations of mathematical and theoretical physics: the Maxwell, D'Alembert, Schrödinger and KdV equations, as well as the equations of heat conduction and acoustics. Exploiting conditional symmetry, we construct a wide class of exact solutions of these equations, which cannot be obtained by the classical method of Sophus Lie.

1. Introduction

The concept and terminology of conditional symmetry and conditional invariance were introduced and developed in the series of articles ([1]-[11]) (see also Mathematical Reviews for the years 1983-1993). Later, this concept was exploited by other authors for the construction of solutions of various non-linear equations of mathematical physics. It turned out that nearly all the basic non-linear equations of mathematical physics have non-trivial conditional symmetry ([2],[9],[10]).

We understand the conditional symmetry of an equation as being a symmetry (local or non-local) of some non-trivial subset of its solution set (the formal definition of the idea of conditional symmetry can be found in Appendix 4 of [2] and in the article [3]). The general definition of conditional symmetry as the symmetry of a subset of the set of solutions is non-constructive and requires further specification: the analytical description of a condition (as an equation) on the solutions of the given equation, which extend or alter the symmetry of the starting equation. Therefore, the basic problem in the investigation of conditional symmetries is that of describing those supplementary equations which increase or change the symmetry of the beginning equation. This is very complex, non-linear problem in general (even in the case of quite simple non-linear equations), which can often be significantly more complicated than constructing solutions of the equation at hand. It is thus meaningful to talk of the conditional symmetry of some class of equations.

Non-trivial conditional symmetries of a PDE (partial differential equation) allows us to obtain in explicit form such solutions which can not be found by using the symmetries of the whole set of solutions of the given PDE. Moreover, conditional symmetries increase significantly the class of PDEs for which we can construct ansatzes which reduce these equations to (systems of) ODEs (ordinary differential equations). As a rule, the reduced equations one obtains from conditional symme-

N. H. Ibragimov et al. (eds.),
Modern Group Analysis: Advanced Analytical and Computational Methods in Mathematical Physics, 231–239.

tries are significantly simpler than those found by reduction using symmetries of the full set of solutions. This allows us to construct exact solutions of the reduced equations.

Looking back, we can say today, that many mathematicians, mechanicians and physicists, such as Euler, D'Alembert, Poincaré, Volterra, Whittaker, Bateman, implicitly used conditional symmetries for the construction of exact solutions of the linear wave equation. Some well-known solutions of this equation can not be obtained by using only Lie symmetries of the full solution set.

2. Conditional symmetry of Maxwell's equation

We shall first consider the first pair of Maxwell's equations

$$\frac{\partial \mathbf{E}}{\partial t} = \text{rot } \mathbf{H}, \quad \frac{\partial \mathbf{H}}{\partial t} = -\text{rot } \mathbf{E} \tag{1}$$

The maximal invariance algebra (in the sense of Lie) of these equations is studied in ([2]). The basis elements of this algebra $< \partial_0, \partial_a, J_{ab}, \mathcal{D} >$ are

$$\begin{gathered} \partial_0 = \frac{\partial}{\partial x^0}, \quad \partial_a = \frac{\partial}{\partial x^a} \\ J_{ab} = x_a\partial_b - x_b\partial_a + s_{ab}, \quad a, b = 1,2,3 \\ \mathcal{D} = x^\mu \partial_\mu + \text{const.} \end{gathered} \tag{2}$$

s_{ab} are 6×6 matrices realizing a representation of the group $O(3)$. Thus the system (1) is invariant under the four-dimensional translations ∂_μ, the rotations J_{ab} and scale transformations $\mathcal{D}$, but it is not invariant under the Lorentz boosts

$$J_{0a} = x_0\partial_a - x_a\partial_{0a} + s_{0a}, \quad x_0 = t \tag{3}$$

the matrices $< s_{0a}, \; s_{ab} >$ realizing a representation of the Lorentz group $O(1,3)$.

Theorem 1 ([2] 1983,[15] 1987). *The system (1) is conditionally invariant under the Lorentz boosts (3) if and only if the solutions of (1) satisfy the conditions*

$$div\, \mathbf{E} = 0, \qquad div\, \mathbf{H} = 0 \tag{4}$$

It is evident from this theorem, that the concept of conditional invariance of a PDE is natural, and leads us, *by purely group-theoretic means*, to the fundamental, overdetermined system of Maxwell's equations.

3. Conditional symmetry of the wave equation

We now examine the non-linear D'Alembert equation

$$\Box u = F(u), \qquad u = u(x_0, x_1, x_2, x_3) \tag{5}$$

$F(u)$ being an arbitrary, smooth function. Equation (5) has conformal symmetry $C(1,3)$ if and only if $F = \lambda u^3$ or $F = 0$ (see for instance [8], [10]). This is the maximal symmetry of all of the solution set of equation (5). For an arbitrary function, (5) admits only the symmetry groups $P(1,3)$ or $C(1,3)$.

Theorem 2 ([5], 1985). *Equation (5), with $F = 0$ is conditionally invariant under the infinite-dimensional algebra with basis elements*

$$X = \xi^\mu(x,u)\frac{\partial}{\partial x^\mu} + \eta(x,u)\frac{\partial}{\partial u} \tag{6}$$

$$\xi^\mu(x,u) = c^{00}(u)\, x^\mu + c^{\mu\nu}(u)\, x_\nu + d^\mu(u), \quad \eta(x,u) = \eta(u) \tag{7}$$

where $c^{00}(u)$, $c^{\mu\nu}(u)$, $d^\mu(u)$, $\eta(u)$ are arbitrary functions of u, if one imposes the condition

$$\frac{\partial u}{\partial x_\mu}\frac{\partial u}{\partial x^\mu} = 0 \tag{8}$$

In this way, the eikonal equation (8), significantly increases the symmetry of the starting equation (5). The system of equations (5), (8), with $F = 0$, is consistent.

Theorem 3 ([10, 15], 1988, 1989). *The equation (5) is conditionally invariant under the conformal group, if*

$$F = \frac{3\lambda}{(u+c)} \tag{9}$$

$$\frac{\partial u}{\partial x_\mu}\frac{\partial u}{\partial x^\mu} = \lambda \tag{10}$$

where λ, c are arbitrary constants. The operators of conformal symmetry are

$$K_\mu = 2x_\mu \mathcal{D} - (x_\alpha x^\alpha - u^2)\frac{\partial}{\partial x^\mu}, \quad \mu = 0,1,2,3. \tag{11}$$

$$\mathcal{D} = x^\mu \frac{\partial}{\partial x^\mu} + u\frac{\partial}{\partial u}$$

Remark. It is important to note, that the operators (11) differ principally from the conformal operators for equation (5), when $F = 0$ or $F = \lambda u^3$. In those cases, the conformal operators are

$$\tilde{K}_\mu = 2x_\mu \mathcal{D} - x_\alpha x^\alpha \frac{\partial}{\partial x^\mu} \tag{12}$$

$$\mathcal{D} = x^\mu \frac{\partial}{\partial x^\mu}$$

The operators (11) are *non-linear*, whereas those in (12) are linear.

Thus the wave equation (5), (9), with non-linear condition (10), has a symmetry possessed by neither the solution set for the linear equation, nor that for the non-linear equation.

4. Criteria for conditional symmetry

Let us consider some PDE

$$L(x, u_{(1)}, u_{(2)}, \ldots, u_{(n)}) = 0 \tag{13}$$

$$u_{(1)} = (u_0, u_1, \ldots, u_{(n)}), \quad u_{(2)} = (u_{01}, u_{02}, \ldots, u_{nn}) \ldots$$

$$u_\mu = \frac{\partial u}{\partial x^\mu}, \quad u_{\mu\nu} = \frac{\partial^2 u}{\partial x^\mu \partial x^\nu}, \ldots$$

Definition 1 (S. Lie, 1884). *Equation (13) is invariant with respect to the operator (6) if*

$$X_s L = \lambda L \tag{14}$$

where X_s is the s-th prolongation of (6), and $\lambda = \lambda(x, u)$ is an arbitrary function.

Let us denote by the symbol

$$Q = < Q_1, Q_2, \ldots, Q_r > \tag{15}$$

some set of operators which does not belong to the invariance algebra (IA) of equation (13).

Definition 2 ([2], 1987). *Equation(13) is said to be conditionally invariant under the operators Q from (15), if there exists a supplementary condition on the solutions of (13) of the form*

$$L_1(x, u, u_{(1)}, \ldots, u_{(n)}) = 0 \tag{16}$$

such that (13) together with (16) is invariant under the Q.

Thus one has the following conditions

$$Q_s L = \lambda_0 L + \lambda_1 L_1 \tag{17}$$

$$Q_s L_1 = \lambda_2 L + \lambda_3 L_1 \tag{18}$$

or

$$Q_s L \Big|_{\substack{L=0 \\ L_1=0}} = 0, \qquad Q_s L_1 \Big|_{\substack{L=0 \\ L_1=0}} = 0 \tag{19}$$

An important class of supplementary conditions (16) is that for which the equation $L_1 = 0$ is a quasi-linear equation of first order

$$L_1(x, u, u_{(1)}) \equiv Qu = 0 \tag{20}$$

$$Q = y^\mu(x, u)\frac{\partial}{\partial x^\mu} + z(x, u)\frac{\partial}{\partial u} \tag{21}$$

with y^μ, z being smooth functions. In this case, we shall say that (13) is Q-conditionally invariant.

In this way, the problem of finding the conditional symmetry of (13) reduces to the solution of the equations (17), (18). The conditions (16), (20) can be considered as equations for the construction of ansatzes for the starting equation (13). The problem of calculating the conditional symmetry is far more complicated than the usual method of Lie for finding the symmetry of the full solution set. In the case of conditional symmetries, the defining equations are, as a rule, non-linear equations which can be solved in only some cases. Fortunately, for most of the equations of non-linear mathematical physics, one can construct partial solutions of the defining equations.

5. A list of equations with non-trivial conditional symmetry

Conditional symmetries began to be exploited only quite recently, and the first publications appeared only in 1983 ([1], [2]). Now, the number of articles in this are is increasing rapidly with each year, and therefore it is difficult to give a complete list (for 1992) of important equations of mathematical physics possessing conditional symmetry. So I shall only give those equations which we have studied and which are interested from our Kievan point of view. We have put in brackets the year(s) when the conditional symmetry of the given equation was found. More detailed information about ansatzes and solutions of the above equations are to be found in the original articles, a list of which are given in ([2], [9], [11]).

1. $$u_0 + u_{11} = F(u) = \begin{cases} \lambda u(u^2-1) \\ \lambda(u^3-3u+2) \\ \lambda u^3 \\ \lambda u(u^3+1) \end{cases} \qquad (1988, 1990)$$

2. $$iu_0 + \Delta u + F(|u|)u = 0 \qquad (1990)$$

$$F(|u|) = \lambda_1 |u|^{4/r} + \lambda_2 |u|^{-4/r}, \quad F(|u|) = \lambda_3 \ln(u^* u)$$

λ_1, λ_2, r arbitrary, real; λ_3 arbitrary, complex

3. $$u_{00} = u\Delta u, \quad u_{00} = c(x, u, u_{(1)})\Delta u \qquad (1987, 1988)$$

4. $$u_{01} - (F(u)u_1)_1 - u_{22} - u_{33} = 0 \qquad (1990)$$

5. $$u_0 + \nabla(F(u)\nabla u) = 0 \qquad (1988)$$

6. $$u_0 + F(u)u_1^{\kappa} + u_{111} = 0 \qquad (1991)$$

7. $$u_0 + (\varphi(u))_{11} + \frac{N}{x_1}(\varphi(u))_1 = F(u)$$

$$u_0 + u_{11} + \frac{3}{2x_1}u_1 = \lambda u^3 \qquad (1992)$$

$$u_0 + uu_{11} + \frac{N}{x_1}uu_1 = \lambda u + \lambda_2$$

8. $$\mathbf{u}_0 + (\mathbf{u}\nabla)\mathbf{u} = -\frac{1}{\rho}\nabla p$$

$$\rho_0 + \mathrm{div}(\rho\mathbf{u}) = 0, \quad p = f(\rho), \ p = \tfrac{1}{2}\rho^2$$

9. $$\gamma^\mu \partial_\mu \Psi + F(\overline{\Psi}\Psi)\Psi = 0 \qquad (1989)$$

10. $$(1 - u_\alpha u^\alpha)\Box u + u^\mu u^\nu u_{\mu\nu} = 0 \qquad (1989)$$

6. Conditional symmetry and exact solutions of KdV type equations

To illustrate the constructive nature of conditional symmetries, we shall examine the equation

$$u_0 + F(u)u_1^\kappa + u_{111} = 0 \tag{22}$$

where $F(u)$ is a smooth function, $\kappa \neq 0$ is an arbitrary, real parameter. When $F(u) = u$, $\kappa = 1$, equation (22) coincides with the standard KdV equation.

Theorem 4 ([11], 1991). *Equation (22) is Q-conditionally invariant with respect to the following operators*

$$Q = x_0^r\, \partial_1 + H(x,u)\, \partial_u \tag{23}$$

with r an arbitrary, real parameter, in the following cases.

$$\begin{aligned} 1.\quad F(u) &= \lambda_1 u^{(2-\kappa)/\kappa} + \lambda_2 u^{(1-\kappa)/2} \\ H(x,u) &= \left(\frac{\lambda_1\kappa}{2}\right)^{-1/\kappa} u^{1/2} \end{aligned} \tag{24}$$

$$\begin{aligned} 2.\quad F(u) &= (\lambda_1 \ln u)^{1-\kappa} \\ H(x,u) &= (\kappa\lambda_1)^{-1/\kappa} \end{aligned} \tag{25}$$

$$\begin{aligned} 3.\quad F(u) &= (\lambda_1 \arcsin u + \lambda_2)(1-u^2)^{(1-\kappa)/2} \\ H(x,u) &= (\kappa\lambda_1)^{-1/\kappa}(1+u^2)^{1/2} \end{aligned} \tag{26}$$

$$\begin{aligned} 4.\quad F(u) &= (\lambda_1 \sinh^{-1} u + \lambda - 2)(1+u^2)^{(1-\kappa)/2} \\ H(x,u) &= (\kappa\lambda_1)^{-1/\kappa}(1+u^2)^{1/2} \end{aligned} \tag{27}$$

$$\begin{aligned} 5. \quad F(u) &= \lambda_1 u \\ H(x,u) &= (\kappa\lambda_1)^{-1/\kappa} \end{aligned} \tag{28}$$

where $r = 1/\kappa$, $\kappa \neq 0$, λ_1, λ_2 are arbitrary, real parameters.

Exploiting the operator of conditional symmetry (23), one can construct ansatzes for the solutions of equation (22), some of which I now exhibit.

The ansatz

$$u = \left(\frac{x_1}{2}(\frac{\kappa\lambda_1 x_0}{2})^{-1/\kappa} + \varphi(x_0)\right)^2$$

gives the solution

$$u = \left(\frac{x_1}{2}(\frac{\kappa\lambda_1 x_0}{2})^{-1/\kappa} + \lambda x_0^{-1/\kappa} - \lambda_2/\lambda_1\right)^2$$

when $F(u)$ is as in (24). The ansatz

$$u = \exp\left(\varphi(x_0 + (\kappa\lambda_1 x_0)^{-1/\kappa}\right)$$

gives the solution

$$u = \exp\left(-\frac{\kappa(\kappa\lambda_1)^{-3\kappa})}{\kappa - 2} x_0^{1-3/\kappa} + \lambda x_0^{-1/\kappa} + (\kappa\lambda_1 x_0)^{-1/\kappa} x_1 - \lambda_2/\lambda_1\right)$$

when $F(u)$ is as in (25) with $\kappa \neq 2$. The ansatz

$$u = \sin(\varphi(x_0) + (\kappa\lambda_1 x_0)^{-1/\kappa} x_1)$$

gives the solution

$$u = \sin\left(\frac{\kappa(\kappa\lambda_1)^{-3/\kappa}}{\kappa - 2} x_0^{1-3/\kappa} + \lambda x_0^{-1/\kappa} + (\kappa\lambda_1 x_0)^{-1/\kappa} x_1 - \lambda_2/\lambda_1\right)$$

for $\kappa \neq 2$.

Theorem 5 ([12], 1990). *The equation*

$$u_{01} - (F(u)u_1)_1 - u_{22} - u_{33} = 0 \tag{29}$$

is invariant under the infinite-dimensional algebra

$$X = a_i(u)R_i, \quad i = 1, \ldots, 12 \tag{30}$$

where the $a_i(u)$ are arbitrary, smooth functions, if one adds to (29) the condition

$$u_0 u_1 - F(u)u_1^2 - u_2^2 - u_3^2 = 0 \tag{31}$$

The operators R_i are given as follows:

$$R_{\mu+1} = \partial_\mu, \quad \mu = 0, \ldots, 3; \; R_5 = x_3\partial_2 - x_2\partial_3$$

$$R_6 = x_2\partial_1 + 2x_0\partial_2, \quad R_7 = x_3\partial_1 + 2x_0\partial_3$$

$$R_8 = x^\mu \partial_\mu, \quad R_9 = x_0\partial_0 + 2x_1\partial_1 + 3x_2\partial_2 + 3x_3\partial_3 - 2\frac{F(u)}{\dot{F}(u)}\partial_u$$

$$R_{10} = \dot{F}(u)x_0\partial_1 - \partial_u, \quad R_{11} = x_2\partial_0 + 2(x_1 + F(u)x_0)\partial_2$$

$$R_{12} = x_3\partial_0 + 2(x_1 + F(u)x_0)\partial_3$$

7. Antireduction

In [10], we have begun work on antireduction. By the term *antireduction* of a PDE we understand the finding of such ansatzes which transform the given PDE into a system of equations for some (unknown, and to be found) functions. In this process, the number of independent variables may remain the same, or be reduced (dimensional reduction), but the number of dependent variables increases. As a rule, one usually exploits the converse of this, that is, one reduces to a system with fewer dependent variables (reduction of components). To illustrate the effectiveness of antireduction, we consider the equation for short waves in gas dynamics

$$2u_{01} - 2(x_1 + u_1)u_{11} + u_{22} + 2\lambda u_1 = 0 \tag{32}$$

We impose the condition

$$\left[u_{111}x_1^{3/2}\right]_1 = 0 \tag{33}$$

on (32). The general solution of (33) is

$$u = x_1^{3/2}\varphi^1 + x_1^2\varphi^2 + x_1\varphi^3 + \varphi^4 \tag{34}$$

with $\varphi^i = \varphi^i(x_0, x_2)$, $i = 1, 2, 3, 4$ being arbitrary functions. Using (34) as an ansatz, equation is reduced to a system with two independent variables.

$$\varphi^3 = 0, \quad \varphi^1_{22} = 0 \quad \varphi^2_{22} = 0$$

$$\varphi^4_{22} = \frac{9}{4}(\varphi^1)^2, \quad \varphi^1_0 = \varphi^1(3\varphi^2 + \frac{1}{2} - \lambda) \tag{35}$$

$$\varphi^2_0 = 2(\varphi_2)^2 - \varphi^2(1 - \lambda)$$

Solving the system (35), we found exact solutions of the starting equation (32) ([11]).

The above results are only a sample of those already obtained. They illustrate the very fruitful nature of conditional symmetry and conditional invariance, and I hope that I have been able to demonstrate that there are new aspects to this concept which are yet to be exploited fully.

Acknowledgements

I would like to express my heartfelt thanks to Professors M. Torrisi and A. Valenti for such a well-organized conference and to Professor M. Boffi for enabling me to attend it. I would also like to thank Dr. P. Basarab-Horwath, Mathematics Department, Linköping University, Sweden, for useful and fruitful discussions during the final preparation of my talk.

References

1. Fushchych, W.: On symmetry and partial solutions of some multi-dimensional equations of mathematical physics. In *Algebraic-theoretical methods in the problems of mathematical physics*, Institute of mathematics, Ukrainian Academy of sciences, Kiev (1983).
2. Fushchych, W., Nikityn, A.: Symmetries of Maxwell's equations. Naukova Dumka, Kiev (1983) and Reidel-Kluwer, Dordrecht (1987).
3. Fushchych, W.: How can one extend the symmetry of differential equations? In *Symmetry and solutions of non-linear equations in mathematical physics*. Institute of mathematics, Ukrainian Academy of Sciences, Kiev (1987).
4. Fushchych, W., Tsyfra, I.: On reduction and solutions of the non-linear wave-equation with broken symmetry. J.Phys. A **20** (1987) 45–48.
5. Shulha, M.: Symmetry and some exact solutions of the D'Alembert equation with a non-linear condition. In *Symmetry and solutions of equations of mathematical physics*. Institute of Mathematics, Ukrainian Academy of Sciences, Kiev (1985).
6. Fushchych, W., Serov, M., Chopyk, W.: Conditional invariance and non-linear heat equations. Proc. Ukr. Acad. Sci. Ser. A. no. 9, (1988) 17–21.
7. Fushchych W., Serov, M.: Conditional invariance and exact solutions of a non-linear equation of acoustics. Proc. Ukr. Acad. Sci. Ser. A. no. 10, (1988) 27–31.
8. Fushchych, W., Shtelen, W., Serov, M.: Symmetry Analysis and Exact Solutions of Non-linear Equations of Mathematical Physics. Naukova dumka, Kiev (1989) and Kluwer, Dordrecht (1993).
9. Fushchych, W., Chopyk, W.: Conditional invariance of a non-linear Schrödinger equation. Proc. Ukr. Acad. Sci. Ser. A. no. 4, (1990) 30–33.
10. Fushchych, W.: Conditional symmetry of the equations of mathematical physics. Ukr. J. Math. **43** no. 11, (1991) 1456–1470.
11. Fushchych, W., Serov, M., Amerov, T.: On the conditional symmetry of the generalized Korteweg-de Vries equation. Proc. Ukr. Acad. Sci. Ser. A. no. 12 (1991) 28–30.
12. Fushchych, W., Chopyk, W., Myronyuk, P.: Conditional invariance and exact solutions of three-dimensional non-linear equations in acoustics. Proc. Ukr. Acad. Sci. Ser. A. no. 9 (1990) 25–28.
13. Fushchych, W., Shtelen, W., Serov, M., Popowych, R.: Conditional symmetry of the linear heat equation. Proc. Ukr. Acad. Sci. Ser. A. no. 12 (1992) 20–26.
14. Fushchych, W., Repeta, W.: Exact solutions of the equations of gas dynamics and non-linear acoustics. Proc. Ukr. Acad. Sci. Ser. A. no. 8 (1991) 35–42.
15. Fushchych, W., Zhdanov, R.: Symmetry and exact solutions of non-linear spinor equations. Phys. Rep. **46** (1989) 325–365.

PSEUDOPOTENTIAL SYMMETRIES FOR INTEGRABLE EVOLUTION EQUATIONS

F. GALAS
Institut für Mathematische Physik,
Technische Universität Braunschweig,
Mendelssohnstr. 3,
W-3300 Braunschweig, Germany

Abstract. The concept of non-local symmetries can be generalized by including pseudopotentials. The KdV-equation has in addition to the two well known hierarchies of Lie-Bäcklund symmetries three hierarchies of pseudopotential symmetries. One of these symmetries generates the solitary wave solution through the transformation of the trivial solution.

We shall consider an integrable evolution equation

$$\frac{\partial u}{\partial t} + K\left(x, t, u, \frac{\partial u}{\partial x}, \ldots, \frac{\partial^k u}{\partial x^k}\right) = 0 \tag{1}$$

with two independent variables x, t and one dependent variable u. Let the variables $\psi = (\psi^1, \ldots, \psi^n)$ denote a number of pseudopotentials of the equation (1). They are defined through the first order differential equations

$$\frac{\partial \psi}{\partial x} = F\left(x, t, u, \frac{\partial u}{\partial x}, \ldots, \frac{\partial^l u}{\partial x^l}, \psi\right), \quad \frac{\partial \psi}{\partial t} = G\left(x, t, u, \frac{\partial u}{\partial x}, \ldots, \frac{\partial^l u}{\partial x^l}, \psi\right), \tag{2}$$

with $F = (F^1, \ldots, F^n), G = (G^1, \ldots, G^n)$, such that the integrability conditions

$$\frac{\partial^2 \psi}{\partial t \partial x} - \frac{\partial^2 \psi}{\partial x \partial t} = \frac{\partial}{\partial t}(F) - \frac{\partial}{\partial x}(G) = 0$$

are satisfied for all solutions u of the equation (1). If the functions F and G depend only on x, t, u and derivatives of u the equations (2) yield potentials ψ and the integrability conditions lead to conservation laws of the equation (1). Pseudopotentials can be calculated by the prolongation method of Estabrook and Wahlquist [1].

The generalization of the concept of non-local symmetries by including pseudopotentials was first proposed by Edelen [2]. We look for pseudopotential symmetries of the equation (1) with infinitesimal generators of the form

$$\mathrm{v} = \xi(x, t, u, \psi)\frac{\partial}{\partial x} + \tau(x, t, u, \psi)\frac{\partial}{\partial t} + \eta(x, t, u, \psi)\frac{\partial}{\partial u}.$$

The prolongation of this vector field

$$pr\,\mathrm{v} = \mathrm{v} + \sum_{\alpha=1}^{n} \phi^\alpha \frac{\partial}{\partial \psi^\alpha}$$

N. H. Ibragimov et al. (eds.),
Modern Group Analysis: Advanced Analytical and Computational Methods in Mathematical Physics, 241–244.

is determined by the prolongation formula

$$\frac{\partial \phi}{\partial x} = pr\,\mathrm{v}(F), \quad \frac{\partial \phi}{\partial t} = pr\,\mathrm{v}(G), \tag{3}$$

with $\phi = (\phi^1, \ldots, \phi^n)$. The integrability conditions $\partial^2\phi/(\partial t \partial x) - \partial^2\phi/(\partial x \partial t) = 0$ are satisfied and so the variables ϕ are also pseudopotentials. If they can be expressed as functions of x, t, u and ψ, that is $\phi = \phi(x, t, u, \psi)$, the prolongation is closed and $pr\,\mathrm{v}$ represents a Lie point symmetry of the enlarged system of differential equations (1) and (2). The symmetry transformations $(\tilde{x}, \tilde{t}, \tilde{u}) = \exp(\varepsilon \mathrm{v}) \cdot (x, t, u)$ with the group parameter ε are obtained by solving the initial value problem

$$\frac{d\tilde{x}}{d\varepsilon} = \xi(\tilde{x}, \tilde{t}, \tilde{u}, \tilde{\psi}), \quad \tilde{x}(\varepsilon = 0) = x; \quad \frac{d\tilde{t}}{d\varepsilon} = \tau(\tilde{x}, \tilde{t}, \tilde{u}, \tilde{\psi}), \quad \tilde{t}(\varepsilon = 0) = t;$$

$$\frac{d\tilde{u}}{d\varepsilon} = \eta(\tilde{x}, \tilde{t}, \tilde{u}, \tilde{\psi}), \quad \tilde{u}(\varepsilon = 0) = u; \quad \frac{d\tilde{\psi}}{d\varepsilon} = \phi(\tilde{x}, \tilde{t}, \tilde{u}, \tilde{\psi}), \quad \tilde{\psi}(\varepsilon = 0) = \psi.$$

Examples of these kind of symmetries in the case of potentials were given by Bluman, Kumei and Reid [3] and by Galas [4]. If $\phi \neq \phi(x, t, u, \psi)$, we can introduce the variables ϕ as new pseudopotentials and repeat the prolongation procedure. The prolongation however may not close in a finite number of steps.

Let $\mathcal{R}$ be a regular recursion operator that maps symmetries of the equation (1) onto new symmetries. Then for any two vector fields $\mathrm{v}_a, \mathrm{v}_b$ the relation $\mathcal{R}^2[\mathrm{v}_a, \mathrm{v}_b] = \mathcal{R}\{[\mathcal{R}\mathrm{v}_a, \mathrm{v}_b] + [\mathrm{v}_a, \mathcal{R}\mathrm{v}_b]\} - [\mathcal{R}\mathrm{v}_a, \mathcal{R}\mathrm{v}_b]$ holds [5]. For $\mathcal{R}\mathrm{v}_b = 0$ this yields $\mathcal{R}^2[\mathrm{v}_a, \mathrm{v}_b] = \mathcal{R}[\mathcal{R}\mathrm{v}_a, \mathrm{v}_b]$. By induction it follows that

$$\mathcal{R}^{n+1}[\mathrm{v}_a, \mathrm{v}_b] = \mathcal{R}[\mathcal{R}^n \mathrm{v}_a, \mathrm{v}_b] \quad \text{for all} \quad n \in N \quad \text{if} \quad \mathcal{R}\mathrm{v}_b = 0. \tag{4}$$

As an example we calculate pseudopotential symmetries for the KdV-equation

$$\frac{\partial u}{\partial t} + 6u\frac{\partial u}{\partial x} + \frac{\partial^3 u}{\partial x^3} = 0. \tag{5}$$

The introduction of the three pseudopotentials ψ^1, ψ^2 and ψ^3 defined by the equations

$$\frac{\partial \psi^1}{\partial x} = \lambda - u - (\psi^1)^2, \quad \frac{\partial \psi^1}{\partial t} = \frac{\partial}{\partial x}[\frac{\partial u}{\partial x} - 2(2\lambda + u)\psi^1] \quad (\lambda \in R), \tag{6}$$

$$\frac{\partial \psi^2}{\partial x} = \psi^1, \quad \frac{\partial \psi^2}{\partial t} = \frac{\partial u}{\partial x} - 2(2\lambda + u)\psi^1, \tag{7}$$

$$\frac{\partial \psi^3}{\partial x} = \exp(-2\psi^2), \quad \frac{\partial \psi^3}{\partial t} = -2(2\lambda + u)\exp(-2\psi^2),$$

leads to three pseudopotential symmetries of the KdV-equation with the infinitesimal generators

$$\mathrm{v}_1 = \psi^1 \exp(2\psi^2)\frac{\partial}{\partial u}, \quad \mathrm{v}_2 = [\psi^1\psi^3 \exp(2\psi^2) + \frac{1}{2}]\frac{\partial}{\partial u},$$

$$v_3 = [\psi^1(\psi^3)^2\exp(2\psi^2) + \psi^3]\frac{\partial}{\partial u}.$$

The operator $\mathcal{R} = D_x^2 + 4(u-\lambda) + 2\frac{\partial u}{\partial x}D_x^{-1}$, where D_x is the total derivative operator with respect to x and D_x^{-1} is the inverse, is a regular recursion operator of the KdV-equation, which differs from the usual one by the constant term -4λ. We have

$$\mathcal{R}v_i = 0 \quad \text{for} \quad i = 1,2,3. \tag{8}$$

Thus the application of the recursion operator does not yield any new symmetries, but we can obtain further symmetries through the commutation relations:

$$[v_1, v_2] = \frac{1}{2}[\psi^1\psi^5\exp(2\psi^2) + \frac{1}{2}\psi^4]\frac{\partial}{\partial u},$$

$$[v_1, v_3] = \frac{1}{2}[\psi^1\left((\psi^3)^2\psi^4 - \psi^6\right)\exp(2\psi^2) + \psi^3\psi^4]\frac{\partial}{\partial u},$$

$$[v_2, v_3] = \frac{1}{2}[\psi^1\psi^3\left((\psi^3)^2\psi^4 - \psi^3\psi^5 - \psi^6\right)\exp(2\psi^2) + (\psi^3)^2\psi^4$$
$$-\psi^3\psi^5 - \frac{1}{2}\psi^6]\frac{\partial}{\partial u},$$

with the pseudopotentials ψ^4, ψ^5 and ψ^6 defined by

$$\frac{\partial\psi^4}{\partial x} = \exp(2\psi^2), \quad \frac{\partial\psi^4}{\partial t} = -2[4\lambda - u - 2(\psi^1)^2]\exp(2\psi^2), \tag{9}$$

$$\frac{\partial\psi^5}{\partial x} = \psi^4\exp(-2\psi^2), \quad \frac{\partial\psi^5}{\partial t} = -2(2\lambda + u)\psi^4\exp(-2\psi^2) - 4\psi^1,$$

$$\frac{\partial\psi^6}{\partial x} = (\psi^3)^2\exp(2\psi^2), \quad \frac{\partial\psi^6}{\partial t} = -2[4\lambda - u - 2(\psi^1)^2](\psi^3)^2\exp(2\psi^2)$$
$$+8\psi^1\psi^3 + 4\exp(-2\psi^2).$$

The application of the recursion operator yields

$$\mathcal{R}[v_1, v_2] = v_1, \quad \mathcal{R}[v_1, v_3] = 2v_2, \quad \mathcal{R}[v_2, v_3] = v_3. \tag{10}$$

We now define inductively

$$v_1^{(0)} = v_1, \quad v_1^{(n)} = [v_1^{(n-1)}, v_2] \quad \text{for} \quad n \in N,$$

$$v_2^{(0)} = v_2, \quad v_2^{(n)} = \frac{1}{2}[v_1^{(n-1)}, v_3] \quad \text{for} \quad n \in N,$$

$$v_3^{(0)} = v_3, \quad v_3^{(n)} = [v_2, v_3^{(n-1)}] \quad \text{for} \quad n \in N.$$

From (4),(8) and (10) one can easily show that $\mathcal{R}^n v_i^{(n)} = v_i$ for $i = 1,2,3$ and all $n \in N$. Thus the KdV-equation has in addition to the two well known hierarchies of Lie-Bäcklund symmetries three hierarchies of pseudopotential symmetries.

We investigate the infinitesimal symmetry generator $v_1 = \psi^1 \exp(2\psi^2)\frac{\partial}{\partial u}$ in detail. The prolongation formula (3) and the equations (6) and (7) yield

$$\frac{\partial \phi^1}{\partial x} = -\psi^1 \exp(2\psi^2) - 2\psi^1\phi^1, \quad \frac{\partial \phi^1}{\partial t} = \frac{\partial}{\partial x}[(\lambda - u - (\psi^1)^2)\exp(2\psi^2) -2(2\lambda + u)\phi^1],$$

$$\frac{\partial \phi^2}{\partial x} = \phi^1, \quad \frac{\partial \phi^2}{\partial t} = (\lambda - u - (\psi^1)^2)\exp(2\psi^2) - 2(2\lambda + u)\phi^1,$$

with the solution $\phi^1 = -\frac{1}{4}\exp(2\psi^2)$ and $\phi^2 = -\frac{1}{4}\psi^4$. From equation (9) it follows

$$\frac{\partial \phi^4}{\partial x} = -\frac{1}{2}\psi^4 \exp(2\psi^2), \quad \frac{\partial \phi^4}{\partial t} = [4\lambda - u - 2(\psi^1)^2]\psi^4 \exp(2\psi^2).$$

The solution is $\phi^4 = -\frac{1}{4}(\psi^4)^2$ and the prolongation is closed.

$$pr\, v_1 = \psi^1 \exp(2\psi^2)\frac{\partial}{\partial u} - \frac{1}{4}\exp(2\psi^2)\frac{\partial}{\partial \psi^1} - \frac{1}{4}\psi^4\frac{\partial}{\partial \psi^2} - \frac{1}{4}(\psi^4)^2\frac{\partial}{\partial \psi^4}$$

is the infinitesimal generator of a Lie point symmetry of the enlarged KdV-system (5),(6),(7) and (9). The symmetry transformations are

$$\tilde{x} = x, \quad \tilde{t} = t, \quad \tilde{u} = u + 2\frac{\partial^2}{\partial x^2}[\ln(4 + \varepsilon\psi^4)], \quad \tilde{\psi}^1 = \psi^1 - \frac{\partial}{\partial x}[\ln(4 + \varepsilon\psi^4)],$$

$$\tilde{\psi}^2 = \psi^2 - \ln(4 + \varepsilon\psi^4) + \ln 4, \quad \tilde{\psi}^4 = \frac{4\psi^4}{4 + \varepsilon\psi^4}. \tag{11}$$

As an example we take the trivial solution $u = 0$. From (6),(7) and (9) with $\lambda>0$ we obtain the following special solutions for the pseudopotentials:

$$\psi^1 = \sqrt{\lambda}, \quad \psi^2 = \sqrt{\lambda}(x - 4\lambda t), \quad \psi^4 = \frac{1}{2\sqrt{\lambda}}\exp[2\sqrt{\lambda}(x - 4\lambda t)].$$

Substitution into (11) leads to the transformed solution

$$\tilde{u} = 2\lambda \mathrm{sech}^2[\sqrt{\lambda}(\tilde{x} - 4\lambda\tilde{t}) + \frac{1}{2}\ln\left(\frac{\varepsilon}{8\sqrt{\lambda}}\right)], \quad \varepsilon \geq 0,$$

which is the solitary wave solution if $\varepsilon>0$ and the trivial solution in the limit $\varepsilon \to 0$.

References

1. H.D. Wahlquist and F.B. Estabrook: J. Math. Phys. **16** (1975) 1
2. D.G.B. Edelen: Isovector Methods for Equations of Balance (Alphen aan den Rijn: Sijthoff and Noordhoff, 1980)
3. G.W. Bluman, S. Kumei and G.J. Reid: J. Math. Phys. **29** (1988) 806
4. F. Galas: Generalized symmetries for the ideal MHD equations, to appear in Physica D, 1993
5. B. Fuchssteiner: Prog. Theor. Phys. **65** (1981) 861

ISOMORPHISM VERIFICATION FOR COMPLEX AND REAL LIE ALGEBRAS BY GRÖBNER BASIS TECHNIQUE

V. P. GERDT
Laboratory of Computing Techniques/Automation
Joint Institute for Nuclear Research
Head Post Office P.O. Box 79, Moscow, Russia

and

W. LASSNER
Fachbereich Mathematik/Informatik, Universität Leipzig
Augustusplatz 10-11,
0-7010 Leipzig, Germany

Abstract. In this paper we present a computer-aided approach to verify the isomorphism between finite-dimensional Lie algebras and to construct an explicit form of a transformation matrix in the case of isomorphism. Our approach is based on the direct investigation of quadratic algebraic equations for matrix elements of a transformation matrix by the Gröbner basis method.

1. Introduction

Lie algebras and groups play an important role in many areas of modern applied mathematics and theoretical physics. One of the most exiting application of Lie algebra methods is symmetry analysis of differential equations, especially of nonlinear ones. Intensive development of computer algebra algorithms and software over last years allows to provide a number of modern computer algebra systems with efficient software packages for finding the explicit form of the classical or point Lie symmetry generators and their commutation relations.

The infinitesimal symmetry generators being linear differential operators form a Lie algebra. The procedure of finding an explicit form of generators is very complicated one, and may lead to different mathematical expressions for generators of a given differential equation or a system of such equations. Therefore, it is very important to have an constructive approach to verifying whether two sets of generators determine essentially the same, i.e. isomorphic Lie algebras, or different ones.

If a symmetry Lie algebra of differential equations can be identified as a member of available tables or databases, one is able to use known properties for further investigations, reduction of the order of ODEs, similarity reduction of PDEs or even their explicit integration [1]. If the symmetry algebra is isomorphic to a known one it may be also useful to find the explicit form of the transformation matrix.

In the present paper we present a straightforward approach (*i*) to decide whether two finite-dimensional Lie algebras are isomorphic or not, (*ii*) to identify a given Lie algebra as a member of known complete tables or databases of Lie algebras and (*iii*) to construct a transformation matrix for two isomorphic Lie algebras. These problems are transformed into commutative algebra as question on the existence

N. H. Ibragimov et al. (eds.),
Modern Group Analysis: Advanced Analytical and Computational Methods in Mathematical Physics, 245–254.

and the construction of the solution of a related system of quadratic equations. This allows to use modern constructive techniques in the theory of polynomial ideals based on construction of Gröbner bases.

2. Basic System of Algebraic Equations

Let $\mathbf{F}$ be the field of real ($\mathbf{R}$) or complex ($\mathbf{C}$) numbers and L and $\tilde{L}$ be Lie algebras over $\mathbf{F}$ of dimension N defined by their structure constants c_{ij}^k and $\tilde{c}_{ij}^k$

$$L: \quad [X_i, X_j] = c_{ij}^k X_k, \qquad \tilde{L}: \quad [\tilde{Y}_i, \tilde{Y}_j] = \tilde{c}_{ij}^k \tilde{Y}_k, \tag{1}$$

in bases $\{X_i\}$ and $\{\tilde{Y}_i\}$ $(i = 1, 2, \ldots, N)$ of the Lie algebras L and $\tilde{L}$, respectively. Here and below we use a convention on summation over the repeated indices.

Assume the existence of an isomorphism $\phi : \tilde{L} \longrightarrow L$ and denote the image of the basis elements $\tilde{Y}_i$ by $Y_i := \phi(\tilde{Y}_i)$. The set $\{Y_i\}$ forms a basis of L with the structure constants $\{\tilde{c}_{ij}^k\}$

$$[Y_i, Y_j] = \tilde{c}_{ij}^k Y_k. \tag{2}$$

So there exists a basis transformation represented by a $N \times N$-matrix $A = (a_i^j)$ with $a_i^j \in \mathbf{F}$.

$$Y_i = a_i^j X_j, \quad det(A) \neq 0. \tag{3}$$

If we replace the $Y's$ in (2) according to (3), we get for the l.h.s. of (2)

$$[Y_i, Y_j] = a_i^k a_j^l [X_k, X_l] = a_i^k a_j^l c_{kl}^n X_n \tag{4}$$

and for the r.h.s. of (2)

$$[Y_i, Y_j] = \tilde{c}_{ij}^m Y_m = \tilde{c}_{ij}^m a_m^n X_n. \tag{5}$$

Comparison of (4) and (5) shows that the elements of the transformation matrix must satisfy the relations

$$a_i^k a_j^l c_{kl}^n - \tilde{c}_{ij}^m a_m^n = 0. \tag{6}$$

For given structure constants $\{c_{ij}^k\}$ of L and $\{\tilde{c}_{ij}^k\}$ of $\tilde{L}$ (6) is a system of at most $N^2(N-1)/2$ quadratic equations in N^2 unknowns a_i^j. It follows from the anti-symmetry $c_{ij}^k = -c_{ji}^k$ of the structure constants.

To use the Gröbner basis technique we add one more variable d and one more equation $d - det(a_i^j) = 0$ to system (6). So we have shown that the answer to the isomorphism problem is given by the following theorem.

Theorem 1 *Two N-dimensional algebras L and $\tilde{L}$ over $\mathbf{F}$ given by the structure constants $\{c_{ij}^k\}$ and $\{\tilde{c}_{ij}^k\}$ are isomorphic if and only if the system of equations*

$$a_i^k a_j^l c_{kl}^n - \tilde{c}_{ij}^m a_m^n = 0, \quad d - det(a_i^j) = 0 \tag{7}$$

in unknowns $\{a_i^j, d\}$, $(i, j = 1, \ldots, N)$ has a solution in $\mathbf{F}$ with $d \neq 0$.

But let us stress the following consequence of Theorem 1. If the Lie algebras L and $\tilde{L}$ are not isomorphic as complex ones, then we must get $d = 0$ from (7). Because of linearity of (7) in d a Gröbner basis G [2] of the ideal generated by the polynomial in the l.h.s. of (7) in the complex case $\mathbf{F} = \mathbf{C}$ shows this fact very explicitly, i.e. $d \in G$. To be more precise, this fact follows from the linearity of (7) in d and the possibility to choose an ordering with d having the highest precedence. It is clear that (7) has always the trivial solution $a_i^j = 0$, $d = 0$. Moreover, the following theorem takes place.

Theorem 2 *If the system (7) has a non-trivial solution, then it has infinitely many solutions.*

Proof. Let matrix $A = (a_i^j)$ be a non-trivial solution of (7). For given two Lie algebras the system of equations (7) is completely determined by their structure constants. Hence, the solution space is invariant under transformations which will not change the structure constants $\{c_{ij}^k\}$ and $\{\tilde{c}_{ij}^k\}$. So, looking at the derivation of our system of equations (6) one immediately sees that one can act first with an automorphism of $\tilde{L}$ represented in the basis Y_i by a matrix $\tilde{T}$ and the structure constants $\{\tilde{c}_{ij}^k\}$ remain unchanged. In the same way one can after the transformation A in (3) apply an automorphism of L represented in the basis X_i by a matrix T and the structure constants $\{c_{ij}^k\}$ remain unchanged.
Therefore, for any solution A

$$B = \tilde{T}^{-1} A T, \tag{8}$$

is also a solution where T and $\tilde{T}$ are automorphism matrices of L and $\tilde{L}$ respectively.

Now, let T be an inner automorphism of the Lie algebra L. It has the following general form

$$T_x = exp(\alpha\, ad_x), \quad ad_x\, y = [x, y], \quad x, y \in L, \tag{9}$$

where α is a parameter with values in $\mathbf{F}$. If L is not an abelian Lie algebra, then there exists at least a one-parametric set of automorphisms and hence infinitely many solutions of (7). The case of abelian L can be neglected here as trivial one. □

With respect to the proposed computer algebra approach to the isomorphism problem it seems of interest to mention the following corollary.

Corollary *Let L be a N-dimensional Lie algebra over* $\mathbf{F}$ *given by the structure constants $\{c_{ij}^k\}$ in an basis $\{X_i\}$. The automorphisms of L are given by the system*

$$a_i^k a_j^l c_{kl}^n - c_{ij}^m a_m^n = 0, \quad d - det(a_i^j) = 0$$

in unknowns $\{a_i^j, d\}$, $(i, j = 1, \ldots, N)$ as solution in $\mathbf{F}$ *with $d \neq 0$.*

3. Gröbner Basis Technique and ASYS Package

To investigate and solve eqs.(7) we need a method which could be applied to non-linear algebraic equations with infinitely many solutions.

For our purposes we select the Gröbner basis technique as an universal, constructive and computer-aided tool of commutative polynomial algebra [2] which allows to deal with polynomial ideals of either zero or positive dimension. All the modern general-purpose computer algebra systems have special built-in modules for a Gröbner basis computation. However, most of them were designed mainly for zero dimensional ideals and have no special facilities for investigation and solving algebraic equations with infinitely many solutions.

In the present paper we use the ASYS package [3] written in REDUCE [4], based on the Gröbner basis technique [2] and especially developed to investigate and solve polynomial algebraic equations with infinitely many solutions. ASYS has a number of special facilities to attack complicated algebraic systems generating positive dimensional ideals:

- **Reduction into subsystems by maximal sets of independent variables**
  In practice, in order to deal with positive dimensional polynomial ideals one needs information on its dimension and on independent sets of polynomial variables [5] which could be treated as free parameters. If any of the maximal independent sets is considered as a parametric set, then the lexicographical Gröbner basis with respect to the remaining variables has a triangular form. Therefore, the problem is reduced to univariate polynomials over rational coefficient fields $Q(a_1, \ldots, a_k)$ where $\{a_1, \ldots, a_k\}$ is one of the maximal sets of independent variables (parameters) according to [5]. ASYS allows to compute all possible maximal sets of independent variables for a given ordering and to execute the corresponding reduction to an equivalent set of triangular subsystems over a rational function coefficient field.
- **Homogeneity reduction**
  If under an appropriate scale transformation of unknowns each monomial of each given polynomial of the initial algebraic system has the same scale factor, the system could be transformed into an equivalent set of subsystems with smaller numbers of variables [3]. Because of double exponential complexity of Buchberger algorithm in the number of variables such a reduction allows a drastically speed up in the computing time of investigating and solving algebraic equations by the Gröbner basis technique. The ASYS package has built-in facilities for homogeneity analysis and the reduction of polynomial equations systems into subsystems.

A user of ASYS may combine homogeneity reduction with the reduction by maximal independent sets to reduce a given system into smaller subsystems with finitely many solutions.

It should be noted that the structure constants in (7) may include parameters, which can be treated as additional unknowns in order to determine their values for providing an isomorphism. This problem arises, first, in symmetry analysis when free parameters are present in differential equations. Second, in known tables [6] different classes of isomorphic Lie algebras are sometimes summarized and represented by parametrized families of classes. The parameters occur in the structure constants and are real in the case of real Lie algebras and sometimes restricted to certain intervals of **R**. In Sect.5 a relevant example of a one-parametric family of four-dimensional Lie algebras is considered which illustrates our method in the presence of parameters. In the next section we describe the computational strategy realized in the package ISOLIE/ASYS. ISOLIE is the project name of a system under development. ISOLIE/ASYS is the adaption of ASYS for the isomorphism problem. ISOLIE contains LIEBASE a database with tables of low-dimensional Lie algebras. Interfaces and preprocessors to ISOLIE/ASYS and to other programs as LIE (REDUCE redlib) have still to be developed.

4. Computational Strategy in ISOLIE/ASYS

A general case, heuristically optimal computational strategy for verifying an isomorphism for given two Lie algebras consists of the following successive steps:

1. Generation of the system of nonlinear algebraic equations (7).
2. Analysis of homogeneity properties of the system and, in the case of their existence, reduction of the system into subsystems with reduced number of variables treating the remaining (homogeneous) variables as free parameters.
3. Construction of the Gröbner basis for the ideals generated by the subsystems of step 2. At this step the lexicographical ordering with d in (7), being of "highest" variable such that the other variables are arranged in heuristically optimal ordering [7], is preferable. If one obtains d as an element of the Gröbner basis this means that the corresponding subsystem does not contribute to an isomorphism. Such subsystems are to be omitted. If there are no other subsystems in the output this means that the Lie algebras are not isomorphic. Otherwise, if at least one subsystem does not contain "d" in the list of the Gröbner basis elements, then the complex Lie algebras are isomorphic. In the case of real Lie algebras and absence of d in the Gröbner basis of a subsystem one has to go to the next step.
4. Because of zero-dimensionality of the ideals for the subsystems obtained at step 3 one has to investigate successively each subsystem starting from their univariate polynomial in the "lowest" variable with respect to the chosen ordering. Generally, these polynomials have rational function coefficients with respect to the parametric set.

We observed that in numerous concrete cases of isomorphism problem the following interesting fact. If the Lie algebras are not isomorphic, then in each subsystem the univariate polynomial in the "lowest" variable is a quadratic one with a negative discriminant for $d \neq 0$. Consequently, if that heuristic criterion is violated in at least one subsystem then the Lie algebras are isomorphic as real ones.

This interesting observation that the nonexistence of real solutions of a subsystem is already expressed by that property of the last equation, could be understood in relation to the paper [8].

5. Demonstration of the Method: Four-dimensional Lie Algebras

In this section we consider two examples of four-dimensional Lie algebras. Initial algebraic systems as well as results of the computations in accordance with the strategy of Sect.4 are given just in the REDUCE-like form.

Example 1

Identification of the Lie Algebra generated by classical (Lie) symmetry analysis of the well-known Korteweg-de Vries equation

$$u_t + u_{xxx} + u_x u = 0. \tag{10}$$

The symmetry Lie algebra of (10) is generated by differential operators

$$\begin{aligned} v_1 &= \partial_x \\ v_2 &= \partial_t \\ v_3 &= t\partial_x + \partial_u \\ v_4 &= x\partial_x + 3t\partial_t - 2u\partial_u \end{aligned}$$

with a commutator table

	v_1	v_2	v_3	v_4
v_1	0	0	0	v_1
v_2		0	v_1	$3v_2$
v_3			0	$-2v_3$

We try to identify this Lie algebra as one of the table [6] of all real four-dimensional Lie algebras up to isomorphism. The explicit form of the algebraic equations (7) for the isomorphism problem of this symmetry Lie algebra and one of the class Lie algebra $A^b_{4,9}$ taken from that table in the classification of paper [6]

$$[e_2, e_3] = e_1, \quad [e_1, e_4] = (1+b)e_1, \quad [e_2, e_4] = e_2, \quad [e_3, e_4] = be_3, \quad (-1 < b \leq 1)$$

with all other commutators being zero, is generated by the separate module of the ISOLIE/ASYS package and consists of 23 equations in 18 unknowns including b.

```
Z1 := 3*(A11*A24 - A14*A21);
Z2 := A11*A23 + A12*A24 - A13*A21 - A14*A22;
Z3 := - 2*(A13*A24 - A14*A23);
Z4 := 3*(A11*A34 - A14*A31);
Z5 := A11*A33 + A12*A34 - A13*A31 - A14*A32;
Z6 := - 2*(A13*A34 - A14*A33);
Z7 := - (B*A11 + A11 - 3*A11*A44 + 3*A14*A41);
Z8 := - (B*A12 + A12 - A11*A43 - A12*A44 + A13*A41 + A14*A42);
Z9 := - (B*A13 + A13 + 2*A13*A44 - 2*A14*A43);
Z10 := - A14*(B + 1);
Z11 := - (A11 - 3*A21*A34 + 3*A24*A31);
Z12 := - (A12 - A21*A33 - A22*A34 + A23*A31 + A24*A32);
Z13 := - (A13 + 2*A23*A34 - 2*A24*A33);
Z14 := - A14;
Z15 := - (A21 - 3*A21*A44 + 3*A24*A41);
Z16 := - (A22 - A21*A43 - A22*A44 + A23*A41 + A24*A42);
Z17 := - (A23 + 2*A23*A44 - 2*A24*A43);
Z18 := - A24;
Z19 := - (B*A31 - 3*A31*A44 + 3*A34*A41);
Z20 := - (B*A32 - A31*A43 - A32*A44 + A33*A41 + A34*A42);
```

```
Z21 := - (B*A33 + 2*A33*A44 - 2*A34*A43);
Z22 := - B*A34;
Z23 := DET MAT ((A11,A12,A13,A14),(A21,A22,A23,A24),
                (A31,A32,A33,A34),(A41,A42,A43,A44)) - D;
```

Then the ISOLIE/ASYS package is applied with the switches **setord, setdim, setgb, scale** on, where

setord generates heuristically optimal ordering [7] for (7).

scale realizes homogeneity reduction. System (7) is split into subsystems such that any solution of the subsystem is a solution of the whole system.

setdim computes all the maximal independent sets of variables (parameters).

setgb makes the further reduction to the subsystems over rational function field and computes their Gröbner bases.

The combination of **scale** and **setgb** is the powerful instrument which leads to the following result of the considered example. The output consists of various subsystems. They are explicitly characterized first by those variables of the original system which has to be zero, second, by those variables of the original system which can be considered as free parameters, and third, by a small number of variables occurring in the Gröbner basis of the subsystem.

As mentioned in Sect. 3, there are two types of parameters in the final subsystems [3]. Parameters coming from homogeneity cannot take the value zero. Parameters resulting from an independent set could take any value in **F**. The first output line gives the heuristically optimal ordering

Order=(D A22 A32 A42 A12 A21 A23 A31 A33 A41 A43 A11 A13 B A44 A24 A34 A14)

If one therefore neglects all output subsystems with $d = 0$, one is focussed immediately to the following output of a subsystem

---

Variables = (D A12 B A44)
Parameters = (A21 A33) % nonzero, from homogeneity
Zeros = (A11 A13 A34 A14 A41 A43 A24 A22 A32 A42 A23 A31)

GROEBNER BASIS

$$\begin{aligned} G(1) &= D + \frac{1}{3} * A21^2 * A33^2 \\ G(2) &= A12 - A21 * A33 \\ G(3) &= B + \frac{2}{3} \\ G(4) &= A44 - \frac{1}{3} \end{aligned} \tag{11}$$

We emphasize the following facts.

- Subsystem (11) determines the parameter b to the value -2/3.
- It shows immediately that the transformation is a real one and so it identifies the KdV symmetry algebra as $A_{4,9}^{-2/3}$.
- One immediately obtains the real isomorphism matrix. The last equation is linear in the single variable $A44$. Already this observation in *one* subsystem with $d \neq 0$ ensures the existence of a real isomorphism. Each variable occurs linearly as leading term in one of the equation. This reflects a general observation on the structure of our subsystems and is in agreement with results of paper [8].

For zero-dimensional ideal generated by (11) it is not necessary to make further reduction by maximal independent sets.

Example 2

Isomorphism analysis of two four-dimensional Lie algebras $A_{4,8}$ and $A_{4,10}$ taken from [6] with the following non-zero commutators

$$A_{4,8}: \quad [e_2, e_3] = e_1, \quad [e_2, e_4] = e_2, \quad [e_3, e_4] = -e_3,$$
$$A_{4,10}: \quad [e_2, e_3] = e_1, \quad [e_2, e_4] = -e_3, \quad [e_3, e_4] = e_2. \tag{12}$$

If one consider (12) as real Lie algebra, they belong to different isomorphic classes [6] and therefore are not isomorphic ones. Let us consider them as complex algebras and verify whether they are still isomorphic or not. In this case the ISOLIE module generates the system of 22 equations in 17 unknowns

```
Z1 := A12*A23 - A13*A22;
Z2 := A12*A24 - A14*A22;
Z3 := - (A13*A24 - A14*A23);
Z4 := A12*A33 - A13*A32;
Z5 := A12*A34 - A14*A32;
Z6 := - (A13*A34 - A14*A33);
Z7 := A12*A43 - A13*A42;
Z8 := A12*A44 - A14*A42;
Z9 := - (A13*A44 - A14*A43);
Z10 := - (A11 - A22*A33 + A23*A32);
Z11 := - (A12 - A22*A34 + A24*A32);
Z12 := - (A13 + A23*A34 - A24*A33);
Z13 := - A14;
Z14 := A31 + A22*A43 - A23*A42;
Z15 := A32 + A22*A44 - A24*A42;
Z16 := A33 - A23*A44 + A24*A43;
Z17 := A34;
Z18 := - (A21 - A32*A43 + A33*A42);
Z19 := - (A22 - A32*A44 + A34*A42);
Z20 := - (A23 + A33*A44 - A34*A43);
Z21 := - A24;
Z22 := DET MAT ((A11,A12,A13,A14),(A21,A22,A23,A24),
                 (A31,A32,A33,A34),(A41,A42,A43,A44)) - D;
```

Using the ISOLIE/ASYS package just in the same way as in previous example, we obtain among output subsystems the following one

---

Parameters = (A32 A33) % non-zero
Zeros = (A34 A24 A14 A41 A13)

SUBBASIS FOR SET (A42 A43) % maximal independent set (arbitrary parameters)

GROEBNER BASIS

$$G(1) = D - 4 * A23 * A33^2 * A32$$
$$G(2) = A11 + 2 * A23 * A32$$
$$G(3) = A21 + A33 * A42 - A32 * A43$$
$$G(4) = \frac{A31 * A33 - A23 * A33 * A42 - A23 * A32 * A43}{A33}$$
$$G(5) = \frac{A44 * A33 + A23}{A33}$$
$$G(6) = A12$$
$$G(7) = \frac{A22 * A33 + A23 * A32}{A33}$$
$$G(8) = A23^2 + A33^2$$

One can immediately see that

- Because d is not an element of the Gröbner basis, the two above Lie algebras are isomorphic as complex ones.
- The fact that they are not isomorphic as real Lie algebras leads to absence of real solutions of the last (univariate) polynomial in $A23$. However, in order to check that real Lie algebras are not isomorphic, one has to look at the last polynomial of *each* subsystem.
- The explicit form of the complex transformation matrix can be easily constructed.

ISOLIE/ASYS needs in its actual (not final) version seconds for Lie algebras of dimension 3, minutes for dimension 4 and 5, hours for dimension 6 and days for higher dimension. At present the method was applied up to eight dimensional Lie algebras.

References

1. Winternitz P. Group Theory and Exact Solutions of Partially Integrable Differential Systems. In: Partially Integrable Nonlinear Evolution Equations and Their Physical Applications (eds. Conte R. and Boccara N.), Kluwer Academic Publishers, Dordrecht, 1990, pp.515-567.
2. Buchberger B. Gröbner Bases: an Algorithmic Method in Polynomial Ideal Theory. In: (Bose N.K., ed.) Recent Trends in Multidimensional System Theory, Reidel, 1985, pp.184-232.
3. Gerdt V.P., Khutornoy N.V., Zharkov A.Yu. Gröbner Basis Technique, Homogeneity and Solving Polynomial Equations. In: (Wu Wen-Tsün, Cheng Min-De, eds.) Proc. of the 1992 Intern. Workshop on Mathematics Mechanization, International Academic Publishers, Beijing, China, 1992, pp.38-51.
4. Hearn A.C. REDUCE User's Manual. Version 3.4. The Rand Corporation, Santa Monica, 1991.
5. Kredel H., Weispfenning V. Computing Dimension and Independent Sets for Polynomial Ideals, *J. Symb. Comp.* 6, 231-247, 1988.

6. Patera J., Sharp R.T., Winternitz P. Invariants of Real Low Dimension Lie Algebras, *J. Math. Phys.* 17, 6, 986-994, 1976.
7. Böge W., Gebauer R., Kredel H. Some Examples for Solving Systems of Algebraic Equations by Calculating Gröbner Bases, *J. Symb. Comp.* 2, 83-98, 1986.
8. Kobayashi H., Moritsugu S., Hogan R.W. On Radical Zero-Dimensional Ideals, *J. Symb. Comp.* 8, 545-552, 1989.

$sl(2,R)$, ERMAKOV SYSTEMS AND THE MAGNETIC MONOPOLE

P G L LEACH
Department of Mathematics and Applied Mathematics
University of Natal, Durban, King George V Avenue, Durban 4001, R S A
Associate member of the Centre for Nonlinear Studies,
University of Witwatersrand, Johannesburg

Abstract. A representation of the group $sl(2,R)$ is the symmetry algebra of the Lewis-Ermakov invariant. Under certain circumstances it is also the symmetry algebra of the associated pair of second order differential equations. The most general pair of sodes possessing the same $sl(2,R)$ symmetry is called a generalized Ermakov system. In three dimensions the system of sodes invariant under $sl(2,R)$ symmetry contains the magnetic monopole as a special case. The general system in three dimensions always admits a Poincaré vector and in fact three such vectors always exist. A special case of weak generalized Ermakov systems in three dimensions is the monopole-Kepler problem.

1. Ermakov systems

In 1880 V P Ermakov [1] put forward a method to find a first integral of

$$\ddot{q}+\omega^2(t)q=0 \tag{1}$$

by introducing the auxiliary equation

$$\ddot{\rho}+\omega^2(t)\rho=\frac{A}{\rho^3}\,. \tag{2}$$

Elimination of $\omega^2(t)$ between (1) and (2) by the combination $\rho\,(1)-(2)q$, multiplication by the integrating factor $(\rho\dot{q}-\dot{\rho}q)$ and integration give the first integral

$$I=\frac{1}{2}(\rho\dot{q}-\dot{\rho}q)^2+\frac{1}{2}A\left(\frac{q}{\rho}\right)^2 \tag{3}$$

known as the Ermakov invariant. It is also known as the Lewis invariant after Ralph Lewis who found it in 1966 by an independent method [2]. An elementary generalisation was provided by Ermakov who replaced the A/ρ^3 term by $\rho^{-3}f(q/\rho)$.

In 1979 Ray and Reid [3] treated the system

$$\begin{aligned}\ddot{x}+\omega^2(t)x &= x^{-3}f(y/x)\\ \ddot{y}+\omega^2(t)y &= y^{-3}g(y/x)\end{aligned} \tag{4}$$

and used the same procedure as Ermakov to determine the first integral

$$I=\frac{1}{2}(x\dot{y}-\dot{x}y)^2+\int^{y/x}\left[uf(u)-u^{-3}g(u)\right]du\,. \tag{5}$$

N. H. Ibragimov et al. (eds.),
Modern Group Analysis: Advanced Analytical and Computational Methods in Mathematical Physics, 255–264.

In fact I exists for any system of the form [4]

$$\begin{aligned} \ddot{x} + \Box x &= x^{-3} f(y/x) \\ \ddot{y} + \Box y &= y^{-3} g(y/x) . \end{aligned} \tag{6}$$

The structure of I suggests angular momentum. In plane polar coordinates the Ermakov system is

$$\begin{aligned} \ddot{r} - r\dot{\theta}^2 + \Box r &= r^{-3} F(\theta) \\ r\ddot{\theta} + 2\dot{r}\dot{\theta} &= r^{-3} G(\theta) \end{aligned} \tag{7}$$

and the Ermakov invariant is

$$I = \frac{1}{2}(r^2\dot{\theta})^2 - \int G(\theta) \mathrm{d}\theta \tag{8}$$

and is manifestly the autonomous first integral of the angular equation in system (7).

2. Algebraic properties of the Ermakov invariant

If a function $I(t, x, \dot{x})$ is invariant under the action of the first extension of a Lie point symmetry $G = \tau(t, x)\partial/\partial t + \eta_i(t, x)\partial/\partial x_i$,

$$G^{[1]} I := \tau \frac{\partial I}{\partial t} + \eta_i \frac{\partial I}{\partial x_i} + (\dot{\eta}_i - \dot{x}_i \dot{\tau}) \frac{\partial I}{\partial \dot{x}_i} = 0 . \tag{9}$$

In the case of

$$I = \frac{1}{2}(r^2\dot{\theta})^2 - \int G(\theta)\dot{\theta} \tag{10}$$

Leach [5] showed that there were three point symmetries, *viz.*

$$\begin{aligned} G_1 &= \frac{\partial}{\partial t} \\ G_2 &= 2t\frac{\partial}{\partial t} + r\frac{\partial}{\partial r} \\ G_3 &= t^2\frac{\partial}{\partial t} + tr\frac{\partial}{\partial r} \end{aligned} \tag{11}$$

with $[G_1, G_2] = 2G_1$, $[G_1, G_3] = G_2$, $[G_2, G_3] = 2G_3$, which is a representation of the Lie algebra $sl(2, R)$. The system of differential equations

$$\mathbf{E}(t, x, \dot{x}, \ddot{x}) = 0 \tag{12}$$

is invariant under the Lie point symmetry $G = \tau\partial/\partial t + \eta_i\partial/\partial x_i$ if

$$G^{[2]} \mathbf{E}(t, x, \dot{x}, \ddot{x}) \,|_{\mathbf{E}=0} = 0 , \tag{13}$$

where

$$G^{[2]} = \tau\frac{\partial}{\partial t} + \eta_i \frac{\partial}{\partial x_i} + (\dot{\eta}_i - \dot{x}_i\dot{\tau})\frac{\partial}{\partial \dot{x}_i} + (\ddot{\eta}_i - 2\ddot{x}_i\dot{\tau} - \dot{x}_i\ddot{\tau})\frac{\partial}{\partial \ddot{x}_i} . \tag{14}$$

The system

$$\begin{aligned} \ddot{r} - r\dot{\theta}^2 &= r^{-3}F(\theta) \\ r\ddot{\theta} + 2\dot{r}\dot{\theta} &= r^{-3}G(\theta) \end{aligned} \tag{15}$$

has the same symmetries as I. This is not the case for the general system (7). However, if the additional term is $\omega^2 r$, another representation of $sl(2,R)$ leaves (7) invariant.

3. Generalized Ermakov systems from the $sl(2,R)$ viewpoint: two dimensions

An inverse problem is created [6] if one takes the $sl(2,R)$ representation, (11), and seeks the form of $f(t,r,\theta,\dot{r},\dot{\theta},\ddot{r},\ddot{\theta}) = 0$ invariant under all three symmetries. The action of G_1 removes t from f. G_2 leads to the five characteristics $\theta, u = r\dot{r}$, $v = r^2\dot{\theta}$, $w = r^3\ddot{r}$ and $x = r^4\ddot{\theta}$. G_3 reduces the number of characteristics to four θ, u, v, w and $y = x + 2uv$ to give the general form

$$f(\theta, r^2\dot{\theta}, r^4\ddot{\theta} + 2r^3\dot{r}\dot{\theta}, r^3\ddot{r}) = 0\,. \tag{16}$$

How is this done? Consider G_2.

$$G_2^{[2]} = 2t\frac{\partial}{\partial t} + r\frac{\partial}{\partial r} + 0\frac{\partial}{\partial \theta} - \dot{r}\frac{\partial}{\partial \dot{r}} - 2\dot{\theta}\frac{\partial}{\partial \dot{\theta}} - 3\ddot{r}\frac{\partial}{\partial \ddot{r}} - 4\ddot{\theta}\frac{\partial}{\partial \ddot{\theta}} \tag{17}$$

and

$$G_2^{[2]} f \,|_{f=0} = 0 \tag{18}$$

leads to the associated Lagrange's system for the characteristics

$$\frac{\mathrm{d}r}{r} = \frac{\mathrm{d}\theta}{0} = \frac{\mathrm{d}\dot{r}}{-\dot{r}} = \frac{\mathrm{d}\dot{\theta}}{-2\dot{\theta}} = \frac{\mathrm{d}\ddot{r}}{-3\ddot{r}} = \frac{\mathrm{d}\ddot{\theta}}{-4\ddot{\theta}} \tag{19}$$

(recall $\partial f/\partial t = 0$). These are easily solved.

The general two-dimensional system invariant under $sl(2,R)$ is

$$\begin{aligned} F(\theta, r^2\dot{\theta}, r^3\ddot{r}, r^4\ddot{\theta} + 2r^3\dot{r}\dot{\theta}) &= 0 \\ G(\theta, r^2\dot{\theta}, r^3\ddot{r}, r^4\ddot{\theta} + 2r^3\dot{r}\dot{\theta}) &= 0\,. \end{aligned} \tag{20}$$

Attention is confined to the forms

$$\begin{aligned} r^3\ddot{r} &= f_1(\theta, r^2\dot{\theta}) \\ r^4\ddot{\theta} + 2r^3\dot{r}\dot{\theta} &= g(\theta, r^2\dot{\theta}) \end{aligned} \tag{21}$$

which can be written in the more physical forms

$$\begin{aligned} \ddot{r} - r\dot{\theta}^2 &= r^{-3}f(\theta, r^2\dot{\theta}) \\ r\ddot{\theta} + 2\dot{r}\dot{\theta} &= r^{-3}g(\theta, r^2\dot{\theta}) \end{aligned} \tag{22}$$

($f_1 = f + (r^2\dot\theta)^2$). Note that (22) differs from (15) by the inclusion of the magnitude of the angular momentum in the arguments of f and g. If new time, $\tau = \int r^{-2}\mathrm{d}t$, and the inverse radial distance $\chi = r^{-1}$, are introduced the system (22) becomes

$$\begin{aligned} \chi'' + \left[\theta'^2 + f(\theta,\theta')\right]\chi &= 0 && (23)\\ \theta'' &= g(\theta,\theta'). && (24) \end{aligned}$$

Under 'appropriate' conditions (24) may be integrated to give

$$I = M(\theta,\theta') \tag{25}$$

which is the generalisation of the Ermakov invariant (5). Assuming that (25) can be inverted (which is always possible locally) as

$$\theta' = N(\theta, I), \tag{26}$$

this can be integrated to give

$$\tau - \tau_0 = \int \frac{\mathrm{d}\theta}{N(\theta,I)}. \tag{27}$$

Finally, with the inversion of (27) to give

$$\theta = J(\tau, I, \tau_0), \tag{28}$$

it is evident that (23) reduces to the time-dependent harmonic oscillator.

4. Generalised Ermakov systems from the $sl(2,R)$ viewpoint: three dimensions

The extension of the results of § 3 to three dimensions is straight forward [7]. For an equation of the form

$$f(t,r,\theta,\phi,\dot r,\dot\theta,\dot\phi,\ddot r,\ddot\theta,\ddot\phi) = 0 \tag{29}$$

to be invariant under the action of the second extensions of the generators of $sl(2,R)$ given in (11) the characteristics are $\theta,\phi,r^2\dot\theta,r^2\dot\phi,r^3\ddot r,r^4\ddot\theta + 2r^3\dot r\dot\theta, r^4\ddot\phi + 2r^3\dot r\dot\phi$. Consideration is restricted to the system of equations

$$\begin{aligned} r^3\ddot r &= f(\theta,\phi,r^2\dot\theta,r^2\dot\phi)\\ r^4\ddot\theta + 2r^3\dot r\dot\theta &= g(\theta,\phi,r^2\dot\theta,r^2\dot\phi)\\ r^4\ddot\phi + 2r^3\dot r\dot\phi &= h(\theta,\phi,r^2\dot\theta,r^2\dot\phi). \end{aligned} \tag{30}$$

In terms of the new time T and the inverse radial distance χ these are

$$\begin{aligned} \chi'' &= f(\theta,\phi,\theta',\phi')\chi\\ \theta'' &= g(\theta,\phi,\theta',\phi')\\ \phi'' &= h(\theta,\phi,\theta',\phi') \end{aligned} \tag{31}$$

which are too general as they stand. This generality is reduced by the imposition of an additional symmetry requirement. A natural one is invariance under the rotation group $so(3)$. Invariance under the algebra $so(3)$ represented by

$$\begin{aligned} G_4 &= \frac{\partial}{\partial\phi} \\ G_5 &= \sin\phi\frac{\partial}{\partial\theta} + \cot\theta\cos\phi\frac{\partial}{\partial\phi} \\ G_6 &= \cos\phi\frac{\partial}{\partial\theta} - \cot\theta\sin\phi\frac{\partial}{\partial\phi} \end{aligned} \tag{32}$$

requires that the system take the form

$$\begin{aligned} r^3\ddot{r} &= A_1(L) \\ r^4\ddot{\theta} + 2r^3\dot{r}\dot{\theta} &= r^4\dot{\phi}^2\sin\theta\cos\theta + B(L)r^2\dot{\theta} - C(L)r^2\dot{\phi}\sin\theta \\ r^4\ddot{\theta} + 2r^3\dot{r}\dot{\phi} &= -2r^4\dot{\theta}\dot{\phi}\cot\theta + \frac{1}{\sin\theta}\left[B(L)r^2\dot{\phi}\sin\theta + C(L)r^2\dot{\theta}\right], \end{aligned} \tag{33}$$

where $L^2 := r^4(\dot{\theta}^2 + \dot{\phi}^2\sin^2\theta)$ is the square of the magnitude of the angular momentum. In vector form (33) takes the compact form

$$\ddot{\mathbf{r}} = \frac{1}{r^3}\left\{A(L)\hat{\mathbf{r}} + B(L)\hat{\omega} + C(L)\hat{\mathbf{L}}\right\}, \tag{34}$$

where $\qquad \hat{\omega} := \hat{\mathbf{L}} \times \hat{\mathbf{r}} \quad \text{and} \quad A = A_1 + L^2 .$

The full symmetry, $sl(2,R) \oplus so(3)$, of (34) can be broken by the addition of an additional radial term without affecting the properties inherent in the $\hat{\omega}$ and $\hat{\mathbf{L}}$ components just as was the case of the two-dimensional system described by (7). Some particular cases are

(i) $B = C = 0$. L is conserved and so $A(L)$ is constant. This represents a Newton-Cotes spiral with an excess/deficit of angular momentum.

(ii) $A = B = 0$. $\quad C(L) = \lambda L$. This describes a particle moving in the field of a magnetic monopole. There is the conserved Poincaré vector [8]

$$\mathbf{P} = \mathbf{L} + \lambda\hat{\mathbf{r}} . \tag{35}$$

The motion is on the surface of a cone with semivertex angle $\arccos(\lambda/P)$. To determine the Poincaré vector for (34) assume a vector of the form

$$\mathbf{P} := I\hat{\mathbf{r}} + J\hat{\omega} + K\hat{\mathbf{L}} .$$

The requirement that $\dot{\mathbf{P}}$ be zero leads to

$$\frac{\mathrm{d}}{\mathrm{d}t}\begin{pmatrix} I \\ J \\ K \end{pmatrix} = r^{-2}\begin{pmatrix} 0 & L & 0 \\ -L & 0 & C/L \\ 0 & -C/L & 0 \end{pmatrix}\begin{pmatrix} I \\ J \\ K \end{pmatrix} . \tag{36}$$

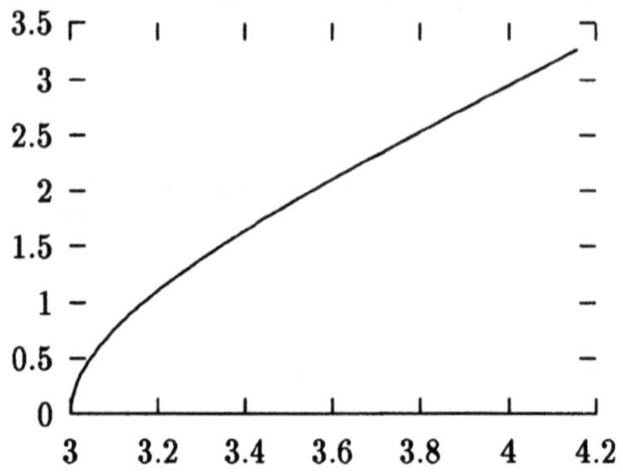

Fig. 1. $B = C = 0$; $A(L)$ is constant. $A(L) > 0$. Newton-Cotes spiral with an excess of angular momentum.

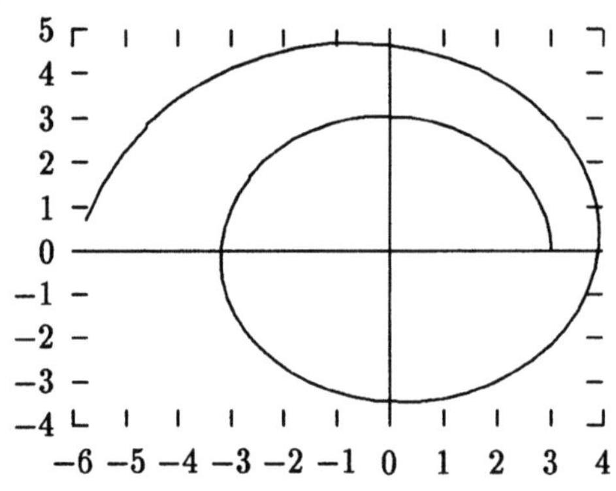

Fig. 2. $B = C = 0$; $A(L)$ is constant. $A(L) < 0$. Newton-Cotes spiral with deficit of angular momentum.

In terms of new time, $T := \int r^{-2}\mathrm{d}t$, (36) is

$$\frac{\mathrm{d}}{\mathrm{d}T}\begin{pmatrix} I \\ J \\ K \end{pmatrix} = \begin{pmatrix} 0 & L & 0 \\ -L & 0 & C/L \\ 0 & -C/L & 0 \end{pmatrix}\begin{pmatrix} I \\ J \\ K \end{pmatrix}. \tag{37}$$

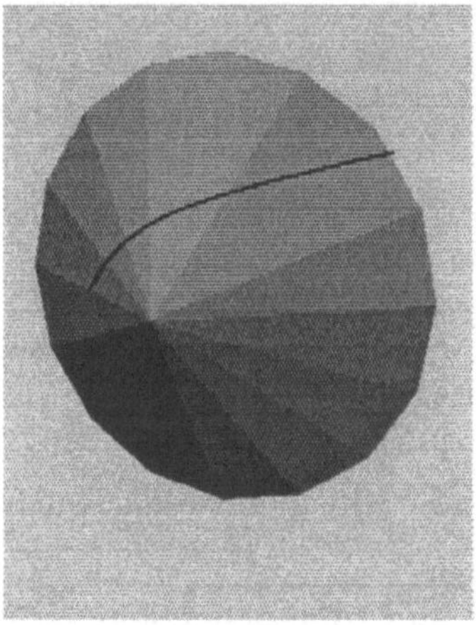

Fig. 3. $A = B = 0$: $C(L) = \lambda L$. Particle moving in the field of a magnetic monopole. Motion is on the surface of a cone with semivertex angle $\arccos(\lambda/P)$.

Returning to eq(34) the vector product with $\mathbf{r}$ is

$$\dot{\mathbf{L}} = r^{-2}(B\hat{\mathbf{L}} - C\hat{\omega}) \tag{38}$$

and the scalar product of (38) with $\mathbf{L}$ gives

$$\begin{aligned} L\dot{L} &= r^{-2}BL \\ \Leftrightarrow \quad L' &= B(L)\,. \end{aligned} \tag{39}$$

Integration of (39) yields $L(T)$ so that (37) has the form

$$\mathbf{I}' = M(T)\mathbf{I}$$

and leads to three vectors of Poincaré type.

Eq (37) can be solved in principle. Normalization of the conserved vector, $\mathbf{P}$, means that there are only two independent variables. Under the Weierstrass transformation [9]

$$\xi = \frac{I + iJ}{1 - K} \qquad \eta = \frac{I + iJ}{1 + K} \tag{40}$$

both ξ and η satisfy

$$\omega' + iL\omega + \frac{iL}{2C}(1 - \omega^2) = 0\,. \tag{41}$$

Under the transformation

$$\omega = C + \frac{2iCu'}{Lu} + i(\frac{C}{L})^2(\frac{L}{C})' \tag{42}$$

(41) becomes

$$u''\left\{\frac{1}{4}\left(\frac{C'}{C}-\frac{L'}{L}+iL\right)^{1/2}-\frac{1}{2}\left(\frac{C'}{C}-\frac{L'}{L}+iL\right)'+\frac{L^2}{4C^2}\right\}u=0 \tag{43}$$

which is just the equation for the time-dependent oscillator. The radial component of (34) is

$$\ddot{r}-r(\dot{\theta}^2+\dot{\phi}^2\sin^2\theta)=r^{-3}A(L)\,. \tag{44}$$

In terms of χ and T (44) is

$$\chi''+[L^2+A(L)]\chi=0 \tag{45}$$

and, since $L=L(T)$, this also is the time-dependent harmonic oscillator.

5. The Pinney equation and time-dependent systems

The time-dependent oscillator

$$\ddot{q}+\omega^2(t)q=0 \tag{46}$$

is transformed to the time-independent oscillator

$$Q''+\Omega^2Q=0 \tag{47}$$

by the transformation

$$Q=\frac{q}{\rho}\qquad T=\int\rho^{-2}(t)\mathrm{d}t\,, \tag{48}$$

where $\rho(t)$ satisfies the equation

$$\ddot{\rho}+\omega^2\rho=\frac{\Omega^2}{\rho^3}\,, \tag{49}$$

which is known as the Pinney equation [10].

The transformation

$$Q=\frac{q-\alpha}{\rho}\qquad T=\int\rho^{-2}(t)\mathrm{d}t \tag{50}$$

with

$$\ddot{\rho}+\omega^2\rho=\frac{\Omega^2}{\rho^3}\,,$$

where Ω^2 is often taken as unity, and its Hamiltonian version

$$Q=\frac{q-\alpha}{\rho},\quad P=\rho(q-\dot{\alpha})-\dot{\rho}(q-\alpha),\quad T=\int\rho^{-2}(t)\mathrm{d}t \tag{51}$$

are used in both classical and quantum mechanics.

The Hamiltonian [11]

$$H=\frac{1}{2}p^2+F(t)q+\frac{1}{2}\omega^2(t)q^2+\frac{1}{\rho^2}U\left(\frac{q-\alpha}{\rho}\right) \tag{52}$$

is transformed to the autonomous form

$$\bar{H} = \frac{1}{2}P^2 + U(Q) \tag{53}$$

provided ρ satisfies (49) and α is a solution of

$$\ddot{\alpha} + \omega^2(t)\alpha = F(t)\,. \tag{54}$$

6. The Pinney equation and $sl(2,R)$

The symmetries of eq(49) are

$$\begin{aligned} G_1 &= u_1(t)\frac{\partial}{\partial t} + \frac{1}{2}\dot{u}_1(t)\rho\frac{\partial}{\partial t} \\ G_2 &= u_2(t)\frac{\partial}{\partial t} + \frac{1}{2}\dot{u}_2(t)\rho\frac{\partial}{\partial t} \\ G_3 &= u_3(t)\frac{\partial}{\partial t} + \frac{1}{2}\dot{u}_3(t)\rho\frac{\partial}{\partial t}\,, \end{aligned} \tag{55}$$

where u_1, u_2 and u_3 are the three linearly independent solutions of

$$\dddot{u} + 4\omega^2\dot{u} + 4\omega\dot{\omega}u = 0 \tag{56}$$

and the Lie algebra of the symmetries is $sl(2,R)$. When (56) is multiplied by u, the resultant equation is easily integrated to give

$$u\ddot{u} - \frac{1}{2}\dot{u}^2 + 2\omega^2u^2 = cst\,. \tag{57}$$

Let $u = \rho^2$, $cst = 2\Omega^2$. Then (57) becomes

$$\rho^3\ddot{\rho} + \omega^2\rho^4 = \Omega^2 \tag{58}$$

or, on division by ρ^3,

$$\ddot{\rho} + \omega^2\rho = \frac{\Omega^2}{\rho^3}\,.$$

The Pinney equation is, in a sense, the second integral of (56).

Note that (56) is a tode which has the maximal number of symmetries, seven (3 + 4), with the algebra $A_1 \oplus 3A_1 \oplus sl(2,R)$. The Pinney equation is the integral associated with the $sl(2,R)$ part of the algebra. The solution of the Pinney equation (49) can be found as follows. Eq(49) has the symmetry

$$G = u(t)\frac{\partial}{\partial t} + \frac{1}{2}\dot{u}(t)\rho\frac{\partial}{\partial t}\,. \tag{59}$$

A new independent variable, $T = \int u^{-1}\mathrm{d}t$, and a new dependent variable, $V = \rho u^{-1/2}$, are introduced. Eq(49) is transformed to

$$V'' + \Omega^2 V = \frac{\Omega^2}{V^3} \tag{60}$$

which is readily solved in terms of the solutions of [10]

$$Y'' + \Omega^2 Y = 0 \tag{61}$$

and is

$$V^2 = A\sin^2\Omega t + 2B\sin\Omega t\cos\Omega t + C\cos^2\Omega t\,, \tag{62}$$

where

$$2AC - B^2 = 1\,. \tag{63}$$

7. Higher order equations and the Pinney connection

Equation (56) has maximal symmetry and is also self-adjoint since

$$\left(-\frac{\mathrm{d}}{\mathrm{d}t}\right)^3 u + \left(-\frac{\mathrm{d}}{\mathrm{d}t}\right)^1 (4\omega^2 u) + \left(-\frac{\mathrm{d}}{\mathrm{d}t}\right)^0 (4\omega\dot{\omega}u) = 0$$

gives

$$-[\dddot{u} + 4\omega^2\dot{u} + 4\omega\dot{\omega}u] = 0\,.$$

Eq(56) is the first of the hierarchy of higher order equations which possesses the maximal symmetry $A_1 \oplus nA_1 \oplus sl(2,R)$. They are all self-adjoint and can be transformed to autonomous form by the transformation

$$T = \int u^{-1}\mathrm{d}t \qquad V = \frac{v}{u^{(n-1)/2}}\,. \tag{64}$$

There should be an equation of order $(n-1)$ which has the essential feature of the Pinney equation, viz. the Lie algebra $sl(2,R)$. It is an open question whether other conditions should be imposed.

Acknowledgements

The Foundation for Research Development of South Africa is thanked for its support.

References

1. V P Ermakov, *Univ Isv Kiev*, Series III, **9** (1880) 1 (translated by A O Harin)
2. H R Lewis Jr., *Phys Rev Lett* **18** (1967) 510; H R Lewis Jr., J Math Phys **9** (1968) 1976
3. J R Ray and J L Reid, *Phys Lett A* **71** (1979) 317
4. J R Ray, *Phys Lett A* **78** (1980) 4; J R Ray and J L Reid, J Math Phys **21** (1980) 1583; J Goedert, *Phys Lett A* **143** (1990) 207
5. P G L Leach, *Phys Lett A* **158** (1991) 102
6. K S Govinder and P G L Leach, Generalized Ermakov systems in terms of $sl(2,R)$ invariants (submitted to *Quest Math*)
7. K S Govinder, C Athorne and P G L Leach, The algebraic structure of generalized Ermakov systems in three systems (submitted to *J Phys A: Math Gen*)
8. H Poincaré, *C R Acad Sci* **123** (1896) 530
9. E Kamke, Differentialgleichungen Lösungsmethoden und Lösungen (Chelsea, New York, 1971) eq 8.50
10. E Pinney, *Proc Amer Math Soc* **1** (1950) 681
11. H Ralph Lewis and P G L Leach, *J Math Phys* **23** (1982) 2371

SYMMETRIES OF DIFFERENTIAL EQUATIONS ON A LATTICE. AN EXAMPLE: THE TODA LATTICE

DECIO LEVI
Dipartimento di Fisica, Universitá di Roma III
I.N.F.N. Sezione di Roma
P.zale A. Moro 2, 00185 Roma Italy

and

PAVEL WINTERNITZ
C.R.M., Université de Montreal
C.P. 6128-A, Montreal, Quebec
Canada H3C 3J7

Abstract. We briefly present some methods which have been introduced to provide Lie symmetries for equations on a lattice, comparing them from the point of view of their applicability. We will apply the introduced techniques to the construction of Lie symmetries of Toda Lattice equations. Moreover, we will show how the group theoretical machinery can be applied also in this case to show the equivalence between nonlinear equations and to find equations of a prescribed form with a well defined symmetry group.

1. Introduction

One of the most efficient methods of obtaining exact analytical solutions of partial differential equations is the method of symmetry reduction [1, 2, 3]. It consists of first finding the Lie group G of local point transformation, leaving the set of solutions of the equation invariant, and then finding solutions invariant under a subgroup G_0 of G. An analysis of the group G allows moreover to classify the equations into classes having the same Lie group G. A condition necessary for being able to transform two equations among each other via a point transformation is that they have the same group G [4].

This essentially continuous technique has been recently extended to the case of discrete equations [5, 6].

In Section 2 we review the passages necessary to get continuous symmetries for differential difference equations (DDE's) and then in Section 3 carry out the explicit calculations in the case of the Toda equation in 1+1 dimensions. In doing so we will present and compare both the results contained in our articles on the subject, which correspond to consider either the *intrinsic method* or the *differential equation method*, and those of the global method introduced by Quispel et. al.[8] .

Apart from the standard Toda Lattice we will consider a Toda Lattice equation with variable coefficients and will show how, by analyzing its symmetry group, one can find the Lie point transformation which maps it into the standard Toda Lattice equation with constant coefficients. At the end, in Section 4, we will present some preliminary results on the class of nonlinear DDE's which have the same symmetry

N. H. Ibragimov et al. (eds.),
Modern Group Analysis: Advanced Analytical and Computational Methods in Mathematical Physics, 265–272.

group as the Toda itself.

2. Lie point symmetries of DDE's

For notational simplicity, let us restrict ourselves to scalar DDE's for one real function $u(n,t)$ depending on just one lattice variable, say n, and one continuous real variable, say t. Moreover we will be only interested in DDE's containing up to second order derivatives as those are the ones of particular interest in applications to dynamical systems. We write such equations as:

$$\begin{gathered}\Delta_n^{(2)} \equiv \Delta(t, n, u(n+k)\,|_{k=-a}^{b}, u_{,t}(n+k)\,|_{k=-a_i}^{b_i}, u_{,tt}(n+k)\,|_{k=-a_{ij}}^{b_{ij}}) = 0 \\ a, b, a_i, b_i, a_{ij}, b_{ij} \in Z^+\end{gathered} \tag{1}$$

Examples of such equations which will be considered in the following are the Toda Lattice

$$\Delta_n^{(2)} = u_{,tt}(n) - e^{u(n-1)-u(n)} + e^{u(n)-u(n+1)} \tag{2}$$

and the inhomogeneous Toda Lattice

$$\Delta_n^{(2)} = u_{,tt}(n) - ne^{u(n-1)-u(n)} + (n+1)e^{u(n)-u(n+1)} - 1 \tag{3}$$

with, in both cases, $u(n) \equiv u(n,t)$.

We are interested in Lie point transformations which leave the solution set of eqs.(1) invariant and which thus have the form:

$$\tilde{t} = \Lambda_g(t, u(n,t)), \tilde{u}(\tilde{n}, \tilde{t}) = \Omega_g(t, n, u(n,t)), \tilde{n} = \nu_g(n) \tag{4}$$

where g represent continuous or discrete group parameters.

Continuous transformations of the form (4) are generated by a Lie algebra of vector fields of the form:

$$\hat{V} = \xi(t, u(n,t))\partial_t + \phi(n, t, u(n,t))\partial_{u(n)} \tag{5}$$

where n is treated as a discrete variable and we have thus set $\tilde{n} = n$.

Let us notice that we have required a priori that ξ does not depend on n as otherwise $\tilde{t}$ would depend on n and thus eq.(1) would be transformed into a nonlocal DDE with $\tilde{u}(n)$ and $\tilde{u}(n+1)$ calculated for different values of $\tilde{t}$.

As in the case of purely differential equations, the following invariance condition

$$pr^{(2)}\hat{V}\Delta_n^{(2)}\,|_{\Delta_n^{(2)}=0} = 0 \tag{6}$$

must be true if $\hat{V}$ is to belong to the Lie symmetry algebra of $\Delta_n^{(2)}$. By the symbol $pr^{(2)}\hat{V}$ we mean:

$$\begin{aligned} pr^{(2)}\hat{V} = \xi(t, u(n))\partial_t + \sum_{k=n-a}^{n+b} \phi(k, t, u(k))\partial_{u(k)} + \sum_{k=n-a_i}^{n+b_i} \phi^t(k, t, u(k))\partial_{u_{,t}(k)} + \\ \sum_{k=n-a_{ij}}^{n+b_{ij}} \phi^{tt}(k, t, u(k))\partial_{u_{,tt}(k)} \end{aligned} \tag{7}$$

with

$$\phi^t(k,t,u(k)) = D_t\phi(k,t,u(k)) - [D_t\xi(t,u(k))]u_{,t}(k) \tag{8}$$

$$\phi^{tt}(k,t,u(k)) = D_t\phi^t(k,t,u(k)) - [D_t\xi(t,u(k))]u_{,tt}(k) \tag{9}$$

where by D_t we mean the total derivative with respect to t .

Let us notice moreover that $\phi^t(n,t,u(n))$ and $\phi^{tt}(n,t,u(n))$ are the prolongation coefficients with respect to the continuous variable while $\phi(k,t,u(k))$, $\phi^t(k,t,u(k))$ and $\phi^{tt}(k,t,u(k))$ for $k \neq n$ are the prolongation coefficients with respect to the discrete variable.

Eq.(6) is to be viewed as just one equation with n as a parametric variable; thus we have a finite algorithm for obtaining the determining equations, a usually overdetermined system of linear partial differential equations for $\xi(t,u(n))$ and $\phi(n,t,u(n))$.

A different approach consists of considering eq.(1) as a system of couple differential equations for the functions $u_n(t)$. Thus, in general we have infinitely many equations for infinitely many functions. In this case the proper ansatz for the vector field $\hat{V}$ would be:

$$\hat{V} = \xi(t,\{u_j(t)\}_j)\partial_t + \sum_k \phi_k(t,\{u_j(t)\}_j)\partial_{u_k(t)} \tag{10}$$

where by $\{u_j(t)\}_j$ we mean the set of all $u_j(t)$ and j and k vary a priori over an infinite range. Calculating the second prolongation $pr^{(2)}\hat{V}$ in a standard manner and imposing

$$pr^{(2)}\hat{V}\Delta_n^{(2)} \mid_{\Delta_j^{(2)}=0} = 0 \qquad \forall n,j \tag{11}$$

we obtain, in general, an infinite system of determining equations for an infinite number of functions.

Conceptually speaking, the *differential equation method* may give rise to a larger symmetry group than the *intrinsic method* we introduced before. In fact the intrinsic method yields purely point transformations while the differential equation method can yield higher transformations with respect to the discrete variable. In practice, it turns out that usually no higher order symmetries with respect to the discrete variable exist; then the two methods give the same result and naturally the intrinsic method is simpler.

A third approach, presented in this Proceedings by Quispel, consists in interpreting the variable n as a continuous variable and consequently the DDE as a differential-delay equation. In such an approach $u(n+k,t) \equiv \exp\left\{[k\frac{d}{dn}]u(n,t)\right\}$ and consequently the DDE reduces to a partial differential equation of ∞ order. In such a case formula (6) is meaningless and the desired Lie symmetries are obtained by returning to first principles, i.e. requiring that the solution of the equation $\Delta_n^{(2)} = 0$ be invariant under the infinitesimal transformation

$$\begin{aligned} t' &= t + \epsilon\xi(t,n,u(t,n)) \\ n' &= n + \epsilon\nu(t,n,u(n,t)) \\ u'(t',n') &= u(t,n) + \epsilon\phi(t,n,u(t,n)) \end{aligned} \tag{12}$$

In such a case the determining equations become very involved and even if in principle this method can provide us also with discrete elements of the symmetry group, often it cannot give more results than those obtained by the intrinsic method. In fact one can easily prove that no discrete element of the symmetry group for DDE's with constant coefficients exists apart from the trivial one generated by $\hat{V} = \frac{\partial}{\partial n}$.

3. Symmetries of the Toda Lattice

Let us now apply the techniques introduced in the previous section to the case of eq.(2). In this case eq.(6) reduces to an overdetermined system of determining equations obtained by equating to zero the coefficients of $[u_t(n)]^k \quad k = 0, 1, 2, 3$ and of $u(n \pm 1)$. They imply:

$$\begin{array}{c} \xi = at + d \\ \phi = b + 2an + ct \end{array} \quad a, b, c, d \quad \text{real constants.} \tag{13}$$

corresponding to a finite dimensional Lie algebra generated by the vector fields:

$$\begin{array}{cc} \hat{D} = t\partial_t + 2n\partial_{u(n)}; & \hat{T} = \partial_t \\ \hat{W} = t\partial_{u(n)}; & \hat{U} = \partial_{u(n)} \end{array} \tag{14}$$

The group transformation which will leave eq.(2) invariant is hence:

$$u'(t', n) = u(t'e^{-\frac{\lambda_4}{2}} - \lambda_3, n) + \lambda_2(t'e^{-\frac{\lambda_4}{2}} - \lambda_3) + \lambda_4 n + \lambda_1 \tag{15}$$

where $\lambda_j \quad j = 1, 2, 3, 4$ are arbitrary real group coefficients.

To the transformation (15) we can add some discrete ones [7]:

$$n' = n + N \qquad N \in Z \tag{16}$$

and

$$(t, u(n)) \to (-t, u(n)); \qquad (t, u(n)) \to (t, -u(-n)). \tag{17}$$

We write the symmetry group of eq.(2)as:

$$G = G_D \rhd G_C \tag{18}$$

where G_D are the discrete transformations (16,17) and the invariant subgroup G_C corresponds to the Lie algebra (14). A complete classification of the one dimensional subgroups of G can be easily obtained and it is partially contained in the literature [6].

In fact, if we complement the Lie algebra (14) with the vector field

$$\hat{Z} = \frac{\partial}{\partial n} \tag{19}$$

and require, at the end of the calculations, that the corresponding group parameter be integer, the symmetry algebra becomes:

$$[\hat{Z}, \hat{D}] = 2\hat{U}; \quad [\hat{D}, \hat{T}] = -\hat{T}; \quad [\hat{D}, \hat{W}] = \hat{W}; \quad [\hat{T}, \hat{W}] = \hat{U} \tag{20}$$

and the one dimensional subalgebras are:

$$\begin{array}{c}\{\hat{Z}+a\hat{D}+b\hat{U}\},\quad \{\hat{Z}+a\hat{T}+k\hat{W}\},\quad \{\hat{Z}+\epsilon\hat{W}\},\quad \{\hat{Z}+a\hat{U}\}\\ \{\hat{T}+a\hat{W}\},\quad \{\hat{D}+a\hat{U}\},\quad \{\hat{U}\},\quad \{\hat{Z}\},\quad \{\hat{W}\}\\ (a,b)\in R;\qquad a\neq 0;\quad k=0,1,-1;\quad \epsilon=\pm 1\end{array} \tag{21}$$

Nontrivial solutions, corresponding to reductions with respect to **continuous subgroups** $G_0 \subset G_C$, are obtained by considering invariance of the Toda Lattice under $\{\hat{T}+a\hat{W}\}$ and $\{\hat{D}+a\hat{U}\}$ and give:

$$u(n,t)=b-\frac{1}{2}at^2-\sum_{j=1}^{n}\log(c-aj) \tag{22}$$

$$u(n,t)=b+2(n+a)\log(t)-\sum_{j=0}^{n}\log[c+(2a-1)j+j^2] \tag{23}$$

where b and c are arbitrary integration parameters.

Reduction by a purely **discrete subgroup**, $G_0 \subset G_D$, implies the invariance of eq(2) under discrete translation of n and makes it possible to impose the periodicity condition

$$u(n+N,t)=u(n,t) \tag{24}$$

In such a case we reduce the DDE (2) to an ordinary differential equation: for example for $N=2$, defining

$$\begin{array}{c}u(n+1)=\frac{1}{2}(\phi(t)+\alpha t+\beta)\\ u(n)=-\frac{1}{2}(\phi(t)-\alpha t-\beta)\end{array} \tag{25}$$

we get

$$\phi_{,tt}=-4\sinh(\phi) \tag{26}$$

while for $N=3$, defining

$$\begin{array}{c}u(n+2)=-q(t)\\ u(n+1)=0\\ u(n)=q(t)\end{array} \tag{27}$$

we get the ordinary Bullough-Dodd-Mikhailov-Shabat differential equation:

$$q_{,tt}=e^{-2q}-e^{q} \tag{28}$$

Let us now consider a subgroup $G_0 \subset G$ that is contained neither in G_C nor in G_D, i.e. a **nonsplitting subgroup** of G. In such a case, considering a reduction invariant under $\hat{Z}+a\hat{D}+b\hat{U}$ we get:

$$F''(y)=e^{-b}[\exp(F(ye^{a})-F(y)+a)-\exp(F(y)-F(ye^{-a})-a)] \tag{29}$$

where the symmetry variables $F(y)$ and y are defined by:

$$\begin{aligned} y &= te^{-an} \\ u(n,t) &= an^2 + bn + F(y) \end{aligned} \tag{30}$$

while under $\hat{Z} + a\hat{T} + k\hat{W}$ we get:

$$F''(y) = e^{F(y+a)-F(y)} - e^{F(y)-F(y-a)} - \frac{k}{a} \tag{31}$$

with

$$\begin{aligned} y &= t - an \\ u(n,t) &= \tfrac{k}{2a}t^2 + F(y) \end{aligned} \tag{32}$$

Eq.(29) can be called a differential "dilation-delay" type equation; it involves one independent variable y, but the function F and its derivatives are evaluated at the point y and at the dilated points ye^a and ye^{-a}. Eq(30) instead is a differential delay equation which have as interesting solutions for $k = 0$ the soliton and periodic solutions of the Toda Lattice.

The other two nonsplitting subgroups give rise to linear delay equations which can be solved explicitly and give trivial solutions.

The same calculation we have carried out up to now for the Toda lattice (2) can be carried out for the inhomogeneous Toda Lattice (3). In such a case we get:

$$\begin{aligned} \tilde{D} = \tilde{t}\partial_{\tilde{t}} + (\tilde{t}^2 + 2n)\partial_{v(n)}; &\qquad \tilde{T} = \partial_{\tilde{t}} + \tilde{t}\partial_{v(n)} \\ \tilde{W} = \tilde{t}\partial_{v(n)}; &\qquad \tilde{U} = \partial_{v(n)} \end{aligned} \tag{33}$$

These vectors have the same symmetry algebra (21) as that of the Toda Lattice. This is a necessary condition for the existence of a point transformation between the two equations. In fact by comparing the two sets of vectors, we get the following transformation which transforms a solution $u(n,t)$ of equation (2) into a solution $v(n,\tilde{t})$ of eq.(3):

$$\begin{aligned} \tilde{t} &= t \\ v(n,\tilde{t}) &= u(n,t) + \tfrac{1}{2}\tilde{t}^2 + \log(n!) \end{aligned} \tag{34}$$

4. Classification of differential equations on a lattice

Group theoretical methods can also be used for finding equations with particular symmetry properties. This has been done in the case of partial differential equations [4] showing, among other things that there is an infinite set of nonlinear partial differential equations in 2+1 dimensions which possess a Kac-Moody-Virasoro algebra of point symmetries of the same kind as the Kadomtsev Petviashvili equation. The same kind of calculations can also be done here.

Let us consider a class of equations of the form:

$$\Delta_n = u_{,tt}(n,t) - F_n(n, u(k,t)) = 0 \tag{35}$$

where F_n is such that $F_{n,u(k)} \neq 0$ for some $k \neq n$. We are interested here just in analyzing the *peculiarity of the Toda Lattice* as an equation on the lattice, differential of second order with respect to t. So we look for F_n such that eq.(35) has the same symmetry group as the Toda Lattice itself.

Under such general assumptions we get that eq.(35) has the same symmetry algebra as the Toda Lattice iff:

$$F_n = K[\{\xi(\alpha, n-1)\}_\alpha]e^{q(n-1)} \tag{36}$$

where α is an arbitrary set of integers, ξ is given by

$$\xi(\alpha, n-1) = u(n-1) - u(n) - \frac{1}{n-\alpha}[u(\alpha) - u(n)] \tag{37}$$

and K is an arbitrary function of its argument.

If we consider just nearest neighbour interactions, as it is the case of the Toda Lattice, we get:

$$F_n = K[u(n+1) - 2u(n) + u(n-1)]e^{u(n-1)-u(n)} \tag{38}$$

a whole class of equations which has the Toda Lattice as a very particular example.

If we fix the form of F_n as polynomial then a dynamical equation which has the same symmetry algebra as the Toda Lattice is, for example

$$\begin{aligned} F_n = &-\tfrac{1}{2}[u(n+1) - 2u(n) + u(n-1)] + [u(n) - u(n-1)]^2 + \\ &+\tfrac{1}{2}[u(n) - u(n+1)][u(n+1) - u(n-1)] \end{aligned} \tag{39}$$

More detailed results on this problems are contained in [9].

5. Conclusions

The main conclusions we can draw are;

1. Symmetry methods work quite well for DDE's and should be further developed.
2. As far as one is interested in point symmetries the **intrinsic method** is quite sufficient. To go over to contact and higher symmetries one should find more elaborated systematic methods since the **differential equation method** is not sufficient for finding higher symmetries in both, continuous and discrete variables.
   One can apply these techniques to a whole range of problems associated with the generic world of classification of DDE's, a very important subject in physics.
3. As in the case of partial differential equations, similarity techniques can be successfully combined with singularity analysis of partial differential equations. In particular differential delay equations like (29) or (31) have the Painlevé property, in the sense of singularity confinement [10, 11].

References

1. Olver , P.J., *Applications of Lie Groups to Differential Equations,* Springer, New York, 1986
2. Ovsiannikov, L.V., *Group Analysis of Differential Equations,* Academic, New York, 1982
3. Winternitz, P., in *Partially Integrable Evolution Equations in Physics,* (R. Conte and N. Boccara, editors), Kluwer, Dordrecht 1990, pp. 515-567
4. David, D., Levi, D. and Winternitz, P., Phys. Lett. **A129** (1988) 161

5. Maeda, S., Math. Japonica **25** (1980) 405
6. Levi, D. and Winternitz, P., Phys. Lett. **A152** (1991) 335; Symmetries and conditional symmetries of differential difference equations, preprint DF-INFN N^0 858 (22/1/1992)
7. Levi, D. and Rodriguez, M.A., J. Phys. A:Math. Gen. **25**(1992) L975
8. Quispel, G.R.W., Capel, H.W. and Sahadevan, R., Phys. Lett. **A170** (1992) 379
9. Levi D. and Winternitz P., Classification of dynamical systems on the lattice, in preparation.
10. Grammaticos, B., Ramani, A. and Papageorgiu, V.G., Phys. Rev. Lett. **67** (1991) 1825
11. Nijhoff, F.W. and Capel, H.W., Phys. Lett. **A163** (1992) 49

SYMMETRIES OF PARTICLE LAGRANGIANS

F M MAHOMED
Centre for Nonlinear Studies and
Department of Computational & Applied Mathematics
University of the Witwatersrand, Johannesburg
P O Wits 2050 - South Africa

A H KARA
Department of Mathematics
University of the Witwatersrand, Johannesburg
P O Wits 2050 - South Africa

and

P G L LEACH
Centre for Nonlinear Studies
University of the Witwatersrand and
Department of Mathematics & Applied Mathematics
University of Natal
King George V Ave
Durban 4001 – South Africa

Abstract. The Lie 'counting' theorem for a second-order ordinary differential equation states that the maximum dimension of its point symmetry algebra is eight. There is an analogous theorem to the Lie 'counting' theorem, viz. the Noether 'counting' theorem, which states that the maximum dimension of the Noether point symmetry algebra of a particle Lagrangian is five. This theorem is discussed. Moreover, a particle Lagrangian admits the maximal $r \in \{0,1,2,3,5\}$-dimensional Noether point symmetry algebra.

1. Introduction

The maximum dimension of the Lie algebra of point symmetries of a scalar second-order ordinary differential equation is eight and the algebra is $sl(3,\mathbb{R})$. This is popularly known as the Lie counting theorem (Lie [4,5]). There are various proofs and generalizations of this famous result (see, e.g. Dickson [2], Anderson and Davison [1], González-Gascón & González-López [3], Mahomed et al. [6]).

One can enquire as to what is the maximum dimension of the (here and in the following the Lie algebra of point symmetries that leave invariant the action integral up to a gauge will be referred to as the Noether point symmetry algebra) Noether point symmetry algebra for a particle Lagrangian? The answer is five, although we have not been able to find an explicit proof of this in the literature. We shall refer to this theorem as the Noether 'counting' theorem in analogy to the Lie 'counting' theorem which we stated above. In the next section we discuss this theorem and its consequences. The maximum dimension of the Noether algebra is attained for the simple Lagrangian, $L = \dot{q}^2/2$, of the simplest equation $\ddot{q} = 0$. The corresponding Lie algebra is $A_{5,40}$ using the classification scheme given in Patera and Winternitz [7].

N. H. Ibragimov et al. (eds.),
Modern Group Analysis: Advanced Analytical and Computational Methods in Mathematical Physics, 273–276.

The knowledge of the maximum Noether algebra is necessary for the classification of first order particle Lagrangians on the line in terms of their Noether point symmetry algebras.

As a consequence of the proof of the Noether 'counting' theorem (see [6]), one can deduce that there does not exist a maximal four-dimensional point symmetry algebra for a first-order particle Lagrangian on the line. Thus, we conclude that a first-order particle Lagrangian on the line admits the maximal $r \in \{0, 1, 2, 3, 5\}$-dimensional Noether point symmetry algebra.

2. The Noether 'counting' theorem

We focus attention on the classical Noether theorem relating to a one-dimensional dynamical system. For a given particle Lagrangian L, the coefficient functions ξ and η of a vector field $G = \xi\partial/\partial t + \eta\partial/\partial q$ that generates a Noether transformation (the transformation which leaves the action functional invariant up to gauge) are found by solving the *Killing* type equation (see Sarlet and Cantrijn [8])

$$G^{[1]}L + \xi^{(1)}L = f^{(1)}, \tag{1}$$

where f is a suitable function of t, q and $G^{[1]}$ is the first prolongation of the symmetry vector field G and is given by $G^{[1]} = G + (\eta^{(1)} - \dot{q}\eta^{(1)})\partial/\partial\dot{q}$. In the absence of the gauge term or strict invariance we shall refer to the transformation that leaves the action functional invariant as a variational transformation.

The classical Noether theorem states that to each Noether transformation generated by G there corresponds a constant of the motion $I(t, q, \dot{q})$ given by

$$I(t, q, \dot{q}) = f(t, q) - [L\xi + (\eta - \dot{q}\xi)\partial L/\partial\dot{q}]. \tag{2}$$

This theorem does possess the beauty of having an explicit formula (2) for the constants of the motion once the Noether symmetries are known. It establishes a precise correspondence between equivalence classes of point symmetries and constants of the motion. It should, however, be pointed out that such a link is strictly with respect to a fixed Lagrangian description, bearing in mind that there are ambiguities in the possible Lagrangian description of a given second order dynamical equation. Therefore, in order to take full advantage of Noether's theorem for a given one-dimensional dynamical system, one needs a suitable Lagrangian that will give rise to the highest number of Noether point symmetries and so the largest number of constants of the motion that is obtainable in this way. Therefore of prime importance is the dimensionality of the Noether point symmetry algebra for a given particle Lagrangian.

We at first find out the maximum dimension of the Noether point symmetry algebra for a variational problem whose Euler-Lagrange equation constitutes a scalar second-order dynamical equation. It is instructive to look firstly at the Noether point symmetries of the free particle equation with respect to the *usual* Lagrangian $L = \dot{q}^2/2$. Substituting this L into the Killing equation (1) and performing simple manipulations we end up with a set of five symmetry vector fields (or any linear combination thereof)

$$G_1 = \frac{\partial}{\partial t} \qquad G_2 = t\frac{\partial}{\partial t} + \frac{1}{2}q\frac{\partial}{\partial q},$$

$$G_3 = t^2\frac{\partial}{\partial t} + tq\frac{\partial}{\partial q} \quad G_4 = \frac{\partial}{\partial q}, \qquad (3)$$
$$G_5 = t\frac{\partial}{\partial q}.$$

The corresponding Lie algebra is isomorphic to $A_{5,40}$ (see [7]).

We introduce the notion of equivalence by means of a point transformation of two particle Lagrangians. Two Lagrangians $L(t,q,\dot{q})$ and $\bar{L}(T,Q,dQ/dT)$ are said to be *equivalent up to gauge* (see [8]) if

$$L(t,q,\dot{q}) = \bar{L}(T,Q,dQ/dT)\frac{dT}{dt} + \frac{df}{dt} \qquad (4)$$

by means of the prescribed coordinate transformation $Q = F(t,q)$, $T = G(t,q)$, where f, the gauge term, is an arbitrary function of t and q. This means that the Euler-Lagrange equations associated with L and $\bar{L}$ are equivalent to each other by the said change of coordinates.

If a second-order equation in q belongs to the equivalence class of the free particle equation $Q'' = 0$ $(' = d/dT)$ via the point transformation $Q = F(t,q)$, $T = G(t,q)$, then a Lagrangian for the equation in q is obtained by invoking (4) where $\bar{L}(T,Q,Q')$ is the Lagrangian of the transformed equation which in this case is that of the free particle equation. We take the usual Lagrangian $\bar{L} = {Q'}^2/2$. With this Lagrangian, the Noether point symmetry algebra associated with Euler-Lagrange second-order linearizable (via a point transformation) equations is a five-dimensional algebra which is isomorphic to $A_{5,40}$. The corresponding Lie group for this algebra of symmetries is $T_2 \otimes_s SL(2,\mathbb{R})$, i.e., the semidirect product of translations T_2 and the special linear group $SL(2,\mathbb{R})$.

The question then is: what is the maximum dimension of the Noether point symmetry algebra with respect to any Lagrangian? The answer is provided by the following theorem, the Noether counting theorem:

Theorem 1. The maximum dimension of the Noether point symmetry algebra for Lagrangians in one-dimensional particle dynamics is five.

The proof of the theorem is algebraic (see [6]) and relies on the maximal dimensionality of the alternative Lagrangian of $\ddot{q} = 0$, viz.

$$L' = \int^{\dot{q}}\int^{r} g(s, q - ts)ds\, dr + H\dot{q} + J, \qquad (5)$$

where g is an arbitrary function of the functionally independent first integrals of $\ddot{q} = 0$, and H and J are functions of t and q such that $H_t = J_q$.

In the case of strict invariance of the action functional, the variational algebra is a subalgebra of the Noether algebra. The variational symmetry algebra of the usual Lagrangian $L = \frac{1}{2}\dot{q}^2$ of the free particle equation is the three-dimensional algebra $A^{\frac{1}{2}}_{3,5}$. Hence we have the following corollary.

Corollary 1. The maximum dimension of the variational point symmetry algebra for Lagrangians in one-dimensional particle dynamics is three.

In view of the above one has to bear in mind that all the symmetries and therefore all the first integrals may not be determined in the absence of a gauge term. This is not to say that they all will be determined using gauge variance since there are alternative Lagrangians for a given problem that may result in fewer (incomplete) symmetries.

Theorem 2. There does not exist a maximal four-dimensional Noether point symmetry algebra for one-dimensional particle Lagrangians.

The proof follows from the proof of Theorem 1 (see [6]).

In conclusion we have the following interesting result:

Theorem 3. A first order particle Lagrangian on the line admits a maximal $r \in \{0, 1, 2, 3, 5\}$ dimensional Noether point symmetry algebra.

References

1. R L Anderson and S M Davison, A generalization of Lie's 'counting' theorem for second-order ordinary differential equations, *J. Math. Anal. Appl.* **48** (1974) 301-315.
2. L E Dickson, Differential Equations from the group standpoint, *Ann. Math.* **25** (1924) 287.
3. F González-Gascón and A González-López, Symmetries of differrential equations, IV, *J. Math. Phys.* **24** (1983) 2006-2021.
4. S Lie, "Differentialgleichungen" (Chelsea, New York, 1967).
5. S Lie, "Vorlesungen über Continuerliche Gruppen mit geometrichen und anderen Anwendungen", Bearbeitet und herausgegben von Dr G Scheffers (Teubner, Leipzig, 1893).
6. F M Mahomed, A H Kara and P G L Leach, Lie and Noether Counting Theorems for one-dimensional systems (to appear in *J. Math. Anal. Appl.*).
7. J Patera and P Winternitz, Subalgebras of real three and four dimensional Lie algebras, *J. Math. Phys* **18** (1977) 1449-1455.
8. W Sarlet and F Cantrijn, Generalization of Noether's Theorem in Classical Mechanics, *SIAM Rev.* **23** (1981) 467-493.

THE GROUP ANALYSIS ALGORITHMS

L. V. OVSIANNIKOV
Lavrentiev Institute of Hydrodynamics
Novosibirsk, Russia

The fundamental notion introduced by S. Lie, i.e., "the system of differential equations admits a group of transformations", is a helpful investigation tool. The admitted group operates on a set of solutions and, therefore, introduces into this set the definite algebraic structure. Searching for this structure and its use to construct or investigate either properties of solutions for a lot of equations of mathematical physics (in particular, continua mechanics) may be effectuated by means of suitable analytical algorithms.

In fact, the presence of such algorithms made the theory capable of living and fruitful. The algorithms under discussion are rather well-known. Their general theory, contents and peculiarities of application are listed, for example, in book [1]. That's why the main aim of this paper is to demonstrate a row of new examples for their realization.

1. Admitted group calculation

Let the system of differential equations E of the order l be defined on the spaces $Z = R^n(x) \times R^m(y)$, $x = (x^1, \ldots, x^n)$, $y = (y^1, \ldots, y^m)$. In prolonged spaces Z_l the point coordinates are introduced

$$\underset{l}{z} = (x, y, \underset{1}{y}, \ldots, \underset{l}{y}), \qquad \underset{l}{y} = \left\{ \frac{\partial^l y^k}{\partial x^{i_1} \ldots \partial x^{i_l}} \right\}$$

for $k = 1, \ldots, m$, $i_1, \ldots, i_l = 1, \ldots, n$, and the equations E as equations of manifold $E \subset Z_l$ are considered:

$$E\colon \quad \psi^\sigma(\underset{l}{z}) = 0 \qquad (\sigma = 1, \ldots, s). \tag{1.1}$$

The calculation algorithm for the group GE admitted by E (Lie pseudogroup) is based on the fact that GE is produced by Lie algebra of the operators

$$X = \xi^i(x, y)\partial_{x^i} + \eta^k(x, y)\partial_{y^k}. \tag{1.2}$$

In turn these operators are found as the solution of the determining equations

$$DE\colon \quad \underset{l}{X}\psi^\sigma(\underset{l}{z})\Big|_E = 0 \qquad (\sigma = 1, \ldots. s),$$

where $\underset{l}{X}$ is the l-th prolongation of X. The system DE is an overdetermined system of linear differential equations for the unknown coordinates ξ^i, η^k in (1.2).

N. H. Ibragimov et al. (eds.),
Modern Group Analysis: Advanced Analytical and Computational Methods in Mathematical Physics, 277–289.

We shall say that the operator X (or Lie algebra LE, or the group GE) possesses the property of x-autonomy if all the coordinates ξ^i are independent of the variables η^k.

If this property is satisfied, the calculations are considerably shortened as to forming and resolving the DE equations. That's why a priori information about the x-autonomy is useful. Moreover, those systems E for which GE is x-autonomic have a special form and because of this deserve separate investigation.

For *quasilinear* systems E the left parts of the DE equations are the second power polynomials from parametric derivatives and, therefore, split into three subsystems: 2-equations, 1-equations and 0-equations. If a conclusion about the x-autonomy can be obtained from 2-equations only, we shall say about the *algebraic 2-autonomy.*

Given below is an example of algebraic 2-autonomy. The first order quasilinear system of m equations for m unknown functions $y = (y^1, \ldots, y^m)$ and independent variables (x, t), where $x = (x^1, \ldots, x^n)$, is considered when $m > 1$ and $n \geq 1$

$$\partial y/\partial t = a^i \partial y/\partial x^i + v. \tag{1.3}$$

Here $a^i = a^i(t, x, y)$ $(i = 1, \ldots, n)$ are $m \times m$ matrices and $v = v(t, x, y)$ are the vectors of "junior terms". The admitted operators are searched in the form

$$X = \tau(t, x, y)\partial_t + \xi^i(t, x, y)\partial_{x^i} + \xi_1^k(t, x, y)\partial_{y^k}. \tag{1.4}$$

With the designations $\xi_k = \partial\tau/\partial y^k$, $\eta_k^i = \partial\xi^i/\partial y^k$, the system of 2-equations is the following linear algebraic system for the unknowns $\xi = (\xi_1, \ldots, \xi_m)$, $\eta^i = (\eta_1^i, \ldots, \eta_m^i)$ $(i = 1, 2, \ldots, n)$:

$$a_q^{ik} a_p^{js} \xi_s + a_p^{jk} a_q^{is} \xi_s - a_s^{jk} a_q^{is} \xi_p - a_s^{ik} a_p^{js} \xi_q + \delta_q^k a_p^{js} \eta_s^i + \delta_p^k a_q^{is} \eta_s^j - a_q^{jk} \eta_p^i - a_p^{ik} \eta_q^j = 0, \tag{1.5}$$

for $i, j = 1, \ldots, n$ and $k, p, q = 1, \ldots, m$, where a_p^{ik} are the elements of matrices a^i, and the summation over $s = 1, \ldots, m$ is supposed.

The algebraic 2-autonomy property is equipollent to the fact that (1.5) has *only* zero solution. This system is further considered in the arbitrary point (t, x, y).

Lemma. *If system* (1.5) *has non-zero solution, then all the matrices* a^i *satisfy the square equations*

$$(a^i)^2 = b^i a^i + c^i I \qquad (i = 1, \ldots, n) \tag{1.6}$$

with the unit $m \times m$ *matrix* I *and scalars* $b^i, c^i \in R$.

Thus, the nonfulfilment of condition (1.6) at least for one matrix a^i results in the x-autonomy of the group GE (1.3).

Let condition (1.6) be valid. The modified matrices a^i are introduced

$$a'^i = a^i - \tfrac{1}{2} b^i I \qquad (i = 1, \ldots, n)$$

and the vector space $\mathfrak{M}_a$ produced by all a'^i over the field R is considered. The symmetric bilinear form $g\colon (a', b') \to R$ is defined on $\mathfrak{M}_a$

$$g(a', b') = \frac{2}{m} \operatorname{tr}(a'b').$$

Let algebra C_a be generated by the matrices a'^i and the unit matrix I.

Theorem. *If system (1.5) has non-zero solution, then C_a is a Clifford's algebra $C_a(g)$, the bilinear form g being degenerate and its rank being ≤ 1.*

Thus, it is sufficient for the x-autonomy GE (1.3) that $C_a \neq C_a(g)$ or the rank of g be > 1.

The set of additional conditions found for the space $\mathfrak{M}_a$ allowed us to formulate a criterion when GE (1.3) is x-autonomic. They remain outside this text since its volume is limited.

2. Group classification

Let the symbol $E(\pi)$ designate the system E containing an "arbitrary element" π which is a set of parameters or functions $Z \to R$.

The intersection

$$G^0 = \bigcap_{\pi} GE(\pi)$$

is called the *kernel* of full groups $GE(\pi)$. The group classification problem consists in searching for the kernel G^0 and all its non-equivalent extensions for specializations of π.

In particular, this problem was solved for gas dynamics equations (see [1], where π is the function of gas state of the form $p = f(\varrho, S)$ (p is the pressure, ϱ is the density and S is the entropy). Here $G^0 = G_{11}$ is the Galilei group G_{10} extended by stretching all the independent variables. The group classification gives 14 non-equivalent forms of the state equation when the extension of G_{11} up to certain groups G_{12}, G_{13}, G_{14} is valid.

A new interesting example of group classification (discovered by Yu. A. Chirkunov and the author) is represented by equations of steady gas flow in the space $R^3(x, y, z)$

$$\varrho \vec{u}\,\nabla \vec{u} + \nabla p = 0, \quad \vec{u}\,\nabla \varrho + \varrho \operatorname{div} \vec{u} = 0, \quad \vec{u}\,\nabla \varrho + A \operatorname{div} \vec{u} = 0. \tag{2.1}$$

Here $\vec{u} = (u, v, w)$ is the velocity vector, p the pressure, ϱ the density, c the speed of sound and $A = c^2$. The arbitrary element is the function $A = A(p, \varrho)$. The kernel of full groups $G^0 = G_7$ is the extension of the Euclidean group G_6 in R^3 by stretching the independent variables.

The possible extensions of G_7 are contained in the following determining equations

$$\sigma p A_p + (\sigma - 2\omega)\varrho A_\varrho = \sigma A, \tag{2.2}$$

$$\varrho A \omega_p + \varrho^2 \omega_\varrho = A \omega_q, \tag{2.3}$$

where $\sigma = const$ and the function $\omega = \omega(p, \varrho, q)$ with $q = \frac{1}{2}|\vec{u}|^2$ is presented. In addition, the extending operators are

$$\bar{X} = \sigma(p\partial_p + \varrho\partial_\varrho) + \omega(u\partial_u + v\partial_v + w\partial_w - 2\varrho\partial_\varrho). \tag{2.4}$$

Among several cases of extension, of special interest are the following.

(a) If $\omega_q \neq 0$, then from (2.2) it follows that $A_\varrho = 0$, i.e., $A = A(p)$. This is the case when the general solution for (2.3) is $\omega = \Omega(S, B)$ with the arbitrary function of the entropy S and the Bernoulli function B (the integral is taken with the constant S)

$$B = q + \int \frac{dp}{\varrho}.$$

It is found, however, that the infinite-dimensional extension (2.4) of the group G_7 is fictitious because system (2.1) with $A = A(p)$ admits the equivalence transformation

$$\vec{u} \to \tilde{\vec{u}} = F(S, B)\vec{u}, \qquad \varrho \to \tilde{\varrho} = F^{-2}(S, B), \qquad p \to \tilde{p} = p, \tag{2.5}$$

which always permits us to make $\omega = const$. That's why it is possible to restrict ourselves to extension of G_7 to G_9 owing to two operators obtained from (2.4) for $(\sigma, \omega) = (1, 0)$ or $(\sigma, \omega) = (0, 1)$.

The transformation (2.5) is known [6] and can be called Munk transformation.

(b) If $\omega_q = 0$, then equation (2.3) can be solved in the form $\omega = \Omega(S)$ with the arbitrary function of the entropy S. Supposing that $\sigma \neq 0$ and changing the independent variables

$$A = pU, \qquad \omega = \frac{1}{2}(\sigma - V) \tag{2.6}$$

results in the system of equations for the functions $U = U(r, t)$ and $V = V(r, t)$, where $r = \log p$, $t = \log \varrho$

$$U_r + VU_t = 0, \qquad UV_r + V_t = 0. \tag{2.7}$$

The evident "hodografwise" linearization of system (2.7) allows us to construct its general solution which is defined implicitly by the equation

$$r = U\varphi'(U) - \varphi(U) + \psi'(V), \qquad t = \varphi'(U) + V\psi'(V) - \psi(V)$$

with the arbitrary functions $\varphi(U)$ and $\psi(V)$, under the condition $\varphi''(U)\psi''(V) \neq 0$.

Here there is an infinitely great number (continuum) of possible variants of concrete state equations $A = A(p, \varrho)$ for which G_7 is extended to G_9 (because the function V is uniquely defined by U from (2.7)).

3. Transformation of solutions

The key property of the admitted group GE is that under the action of any transformation $T \in GE$ each solution $y = u(x)$ for E passes into the solution $y = u'(x)$ for the same equations E. Under the transformation given by formulae

$$T: \quad x' = f(x, y), \quad y' = g(x, y) \tag{3.1}$$

the functions $u'(x)$ are uniquely obtained from identity $[y' = u'(x')]_{y=u(x)}$, or in detail

$$g(x, u(x)) = u'(f(x, u(x))). \tag{3.2}$$

With T running admitted group GE the set of solutions obtained by means of algorithm (3.1), (3.2) is called the *orbit* of the solution $y = u(x)$. This algorithm is evidently simple and we shall make only one notation.

Let $\mathcal{L}_u E$ be the system of differential equations obtained by *linearization* of the equations E over its solution $y = u(x)$. Then by differentiation of (3.2) with respect to parameters contained in the transformation $T \in GE$ we obtain a family of solutions $y = v(x)$ for the system $\mathcal{L}_u E$. Namely, with operators from the Lie algebra LE of the form (1.2) the discussed solutions for $\mathcal{L}_u E$ are given by formulae

$$v^k(x) = \eta^k(x, u(x)) - \frac{\partial u^k(x)}{\partial x^i}\xi^i(x, u(x)) \qquad (k = 1, \ldots, m). \tag{3.3}$$

This allows us to predict some definite particular solutions for the linearized equations algorithmically.

4. Invariant solutions

It is convenient to consider the solution $y = u(x)$ for the system E as the manifold $U \subset Z$ and to say simply about the "solution U". The notion of invariant H-solution is defined for any subgroup $H \subset GE$.

The solution U is called the *invariant H-solution* if U is invariant manifold of the group H.

The necessary condition for the existence of the invariant H-solutions is formulated in terms of the r-parametric group H with basic operators of its Lie algebra

$$X_s = \xi_s^i \partial_{x^i} + \eta_s^k \partial_{y^k} \qquad (s = 1, \ldots, r). \tag{4.1}$$

In general, let $r_*(a_q^p)$ designate the *general rank* of the matrix with elements a_q^p. For nonsingular invariant H-solution to exist, it is necessary that relations be satisfied

$$r_*(\xi_s^i) = r_*(\xi_s^i, \eta_s^k) \leq n. \tag{4.2}$$

The following theorem [1] about the conditional existence of the invariant H-solutions is valid. In the below formulation r_* is the common value of ranks in (4.2).

Theorem. *Let the subgroup $H \subset GE$ satisfy the conditions* (4.2). *Then there exist a factor system E/H over the space $Z_H = R^{n-r_*}(t) \times R^m(w)$ and a mapping $Z \to Z_H$ defined by formulae*

$$t^l = t^l(z), \quad w^k = w^k(z) \qquad (l = 1, \ldots, n - r_*, \ k = 1, \ldots, m) \tag{4.3}$$

with the following property: any nonsingular invariant H-solution $U = U\{y = u(x)\}$ is mapped into the solution $W = \{w = W(t)\}$ for the system E/H satisfying identity

$$w(x, u(x)) = W(t(x, u(x))) \tag{4.4}$$

and inversely if the solution W for the system E/H produces the manifold in Z

$$w(x, y) = W(t(x, y)) \tag{4.5}$$

which can be written in the form $y = u(x)$, then U is invariant H-solution for E.

The number $\varrho = n - r_*$ equal to that of independent variables in the factor system E/H is called the *rank* of the invariant H-solution.

The factor system E/H is built algorithmically by means of *inversal invariants* $J = (J^1, \ldots, J^q)$ of the group H, where $q = m + n - r_* = m + \varrho$ and J^l $(l = 1, \ldots, q)$ are functionally independent scalar invariant of the group H.

NOTE 1. If the system E is linearized on its invariant H-solution $y = u(x)$, then the system $\mathcal{L}_u E$ admits the group H. This is the case when formulae (3.3) for $T \in H$ give zero solution for $\mathcal{L}_u E$ only. Therefore, the non trivial solutions can be obtained from (3.3) with operators of the form (4.1) which do not belong to Lie algebra LH.

NOTE 2. A subgroup $H_1 \subset GE$ is equivalent to a subgroup $H_2 \subset GE$ if there exists such $T \in GE$ that $H_2 = TH_1T^{-1}$. In this case the subgroups H_1 and H_2 are connected by inner automorphism of the group GE. This is the case when the invariant H_1-solutions U_1 and invariant H_2-solutions U_2 are *similar*: they are connected by formula $U_2 = TU_1$. Therefore, the classification problem for invariant H-solutions up to similarity is exactly the classification problem for subgroups of the group GE up to inner automorphisms.

5. Partial invariance

If condition (4.2) is not fulfilled, i.e., if $r_*(\xi_s^i) < r_*(\xi_s^i, \eta_s^k)$ or $r_*(\xi_s^i, \eta_s^k) > n$, then there are no invariant H-solutions. The reason is that the group H has not enough invariants for constructing the system E/H. Nevertheless, in this case the algorithm exists for constructing some classes of partial solutions called *partially invariant*. This algorithm is described in [1] in detail and only one example, which is interesting in our opinion, is given below.

Full gas dynamics equations E with an arbitrary state equation are considered. The basic space is $Z = R^4(t, x, y, z) \times R^5(u, v, w, \varrho, S)$. The four-dimensional group H_4 admitted by E is taken, which has the basis of operators in LH_4

$$X_1 = \partial_x, \quad X_2 = \partial_y, \quad X_3 = t\partial_x + \partial_u, \quad X_4 = t\partial_y + \partial_v.$$

Here $r_*(\xi_s^i) = 2$, however, $r_*(\xi_s^i, \eta_s^k) = 4 = n$. Therefore, the defect of partially invariant H_4-solution will be $\delta = 2$.

The universal invariant for H_4 is $J = (t, z.w, p, \varrho)$ and the parametric variables are u and v. Then the representation of partially invariant H_4-solution is

$$u = u(t, x, y, z), \quad v = v(t, x, y, z), \quad w = w(t, z), \quad \varrho = \varrho(t, z), \quad S = S(t, z). \quad (5.1)$$

With the operators $D = \partial_t + u\partial_x + u\partial_y + w\partial_z$, $D_0 = \partial_t + w\partial_z$ and the auxiliary function $h = h(t, z)$, the gas dynamics equations for the solutions of the form (5.1) are divided into the *invariant subsystem*

$$D_0 w + \varrho^{-1} p_z = 0, \quad D_0 \varrho + \varrho w_z = 2h\varrho, \quad D_0 S = 0 \quad (5.2)$$

and the auxiliary subsystem

$$Du = 0, \qquad Dv = 0, \qquad u_x + v_y = -2h. \quad (5.3)$$

The subsystem (2.5) is *active*. It turns out that this subsystem can be fully solved by the process of reduction to the passive form. The result depends on the sign $\varepsilon = \text{sign}(h_t - h^2)$. Such a result will be shown below for $\varepsilon = -1$ (*elliptical* case).

The system (5.3) is equipollent to the passive system

$$u_x v_y - u_y v_x = 2h^2 - h_t, \qquad u_x + v_y = -2h, \tag{5.4}$$

where

$$h = -\frac{\alpha + \beta t}{1 + 2\alpha t + \beta t^2},$$

herewith α, β are arbitrary functions of the entropy S satisfying the inequality $\beta > \alpha^2$. Under this condition system (5.4) in its turn is equivalent to the Cauchy-Riemann system and its general solution (in the implicit form) depends on one arbitrary analytical function.

From the viewpoint of the group analysis the solution obtained is partially invariant H_4-solution of rank 2 and defect 2.

From the viewpoint of physics the gas motion described is divided into one-dimensional longitudinal motion along the z-axis described by system (5.2) and two-dimensional crosswise motion in planes orthogonal to the z-axis described by system (5.4). In this case the longitudinal motion goes on with mass inflow (the second equation (5.2)) which is uniformly taken away from two-dimensional crosswise motion as is shown by the second equation (5.4).

The question arises just without an answer: does the solution of system (5.4) exist for all $t > 0$ within a *fixed* domain $\Omega \subset R^2(x, y)$ with physically appropriate boundary conditions on $\partial\Omega$?

The notion of partially invariant solution induces the following statement of *reduction problem*. It is required to ascertain whether for the given partially invariant H-solution U of the rank ϱ and defect δ the subgroup $H' \subset H$ exists, relative to which U as the partially invariant H' solution has the same rank ϱ, however, smaller defect $\delta' < \delta$. If such a subgroup exists, then we say that the partially invariant solution U is reducible to a smaller defect. In the case $\delta' = 0$, the reduction leads to invariant solution.

Recently this problem, in the case of reduction to the invariant solution, was considered in a new fashion by Vorob′ev [3]. He noted that our known condition of such reducibility (see [1], ch. VI) can be generalized to derivatives of higher order.

6. Subalgebras classification

In what follows we shall say, restricting ourselves to a local aspect of the classification problem for subgroups of Lie groups, about the classification for subalgebras of the given finite-dimensional Lie algebra L only.

A number of works is devoted to this problem, among them the achievements of Patera and Winternitz [4] are distinguished. The subsequent descriptions of Fushich [5] are also worth of noting.

The interest to this problem is stimulated by application to the equations of mathematical physics for which the admitted groups allow us to construct the *submodels* of various physical processes described by the equations E. In this case the submodel is realized as the factor system E/H relative to the subgroup $H \subset GE$. That's why the classification of the subgroups (subalgebras) induces, as a minimum, the classification of essentially different (see note 2 in section 4) invariant and partially invariant submodels.

The problem consists in enumerating the subalgebras $N \subset L$ up to inner automorphisms of the algebra or, as we shall say, up to *similarity*.

The set $\Theta_L = \{N\}$ of the subalgebras $N \subset L$ is called the *optimal system* of subalgebras if Θ_L is a *minimal* (under inclusion) set among all the sets of subalgebras possessing the following property: any subalgebra $N \subset L$ is similar to one of subalgebras in Θ_L. Equipollent formulation: Θ_L is a *maximal* (under inclusion) set among all the sets of the subalgebras $N \subset L$ possessing the property: not any of two subalgebras in Θ_L are similar.

Generally speaking, for the given finite-dimensional Lie algebra there exists an infinitely great number of optimal systems Θ_L. That's why the use of different algorithms of their construction leads to the different systems Θ_L. Certainly, all these systems are equivalent in the following sense: for two optimal systems Θ_L^1 and Θ_L^2 such collection of inner automorphisms $\{A\} \subset \operatorname{Int} L$ exists that for any $N^1 \in \Theta_L^1$ there is such automorphism $A^1 \in \{A\}$ with which $A^1 N^1 \in \Theta_L^2$ and inversely.

The problem of isolation of a certain Θ_L is complicated due to the fact that optimal systems contain, as a rule, an infinitely great number of subalgebras. Herewith, *subalgebra series* contained in Θ_L are usually described by the *group parameters* incoming linearly and are *invariants* of some subgroups of automorphisms from $\operatorname{Int} L$.

The essential invariant is the subalgebra *dimension*. If L has the dimension $r > 1$, then let $\Theta_s = \Theta_{L,s}$ designate the optimal system of subalgebras $N_s \in \Theta_L$ whose dimension is s, where $1 \leq s \leq r$. Then the subsystems Θ_s for each s will be equivalent (in the above-mentioned sense).

All the subalgebras (series) which belong to Θ_L can be placed in the form of "layers" Θ_s. If the subalgebras $N_s \in \Theta_s$ and $N_{s+k} \in \Theta_{s+k}$ are connected by the inclusion $N_s \subset N_{s+k}$, where $k > 0$ is minimal for the given N_s, then in the above-mentioned arrangement they are joined by the arrow $N_s \to N_{s+k}$. After this Θ_L takes the form of a connected graph called a *tree of subalgebras*. In this case the number $k > 0$ is called the *length of the arrow*.

For unification of the optimal systems Θ_L, use can be made of the below-described algorithm for construction of Θ_L, which results from the following considerations.

Let G be a group of linear transformations of the vector space L (representation $G|L$). The orbit of the vector $x \in L$ is a vector set $Gx = \{\, gx \mid g \in G \,\}$. Let V be a subset in L. Then the set $\Theta_V \subset V$ is called the optimal GV-system if Θ_V is such a *minimal* (under inclusion) subset in V that for any $x \in V$ the intersection $Gx \cap \Theta_V$ is not empty.

Suppose that the following condition are satisfied:

1. There exists a decomposition of L into a direct sum of proper subspaces: $L = J \oplus N$;
2. There exists a decomposition of G into a semidirect product $G = G_J G_N$ of the proper normal subgroup G_J and subgroup G_N, which is isomorphic to the factor group G/G_J;
3. If $x = x_J \oplus x_N$ is a decomposition of the vector $x \in L$ into components $x_J \in J$

and $x_N \in N$, then the action of transformations $g = g_J g_N$, where $g_J \in G_J$, $g_N \in G_N$, by formula $\bar{x} = gx$ is described in the following table.

	$\bar{x}_J$	$\bar{x}_N$
G_J	$g_J x_J + (g_J x_N)_J$	x_N
G_N	$g_N x_J$	$g_N x_N$

This means that not only J is an invariant subspace of the action G (the representation $G|J$ is reducible) but also that the contraction of G_J to N is trivial: $(g_J x_N)_N = x_N$. We call the representation $G|L$ with properties 1–3 *decomposable.*

For decomposable representation $G|L$, it is possible to perform the following construction in three stages: (a), (b), and (c).
(a) The optimal $G_N N$-system Θ_N constructed of the vectors x_{Ni} $(i = 1, 2, \ldots)$;
(b) For each i the *stabilizer* (subgroup) $S_i \subset G$ of the vector x_{Ni} is found;
(c) The $S_i J$-system Θ_{Ji} of the known x_{Jij} $(j = 1, 2, \ldots)$ is constructed for each i.

Then the statement is valid.

Theorem 1. *The set Θ_L of vectors x_{ij} where*

$$x_{ij} = x_{Jij} \oplus x_{Ni} \qquad (i = 1, 2, \ldots;\ j = 1, 2, \ldots)$$

is the optimal GL-system.

This construction is applied to the subalgebras classification problem on the following basis. Let us call Lie algebra L decomposable if L contains such proper ideal J and subalgebra N that the factor algebra L/J is isomorphic to N.

Theorem 2. *If Lie algebra L is decomposable, then the group of its inner automorphisms* Int L *produces the decomposable representation* Int $L|L$.

It is worth of noting that every *non-simple* Lie algebra is decomposable. Therefore, the algorithm founded on the above-mentioned construction refuses only in the case of the non-simple Lie algebras. For such algebras the problem can be solved on the ground of the well-known simple Lie algebras classification ascenting to Killing and Cartan.

As an example of application, we demonstrate the result of construction Θ for Lie algebra L_6 of the plane Galilei group G_6 (see Appendix). Here the realization of L_6 is used as Lie algebra of operators acting on the space $R^3(t, x, y)$ with the basis

$$X_1 = \partial_x, \quad X_2 = \partial_y, \quad X_3 = t\partial_x, \quad X_4 = t\partial_y, \quad X_5 = y\partial_x - x\partial_y, \quad X_6 = \partial_t. \quad (6.1)$$

The optimal system Θ_{L_6} is presented in the form of the list of subalgebras $L_{s,i}$ $(s = 6, 5, 4, 3, 2, 1)$ having the dimension s and the tree of subalgebras for L_6.

The following fact can be pointed out. Despite Lie algebra L_6 is solvable, its tree of subalgebras contains the arrows which length is more than unity.

In this example as well as in the general case the question concerning the optimization of the tree of subalgebras remains open.

7. Group stratification

The set of all solutions for the system E admitting the group G is naturally stratified into orbits of the solutions. S. Lie posed the problem of representation of this set by describing the set of orbits and solutions inside each orbit by means of differential equations.

The solution of Lie's problem called *group stratification* of the system E by means of the group G can be represented symbolically as

$$(E, G) = RE \cup AG. \tag{7.1}$$

Here the *resolving* system RE connects only invariants of the group G (thereby describing the set of orbits), and *automorphic* system AG is determined by the group itself. The automorphic property means that the group G is transitive on a set of all solutions for the system AG.

In practice, group stratification is especially interesting when the admitted group GE is infinite-dimensional, i.e., it is Lie pseudogroup which general element depends on some arbitrary functions. The group stratification allows us as if to "clean" the system E from these elements and to lay the responsibility for arbitrary functions upon the automorphic system.

The group stratification is performed algorithmically by introducing auxiliary connections for the differential invariants of the group G (see [1]).

The example of hydrodynamics equations for ideal incompressible fluid is considered:

$$E\colon\quad \vec{u} + \vec{u}\nabla\vec{u} + \nabla p = 0, \quad \operatorname{div}\vec{u} = 0. \tag{7.2}$$

Here the velocity vector $\vec{u} = (u, v, w)$ and the pressure p are the functions of time t and Descartes coordinates $\vec{x} = (x, y, z)$.

System (7.2) admits, with the exception of evident transformations such as translations for t, dilatations and rotations, the infinite-dimensional Lie pseudogroup $G_{\varphi,\sigma}$ of transformations $(\vec{x}, \vec{u}, p) \to (\vec{x}^{\,n}, \vec{u}^{\,n}, p^n)$ given by formulae

$$\vec{x}^{\,n} = \vec{x} + \vec{\varphi}(t), \quad \vec{u}^{\,n} = \vec{u} + \vec{\varphi}\,'(t), \quad p^n = \vec{\varphi}\,''(t) \cdot \vec{x} - \sigma(t). \tag{7.3}$$

Here the vector function $\vec{\varphi}$ and the function σ are arbitrary versus time t (primes designate derivatives by t).

Group stratification of system (7.2) relative to the group $G_{\varphi,\sigma}$ is obtained by introducing the invariant independent variables t, $\vec{\xi} = (\xi, \eta, \zeta)$ and their functions: a matrix $M = (a^i_k)$ and vector $\vec{f}$. Let $\vec{a}^{\,i} = (a^i_1, a^i_2, a^i_3)$ $(i = 1, 2, 3)$ be the lines of the matrix M, $|M| = \det M$ and M^* be the transposed matrix of M. The matrix $Q(\vec{f})$ corresponds to the vector $\vec{f} = (f, g, h)$

$$Q(\vec{f}) = \begin{pmatrix} 0 & -h & g \\ h & 0 & -f \\ -g & f & 0 \end{pmatrix}.$$

The index ξ designates differential operations with respect to the variables $\vec{\xi}$. The following statement is valid with introduced designations.

Theorem. *The group stratification of system* (7.2) *relative to the admitted group* $G_{\varphi,\sigma}$ *is given by the resolving system*

$$\left.\begin{array}{c} |M| = 1, \quad M^* M_t - M_t^* M = Q(\vec{f}); \\ \mathrm{rot}_\xi\, \vec{a}^{\,i} = 0, \quad (i = 1, 2, 3); \\ \vec{f}_t = 0, \quad \mathrm{div}_\xi\, \vec{f} = 0 \end{array}\right\} RE$$

and the automorphic system for the functions $\vec{x}$, $\vec{u}$, p *of the variables* t, $\vec{\xi}$

$$\left.\partial\vec{x}/\partial\vec{\xi} = M, \quad \vec{u} = \vec{x}_t, \quad \nabla_\xi p = -M^* \vec{x}_{tt}\right\} AG_{\varphi,\sigma}.$$

Evidently, the variables are the Lagrangian coordinates of liquid particles. The system RE admits some equivalence transformations changing the vector $\vec{f}$ because these coordinates are defined ambiguously. It turns out that the choice of $\vec{\xi}$ is always possible to give $\vec{f} = (1, 0, 0)$. This is the case when integration of RE results in the known Clebsh representation for the vortex velocity field and in Weber integral (see [6]).

8. Other algorithms

In conclusion, let us pay attention to some algorithms which are useful for the DE group analysis.

The classification of *differential invariants* is meaningful for the transformation groups admitted by DE. It is based on the notion of invariant differentiation. Here the theorem is valid which is about the *basis* of the differential invariants. This is a finite collection of invariants that allows us to produce any differential invariant by function operations and invariant differentiation. The theorem concerning the basis is also valid for infinite-dimensional Lie pseudogroups. Besides, the basis searching is carried out algorithmically (for example see discussion of this question in [1]).

The algorithm for finding the *conservation laws* has the fundamental base, i.e., the well-known Noether theorem. Its further development and a lot of examples are presented in the book of N. H. Ibragimov [2]. In particular, an important notion of the *basis* of the conservation laws introduced there engages our attention.

We mention also the algorithmical construction of normal extensions of Lie algebra by means of searching for its tower of differentiations. The task is reduced to solving a linear algebraic system of equations. Those from differentiations which are not inner produce the groups of *outer* automorphisms of Lie algebra. Nevertheless, we can't thus far say definitely what role is played by outer automorphisms of Lie algebra LE admitted by the system E in questions concerned with the structure of its solutions.

Thus, the algorithmic character of Lie theory gives not only a possibility of solving the differential problems in group analysis of differential equations and also serves to the development of the theory itself. We can do no more than wish the same to the so-called "non-Lie" theories arising recently in the field of group analysis.

Appendix

TABLE 1: optimal system Θ_{L_6} of subalgebras $L_{s,i}$, where s is the dimension and t the order number in the optimal system Θ_s. The table gives the basis $(Y_1, \ldots, Y_s)$ of subalgebras $L_{s,i} \in \Theta_s$ written through operators (6.1). The arbitrary parameters α, β are invariants.

s	t	Y_1	Y_2	Y_3	Y_4	Y_5	Y_6
6	1	X_1	X_2	X_3	X_4	X_5	X_6
5	1	X_1	X_2	X_3	X_4	X_5	
	2	X_1	X_2	X_3	X_4	X_6	
4	1	X_1	X_2	X_3	$\alpha X_4 + X_6$		
	2	X_1	X_2	X_3	X_4		
3	1	$\alpha X_1 + X_4$	$-\alpha X_2 + X_3$	X_5			
	2	X_1	X_2	X_5			
	3	X_1	$\alpha X_2 + X_3$	$\beta X_4 + X_6$			
	4	X_1	X_2	$\alpha X_3 + X_6$			
	5	X_2	X_3	$\alpha X_1 + X_4$			
	6	X_1	X_2	X_3			
2	1	X_5	X_6				
	2	$\alpha X_1 + X_2$	$\beta X_3 + X_6$				
	3	X_1	$\alpha X_3 + X_6$				
	4	$\alpha X_1 + X_4$	$\beta X_2 + X_3$				
	5	X_1	X_3				
	6	X_1	X_2				
1	1	$\alpha X_5 + X_6$					
	2	X_5					
	3	$\alpha X_3 + X_6$					
	4	$\alpha X_1 + X_4$					
	5	X_1					

TABLE 2: tree of subalgebras for L_6. The line with the number s contains the numbers i of subalgebras $L_{s,i}$ in accordance with Table 1.

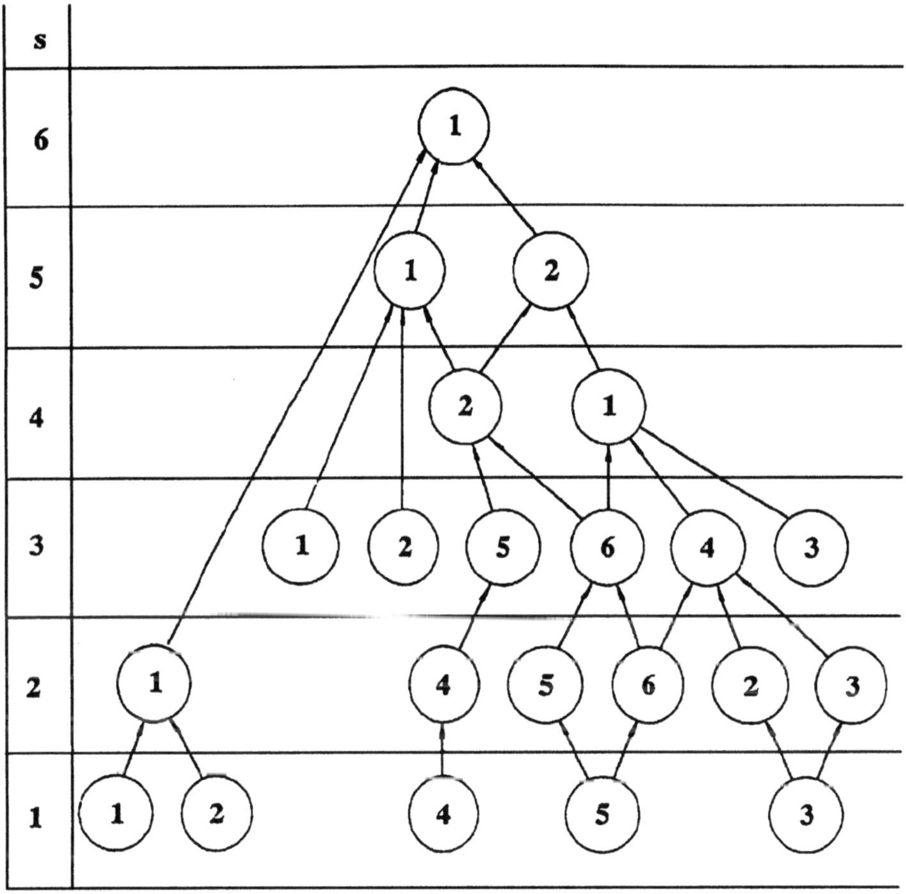

References

1. L. V. Ovsiannikov. Group Analysis of Differential Equations. Academic Press, 1982.
2. N. H. Ibragimov. Transformation Group In Mathematical Physics. Moscow, "Nauka", 1983 (in Russian).
3. E. M. Vorob'ev. Reduction and Quotient Equations for Differential Equations With Symmetries. Acta Applicandae Mathematicae, Kl. Acad. Publ. (Netherlands), 23, 1991.
4. J. Patera, P. Winternitz, H. Zassenhaus. Continuous Subgroups of the Fundamental Groups of Physics, I. General Method and the Poincaré Group. J. Math. Phys., 1974, 15, N 8.
5. V. I. Fushich, L. F. Barannik, A. F. Barannik. Podgruppovoi analiz grupp Galilei, Puancare i reduktsiya nelineinykh uravnenii (in Russian). Kiev, Naukova Dumka, 1991.
6. J. Serrin. Mathematical Principles of Classical Fluid Mechanics. Berlin, 1959.

POTENTIAL SYMMETRIES OF FOKKER-PLANCK EQUATIONS

EDVIGE PUCCI and GIUSEPPE SACCOMANDI
Istituto di Energetica, Università degli Studi di Perugia
06100 Perugia, ITALY

Abstract. The characterization of potential systems derived from first order conservation laws and of the related potential symmetries is carried out for the Fokker-Planck equation $u_t = u_{xx} + (a(x)u)_x$. The potential symmetries for the *natural* potential system are classified. Via a generalization of the classical reduction method, we obtain classes of exact solutions which contain invariant solutions as a particular case.

1. Introduction

Lie point symmetries for the family of Fokker-Planck equations (FPEs)

$$\Delta \equiv u_t - u_{xx} - (a(x)u)_x = 0 \ , \tag{1.1}$$

where $a(x)$ is a smooth function of x, have been considered in [1, 2]. Besides the trivial symmetries when $a(x)$ satisfies the equation

$$\text{I)} \qquad 2a' - a^2 = \hat{s}_1 x^2 + \hat{s}_2 x + \hat{s}_3 \ , \tag{1.2}$$

where $\hat{s}_h$ are arbitrary constants and $(\cdot)'$ denotes the ordinary derivative, there are four symmetry groups $\mathcal{G}_i^I$, $i = 1, \ldots, 4$. On the other hand, when $a(x)$ satisfies the equation

$$\text{II)} \qquad 2a' - a^2 = \frac{\hat{\nu}}{4}(x + \hat{\nu}_0)^2 + \hat{\nu}_1(x + \hat{\nu}_0)^{-2} + \hat{\nu}_2 \ , \tag{1.3}$$

where $\hat{\nu}$ and the $\hat{\nu}_k$ are arbitrary constants, there are two non-trivial symmetry groups $\mathcal{G}_i^{II}$, $i = 1, 2$. If $\hat{s}_1 \neq 0$ we can rewrite I) as a particular case of II) with $\hat{\nu}_1 = 0$. Then for the functions $a(x)$ which satisfy I with $\hat{s}_1 \neq 0$, two of the symmetry groups $\mathcal{G}_i^I$ can be obtained from the symmetry groups $\mathcal{G}_i^{II}$ setting $\hat{\nu}_1 = 0$.

Non local symmetries for the FPE (1.1) with $a(x) \equiv x$ have been found in [3], just like Lie point symmetries of the system obtained from (1.1) introducing the potential $v(x, t)$ as new unknown function

$$v_x = u \ , \qquad v_t = u_x + xu \ . \tag{1.4}$$

If for equation (1.1) it is possible to determine two functions F and G of (x, t, u, u_x, u_t) so that

$$D_t F - D_x G \equiv \lambda \Delta \ , \tag{1.5}$$

where D_x and D_t are the total derivative operators with respect to x and t, and $\lambda(x, t, u, u_x, u_t)$ is the nonzero Lagrange multiplier, then the Lie point symmetries of the potential system Σ

$$v_x = F \ , \qquad v_t = G \ , \tag{1.6}$$

N. H. Ibragimov et al. (eds.),
Modern Group Analysis: Advanced Analytical and Computational Methods in Mathematical Physics, 291–298.

with infinitesimal operator:

$$\mathcal{X} = \xi(x,t,u,v)\frac{\partial}{\partial x} + \tau(x,t,u,v)\frac{\partial}{\partial t} + \eta(x,t,u,v)\frac{\partial}{\partial u} + \phi(x,t,u,v)\frac{\partial}{\partial v} ,$$

if $(\partial\xi/\partial v)^2 + (\partial\tau/\partial v)^2 + (\partial\eta/\partial v)^2 \neq 0$ are *potential symmetries (P.S.)* of (1.1), which, as it is well known, are particular nonlocal symmetries [4]. The pair of functions F and G are the density and the flux of a first order conservation law. If these functions satisfy (1.5) with $\lambda = 1$ their potential system will be called *natural*. It is important to note that, for every F and G satisfying (1.5), $\hat{F}$ and $\hat{G}$ defined as $\hat{F} = F + D_x K$ and $\hat{G} = G + D_t K$, where K is an arbitrary regular function, satisfy again (1.5). However the two potential systems associated with these pairs are equivalent. Since the choice of the functions F and G is not unique and non-equivalent pairs of functions may exist, the possibility of finding P.S. of a given equation is closely related to the choice of the particular potential system [3].

In section 2, we determine the set of conservation laws (1.5) and the conditions which ensure that P.S. could exist. In section 3, we study the Lie point symmetries of the natural potential system

$$v_x = u , \qquad v_t = u_x + a(x)u \tag{1.7}$$

so that all the potential symmetry for (1.1), related to (1.7), are classified. In the last section, we determine some solutions of (1.1) by using reduction methods, applying a generalization of the classical procedure [3].

2. Conservation laws and related potential symmetries

Using the direct method, we find that the general form for F and G is

$$F = F_1 u_x + F_2 , \qquad G = F_1 u_t + (F_{2,u} - F_{1,x})u_x + G_1 \tag{2.1}$$

where $F_1(x,t,u)$, $F_2(x,t,u)$ and $G_1(x,t,u)$ are solutions of the system

$$F_{1,x} - F_{2,u} + F_3 = 0 , \quad G_{1,x} - F_{2,t} - a'uF_3 = 0 , \quad G_{1,u} + F_{3,x} - aF_3 - F_{1,t} = 0 , \tag{2.2}$$

and $F_3(x,t)$, since $\lambda = F_3$, is a non-zero solution of the compatibility condition of $(2.2)_2$ and $(2.2)_3$

$$F_{3,xx} - aF_{3,x} + F_{3,t} = 0 . \tag{2.3}$$

Without loss of generality, we introduce $H(x,t,u)$ and we assume $F_1 = H_u$; thus from $(2.2)_1$ and $(2.2)_2$ we have

$$F_2 = F_3 u + H_x - K_1 , \qquad G_1 = aF_3 u - F_{3,x}u + H_t + K_2 \tag{2.4}$$

where $K_1(x,t)$ and $K_2(x,t)$, from $(2.2)_3$, must satisfy

$$K_{1,t} + K_{2,x} = 0 . \tag{2.5}$$

Introducing these results in (2.1), we obtain

$$F = H_x + H_u\, u_x - K_1 + F_3\, u , \qquad G = H_t + H_u u_t + K_2 + F_3 u_x - F_{3,x}u + a\, uF_3 .$$

We point out that H, K_1 and K_2 can be dropped because they are divergence terms; then every conservation law of the first order for the equation (1.1) is defined by

$$F = F_3 u \,, \qquad G = F_3 u_x - F_{3,x} u + a \; u \; F_3 \,. \tag{2.6}$$

The related potential systems are

$$v_x = F_3 u \,, \qquad v_t = F_3 u_x - F_{3,x} u + a \; u \; F_3 \,. \tag{2.7}$$

The standard method [5] for the determination of the Lie point symmetries applied to (2.7) gives the following generators:

$$\begin{aligned}
\tau &= \tau(t)\,,\ \xi = \tau_t \frac{x}{2} + q_1(t)\,,\ \phi = w_2 - v\{F_3 \tau_{tt} x^2 + 2[(aF_3 - 2F_{3,x})\tau_t + \\
&+ 2F_3 q_{1,t}]x + 8(F_{3,xx} - aF_{3,x})\tau + 4(aF_3 - 2F_{3,x})q_1 - 8F_3 q_2(t)\}/(8F_3)\,, \\
\eta &= -\{uF_3^3[\tau_{tt}x^2 + 2(a\tau_t + 2q_{1,t})x + 4\tau_t - 8q_2 + 4aq_1] + \\
&+ v[(2F_3^2\tau_{tt} + 2(-2F_3F_{3,xx} + \\
&+ 2(F_{3,x})^2 + a'F_3^2)\tau_t)x + 2(aF_3 - 2F_{3,x})F_3\tau_t + \\
&+ 8(F_3F_{3,xxx} - (F_{3,x} + aF_3)F_{3,xx} + a(F_{3,x})^2 + \\
&- a'F_3F_{3,x})\tau + 4F_3^2 q_{1,t} + 4(-2F_3F_{3,xx} + 2(F_{3,x})^2 + a'F_3^2)q_1] - \\
&- 8F_3^2 w_{2,x}\}/(8F_3^3)\,,
\end{aligned} \tag{2.8}$$

where $\tau(t)$, $q_1(t)$, $q_2(t)$, and $w_2(x,t)$ must satisfy

$$F_3 w_{2,t} - F_3 w_{2,xx} + w_{2,x}(2F_{3,x} - aF_3) = 0 \tag{2.9}$$

and

$$\begin{aligned}
F_3^3\tau_{ttt}x^2 &+ [(8F_3^2F_{3,xxx} - 24F_3F_{3,x}F_{3,xx} + 16(F_{3,x})^3 - 2(a'' + aa')F_3^3)\tau_t + \\
&+ 4F_3^3 q_{1,tt}]x - 2F_3^3\tau_{tt} + \\
&+ [16F_3^2F_{3,xx} - 16F_3(F_{3,x})^2 - 2(2a' + a^2)F_3^3]\tau_t + \\
&+ 16[-F_3^2F_{3,xxxx} + (2F_3F_{3,x} + aF_3^2)F_{3,xxx} + \\
&+ F_3(F_{3,xx})^2 - (2(F_{3,x})^2 + 3aF_3F_{3,x} - 2a'F_3^2)F_{3,xx} + \\
&+ 2a(F_{3,x})^3 - 2a'F_3(F_{3,x})^2 + \\
&+ a''F_3^2F_{3,x}]\tau - 8F_3^3 q_{2,t} + 4[4F_3^2F_{3,xxx} - 12F_3F_{3,x}F_{3,xx} + \\
&+ 8(F_{3,x})^3 - (a'' + aa')F_3^3]q_1 = 0\,.
\end{aligned} \tag{2.10}$$

For a function $a(x)$ and a chosen $F_3(x,t)$ solution of (2.3), P.S. will be admitted if it is possible to satisfy (2.10) with some $\tau(t)$, $q_1(t)$ and $q_2(t)$ such as $\eta_v \neq 0$. In [3], it has been show that by choosing $a \equiv x$ and $F_3 = 1$ we can find τ, q_1 and q_2; on the other hand always for $a \equiv x$, but for $F_3 = \int \exp(x^2/2)dx$, it is not possible to satisfy (2.14) and therefore P.S. do not exist.

3. Natural system and potential symmetries

The system (1.7) is obtained by taking $F_3 = 1$; then (1.7) is the *natural* potential system associated to (1.1). In this case, the generators (2.8) assume the simplified form

$$
\begin{aligned}
\tau &= \tau(t)\ , \quad \xi = \frac{1}{2} x\tau_t + q_1(t)\ , \\
\phi &= w_2 - v\left\{\frac{1}{8}\left[4\, xq_{1,t} + 4aq_1 + \tau_{tt}x^2 + 2xa\tau_t\right] - q_2\right\}\ , \\
\eta &= -\frac{1}{8}\{u[\tau_{tt}x^2 + (2a\tau_t + 4q_{1,t})x + 4\tau_t - 8q_2 + 4aq_1] + \\
&+ v[(2\tau_{tt} + 2a'\tau_t)x + 2a\tau_t + 4q_{1,t} + 4a'q_1] - 8w_{2,x}\}\ ,
\end{aligned}
\tag{3.1}
$$

where $w_2(x,t)$ satisfies

$$w_{2,xx} + aw_{2,x} - w_{2,t} = 0\ . \tag{3.2}$$

By setting $B \equiv 2a' + 2a^2$ the equation (2.10) becomes

$$B'(2q_1 + x\tau_t) + 2B\tau_t - \tau_{ttt}x^2 - 4q_{1,tt}x + 2\tau_{tt} + 8q_{2,t} = 0\ . \tag{3.3}$$

The function $w_2(x,t)$ is the generator of the ∞-dimensional group typical for every linear equation; since we are not interested in this group we will set $w_2 \equiv 0$. By considering (3.3) as an ordinary differential equation in the unknown function $B(x)$, we have that

$$B = (\bar{s}_0x^4 + \bar{s}_1x^3 + \bar{s}_2x^2 + \bar{s}_3x + \bar{s}_4)/(12(\tau_tx + 2q_1)^2)\ , \tag{3.4}$$

where:

$$
\begin{aligned}
\bar{s}_0 &= 3\tau_t\tau_{ttt}\ , \ \bar{s}_1 = 8q_1\tau_{ttt} + 16q_{1,tt}\tau_t\ , \ \bar{s}_2 = -12(\tau_t\tau_{tt} + 4q_{2,t}\tau_t - 4q_1q_{1,tt})\ , \\
\bar{s}_3 &= -48q_1(\tau_{tt} + 4q_{2,t})\ , \ \bar{s}_4 = 12c
\end{aligned}
$$

and $c(t)$ is an arbitrary function introduced by integration.

The relation (3.4) shows that B is a rational function of x obtained as $P_4(x)/P_2(x)$ where P_4 and P_2 are polynomials at most of the fourth and second degree respectively, and the time is considered as a parameter. The function B depends only on x and then only two possibilities can occur

i) The quantities $\bar{s}_i/\tau_i^2$ for $i = 0,\ldots,4$ and q_1/τ_t do not depend on t.

ii) P_4 contains P_2 as a factor and B is a polynomial with constant coefficients.

In i) we have assumed $\tau_t \neq 0$ since the case $\tau_t = 0$ falls within ii). Then we have the following cases

A) The (3.4) can be rewritten as

$$2a' + a^2 = \frac{\nu}{4}(x + \nu_0)^2 + \nu_1(x + \nu_0)^{-2} + \nu_2\ , \tag{3.5}$$

where ν, ν_1, ν_2 are arbitrary constants ($\nu \neq 0, \nu_1 \neq 0$); and $\tau(t)$, $q_1(t)$, $q_2(t)$ are solutions of

$$\tau_{ttt} - \nu\tau_t = 0\ , \quad q_1 - \frac{\nu_0}{2}\tau_t = 0\ , \quad 8q_{2,t} + 2\tau_{tt} + (\nu\nu_0^2 + 2\nu_2)\tau_t = 0\ . \tag{3.6}$$

B) The (3.4) can be rewritten as

$$2a' + a^2 = \frac{\nu}{4}(x+\nu_0)^2 + \nu_2 \, , \tag{3.7}$$

where $\nu \neq 0$; $\tau(t)$, $q_1(t)$, $q_2(t)$ are solutions of

$$\tau_{ttt} - \nu\tau_t = 0 \, , \quad 8q_{1,tt} - 2\nu q_1 - 3\nu\nu_0\tau_t = 0 \, , \tag{3.8}$$
$$16q_{2,t} + 4\tau_{tt} + 2\nu\nu_0 q_1 + (\nu\nu_0^2 + 4\nu_2)\tau_t = 0 \, .$$

C) The (3.4) can be rewritten as

$$2a' + a^2 = \nu_1(x+\nu_0)^{-2} + \nu_2 \, , \tag{3.9}$$

where $\nu_1 \neq 0$; $\tau(t)$, $q_1(t)$, $q_2(t)$ are solutions of

$$\tau_{ttt} = 0 \, , \quad 2q_1 - \nu_0\tau_t = 0 \, , \quad 4q_{2,t} + \tau_{tt} + \nu_2\tau_t = 0 \, . \tag{3.10}$$

D) The (3.4) can be rewritten as

$$2a' + a^2 = 4\mu_1 x + \mu_2 \, , \tag{3.11}$$

and μ_1 and μ_2 are arbitrary parameters; and $\tau(t)$, $q_1(t)$, $q_2(t)$ are solutions of

$$\tau_{ttt} = 0 \, , \quad q_{1,tt} - 3\mu_1\tau_t = 0 \, , \quad 4q_{2,t} + 4\mu_1 q_1 + \tau_{tt} - \mu_2\tau_t = 0 \, . \tag{3.12}$$

The generators of the P.S. are reported in table 1.

4. Solutions by reduction

In this section, we determine solutions of the FPE through reduction methods related to the P.S. of section 3. For lack of space we will report here only the results related to the generator χ_A^1. Solutions related to the others generators and more details can be found in [6]. In the classical procedure [4,5], we determine invariant solutions of the potential system (1.7) which, in turn, lead to solutions of (1.1) because this equation is a differential consequence of the system.

Let be

$$\mu(x) = -(\nu x^2 + 2(a\sqrt{\nu} + \nu\nu_0)x + 2\sqrt{\nu}(a\nu_0 + 1) + 2\nu_2 + \nu\nu_0^2)/(4\sqrt{\nu}(x+\nu_0)) \, , \tag{4.1}$$

we have that the invariant solutions of (1.7) are

$$v = h_0(z)\exp\left(\int \mu(x)\,\mathrm{d}x\right) \, , \quad u = \exp\left(\int \mu(x)\,\mathrm{d}x\right) \tag{4.2}$$
$$\left(h_1(z) - \frac{1}{4}h_0(z)[x\sqrt{\nu}(x+2\nu_0) + 2a(x+\nu_0)]\right)/(x+\nu_0) \, ,$$

where the similarity variable is $z = \exp(-\sqrt{\nu}t/2)(x+\nu_0)$ and the functions $h_0(z)$, $h_1(z)$ satisfy the system

$$\mathcal{E}_1 \equiv 4\nu z h_0' - h_0[\sqrt{\nu}(2\nu_2 + \nu\nu_0^2) + 2\nu] - 4\nu h_1 = 0 \, , \tag{4.3}$$
$$\mathcal{E}_2 \equiv 4\nu h_0'' z^2 - 4h_0' z(\sqrt{\nu}\nu_2 + \nu) + h_0(\nu_2^2 + 4\sqrt{\nu}\nu_2 - \nu\nu_1 + 3\nu) = 0 \, .$$

The $(4.3)_2$ is a Euler equation; the $(4.3)_1$ defines h_1 when h_0 is known. The function $(4.2)_2$ where we have introduced the solutions of (4.3) is a two parameter family of solutions of $\Delta = 0$. This algorithm can be generalized, as pointed out in [3], by introducing $(4.2)_2$ directly into (1.1). By doing this substitution and by setting

$$k_1(z) = \mathcal{E}_1 \ , \quad k_2(z) = \mathcal{E}_2 \ , \tag{4.4}$$

we obtain the relation

$$\Lambda(z, k_1', k_2', k_1'', k_2'') + k_1\Lambda_1(x) - k_2\Lambda_2(x) = 0 \ , \tag{4.5}$$

where

$$\begin{aligned}
\Lambda &= 8\nu k_1'' z^2 - 8k_1' z(\sqrt{\nu}\nu_2 - 3\nu) - 8\nu k_2' z \ , \\
\Lambda_1 &= 4\sqrt{\nu}(\nu x^2 + 2\nu\nu_0 x + 4\nu_2 + \nu\nu_0^2) + \nu^2 x^4 + 4\nu^2\nu_0 x^3 + \\
&+ 2\nu(2\nu_2 + 3\nu\nu_0^2 - 2a^2)x^2 + \\
&+ 4\nu\nu_0(2\nu_2 + \nu\nu_0^2 - 2a^2)x + 2\nu_2^2 + 4\nu\nu_2\nu_0^2 + 2\nu\nu_1 + \nu^2\nu_0^4 - 4a^2\nu\nu_0^2 + 30\nu \ , \\
\Lambda_2 &= 2\sqrt{\nu}(\nu x^2 + 2\nu\nu_0 x + 2\nu_2 + \nu\nu_0^2) + 4a\nu x + 4a\nu\nu_0 + 20\nu \ .
\end{aligned}$$

The (4.5) must be satisfied for every value of the parameter x; then it must be compatible with

$$k_1\frac{d^n\Lambda_1}{dx^n} - k_2\frac{d^n\Lambda_1}{dx^n} = 0 \ , \quad \forall n \in \mathbb{N} \ . \tag{4.6}$$

Besides the case $k_1 \equiv k_2 \equiv 0$ (the case of invariant solutions), this compatibility can be obtained in the following eventuality:

$$\begin{aligned}
&\nu_1 = \frac{\nu_2^2 - 4\sqrt{\nu}\nu_2 + 2\nu}{\nu} \ , \qquad a(x) = \frac{\nu(x+\nu_0)^2 + 2(\nu_2 - \sqrt{\nu})}{2\sqrt{\nu}(x+\nu_0)} \ ; \\
&k_2 = -2k_1 \ , \qquad k_1''\sqrt{\nu}z - k_1'(\nu_2 + \sqrt{\nu}) = 0 \ .
\end{aligned}$$

In correspondence to every $k_1(z)$ and $k_2(z)$ compatible, we can solve the equations (4.4) in the unknown functions $h_0(z)$ and $h_1(z)$. The equations (4.3) are the homogeneous system associated with (4.4); therefore we obtain a four parameter family of solutions, given by the sum of solutions of (4.3) with a particular integral of (4.4).

5. Concluding remarks

In this work we have classified all the P.S. of the FPEs (1.1) and by using these symmetries, we have found some solutions by reduction methods. P.S. exist iff the function $a(x)$ is a solution of one of the Riccati equations (3.5), (3.7), (3.9), (3.11); then the more general $a(x)$ admitting P.S. can be expressed by means of hypergeometric functions. A solution of function (3.5), (3.7), (3.9), (3.11) is also a solution of (1.2) of (1.3) iff its derivative is a rational function of x. Therefore some $a(x)$ exist so that the corresponding FPE allows P.S. but their point symmetries are only trivial.

It is well known that point symmetries can be also used to study BVP; in particular it has been done by Bluman in [1] for the FPEs (1.1) when $a(x) = -a(-x)$.

Also P.S. can be used to solve BVP [5]; obviously it is necessary to embed the given BVP into a new BVP posed for the potential system. Since $a(x)$ exists as a solution of (3.5), (3.7), (3.9), (3.11) which is not a solution of (1.2) or (1.3), and the Lie algebra of the potential system is large enough to ensure the invariance of boundary conditions of (1.1) which are interesting, then the results of this work can be used to solve new BVP for (1.1) by following the guidelines of [1].

References

1. G.W. Bluman, *Int. J. Non-Linear Mechanics*, **6** (1971) pp. 143–153.
2. G. Cicogna, D. Vitali, *J. Phys. A: Math. Gen.*, **23** (1990) L85–L88.
3. E. Pucci, G. Saccomandi, to appear *J. Phys. A: Math. Gen.*, (1993).
4. G.W. Bluman, G.J. Reid, S. Kumei, *J. Math. Phys.*, **29** (1988) pp. 806–811.
5. G.W. Bluman, S. Kumei, "Symmetries and Differential Equations", *Appl. Math. Sci.*, **81**, Springer (1989).
6. E. Pucci, G. Saccomandi, *Technical Report FM2/92*, Istituto di Energetica, Università di Perugia (1992).

Table I

χ_1^A: $\tau = \exp(\sqrt{\nu t})$; $\xi = \frac{1}{2}\exp(\sqrt{\nu t})\sqrt{\nu}(x+\nu_0)$; $\phi = -\frac{1}{8}\exp(\sqrt{\nu t})v[\nu x^2 + 2\nu_0\nu x + 2\nu_2 + \nu\nu_0^2 + 2(ax + a\nu_0 + 1)\sqrt{\nu}]$; $\eta = -\exp(\sqrt{\nu t})\{v[\nu^2x^4 + 4\nu^2\nu_0x^3 + 8\sqrt{\nu}(\nu x^2 + 2\nu\nu_0x + \nu\nu_0^2) + (4\nu\nu_2 + 6\nu^2\nu_0^2 - 4a^2\nu)x^2 + 4(2\nu\nu_0\nu_2 + \nu^2\nu_0^3 - 2a^2\nu\nu_0 + 2a\nu)x + 4\nu\nu_0^2\nu_2 + 4\nu\nu_1 + \nu^2\nu_0^4 - 4a^2\nu\nu_0^2 + 8a\nu\nu_0] + u[\sqrt{\nu}(4\nu x^3 + 12\nu\nu_0x^2 + (8\nu_2 + 12\nu\nu_0^2)x + 8\nu_0\nu_2 + 4\nu\nu_0^3) + 8a\nu x^2 + (16a\nu\nu_0 + 24\nu\nu_0)x + 8a\nu\nu_0^2 + 24\nu\nu_0]\}/[32\sqrt{\nu}(x+\nu_0)]$

χ_2^A: $\tau = \exp(-\sqrt{\nu t})$; $\xi = -\frac{1}{2}\exp(-\sqrt{\nu})\sqrt{\nu}(x+\nu_0)$; $\phi = -\frac{1}{8}\exp(-\sqrt{\nu t})\cdot v[\nu x^2 + 2\nu_0\nu x + 2\nu_2 + \nu\nu_0^2 - 2(ax + a\nu_0 + 1)\sqrt{\nu}]$; $\eta = \exp(-\sqrt{\nu t})\{v[\nu^2x^4 + 4\nu^2\nu_0x^3 - 8\sqrt{\nu}(\nu x^2 + 2\nu\nu_0x + \nu\nu_0^2) + (4\nu\nu_2 + 6\nu^2\nu_0^2 - 4a^2\nu)x^2 + 4(2\nu\nu_0\nu_2 + \nu^2\nu_0^3 - 2a^2\nu\nu_0 + 2a\nu)x + 4\nu\nu_0^2\nu_2 + 4\nu\nu_1 nu^2\nu_0^4 - 4a^2\nu\nu_0^2 + 8a\nu\nu_0] - u[\sqrt{\nu}(4\nu x^3 + 12\nu\nu_0x^2 + (8\nu_2 + 12\nu\nu_0^2)x + 8\nu_0\nu_2 + 4\nu\nu_0^3) - 8a\nu x^2 - (16a\nu\nu_0 + 24\nu)x - 8a\nu\nu_0^2 - 24\nu\nu_0]\}/[32\sqrt{\nu}(x+\nu_0)]$

χ_1^B: $\tau = \exp(\sqrt{\nu t})$; $\xi = \frac{1}{2}\sqrt{\nu}\exp(\sqrt{\nu t})(x+\nu_0)$; $\phi = -\frac{1}{8}\exp(\sqrt{\nu t})\cdot v[2\sqrt{\nu}(ax + a\nu_0 + 1) + \nu x^2 + 2\nu\nu_0x + 2\nu_2 + \nu\nu_0^2]$; $\eta = -\exp(\sqrt{\nu t})\{v[t\sqrt{\nu}(\nu x^3 + 3\nu\nu_0x^2 + (4\nu_2 + 3\nu_0^2\nu - 4a^2)x + 4\nu_0\nu_2 + \nu_0^3\nu - 4a^2\nu_0 + 8a) + 8\nu x + 8\nu\nu_0] + u[\sqrt{\nu}(8ax + 8a\nu_0 + 24) + 4\nu x^2 + 8\nu\nu_0x + 8\nu_2 + 4\nu\nu_0^2]\}/32$

χ_2^B: $\tau = \exp(-\sqrt{\nu t})$; $\xi = -\frac{1}{2}\sqrt{\nu}\exp(-\sqrt{\nu t})(x+\nu_0)$; $\phi = \frac{1}{8}\exp(-\sqrt{\nu t})\cdot v[2\sqrt{\nu}(ax + a\nu_0 + 1) - \nu x^2 - 2\nu\nu_0x - 2\nu_2 - \nu\nu_0^2]$; $\eta = \exp(-\sqrt{\nu t})\cdot\{v[\sqrt{\nu}(\nu x^3 + 3\nu\nu_0x^2 + (4\nu_2 + 3\nu_0^2\nu - 4a^2)x + 4\nu_0\nu_2 + \nu_0^3\nu - 4a^2\nu_0 + 8a) - 8\nu x - 8\nu\nu_0] + u[\sqrt{\nu}(8ax + 8a\nu_0 + 24) - 4\nu x^2 - 8\nu\nu_0x - 8\nu_2 - 4\nu\nu_0^2]\}/32$

χ_3^B: $\tau = 0$; $\xi = \exp\left(\frac{1}{2}\sqrt{\nu t}\right)$; $\phi = -\exp\left(\frac{1}{2}\sqrt{\nu t}\right)v[\sqrt{\nu}(x+\nu_0) + 2a]/4$; $\eta = -\exp\left(\frac{1}{2}\sqrt{\nu t}\right)\{v(\nu x^2 + 2\nu\nu_0x + 4\nu_2 + \nu\nu_0^2 + 4\sqrt{\nu} - 4a^2) + u[4\sqrt{\nu}(x+\nu_0) + 8a]\}/16$

χ_4^B: $\tau = 0$; $\xi = \exp\left(-\frac{1}{2}\sqrt{\nu t}\right)$; $\phi = \exp\left(-\frac{1}{2}\sqrt{\nu t}\right)v[\sqrt{\nu}(x+\nu_0) - 2a]/4$; $\eta = -\exp\left(-\frac{1}{2}\sqrt{\nu t}\right)\{\nu(\nu x^2 + 2\nu\nu_0x + 4\nu_2 + \nu\nu_0^2 - 4\sqrt{\nu} - 4a^2) + u[8a - 4\sqrt{\nu}(x+\nu_0)]\}/16$

χ_1^C: $\tau = t$; $\xi = \frac{1}{2}(x+\nu_0)$; $\phi = -v[a(x+\nu_0) + \nu_2t]/4$; $\eta = -\{v[(\nu_2 - a^2)x^2 + 2(\nu_0\nu_2 - a^2\nu_0 + a)x + \nu_0^2\nu_2 + \nu_1 - a^2\nu_0^2 + 2a\nu_0] + u[2ax^2 + 2(\nu_2t + 2a\nu_0 + 2)x + 2\nu_0\nu_2t + 2a\nu_0^2 + 4\nu_0]\}/[8(x+\nu_0)]$

χ_2^C: $\tau = t^2$; $\xi = t(x+\nu_0)$; $\phi = -v[x^2 + 2(at+\nu_0)x + \nu_2t^2 + 2(a\nu_0+1)t]/4$; $\eta = -\{v[((\nu_2 - a^2)t + 2)x^2 + 2((\nu_0\nu_2 - a^2\nu_0 + a)t + 2\nu_0)x + (\nu_0^2\nu_2 + \nu_1 - a^2\nu_0^2 + 2a\nu_0)t + 2\nu_0^2] + u[x^3 + (2at + 3\nu_0)x^2 + (\nu_2t^2 + (4a\nu_0 + 6)t + 2\nu_0^2)x + \nu_0\nu_2t^2(2a\nu_0^2 + 6\nu_0)t]\}/[4(x+\nu_0)]$

χ_1^D: $\tau = t^2$; $\xi = tx + \mu_1t^3$; $\phi = -v[x^2 + (6\mu_1t^2 + 2at)x + \mu_1^2t^4 + 2a\mu_1t^3 + \mu_2t^2 + 2t]/4$; $\eta = -\{v[4\mu_1tx^2 + (4\mu_1^2t^3 + \mu_2t - a^2t + 2)x + (\mu_1\mu_2 - a^2\mu_1)t^3 + 6\mu_1t^2 + 2at] + u[x^2 + (6\mu_1t^2 + 2at)x + \mu_1^2t^4 + 2a\mu_1t^3 + \mu_2t^2 + 6t]\}/4$

χ_2^D: $\tau = t$; $\xi = \frac{1}{2}(x + 3\mu_1t^2)$; $\phi = -v[(6\mu_1t + a)x + 2\mu_1^2t^3 + 3a\mu_1t^2 + \mu_2t]/4$; $\eta = -\frac{1}{8}\{v[4\mu_1x^2 + (12\mu_1^2t^2 + \mu_2 - a^2)x + 3\mu_1t^2(\mu_2 - a^2) + 12\mu_1t + 2a] + u[(2a + 12\mu_1t)x + 4\mu_1^2t^3 + 6a\mu_1t^2 + 2\mu_2t + 4]\}$

χ_3^D: $\tau = 0$; $\xi = t$; $\phi = -\frac{1}{2}v(x + \mu_1t^2 + at)$; $\eta = -\{v[4\mu_1tx + (\mu_2 - a^2)t + 2] + u(2x + 2\mu_1t^2 + 2at)\}/4$

χ_4^D: $\tau = 0$; $\xi = 1$; $\phi = -\frac{1}{2}v(2\mu_1t + a)$; $\eta = -[v(4\mu_1x + \mu_2 - a^2) + u(4\mu_1t + 2a)]/4$

CONTINUOUS SYMMETRIES OF DIFFERENCE EQUATIONS

G. R. W. QUISPEL and R. SAHADEVAN*
Department of Mathematics
La Trobe University
Melbourne 3083
Australia

Abstract. A method is given to find the continuous symmetries of ordinary difference equations. It is shown how the Lie symmetries can be used to find the general solution of difference equations. The method can also be applied to mappings.

In the second half of the nineteenth century Sophus Lie unified and extended the solution methods for ordinary differential equations. He discovered that the existing special techniques used for solving ordinary differential equations were all special cases of a general integration procedure based on the invariance of the differential equation under a continuous group of symmetries. In partial differential equations a continuous group can be used to find so-called "group invariant" or "similarity" solutions [1].

It has recently become clear that the same ideas work for difference equations. In refs. [2-4] Lie's method for finding symmetries was extended to differential-difference equations. It was also shown how Lie symmetries can be used to obtain similarity solutions of *partial* differential-difference equations [3,4]. In this paper we will show that the same method can be applied to *ordinary* difference equations.[1] To simplify the presentation we give an intuitive treatment based on an example, rather than presenting the method in full mathematical detail and generality. We note here that the method of [6], though superficially somewhat similar, in general does not yield Lie symmetries, and, in contrast to the method presented here, cannot be applied to nonautonomous difference equations.

The example we will treat in this paper is the difference equation

$$u(x+1) = 4u(x)^3 + 3u(x), \tag{1}$$

where u is a function: $\mathbb{R} \to \mathbb{R}$. We will find the Lie symmetries of this equation, and use them to derive its general solution. Consider the one parameter (ϵ) group of infinitesimal point transformations in $(x, u(x))$ given by

$$x^* = x + \epsilon\xi(x, u(x)) + O(\epsilon^2), \tag{2a}$$
$$u^*(x^*) = u(x) + \epsilon v(x, u(x)) + O(\epsilon^2). \tag{2b}$$

* On leave of absence from Ramanujan Institute for Advanced Study in Mathematics, University of Madras, Chepauk, Madras-600 005, Tamilnadu, India.

[1] For an alternative approach see [5].

N. H. Ibragimov et al. (eds.),
Modern Group Analysis: Advanced Analytical and Computational Methods in Mathematical Physics, 299–302.

Equation (1) is invariant under the transformation (2) if

$$u^*(x^*+1) = 4u^*(x^*)^3 + 3u^*(x^*). \tag{3}$$

Therefore, we need to know $u^*(x^*+1)$. The expression for this quantity was given in [2-4]

$$u^*(x^*+1) = u(x+1) + \epsilon v(x+1, u(x+1)) \\ + \epsilon[\xi(x,u(x)) - \xi(x+1,u(x+1))]\frac{du}{dx}(x+1). \tag{4}$$

Using (2) and (4) in (3) we obtain the equation to be satisfied by the symmetries:

$$v(x+1,u(x+1)) + [\xi(x,u(x)) - \xi(x+1,u(x+1))]\frac{du}{dx}(x+1) \\ = [12u^2(x)+3]v(x,u(x)). \tag{5}$$

Equation (5) implies

$$\xi(x,u(x)) - \xi(x+1,u(x+1)) = 0, \tag{6a}$$

$$v(x+1,u(x+1)) = 3[4u(x)^2+1]v(x,u(x)). \tag{6b}$$

Looking at equation (6a), it is clear that any unit periodic function $\alpha(x)$, i.e. a function defined by $\alpha(x) = \alpha(x+1)$ [3, 4, 7], is a solution, and substituting the equation of motion (1) in (6b) we obtain

$$\xi(x,u(x)) = \alpha(x) \tag{7}$$

$$v(x+1, 4u(x)^3+3u(x)) = 3[4u^2(x)+1]v(x,u(x)). \tag{8}$$

Equation (8) is a functional equation, that we solve by expanding $v(x,u(x))$ in a Laurent series in $\frac{1}{u(x)}$

$$v(x,u(x)) = a_0(x)u^b(x) + a_1(x)u^{b-1}(x) + \dots , \tag{9}$$

where $a_i(x)$, $i = 0,1,2,\dots$ are unknown functions of x and b is an unknown constant, all to be determined. Inserting equation (9) in (8), and equating the leading terms, it follows that

$$b = 1 \text{ and } a_0(x) = 3^x\beta(x), \tag{10}$$

where $\beta(x)$ is another arbitrary unit periodic function. Equating coefficients in the expansion of the right hand and left hand sides of (8), we get

$$v(x,u) = 3^x\beta(x)u[1 + \frac{1}{2}(\frac{1}{u})^2 - \frac{1}{8}(\frac{1}{u})^4 + \frac{1}{16}(\frac{1}{u})^6 - \frac{5}{128}(\frac{1}{u})^8 + \frac{7}{256}(\frac{1}{u})^{10} - \dots]. \tag{11}$$

With a little experience we recognize the terms in the square bracket as the first terms in the expansion of $(1+\frac{1}{u^2})^{\frac{1}{2}}$. Indeed, substituting this in (11) it is easy to verify that

$$v(x,u(x)) = 3^x\beta(x)(1+u^2(x))^{\frac{1}{2}} \tag{12}$$

is a solution of (8) and hence an infinitesimal symmetry of (1). [Note that even if a difference equation is autonomous, its symmetries are in general non-autonomous.]

The finite symmetry transformation associated with (12) is given by

$$\overline{u} = [\exp\{\epsilon\ v(x, u(x))\frac{\partial}{\partial u}\}]u. \tag{13}$$

In this case, the right hand side of (13) equals

$$[\exp\{\epsilon\ 3^x \beta(x)(1+u^2)^{\frac{1}{2}} \frac{\partial}{\partial u}\}]u. \tag{14}$$

We now look for a homogenizing change of variables $u \to z$ such that exponential operator in (14) becomes

$$\exp\{\epsilon\ 3^x \beta(x)\frac{\partial}{\partial z}\}. \tag{15}$$

To this end we must choose

$$z = \int \frac{du}{\sqrt{1+u^2}} = \text{arcsinh } u. \tag{16}$$

Thus (14) becomes

$$[\exp \epsilon\ 3^x \beta(x)\frac{\partial}{\partial z}] \sinh z = \sinh(z + \epsilon\ 3^x \beta(x)). \tag{17}$$

Hence a finite symmetry group leaving equation (1) invariant is

$$\overline{u}(x) = \sinh[\text{arcsinh } u(x) + \epsilon\ 3^x \beta(x)]. \tag{18}$$

The finite symmetry group (18) can be used to generate infinitely many solutions from a given solution. For example, to get the general solution of the difference equation (1), we apply the group (18) to a particular solution of (1), say $u \equiv 0$. Thus, taking $\epsilon = 1$ without loss of generality, the general solution of (1) is

$$\overline{u}(x) = \sinh(3^x \beta(x)), \tag{19}$$

where, as was mentioned, $\beta(x)$ is an arbitrary unit periodic function determined by the initial conditions.[2] For applications of this method to other autonomous and non-autonomous difference equations (with one or more spans) we refer the interested reader to [8].

Acknowledgement

This investigation is part of the research programme of the Australian Research Council.

---

[2] Those readers who are interested in mappings rather than difference equations can take $\beta(x)$ equal to a constant.

References

1. P.J. Olver, *Applications of Lie Groups to Differential Equations*, (Springer, Berlin, 1986).
2. J.M. Hill, Solution of Differential Equations by means of One-parameter Groups, Pitman Research Notes in Mathematics, **63** (Pitman, Melbourne, 1982).
3. G.R.W. Quispel, H.W. Capel and R. Sahadevan, Phys. Let. **A170** (1992) 379.
4. G.R.W. Quispel, H.W. Capel and R. Sahadevan, to appear in Proceedings of the NATO Workshop on Applications of Analytic and Geometric methods in Nonlinear Differential Equations, (ed.) P.A. Clarkson, University of Exeter, U.K. (1992).
5. V.A. Dorodnitsyn, Journal of Soviet Mathematics, **55** (1991) 1490.
6. S. Maeda, Math. Japonica **25** (1980) 405; IMA J. Appl. Maths. **35** (1987) 129.
7. T.L. Saaty, Modern Nonlinear Equations, (McGraw-Hill, New York, 1967).
8. G.R.W. Quispel and R. Sahadevan, to be published.

INTEGRABLE MECHANICAL SYSTEMS INVARIANT WITH RESPECT TO THE ACTION OF THE KDV HIERARCHY

S. RAUCH-WOJCIECHOWSKI
Department of Mathematics, Linköping University
S-58183 Linköping, Sweden

1. Introduction

In this note we show how to construct integrable mechanical systems invariant with respect to the action of all flows of the KdV hierarchy [1, 2]. We illustrate on a very simple example how these systems inherit from the KdV hierarchy a bihamiltonian formulation and a spectral parameter dependent Lax representation which may be used for solving these systems.

The main steps of the construction are of general character and apply to other soliton hierarchies as well. Many new integrable systems await to be discovered this way. The method and results presented in this paper lead to unification of soliton theory with integrability theory in classical mechanics.

2. The KdV hierarchy. Notation and conventions

Let us denote the KdV hierarchy of equations as

$$\frac{\partial u}{\partial t_m} = K_m[u] = B_0 P_m = B_1 P_{m-1} \qquad m = 0, 1, 2, \ldots \tag{1}$$

where the vector fields $K_m[u]$ are differential functions of u and its derivatives u_x, u_{xx}, ...; $P_m = \frac{\delta \mathcal{H}_m}{\delta u}$ are variational derivatives of hamiltonian densities $\mathcal{H}_m[u]$ and $B_0 = \partial = \frac{\partial}{\partial x}$, $B_1 = \frac{1}{4}\partial^3 + u\partial + \frac{1}{2}u_x$ are hamiltonian operators. Equations (1) follow from the Schrödinger spectral problem

$$0 = (\partial^2 + u - \lambda)\phi \tag{2}$$

as the condition of compatibility with the auxiliary linear problem of the form

$$\frac{\partial \phi}{\partial t_m} = (\partial^{2m-1} + a_{2m-3}[u]\partial^{2m-3} + \ldots + a_0[u])\phi = A_m \phi$$

The first operators A_m read: $A_0 = 0, A_1 = \partial, A_2 = \partial^3 + \frac{3}{2}u\partial + \frac{3}{4}u_x, \ldots$ and the corresponding vector fields are: $K_0 = 0, K_1 = u_x, K_2 = \frac{1}{4}u_{xxx} + \frac{3}{2}uu_x, K_3 = \frac{1}{16}(u_{xxxxx} + 10uu_{xx} + 5u_x^2 + 10u^3), \ldots$. All P_m in (1) can be calculated from the

N. H. Ibragimov et al. (eds.),
Modern Group Analysis: Advanced Analytical and Computational Methods in Mathematical Physics, 303–309.

recursion relation [3]

$$0 = [\frac{1}{4}\partial^3 + (u-\lambda)\partial + \frac{1}{2}u_x]\mathcal{P} \qquad \text{where} \qquad \mathcal{P} = \sum_{m=0}^{\infty} P_m \lambda^{-m} \tag{3}$$

or from its integrated form

$$0 = \frac{1}{2}\mathcal{P}\mathcal{P}_{xx} - \frac{1}{4}{\mathcal{P}_x}^2 + (u-\lambda)\mathcal{P}^2 = C(\lambda) = -4\lambda \tag{4}$$

where the integration constant $C(\lambda)$ has been chosen as -4λ. Every pair of functions u, ϕ solving (2) satisfies the square eigenfunction relation (SER)

$$0 = [\frac{1}{4}\partial^3 + (u-\lambda)\partial + \frac{1}{2}u_x](\phi\phi) = (B_1 - \lambda B_0)(\phi\phi) \tag{5}$$

which is a differential consequence of (2).

Restricted flows of the KdV hierarchy are defined [1, 5] as the set of $N+1$ equations

$$\begin{aligned} &0 = (\partial^2 + u - \lambda_k)\phi_k \qquad k = 1, \ldots, N \\ &K_s[u] = \alpha B_0(\textstyle\sum_{k=1}^{N} \phi_k^2) \ \ (\text{for fixed } s = 0, 1, 2, \ldots \text{ and } \alpha = \text{constant}) \end{aligned} \tag{6}$$

Equations (6) are invariant with respect to the action of all flows of KdV hierarchy since these flows preserve the form of the spectral problem (2), of the vector fields $K_s[u]$, of the $(\phi\phi) = \frac{\delta\lambda\,[u]}{\delta u}$ and also of the hamiltonian operator B_0. Integrals of motion corresponding to the hamiltonian action of each flow of the hierarchy render equation (6) to be completely integrable. We shall show below that all restricted flows of the KdV hierarchy are equivalent to a certain set of Newton equations.

3. Trivial example

The stationary KdV equation

$$0 = K_2 = \frac{1}{4}u_{xxx} + \frac{3}{2}uu_x \tag{7}$$

Physical representations

There are two ways of turning (7) into a Newton equation.
(a) By direct integration we get

$$0 = \frac{1}{4}u_{xx} + \frac{3}{4}u^2 - c = r_{xx} + 12r^2 - c = \frac{d}{dx}\frac{\partial \mathcal{L}}{\partial r_x} - \frac{\partial \mathcal{L}}{\partial r} = \frac{\delta \mathcal{L}}{\delta r} \tag{8}$$

where $r = \frac{1}{4}u$, c is the integration constant and $\mathcal{L} = \frac{1}{2}r_x^2 - 4r^3 + cr$ is the lagrangian function. Equation (8) has the hamiltonian form

$$r_x = \frac{\partial \mathcal{H}}{\partial s} = s, \qquad s_x = -\frac{\partial \mathcal{H}}{\partial r} = -12r^2 + c \tag{9}$$

with $\mathcal{H} = \frac{1}{2}s^2 + 4r^3 - cr$ where $s = \frac{\partial \mathcal{L}}{\partial r_x} = r_x$ is the canonically conjugate momentum.
(b) By multiplication of (7) with u and by subsequent integration we get

$$0 = \frac{1}{4}uu_{xx} - \frac{1}{8}u_x^2 + \frac{1}{2}u^3 - d \tag{10}$$

where d is another (different from c) integration constant. Equation (10) has a Newton equation form if we divide it by u, however the force depends on the velocity u_x. The substitution $u = q^2$ removes the velocity dependent term and then (10) becomes

$$0 = q_{xx} + q^3 - \frac{2d}{q^3} = \frac{\delta \bar{\mathcal{L}}}{\delta q} \tag{11}$$

with $\bar{\mathcal{L}} = \frac{1}{2}q_x^2 - \frac{1}{4}q^4 - \frac{d}{q^2}$. Equation (11) has the hamiltonian form

$$q_x = \frac{\partial \mathcal{H}}{\partial p} = p, \qquad p_x = -\frac{\partial \mathcal{H}}{\partial q} = -q^3 + \frac{2d}{q^3} \tag{12}$$

with $\bar{\mathcal{H}}(q,p,d) = \frac{1}{2}p^2 + \frac{1}{4}q^4 + \frac{d}{q^2}$ where $p = \frac{\partial \bar{\mathcal{L}}}{\partial q_x} = q_x$.

Bihamiltonian formulation

Equations (9) and (12) are equivalent (since they both follow from (7)) in the extended phase space of variables (r,s,c) and (q,p,d). If we substitute $r = \frac{1}{4}u = \frac{1}{4}q^2$ into (8) then

$$0 = r_{xx} + 12r^2 - c = \frac{1}{2}q[q_{xx} + q^3 - \frac{2d}{q^3}] + [\frac{1}{2}q_x^2 + \frac{1}{4}q^4 + \frac{d}{q^2} - c] \tag{13}$$

It says that if q satisfies equation (11) then r satisfies (8) with $c = \bar{\mathcal{H}}(q,p,d)$ calculated from the energy integral and vice versa. Thus equations (9) and (12) are connected through the map

$$r = \tfrac{1}{4}u = \tfrac{1}{4}q^2, \; s = r_x = \tfrac{1}{4}u_x = \tfrac{1}{2}qq_x = \tfrac{1}{2}qp, \; c = \bar{\mathcal{H}}(q,p,d) = \tfrac{1}{2}p^2 + \tfrac{1}{4}q^4 + \tfrac{d}{q^2}.$$

By using this map we calculate new Poisson brackets

$$\{r,s\} = \{\tfrac{1}{4}q^2, \tfrac{1}{2}qp\} = r, \; \{r,c\} = s, \; \{s,c\} = 12r^2 - c$$

and obtain the bihamiltonian formulation

$$\begin{bmatrix} r \\ s \\ c \end{bmatrix}_x = \begin{bmatrix} 0 & 1 & 0 \\ -1 & 0 & 0 \\ 0 & 0 & 0 \end{bmatrix} \begin{bmatrix} 12r^2 - c \\ s \\ -r \end{bmatrix} = \begin{bmatrix} 0 & r & s \\ -r & 0 & -12r^2 + c \\ -s & 12r^2 - c & 0 \end{bmatrix} \begin{bmatrix} 0 \\ 0 \\ 1 \end{bmatrix} = \tag{14}$$

$$= \Pi_0 d\mathcal{H}(r,s,c) = \Pi_1 dc$$

for equations (9). We stress the fact that the Newton representations (8), (11) and the Poisson operators Π_0, Π_1 are a consequence of the bihamiltonian formulation (in terms of B_0 and B_1) for the KdV hierarchy.

Lax representation

The Lax representation of the KdV equation naturally provides a commutative representation for its stationary flow (7)

$$0 = u_t = K_2 = [A_2, \mathbb{L}], \qquad \mathbb{L} = (\partial^2 + u - \lambda) \tag{15}$$

Equation (15) is a compatibility condition for

$$0 = (\partial^2 + u - \lambda)\phi = \mathbb{L}\phi, \qquad A_2\phi = (\partial^3 + \frac{3}{2}u\partial + \frac{3}{4}u_x)\phi = \mu\phi \tag{16}$$

which has the following matrix form

$$\begin{bmatrix} \phi \\ \phi_x \end{bmatrix}_x = \begin{bmatrix} 0 & 1 \\ \lambda - u & 0 \end{bmatrix} \begin{bmatrix} \phi \\ \phi_x \end{bmatrix} = L \begin{bmatrix} \phi \\ \phi_x \end{bmatrix} \tag{17}$$

$$\mu \begin{bmatrix} \phi \\ \phi_x \end{bmatrix}_x = \begin{bmatrix} -\frac{1}{4}u_x & \lambda + \frac{1}{2}u \\ (\lambda + \frac{1}{2}u)(\lambda - u) - \partial\frac{1}{4}u_{xx} & \frac{1}{4}u_x \end{bmatrix} \begin{bmatrix} \phi \\ \phi_x \end{bmatrix} = M \begin{bmatrix} \phi \\ \phi_x \end{bmatrix} \tag{18}$$

where the second row in (18) follows by x-differentiation of $A_2\phi = \mu\phi$. The compatibility condition for (17),(18) is the Lax equation $M_x = [L, M]$. Lax representations for (9) and (12) are provided by the same matrices M, L when we express u, u_x, u_{xx} through the canonical variables (r, s, c) or (q, p, d).

4. Less trivial example [5]

The next stationary KdV flow

$$0 = K_3 = \frac{1}{16}\partial(u_{xxxx} + 10uu_{xx} + 5u_x^2 + 10u^3) = (\frac{1}{4}\partial^3 + u\partial + \frac{1}{2}u_x)(\frac{1}{4}u_{xx} + \frac{3}{4}u^2) \tag{19}$$

is equivalent to the one parameter family of Newton equations

$$0 = r_{1xx} - 4r_2 + 10r_1^2 = \frac{\delta \mathcal{L}_3}{\delta r_2}, \qquad 0 = r_{2xx} + 20r_1r_2 - 10r_1^3 - c = \frac{\delta \mathcal{L}_3}{\delta r_1} \tag{20}$$

if we integrate the first hamiltonian operator $B_0 = \partial$. Equations (20) follow from the lagrangian $\mathcal{L}_3 = r_{1x}r_{2x} - W(r_1, r_2, c)$, $W(r_1, r_2, c) = 10r_1^2r_2 - 2r_2^2 - \frac{5}{2}r_1^4 - cr_1$ and can be written in a hamiltonian form with $\mathcal{H}_3 = s_1s_2 + W(r_1, r_2, c)$, $s_1 = r_{2x}$, $s_2 = r_{1x}$. By multiplying the B_1 form of (19) with $\Delta = \frac{1}{4}u_{xx} + \frac{3}{4}u^2$ and by subsequent integrating we reduce (19) to two equations

$$0 = \frac{1}{4}\Delta\Delta_{xx} - \frac{1}{8}\Delta_x^2 + \frac{1}{2}u\Delta^2 - \frac{1}{128}\alpha, \qquad \Delta = \frac{1}{4}u_{xx} + \frac{3}{4}u^2$$

which after the change of variables $u = q_1$, $\Delta = -\frac{1}{8}q_2^2$ yield the Henon-Heiles system [5, 6]

$$0 = q_{1xx} + 3q_1^2 + \frac{1}{2}q_2^2, \qquad 0 = q_{2xx} + q_1q_2 - \frac{\alpha}{q_2^3}$$

This system follows from the hamiltonian $H_q(q,p,,\alpha) = \frac{1}{2}p_1^2 + \frac{1}{2}p_2^2 + \frac{1}{2}q_1q_2^2 + q_1^3 + \frac{\alpha}{2q_2^2}$. Similarly as in section 3 we get the map

$$r_1 = \frac{1}{4}q_1,\ s_2 = \frac{1}{4}p_1,\ r_2 = -\frac{1}{32}(q_1^2+q_2^2,\ s_1 = -\frac{1}{16}(q_1p_1+q_2p_2),\ c = -\frac{1}{8}H_q(q,p,\alpha)$$

connecting the (r,s,c) and (q,p,α) variables and equations (20) acquire the bihamiltonian formulation $(r_1,r_2,s_1,s_2,c)^t = \Pi_0 d\mathcal{H}_3 = \Pi_1 dc$ where

$$\Pi_0 = \begin{bmatrix} 0 & I & 0 \\ -I & 0 & 0 \\ 0 & 0 & 0 \end{bmatrix}, \quad \Pi_1 = \begin{bmatrix} 0 & 0 & \frac{1}{2}r_1 & -\frac{1}{2} & s_2 \\ 0 & 0 & r_2 & \frac{1}{2}r_1 & s_1 \\ -\frac{1}{2}r_1 & -r_2 & 0 & \frac{1}{2}s_2 & -\frac{\partial W}{\partial r_1} \\ \frac{1}{2} & -\frac{1}{2}r_2 & -\frac{1}{2}s_2 & 0 & -\frac{\partial W}{\partial r_2} \\ -s_2 & -s_1 & \frac{\partial W}{\partial r_1} & \frac{\partial W}{\partial r_2} & 0 \end{bmatrix}$$

are Poisson operators mapping differentials $d\mathcal{H}_3$ and dc into vector fields and

$$W(r_1,r_2,c) = 10r_1^2r_2 - 2r_2^2 - \frac{5}{2}r_1^4 - cr_1.$$

5. Newton representation for stationary flows of the KdV hierarchy [2]

Theorem

Each stationary KdV equation $c = P_{s+1}[u]$ is, under substitution $u = 4r_1$, equivalent to the following set of Newton equations

$$\begin{aligned} 0 &= Res[\mathcal{R}_{xx} + 4r_1\mathcal{R} - \lambda(\mathcal{R}-\mathcal{R}^{-3})] = \tfrac{\delta\mathcal{L}_{s+1}}{\delta r_s} \\ &\vdots \\ 0 &= Res\lambda^{s-2}[\mathcal{R}_{xx} + 4r_1\mathcal{R} - \lambda(\mathcal{R}-\mathcal{R}^{-3})] = \tfrac{\delta\mathcal{L}_{s+1}}{\delta r_2} \\ 0 &= Res\lambda^{s-1}[\mathcal{R}_{xx} + 4r_1\mathcal{R} - \lambda(\mathcal{R}-\mathcal{R}^{-3})] + 2Res(\lambda^s\mathcal{R}^2) - c = \tfrac{\delta\mathcal{L}_{s+1}}{\delta r_1} \end{aligned} \tag{21}$$

where $\mathcal{R} = \sum_{m=0}^{\infty} r_m\lambda^{-m}$. Equations (21) follow from the lagrangian

$$\mathcal{L}_{s+1} = Res[\frac{1}{2}\lambda^s\mathcal{R}_x^2 - 2\lambda^s r_1\mathcal{R}^2 + \lambda^{s+1}\frac{1}{2}(\mathcal{R}^2 + \mathcal{R}^{-2})] - cr_1$$

Proof

The stationary KdV equations are $(2s+1)st$ order ODE's for u which are defined as

$$K_{s+1}[u] = 0 \quad \text{or} \quad P_{s+1}[u] = c = \text{constant}, \qquad s = 1,2,\ldots \tag{22}$$

They can alternatively be parametrized by P_m which satisfy a set of 2nd order nonlinear ODE's which follow from (4). But a really interesting parametrization follows from the simple formal substitution $\mathcal{P} = 2\mathcal{R}^2$, $\mathcal{R} = \sum_{m=0}^{\infty} r_m\lambda^{-m}$ (and hence $u = 4r_1$) which turns equation (4) (and therefore (22) into $0 = \mathcal{R}_{xx} + 4r_1\mathcal{R} - \lambda(\mathcal{R}-\mathcal{R}^{-3})$. Then the first s equations, together with the condition $c = P_{s+1}[u] = 2Res(\lambda^s\mathcal{R}^2)$, become an autonomous set of Newton equations for $r_1,\ldots,r_s$. The symbol *Res* means here residuum at $\lambda = \infty$. Equations (21) follow directly from the change of variables $\mathcal{P} = 2\mathcal{R}^2$. The additional term $c - 2Res(\lambda^s \mathcal{R}^2)$ in the last equation (21) contains the information that we consider the $(2s+1)st$ stationary KdV equation. It is also necessary for cancelling the r_{s+1} variable which appears in the expression $Res\lambda^{s-1}[\mathcal{R}_{xx} + 4r_1\mathcal{R} - \lambda(\mathcal{R} - \mathcal{R}^{-3})]$. The proof of the lagrangian form of (21) is a simple exercise in deriving the formal series for $\mathcal{R}$.

Corollary

All restricted flows (6) of the KdV hierarchy written as

$$0 = \phi_{kxx} + (4r_1 - \lambda_k)\phi_k \quad k = 1, \ldots, N \qquad P_{s+1} = (\sum_{k=1}^{N} \phi_k^2) + c$$

admit the Newton representation (21) if in the last equation c is substituted by $(\sum_{k=1}^{N} \phi_k{}^2) + c$. These Newton equations follow from the lagrangian

$$\bar{\mathcal{L}}_{s+1} = \mathcal{L}_{s+1} + \frac{1}{4}(\sum_{k=1}^{N} \phi_{kx}^2) - r_1(\sum_{k=1}^{N} \phi_k{}^2) + \frac{1}{4}(\sum_{k=1}^{N} \lambda_k \phi_k{}^2)$$

6. Conclusions

We have shown that all stationary and restricted flows of the KdV hierarchy can be parametrized as a set of Newton equations. These Newton equations are completely integrable and inherit Lax representation and bihamiltonian formulation from the KdV hierarchy. The method of endowing stationary flows of the KdV hierarchy with Newton parametrization relies essentially on the structure of the recursion relation (3) and may be applied to other soliton hierarchies in order to provide interesting parametrizations for their stationary and restricted flows as sets of Newton equations or as dynamical systems. The main elements of structure of soliton hierarchy, such as Lax representation, bihamiltonian formulation, Miura map and Darboux transformation can be carried to these systems and used for their solving. This way one obtains new Lax representations and bihamiltonian structures for classical mechanical systems which are worth of further study on their own. This area of research is still to be explored.

It is necessary to stress here that we have presented a method where we can first (and easily) construct interesting sets of integrable equations and then to derive, from the underlying soliton hierarchy, all the elements of structure necessary to solve these equations. It is in contrast with the Kac-Moody and R-matrix approaches which provide nice theorems on an abstract level but give no clue to what would be physically interesting parametrizations.

Acknowledgements

I would like to thank Krzysztof Marciniak for help with LATEX.

References

1. Antonowicz M., Rauch-Wojciechowski S. *Phys.Lett.* **147A**, 455 (1990)
2. Rauch-Wojciechowski S. *Phys.Lett.* **173A**, 91 (1992)
3. Dickey L.A., Soliton equations and hamiltonian systems, *World Scientific*, (Singapore 1991) ch.12
4. Antonowicz M., Rauch-Wojciechowski S. *J.Math.Phys.* **33**, 2115 (1992)
5. Antonowicz M., Rauch-Wojciechowski S. *Phys.Lett.* **163A**, 167 (1992)
6. Fordy A. *Physica D* **52**, 204 (1991)

A POINT SYMMETRY GROUP OF A DIFFERENTIAL EQUATION WHICH CANNOT BE FOUND USING INFINITESIMAL METHODS

G. J. REID, D. T. WEIH
Mathematics Department, University of British Columbia,
Vancouver B.C., Canada V6T 1Z2.
email: reid@math.ubc.ca

and

A. D. WITTKOPF
Mathematics Department, Simon Fraser University,
Vancouver B.C., Canada V5A 1S6.

Abstract. A point symmetry group of a differential equation is a group of invertible transformations of its dependent and independent variables which leaves the differential equation invariant. Lie's infinitesimal method recovers the connected components of Lie symmetry groups of differential equations, and, by analytic continuation, sometimes yields disconnected components.

We show that Lie's infinitesimal method and its analytic continuation fails to yield the full group of point symmetries of the ODE $u_{xx} = \frac{1}{x}u_x + \frac{4}{x^3}u^2$. To find the full point symmetry group of this ODE we apply a differential analogue of Buchberger's algorithm to simplify its *nonlinear* determining equations. The reduced form of these equations is easily solved to yield the explicit form of the group transformations.

1. The group and infinitesimal determining systems of $u_{xx} = \frac{1}{x}u_x + \frac{4}{x^3}u^2$

We find the full group of point symmetries $(x, u) \mapsto (\hat{x}, \hat{u})$ of the form

$$\hat{x} = X(x, u), \quad \hat{u} = U(x, u), \tag{1}$$

for the ODE

$$u_{xx} = \frac{1}{x}u_x + \frac{4}{x^3}u^2, \quad u = u(x). \tag{2}$$

Invertibility of the group transformations implies that their Jacobian does not vanish:

$$\frac{\partial(X, U)}{\partial(x, u)} = X_x U_u - X_u U_x \not\equiv 0. \tag{3}$$

To determine symmetries of the form (1) of (2) we substitute the transformations (1) into $\hat{u}_{\hat{x}\hat{x}} = \hat{u}_{\hat{x}}/\hat{x} + 4\hat{u}^2/\hat{x}^3$. In the (x, u) coordinates

$$\hat{u}_{\hat{x}} = \frac{dU}{dX} = \frac{U_x + u_x U_u}{X_x + u_x X_u} \tag{4}$$

where $X_x + u_x X_u \not\equiv 0$ as a consequence of the invertibility requirement (3) (since $X_x + u_x X_u \equiv 0$ implies $X_x \equiv X_u \equiv 0$ and the violation of (3)). Similarly we obtain an expression for $\hat{u}_{\hat{x}\hat{x}} = \frac{d\hat{u}_{\hat{x}}}{dX}$ in the (x, u) coordinates. General formulae for

N. H. Ibragimov et al. (eds.),
Modern Group Analysis: Advanced Analytical and Computational Methods in Mathematical Physics, 311–316.

transforming systems of PDEs are given in [1], and we have automated this process using the symbolic language Maple. The net result is that the transformed equation is obtained in terms of x, u, u_x, u_{xx}, X, U and derivatives of X, U.

The condition that the point transformations (1) are symmetries is then imposed by using (2) to eliminate u_{xx} from the transformed equation. Polynomial decomposition of the resulting equation with respect to u_x yields the following *group determining equations* for $X(x, u)$ and $U(x, u)$:

$$
\begin{aligned}
&4U^2X_u^3 - X^3X_uU_{uu} + X^3U_uX_{uu} + X^2U_uX_u^2 = 0,\\
&X^2U_xX_u^2 + 2X^3U_uX_{xu} - X^3X_xU_{uu} - 2X^3X_uU_{xu} + 2X^2U_uX_xX_u\\
&\qquad\qquad +12U^2X_xX_u^2 + X^3U_xX_{uu} = 0,\\
&x^3X^3U_xX_{xx} - 4u^2X^3U_uX_x - x^3X^3U_{xx}X_x + x^3X^2U_xX_x^2\\
&\qquad\qquad +4u^2X^3U_xX_u + 4x^3U^2X_x^3 = 0,\\
&2xX^3U_xX_{xu} + X^3U_xX_u - xX^3U_{xx}X_u + 2xX^2U_xX_xX_u - X^3U_uX_x + xX^2U_uX_x^2\\
&\qquad\qquad -2xX^3U_{xu}X_x + xX^3U_uX_{xx} + 12xU^2X_x^2X_u = 0,
\end{aligned} \tag{5}
$$

which are restricted only by the invertibility condition (3). The group determining system (5) is polynomially nonlinear in its dependent variables and their derivatives.

The nonlinearity of group determining equations such as those above was partially circumvented by Lie's work [1],[2]. He considered multi-parameter local groups of point transformations (now called local Lie point groups) that are analytic at their identities. He showed that the subgroup formed by the connected component of the identity of such groups could be characterized infinitesimally and linearly by expanding the group transformations about their identities.

To apply Lie's method to the ODE (2) one seeks a continuous local point group of symmetries of (2), analytic at its identity, depending on a single continuous parameter (ϵ) of the form

$$
\begin{aligned}
\hat{x} &= X(x, u; \epsilon) = x + \epsilon\xi(x, u) + O(\epsilon^2)\,,\\
\hat{u} &= U(x, u; \epsilon) = u + \epsilon\eta(x, u) + O(\epsilon^2)\,,
\end{aligned} \tag{6}
$$

where $\xi(x, u)$ and $\eta(x, u)$ are the *infinitesimals* of the group. Lie showed that the subgroup formed by the connected component of the identity of such groups could be uniquely characterized in terms of its infinitesimals and that these satisfied a system of linear homogeneous PDEs called *infinitesimal determining equations.* These equations are derived by an explicit algorithm [1,2], for which many computer implementations are available (see the references cited in [3]).

For the ODE (2) the infinitesimal determining equations are:

$$
\begin{aligned}
&\xi_{uu} = 0, \;\; x\eta_{uu} - 2x\xi_{xu} - 2\xi_u = 0,\\
&x^4\eta_{xx} - x^3\eta_x + 4u^2x\eta_u - 8u^2x\xi_x - 8ux\eta + 12u^2\xi = 0,\\
&2x^3\eta_{xu} - x^3\xi_{xx} - x^2\xi_x - 12u^2\xi_u + x\xi - 0\,.
\end{aligned} \tag{7}
$$

They are linear and much simpler than the group determining system (5).

2. Reduction of the Group and Infinitesimal Determining Systems to Differential Gröbner Bases

One of us (ADW) is developing an algorithm **diff_reduce** which attempts to reduce polynomially nonlinear systems of PDEs to the form of a differential Gröbner basis. The algorithm **diff_reduce** is a differential analogue of Buchberger's elimination algorithm for polynomial equations (also see [4],[5]). It takes as its input the system of PDEs and a total ordering on monomials of derivatives. The most important property of these orderings is that they are preserved under differentiation and multiplication by differential monomials. A key feature of Wittkopf's algorithm is that it uses reduction instead of the pseudo-reduction used in [4].

We reduced the infinitesimal determining system (7) to a differential Gröbner basis subject to the lexicographical ordering which is graded first by function $\xi \prec \eta$ (so that any derivative of ξ is smaller than any derivative of η), next by total order of derivative, and finally lexicographically by derivative ($\partial_x \prec \partial_u$):

$$\xi \prec \xi_x \prec \xi_u \prec \xi_{xx} \prec \xi_{xu} \prec \xi_{uu} \prec \ldots \prec \eta \prec \eta_x \prec \eta_u \prec \eta_{xx} \prec \eta_{xu} \prec \ldots \tag{8}$$

An extension of the same ordering with ξ, η replaced by X, U respectively, was used for the reduction of the group determining system.

The reduced form of the infinitesimal determining system using Wittkopf's algorithm is

$$\xi_u = 0, \quad x\xi_x - \xi = 0, \quad x\eta - u\xi = 0. \tag{9}$$

The reduced form of the group determining system is

$$\begin{aligned} &X_u = 0,\ x^4X_x^4 - X^4 = 0,\ 8XU + X^2 - x(8u + x)X_x^2 = 0, \\ &8x^3X_x^2U + x^3XX_x^2 - (8u + x)X^3 = 0. \end{aligned} \tag{10}$$

Both of the above systems are auto-reduced algebraic Gröbner bases and also differential Gröbner bases. The reduced form of the group determining system is a differential Gröbner basis in the sense that it solves the differential ideal membership problem[1] with the restriction that $\frac{\partial(X,U)}{\partial(x,u)}$ does not lie in the differential ideal generated by the system. The last equation of (10) is a consequence of the first three. Despite recent progress [4],[5] all the existing differential-Buchberger type algorithms suffer a drawback. None of them can always guarantee reduction of polynomially nonlinear systems of PDEs to differential Gröbner bases in a finite number of steps.

3. Explicit Form of the Group

The reduced forms of the infinitesimal and group determining equations are easy to solve.

Solving the infinitesimal determining system (9), and exponentiating its infinitesimal operator, shows that the corresponding connected[2] Lie group is the simple

[1] A differential ideal generated by a system of differential polynomials is the infinite set of sums of products of derivatives of these polynomials.

[2] By analytic continuation (i.e. by considering α complex) the two disconnected components of the real case corresponding to $\alpha < 0$ and $\alpha > 0$ become a single connected component of the identity.

scaling group H:

$$(\hat{x}, \hat{u}) = P_\alpha(x, u) = (\alpha x, \alpha u). \tag{11}$$

This scaling group is also obvious by a direct inspection of ODE (2).

To solve the group determining system we note that the second equation of its reduced form $x^4 X_x^4 - X^4 = 0$ has four solutions for X_x: X/x, iX/x, $-X/x$ and $-iX/x$. The corresponding ODEs are easily integrated for $X(x)$. $U(x, u)$ is then obtained by substituting the corresponding solutions into the third equation of (10). We obtain four one parameter classes of transformations:

$$P_\alpha(x, u) = (\alpha x, \alpha u), \tag{12}$$

$$Q_\beta(x, u) = (\beta x^i, -\beta u x^{i-1} - \frac{\beta x^i}{4}), \tag{13}$$

$$R_\gamma(x, u) = (\gamma x^{-1}, \gamma u x^{-2}), \tag{14}$$

$$S_\delta(x, u) = (\delta x^{-i}, -\delta u x^{-i-1} - \frac{\delta x^{-i}}{4}), \tag{15}$$

where $i^2 = -1$. Also x, u take complex values, and α, β, γ, δ are nonzero complex parameters. There is no loss in defining x^i by its principal branch ($x^i = e^{-Arg\, x}(\cos \log|x| + i \sin \log|x|)$) since the other branches are equivalent by rescalings of the group parameters. The identity for the point group defined by the above transformations is P_α with $\alpha = 1$. The four classes of transformations are disjoint in that there are no values of the parameters $\alpha, \beta, \gamma, \delta$ such that the action of any two of these transformations agrees on some open neighbourhood in the complex plane. In particular there are no values of β, γ, δ such that operators Q_β, R_γ, S_δ yield the identity. These transformations satisfy the group composition laws given in the table below.

$\circ$	P_β	Q_β	R_β	S_β
P_α	$P_{\alpha\beta}$	$Q_{\alpha\beta}$	$R_{\alpha\beta}$	$S_{\alpha\beta}$
Q_α	$Q_{\alpha\beta^i}$	$R_{\alpha\beta^i}$	$S_{\alpha\beta^i}$	$P_{\alpha\beta^i}$
R_α	$R_{\alpha\beta^{-1}}$	$S_{\alpha\beta^{-1}}$	$P_{\alpha\beta^{-1}}$	$Q_{\alpha\beta^{-1}}$
S_α	$S_{\alpha\beta^{-i}}$	$P_{\alpha\beta^{-i}}$	$Q_{\alpha\beta^{-i}}$	$R_{\alpha\beta^{-i}}$

Table: Point symmetry group of $\mathbf{u_{xx} = u_x/x + 4u^2/x^3}$

In the above table the (1,1) entry represents $P_\alpha \circ P_\beta$, the (1,2) entry represents $P_\alpha \circ Q_\beta$ and $(P_\alpha \circ Q_\beta)(x, u) = P_\alpha(Q_\beta(x, u))$ etc. Also $(P_\alpha)^{-1} = P_{\alpha^{-1}}$, $(Q_\alpha)^{-1} = S_{\alpha^i}$, $(R_\alpha)^{-1} = R_\alpha$, $(S_\alpha)^{-1} = Q_{\alpha^{-i}}$. The full group is generated by the transformations P_α and the discrete symmetries Q_1, R_1, S_1. The four disjoint components P_α, Q_β, R_γ, S_δ are left invariant under the action of P_α and the subgroup $H = \{P_\alpha : \alpha \neq 0\}$ is a normal subgroup of the full point group $G = \{P_\alpha, Q_\beta, R_\gamma, S_\delta : \alpha, \beta, \gamma, \delta \neq 0\}$ of the ODE. The quotient group G/H is isomorphic to the cyclic group on four elements. The only subgroup which can be found by Lie's infinitesimal method is H.

4. Discussion

We showed that the application of Lie's infinitesimal method to the ODE $u_{xx} = \frac{1}{x}u_x + \frac{4}{x^3}u^2$ only reveals a simple one parameter scaling group $H = \{P_\alpha : \alpha \neq 0, P_\alpha(x,u) = (\alpha x, \alpha u)\}$ of the ODE. To find the full group of point symmetries of the ODE we solved its nonlinear group determining equations. In addition to H this group consisted of three disjoint one parameter classes of transformations whose existence could not be predicted by Lie's infinitesimal method.

At first sight this is surprising since the full point symmetry group of the ODE must itself be a Lie group[3]. The surprise is lessened when we realize that Lie's infinitesimal method is only guaranteed to uncover the component of such groups that is connected to the identity and our group consists of four disjoint components. There are many known examples of disconnected Lie symmetry groups of differential equations. As far as we know, however, all of these examples can be obtained by either inspection or analytic continuation of the connected component determined by Lie's infinitesimal method. The main point of our article is that the full point group of ODE (2) cannot be obtained by such methods and we needed to solve the nonlinear group determining equations to find this group.

We were motivated to completely classify the point symmetries of ODE (2) by some remarkable transformations appearing in the work of Zaitsev [6]. Zaitsev classified equivalence transformations of a class of generalized Emden-Fowler ODEs of the form $u_{xx} = Ax^n u^m u_x^l$ where A, n, m, l are constants. Each such transformation mapped one member of the class to another member. By a simple change of coordinates our ODE is a fixed point of some of Zaitsev's transformations. For this ODE Zaitsev's transformations are contained in the point group that we calculated.

Elsewhere [7] we will show that structural characteristics of point groups of differential equations (e.g. their dimension and group table structure) can be calculated by manipulating the local initial data of their group determining equations. Such data is often explicitly calculable without solving group determining equations. This data is in one to one correspondence with the solutions of its corresponding group determining system. This approach is an extension to general point groups of differential equations, of our previous successful algorithms [8,9,10], for calculating the dimension and structure of Lie symmetry groups. Consequently we can often avoid the difficulty of integrating nonlinear systems of PDEs in determining structural aspects of the full point groups of differential equations.

Acknowledgements

GJR and DTW gratefully acknowledge the assistance of grants from the Natural Sciences and Engineering Research Council of Canada. We also thank George Bluman and Ian Lisle for discussions about this work. Information on how to obtain the Maple programs automating the algorithms described in [8,9,10] can be obtained by emailing GJR (email address: reid@math.ubc.ca).

[3] We take the definition of a finite (r) parameter Lie group to be a group which carries the structure of an r-dimensional (not necessarily connected) smooth manifold so that the group operation and inversion are smooth maps between manifolds.

References

1. Bluman G. W. and Kumei S. (1989). *Symmetries and Differential Equations* (Springer-Verlag, New York).
2. Ovsiannikov L. V. (1982). *Group Analysis of Differential Equations* (Academic Press, New York).
3. Champagne B., Hereman W. and Winternitz P. (1991). The computer calculation of Lie point symmetries of large systems of differential equations. Comp. Phys. Comm. **66** 319–340.
4. Mansfield E. (1991). *Differential Gröbner Bases.* Ph. D. Thesis (University of Sydney).
5. Carrà-Ferro G. (1987). Gröbner Bases and Differential Algebra. Lec. Notes in Comp. Sci. **356** 128–140.
6. Zaitsev V. F. (1988). On Discrete-Group Analysis of Ordinary Differential Equations. Sov. Math. Dokl. **37** No. 2, 403–406.
7. Reid G. J., Weih D. T. and Wittkopf A. D. *Point symmetries of differential equations which are not Lie symmetries* (in preparation).
8. Reid G. J. (1991). Algorithms for reducing a system of PDEs to standard form, determining the dimension of its solution space and calculating its Taylor series solution. Eur. J. of Appl. Math. **2**, 293–318.
9. Reid G. J. (1991). Finding abstract Lie symmetry algebras of differential equations without integrating determining equations. Eur. J. of Appl. Math. **2**, 319–340.
10. Reid G. J., Lisle I. G., Boulton A. and Wittkopf A. D. (1992). Algorithmic determination of commutation relations for Lie symmetry algebras of PDEs. In Proceedings of the 1992 International Symposium on Symbolic and Algebraic Computation, edited by P. S. Wang, 63–68 (acm press, New York).

ERMAKOV STRUCTURE IN 2 + 1-DIMENSIONAL SYSTEMS. CANONICAL REDUCTION

C. ROGERS,* C. HOENSELAERS and U. RAMGULAM
Department of Mathematical Sciences
Loughborough University of Technology, U.K.

Abstract. Ermakov-type systems in 2 + 1-dimensions are introduced. Multi-wave solutions of a 2 + 1-dimensional Pinney equation and a modulated 2 + 1-dimensional sine-Gordon equation are constructed via the standard Ermakov system. Lie group analysis of the latter is undertaken to reveal its underlying linear structure in terms of canonical coordinates. In conclusion, a 2 + 1-dimensional nonlinear system due to Stoker which arises in two-layer shallow water theory is shown to admit a symmetry reduction to a Ermakov system.

1. Introduction

The analysis of the coupled nonlinear ordinary differential equations now known as Ermakov systems originated in 1880 [1]. There has since been an extensive literature devoted to their study [2]–[30]. In terms of applications, Ermakov systems arise, in particular, in nonlinear elasticity [31, 32] and nonlinear optics [33]–[37]. The main theoretical interest in such systems resides in the fact that they admit a constant of motion, the Lewis-Ray-Reid invariant whereby a nonlinear superposition principle may be constructed.

Here, 2+1-dimensional Ermakov systems are introduced and multi-wave solutions constructed for a 2+1-dimensional modulated sine-Gordon equation via the standard Ermakov system [30]. Lie group analysis of the latter is used to reveal its underlying linear structure in terms of canonical coordinates. This linearization was obtained in an 'ad hoc' manner in [24].

In conclusion, a nonlinear 2 + 1-dimensional two-layer shallow water system due to Stoker [38] is shown to admit a symmetry reduction to a special Ermakov system. The linearization of the latter has recently been exploited by Athorne [29] to analyse stability and periodicity.

2. 2+1-dimensional Ermakov systems

Here, we consider 2 + 1-dimensional Ermakov-type systems [30]

$$\begin{aligned} \Box\phi + \omega^2\phi - \rho^{-3}H(\phi/\rho) &= 0, \\ \Box\rho + \omega^2\rho - \phi^{-3}J(\rho/\phi) &= 0, \end{aligned} \tag{2.1}$$

* Present address: Department of Applied Mathematics, University of New South Wales, Australia.

N. H. Ibragimov et al. (eds.),
Modern Group Analysis: Advanced Analytical and Computational Methods in Mathematical Physics, 317–328.

where $\Box := \partial^2/\partial x^2 + \partial^2/\partial y^2 - \partial^2/\partial t^2$ and ω^2 may, in general, depend in an arbitrary manner on ϕ, ρ and their partial derivatives.

A dimensional reduction of the system (2.1) is sought whereby it becomes the classical Ermakov system. Thus, it is required that $\phi = \phi(g,h), \rho = \rho(g,h)$ where g, h are subject to the conditions[1]

$$\begin{gathered} \Box g = g, \ \Box h = h, \\ g_x^2 + g_y^2 - g_t^2 = g^2, \ h_x^2 + h_y^2 - h_t^2 = h^2, \\ g_x h_x + g_y h_y - g_t h_t = gh. \end{gathered} \tag{2.2}$$

This approach corresponds to that adopted in another context in [39]–[42].

Introduction of the new variables ξ, η according to

$$\xi = \frac{1}{2}\ln(gh), \tag{2.3}$$

$$\eta = \frac{1}{2}\ln\left(\frac{g}{h}\right) \tag{2.4}$$

now reduces the $2+1$-dimensional system (2.1) to

$$\begin{aligned} \phi_{\xi\xi}|_\eta + \omega^2\phi &= \rho^{-3}H(\phi/\rho), \\ \rho_{\xi\xi}|_\eta + \omega^2\rho &= \phi^{-3}J(\rho/\phi). \end{aligned} \tag{2.5}$$

This is the classical Ermakov system except that the quantity η may enter as a parameter.

Now, (2.5) yields

$$\partial_\xi(\rho\phi_\xi - \phi\rho_\xi) = \rho^{-2}H(\phi/\rho) - \phi^{-2}J(\rho/\phi),$$

whence,

$$r_{\tau\tau} = H(r) - r^{-2}J(r^{-1}), \tag{2.6}$$

where

$$r = \phi/\rho, \tag{2.7}$$

$$\tau = \int \rho^{-2}|_\eta d\xi + \alpha(\eta) \tag{2.8}$$

where $\alpha(\eta)$ is arbitrary. Integration of (2.6) yields

$$r_\tau = \sqrt{2\{I(\eta) + \textstyle\int [H(r) - r^{-2}J(r^{-1})]\, dr}$$

whence

$$\frac{\rho^4}{2}\left[g\partial_g(\phi/\rho) + h\partial_h(\phi/\rho)\right]^2 - \int^{\phi/\rho} \left[H(r) - r^{-2}J(r^{-1})\right]\, dr = \bar{I}(g/h). \tag{2.9}$$

[1] It is noted that dimensional reduction of the $2+1$-dimensional Ermakov system (2.1) may be sought systematically by Lie group methods.

$$(\bar{I}(g/h) := I(\frac{1}{2}\ln(g/h)).$$

Here, $\bar{I}(g/h)$ corresponds to the Lewis-Ray-Reid invariant of standard Ermakov theory.

Let us now turn to the conditions (2.2) on g and h and introduce the Ansatz

$$g = \sum_{i=1}^{N} e^{\theta_i+\delta_i}, \tag{2.10}$$

$$h = \sum_{i=1}^{M} e^{\theta_i+\Delta_i}, \tag{2.11}$$

where

$$\theta_i = p_i x + q_i y + w_i t, \tag{2.12}$$

and δ_i, Δ_i are arbitrary phase constants. Insertion of (2.10), (2.11) into (2.2) produces the restrictions

$$p_i^2 + q_i^2 - w_i^2 = 1, \tag{2.13}$$

$$(p_i - p_j)^2 + (q_i - q_j)^2 - (w_i - w_j)^2 = 0. \tag{2.14}$$

The condition (2.14) is satisfied if

$$(p_i - p_j, q_i - q_j, w_i - w_j) = \alpha_{ij}\ (\lambda, \mu, \nu), \tag{2.15}$$

where

$$\lambda^2 + \mu^2 = \nu^2, \tag{2.16}$$

and (2.13) requires that

$$\lambda p_j + \mu q_j - \nu w_j = 0, \quad \alpha_{ij} \neq 0. \tag{2.17}$$

In the above, the α_{ij} are arbitrary constants.

Once λ, μ, ν are chosen in accordance with (2.16), if p_1 is chosen arbitrarily, then q_1 and w_1 are determined via the constraints

$$p_1^2 + q_1^2 - w_1^2 = 1, \tag{2.18}$$

$$\lambda p_1 + \mu q_1 - \nu w_1 = 0. \tag{2.19}$$

The remaining (p_i, q_i, w_i) are then given by

$$p_i = p_1 + \alpha_{i1}\lambda, \quad q_i = q_1 + \alpha_{i1}\mu, \quad w_i = w_1 + \alpha_{i1}\nu. \tag{2.20}$$

Accordingly, multi-wave solutions of the constraints (2.2) are delivered in the form

$$g = e^{p_1 x + q_1 y + w_1 t} \sum_{i=1}^{N} e^{\alpha_{i1}(\lambda x + \mu y + \nu t) + \delta_i}, \tag{2.21}$$

$$h = e^{p_1 x + q_1 y + w_1 t} \sum_{i=1}^{M} e^{\alpha_{i1}(\lambda x + \mu y + \nu t) + \Delta_i}, \tag{2.22}$$

where λ, μ, ν and p_1, q_1, w_1 are subject to the conditions (2.16), (2.18) and (2.19).

3. The 2+1-dimensional Pinney equation

Here, attention is turned to the particular Ermakov system

$$\Box\phi + \omega^2\phi - k\phi^{-3} = 0, \tag{3.1}$$

$$\Box\rho + \omega^2\rho = 0, \tag{3.2}$$

coupled through ω^2. Thus, (3.1)–(3.2) represents a 2+1-dimensional Pinney system. It is noted that higher-dimensional Pinney equations arise in quantum mechanical systems (Lee [33]).

In this case, the reduced Ermakov system is

$$\phi_{\xi\xi}|_\eta + \omega^2\phi - k\phi^{-3} = 0, \tag{3.3}$$

$$\rho_{\xi\xi}|_\eta + \omega^2\rho = 0, \tag{3.4}$$

with associated first integral

$$\rho\phi_\xi - \phi\rho_\xi = \sqrt{2I(\eta) - k(\rho/\phi)^2}. \tag{3.5}$$

Let ρ_1, ρ_2 be two linearly independent solutions of (3.4). The relation (3.5) gives

$$\begin{aligned} \rho_1\phi_\xi - \phi\rho_{1\xi} &= \sqrt{2I_1(\eta) - k(\rho_1/\phi)^2}, \\ \rho_2\phi_\xi - \phi\rho_{2\xi} &= \sqrt{2I_2(\eta) - k(\rho_2/\phi)^2}, \end{aligned} \tag{3.6}$$

whence, on elimination of ϕ_ξ,

$$\phi W(\rho_1, \rho_2) = \rho_2\sqrt{2I_1(\eta) - k(\rho_1/\phi)^2} - \rho_1\sqrt{2I_2(\eta) - k(\rho_2/\phi)^2} \tag{3.7}$$

where $W(\rho_1, \rho_2)$ is the Wronskian of ρ_1, ρ_2. Explicit solution of (3.7) for ϕ produces the nonlinear superposition

$$\phi(\xi, \eta) = \sqrt{\bar\lambda(\eta)\rho_1^2 + 2\bar\mu(\eta)\rho_1\rho_2 + \bar\nu(\eta)\rho_2^2} \tag{3.8}$$

where

$$\bar\lambda\bar\nu - \bar\mu^2 = k/W(\rho_1, \rho_2). \tag{3.9}$$

If $\omega = \omega(\eta)$ and we take

$$\rho_1 = \sin\omega\xi, \tag{3.10}$$

$$\rho_2 = \cos\omega\xi \tag{3.11}$$

then (3.8) gives the solution of (3.1), namely

$$\phi(\xi, \eta) = \sqrt{\bar\alpha(\eta) + \bar\beta(\eta)\sin(2\omega(\eta)\xi + \bar\alpha(\eta))} \tag{3.12}$$

where

$$\bar\alpha^2 - \bar\beta^2 = k/W(\rho_1, \rho_2) = k/\omega(\eta) \tag{3.13}$$

and

$$\begin{aligned} \xi &= p_1 x + q_1 t + w_1 t + \tfrac{1}{2}\ln[\textstyle\sum_{i=1}^N e^{\alpha_{i1}(\lambda x + \mu y + \nu t) + \delta_i} \sum_{i=1}^M e^{\alpha_{i1}(\lambda x + \mu y + \nu t) + \Delta_i}] \\ \eta &= \tfrac{1}{2}\ln[\textstyle\sum_{i=1}^N e^{\alpha_{i1}(\lambda x + \mu y + \nu t) + \delta_i} / \sum_{i=1}^M e^{\alpha_{i1}(\lambda x + \mu y + \nu t) + \Delta_i}]. \end{aligned} \tag{3.14}$$

It is noted that, for constant ω and k, there exists a Lagrangian for Pinney's equation.

4. A 2+1-dimensional modulated sine-Gordon equation

The 2 + 1-dimensional Ermakov system

$$\Box\phi + \omega^2\phi - k\rho^{-3}\sin(\phi/\rho) = 0, \tag{4.1}$$

$$\Box\rho + \omega^2\rho = 0. \tag{4.2}$$

is considered next. This represents a 2 + 1-dimensional sine-Gordon equation (4.1) modulated by a function $\rho(x, y, t)$ which, in turn, is governed by (4.2).

The unmodulated 2 + 1-dimensional sine-Gordon equation (4.1) corresponds to $\rho = 1, \omega = 0$ and has been investigated in particular by Leibbrandt et al [43]. It is nonintegrable and admits only a restricted auto-Bäcklund transformation [43],[44]. It is only recently that the integrable 2+1-dimensional sine-Gordon equation which like the Davey-Stewartson and Vizhnik-Novikov-Veselov equation contains the spatial variables on an equal basis has been discovered [34]. This equation has been shown to be amenable to the $\bar{\partial}-\partial$ scattering method and to admit an auto-Bäcklund transformation together with soliton, breather and coherent structure solutions [46]–[50]. By contrast, the non-integrable sine-Gordon equation as considered in [43] has limited structure. However, restricted multi-wave solutions may be constructed in the manner described in [39]–[42]. It is noted that 1 + 1-dimensional modulated sine-Gordon systems of the type (4.1)–(4.2) have been investigated by Ray [17] and Saermark [19]. Therein, travelling wave solutions were obtained.

In this case, $H = k\sin(\phi/\rho)$, $J = 0$ so that the canonical equation (2.6) becomes

$$r_{\tau\tau} = k\sin r. \tag{4.3}$$

Thus, if $r = r(\tau)$ is any solution of the nonlinear pendulum equation (4.3) then the nonlinear superposition

$$\phi = \rho(\xi)r\{\int \rho^{-2}|_\eta d\xi + \alpha(\eta)\} \tag{4.4}$$

where

$$\rho_{\xi\xi}|_\eta + \omega^2\rho = 0 \tag{4.5}$$

and ξ, η are given by (3.14) provides a multi wave solution of the modulated 2 + 1-dimensional sine-Gordon equation (4.1).

If, as in the 1 + 1-dimensional case considered by Ray [17], the base solution

$$r = 4\tan^{-1}\exp[\tau/k^{\frac{1}{2}}] \tag{4.6}$$

of (4.3) is taken along with the particular solution

$$\rho = \sin\xi \tag{4.7}$$

of (4.5) with $\omega = 1$ then the nonlinear superposition (4.4) of (4.6) and (4.7) produces the multi-wave solution

$$\phi = 4\sin\xi\tan^{-1}\exp\frac{1}{k^{\frac{1}{2}}}(-\cot\xi + \alpha(\eta)) \tag{4.8}$$

of the modulated 2 + 1-dimensional sine-Gordon equation

$$\Box\phi - \phi + k(\sin\xi)^{-3}\sin(\phi/\sin\xi) = 0 \tag{4.9}$$

where ξ, η are given by (3.14)

5. Canonical reduction of the standard Ermakov systems

In the preceding, dimensional reductions of the 2 + 1-dimensional Ermakov systems have been noted. Here, Lie group methods are adduced to reveal the underlying linear structure of the standard Ermakov system in terms of canonical variables. In recent work, it is noted that Leach [28] has used Lie group methods to explain the nature of generalized Ermakov systems in terms of the symmetry algebra $sL(2, \mathbb{R})$.

The standard Ermakov system

$$\begin{aligned} \ddot{u} + \omega^2(t)u &= \tfrac{f(v/u)}{u^2 v}, \\ \ddot{v} + \omega^2(t)v &= \tfrac{g(u/v)}{v^2 u}, \end{aligned} \tag{5.1}$$

is considered where w, f, g are arbitrary functions of their indicated arguments.

The admitted Lie group of invariant transformations is sought with associated operator

$$\Gamma = T(t,u,v)\partial_t + U^1(t,u,v)\partial_u + U^2(t,u,v)\partial_v. \tag{5.2}$$

The invariance requirement produces the following system of determining equations (Ramgulam [51]).

$$\begin{aligned} &U^1_{vv} = T_{vv} = T_{uv} = T_{uu} = U^2_{uu} = 0, \\ &U^1_{uu} - 2T_{ut} = 0,\ U^2_{vv} - 2T_{vt} = 0, \\ &U^1_{uv} - T_{vt} = 0,\ U^2_{uv} - T_{ut} = 0, \\ &U^1_{vt} + (\omega^2 u - \tfrac{f}{u^2 v})T_v = 0,\ U^2_{ut} + (\omega^2 v - \tfrac{g}{uv^2})T_u = 0, \\ &2U^1_{ut} - T_{tt} + 3(\omega^2 u - \tfrac{f}{u^2 v})T_u + (\omega^2 v - \tfrac{g}{uv^2})T_v = 0, \\ &2U^2_{vt} - T_{tt} + 3(\omega^2 v - \tfrac{g}{uv^2})T_v + (\omega^2 u - \tfrac{f}{u^2 v})T_u = 0, \\ &U^1_{tt} + 2\omega\dot{\omega}uT + \omega^2 U^1 + \left(\tfrac{f'}{u^4} + \tfrac{2f}{u^3 v}\right)U' + \left(\tfrac{-f'}{u^3 v} + \tfrac{f}{u^2 v^2}\right)U^2 \\ &\qquad - \left(\omega^2 v - \tfrac{g}{uv^2}\right)U^1_v + (\omega^2 u - \tfrac{f}{u^2 v})(2T_t - U^1_u) = 0, \\ &U^2_{tt} + 2\omega\dot{\omega}vT + \omega^2 U^2 + \left(-\tfrac{g'}{uv^3} + \tfrac{g}{u^2 v^2}\right)U^1 + \left(\tfrac{g'}{v^4} + \tfrac{2g}{uv^3}\right)U^2 \\ &\qquad -(\omega^2 u - \tfrac{f}{u^2 v})U^2_u + (\omega^2 v - \tfrac{g}{uv^2})(2T_t - U^2_v) = 0. \end{aligned} \tag{5.3}$$

The solution of the system (5.3) yields

$$T(t) = \tau^2(t), \quad U^1 = \tau\dot{\tau}u, \quad U^2 = \tau\dot{\tau}v \tag{5.4}$$

where τ is governed by the standard Pinney equation

$$\ddot{\tau} + w^2(t)\tau - k\tau^{-3} = 0, \tag{5.5}$$

where k is an arbitrary constant of integration.

The reduction of the Ermakov system (5.1) is now undertaken in terms of canonical variables x, y and z given by

$$\begin{aligned} \tau^2\frac{\partial x}{\partial t} + \tau\dot{\tau}u\frac{\partial x}{\partial u} + \tau\dot{\tau}v\frac{\partial x}{\partial v} &= 0, \\ \tau^2\frac{\partial y}{\partial t} + \tau\dot{\tau}u\frac{\partial y}{\partial u} + \tau\dot{\tau}v\frac{\partial y}{\partial v} &= 0, \\ \tau^2\frac{\partial z}{\partial t} + \tau\dot{\tau}u\frac{\partial z}{\partial u} + \tau\dot{\tau}v\frac{\partial z}{\partial v} &= 1. \end{aligned} \tag{5.6}$$

Thus, x and y are group invariants and z is the variable of translation. Here, we proceed with the solution set

$$\begin{gathered} x = v^2/u^2, \quad y = \tau/v, \\ z = \int \tau^{-2}\, dt \end{gathered} \tag{5.7}$$

with τ governed by (5.5) with $k = 0$. The invariant x is taken as the new independent variable and y and z as the new dependent variables.

The Ermakov system (5.1), in terms of the canonical variables, remarkably reduces to the linear equation [24]

$$xh\frac{d^2y}{dx^2} + \frac{d}{dx}(xh)\frac{dy}{dx} + \frac{1}{x^{\frac{1}{2}}h}g(x^{-\frac{1}{2}})y = 0, \tag{5.8}$$

together with the first integral

$$\frac{dz}{dx} = \frac{1}{xy^2h}, \tag{5.9}$$

where

$$h = 2x^{\frac{1}{2}}[2I + \int^x \sigma^{-\frac{3}{2}}g(\sigma^{-\frac{1}{2}})\, d\sigma - \int^x \sigma^{-\frac{1}{2}}f(\sigma^{\frac{1}{2}})\, d\sigma]^{\frac{1}{2}}. \tag{5.10}$$

The relation (5.9) together with (5.10) provides the Lewis-Ray-Reid invariant

$$I = \frac{(u\dot{v} - v\dot{u})^2}{2} + \int^{u/v} g(\lambda)\, d\lambda + \int^{v/u} f(\mu)\, d\mu. \tag{5.11}$$

of the original Ermakov system (5.2).

Nonlinear superposition principles for u and v in terms of linearly independent solutions $X_1(t), X_2(t)$ of the linear oscillator equation

$$\ddot{X} + w^2(t)X = 0 \tag{5.12}$$

may now be constructed [24]. Thus, if $X_1(t), X_2(t)$ have unit Wronskian then the canonical variable z can be taken as

$$z = X_1(t)/X_2(t) \tag{5.13}$$

where τ is identified with $X_2(t)$. Let y_1, y_2 be linearly independent solutions of (5.8) for specified f, g so that

$$y = y(x; I, c_1, c_2) = c_1y_1 + c_2y_2. \tag{5.14}$$

On substitution of (5.14) into (5.9) and integration there results, in principle,

$$z = z(x; I, c_1, c_2, c_3) \tag{5.15}$$

where c_3 is an arbitrary constant of integration. Under appropriate conditions as discussed in [24], the relation (5.15) may be inverted to yield $x = x(X_1/X_2, I, c_1, c_2, c_3)$. The relations (5.7) $_{1,2}$ now provide the nonlinear superposition principles

$$\begin{gathered} u = X_2[x^{\frac{1}{2}}(c_1y_1(x, I) + c_2y_2(x, I))], \\ v = X_2/(c_1y_1(x, I) + c_2y_2(x, I)). \end{gathered} \tag{5.16}$$

The subject of explicit algebraic nonlinear superposition principles of the type (5.16) has been recently addressed by Athorne [26].

6. A 2+1-dimensional nonlinear shallow water system. Symmetry reduction to a Ermakov-system.

The linearization via reciprocal-type transformations of the 1+1-dimensional nonlinear shallow water equations descriptive of motion on a beach of constant inclination is well-known (Carrier and Greenspan [52]). The result is important, in particular, because it facilitates the moving shoreline analysis and the calculation of run-up or run-down [32],[53]. In 2 + 1-dimensional shallow water theory no such linearization is known. However, Lie group analysis may be applied to reveal important structure inherent in the system. Indeed, this has been conducted in [54],[55] for the motion of rotating shallow water in a rigid container. In particular, Ball-type theorems [56],[57] were retrieved therein which describe the temporal evolution of certain physically important quantities, Moreover, in [58], group analysis was used to construct an analytic solution of the 2 + 1-dimensional rotating shallow water system wherein the moving shoreline in a paraboloidal basin is a circle with radius oscillating in accordance with Pinney's equation.

Here, the Lie group analysis of a 2 + 1-dimensional two-layer shallow water system is conducted with a view to revealing a Ermakov-type system as a symmetry reduction.

The following two-layer nonlinear shallow water system is considered (Stoker [38]):

$$\begin{gathered}
u_t^{(0)} + u^{(0)}u_x^{(0)} + v^{(0)}u_y^{(0)} - fv^{(0)} + h_x^{(0)} + \frac{\varrho'}{\rho}(h'^{(0)} - h^{(0)})_x = 0,\\
v_t^{(0)} + u^{(0)}v_x^{(0)} + v^{(0)}v_y^{(0)} + fu^{(0)} + h_y^{(0)} + \frac{\varrho'}{\rho}(h'^{(0)} - h^{(0)})_y = 0,\\
h_t^{(0)} + (u^{(0)}h^{(0)})_x + (v^{(0)}h^{(0)})_y = 0,\\
u_t'^{(0)} + u'^{(0)}u_x'^{(0)} + v'^{(0)}u_y'^{(0)} - fv'^{(0)} + h_x'^{(0)} = 0,\\
v_t'^{(0)} + u'^{(0)}v_x'^{(0)} + v'^{(0)}v_y'^{(0)} + fu'^{(0)} + h_y'^{(0)} = 0,\\
(h'^{(0)} - h^{(0)})_t + (u'^{(0)}(h'^{(0)} - h^{(0)}))_x + (v'^{(0)}(h'^{(0)} - h^{(0)}))_y = 0.
\end{gathered} \tag{6.1}$$

In the above, the quantities $\{u^{(0)}, v^{(0)}, p^{(0)}, h^{(0)}\}$, $\{u'^{(0)}, v'^{(0)}, p'^{(0)}, h'^{(0)}\}$ represent, in turn, the leading order terms in shallow water approximations to the lower and upper layer horizontal velocity components, pressures and layer elevations with respect to the impervious boundary $z = 0$. The derivation of the 2 + 1-dimensional nonlinear system (6.1) is analogous to that conducted for a single layer shallow water system by Keller [59]. If the Coriolis parameter f is non-zero, then the change of independent variables $t \to ft$, $x \to fx, y \to fy$ allow us to set $f = 1$ in the system (6.1). In addition, the superscript (0) is dropped in the sequel.

The Lie group of transformations is sought with associated operator

$$\Gamma = T\partial_t + X\partial_x + Y\partial_y + U\partial_u + V\partial_v + H\partial_h + U'\partial_{u'} + V'\partial_{v'} + H'\partial_{h'} \tag{6.2}$$

which leaves invariant the nonlinear system (6.1). The generators $T, X, Y, \ldots H'$ are all taken as functions of the variables $t, x, y, u, v, h, u', v', h'$. The invariance requirement produces, with the aid of MACSYMA, a linear system of determining equations for $T, X, Y, \ldots, H'$. These can be duly solved in generality to yield

$$
\begin{aligned}
T &= \alpha(t), \\
X &= (\dot{\alpha}/2 + a_0)x + (\alpha/2 - a_1)y + \beta(t), \\
Y &= (\dot{\alpha}/2 + a_0)y - (\alpha/2 - a_1)x + \gamma(t), \\
U &= (-\dot{\alpha}/2 + a_0)u + (\alpha/2 - a_1)v + \ddot{\alpha}x/2 + \dot{\alpha}y/2 + \dot{\beta}, \\
V &= (-\dot{\alpha}/2 + a_0)v - (\alpha/2 - a_1)u + \ddot{\alpha}y/2 - \dot{\alpha}x/2 + \dot{\gamma}, \\
H &= (-\dot{\alpha} + 2a_0)h, \\
U' &= (-\dot{\alpha}/2 + a_0)u' + (\alpha/2 - a_1)v' + \ddot{\alpha}x/2 + \dot{\alpha}y/2 + \dot{\beta}, \\
V' &= (-\dot{\alpha}/2 + a_0)v' - (\alpha/2 - a_1)u' + \ddot{\alpha}y/2 - \dot{\alpha}x/2 + \dot{\gamma}, \\
H' &= (-\dot{\alpha} + 2a_0)h',
\end{aligned} \tag{6.3}
$$

where a_0, a_1 are arbitrary constants and α, β, γ are functions of t which satisfy the constraints

$$\dddot{\alpha} + \dot{\alpha} = 0, \quad \ddot{\beta} - \dot{\gamma} = 0, \quad \ddot{\gamma} + \dot{\beta} = 0. \tag{6.4}$$

The latter system is readily solved to give

$$\alpha = a_2 \cos t + a_3 \sin t + a_4, \quad \beta = a_5 \cos(t + a_8) + a_6$$

$$\gamma = -a_5 \sin(t + a_8) + a_7$$

where the a_i are arbitrary constants.

Symmetry reduction of the nonlinear system (6.1) may now be sought via the admitted group. In this connection, it is necessary to calculate the invariants of the group as given by the characteristic equations

$$dt/T = dx/X = dy/Y = \ldots = dh'/H'. \tag{6.5}$$

Here, attention is restricted to the case in which all the parameters a_i except a_0 are set zero; without loss of generality, we take $a_0 = 1$. As a result, we obtain the similarity representations

$$
\begin{aligned}
u &= y\tilde{u}(\xi, \eta), & u' &= y\tilde{u}'(\xi, \eta), \\
v &= y\tilde{v}(\xi, \eta), & v' &= y\tilde{v}'(\xi, \eta), \\
h &= y^2\tilde{h}(\xi, \eta), & h' &= y^2\tilde{h}'(\xi, \eta),
\end{aligned} \tag{6.6}
$$

with similarity variables ξ, η given by

$$\xi = t, \quad \eta = x/y. \tag{6.7}$$

Insertion of the expressions (6.6) into the original system (6.1) produces the 1 + 1-dimensional system

$$
\begin{gathered}
\tilde{u}_\xi + \tilde{u}\tilde{u}_\eta + \tilde{u}\tilde{v} - (\eta\tilde{u}_\eta + 1)\tilde{v} + \tfrac{\varrho'}{\rho}\tilde{h}'_\eta + (1 - \tfrac{\varrho'}{\rho})\tilde{h}_\eta = 0, \\
\tilde{v}_\xi + \tilde{u}\tilde{v}_\eta + \tilde{v}^2 - \eta\tilde{v}\tilde{v}_\eta + \tilde{u} + \tfrac{\varrho'}{\rho}(2\tilde{h}' - \eta\tilde{h}'_\eta) + (1 - \tfrac{\varrho'}{\rho})(2\tilde{h} - \eta\tilde{h}_\eta) = 0, \\
\tilde{h}_\xi + (\tilde{u}_\eta + 3\tilde{v} - \eta\tilde{v}_\eta)\tilde{h} + (\tilde{u} - \eta\tilde{v})\tilde{h}_\eta = 0, \\
\tilde{u}'_\xi + \tilde{u}'\tilde{u}_\eta + \tilde{u}'\tilde{v}' - (\eta\tilde{u}'_\eta + 1)\tilde{v}' + \tilde{h}'_\eta = 0, \\
\tilde{v}'_\xi + \tilde{u}'\tilde{v}'_\eta + \tilde{v}'^2 - \eta\tilde{v}'\tilde{v}'_\eta + \tilde{u}' + 2\tilde{h}' - \eta\tilde{h}'_\eta = 0, \\
(\tilde{h}' - h)_\xi + (\tilde{u}'_\eta + 3\tilde{v}' - \eta\tilde{v}'_\eta)(\tilde{h}' - \tilde{h}) + (\tilde{u}' - \eta\tilde{v}')(\tilde{h}' - \tilde{h})_\eta = 0.
\end{gathered} \tag{6.8}
$$

The above set of equations is not tractable as it stands. Accordingly, Lie group methods are applied once again in order to obtain a second symmetry reduction. It turns out that the system (6.8) is invariant under the Lie group whose infinitesimal form is

$$\begin{aligned}
\xi^* &= \xi + \epsilon\delta(\xi) + 0(\epsilon^2),\\
\eta^* &= \eta + \epsilon[\eta^2(\delta/2 + b_0) + \delta/2 + b_0] + 0(\epsilon^2),\\
\tilde{u}^* &= \tilde{u} + \epsilon[\tilde{u}\{\eta(\delta/2 + b_0) - \dot{\delta}\} + \tilde{v}(\delta/2 + b_0) + \eta\ddot{\delta}/2 - \dddot{\delta}/2] + 0(\epsilon^2),\\
\tilde{v}^* &= \tilde{v} + \epsilon[\tilde{v}\{\eta(\delta/2 + b_0) - \dot{\delta}\} - \tilde{u}(\delta/2 + b_0) + \ddot{\delta}/2 - \eta\dot{\delta}/2] + 0(\epsilon^2),\\
\tilde{h}^* &= \tilde{h} + \epsilon[\tilde{h}\{2\eta(\delta/2 + b_0) - 2\dot{\delta}\}] + 0(\epsilon^2),\\
\tilde{u}'^* &= \tilde{u}' + \epsilon[\tilde{u}'\{\eta(\delta/2 + b_0) - \dot{\delta}\} + \tilde{v}'(\delta/2 + b_0) + \eta\ddot{\delta}/2 - \dddot{\delta}/2] + 0(\epsilon^2),\\
\tilde{v}'^* &= \tilde{v}' + \epsilon[\tilde{v}'\{\eta(\delta/2 + b_0) - \dot{\delta}\} - \tilde{u}'(\delta/2 + b_0) + \ddot{\delta}/2 - \eta\dot{\delta}/2] + 0(\epsilon^2),\\
\tilde{h}'^* &= \tilde{h}' + \epsilon[\tilde{h}'(2\eta(\delta/2 + b_0) - 2\dot{\delta}] + 0(\epsilon^2),
\end{aligned} \tag{6.9}$$

where

$$\dddot{\delta} + \dot{\delta} = 0, \tag{6.10}$$

so that

$$\delta = b_1 \cos t + b_2 \sin t + b_3. \tag{6.11}$$

In the above, $b_i, i = 0, 1, 2, 3$ are arbitrary constants.

If attention is restricted to the subgroup determined by $b_0 = 1,\ \delta(t) = 0$ then we obtain the similarity forms

$$\begin{aligned}
\tilde{u} &= \eta\tilde{U}(\sigma) + \tilde{V}(\sigma), & \tilde{u}' &= \eta\tilde{U}'(\sigma) + \tilde{V}'(\sigma),\\
\tilde{v} &= -\eta\tilde{V}(\sigma) + \tilde{U}(\sigma), & \tilde{v}' &= -\eta\tilde{V}'(\sigma) + \tilde{U}'(\sigma),\\
\tilde{h} &= (\eta^2 + 1)\tilde{H}(\sigma), & \tilde{h}' &= (\eta^2 + 1)\tilde{H}'(\sigma),
\end{aligned} \tag{6.12}$$

where $\sigma = \xi = t$. Insertion of the representations (6.12) into (6.8) produces the reduced system

$$\begin{aligned}
&\dot{\tilde{V}} + 2\tilde{U}\tilde{V} - \tilde{U} = 0,\\
&\dot{\tilde{U}} - \tilde{V}^2 + \tilde{U}^2 + \tilde{V} + \tfrac{2\rho'}{\rho}\tilde{H}' + 2(1 - \tfrac{\rho'}{\rho})\tilde{H} = 0,\\
&\dot{\tilde{H}} + 4\tilde{H}\tilde{U} = 0,\\
&\dot{\tilde{V}}' + 2\tilde{U}'\tilde{V}' - \tilde{U}' = 0,\\
&\dot{\tilde{U}}' - \tilde{V}'^2 + \tilde{U}'^2 + \tilde{V}' + 2\tilde{H}' = 0,\\
&(\dot{\tilde{H}}' - \dot{\tilde{H}}) + 4(\tilde{H}' - \tilde{H})\tilde{U}' = 0.
\end{aligned} \tag{6.13}$$

If we now set

$$\Omega = (-\tilde{V} + \frac{1}{2})^{-\frac{1}{2}}, \tag{6.14}$$

$$\Omega' = (-\tilde{V}' + \frac{1}{2})^{-\frac{1}{2}} \tag{6.15}$$

then the solution of the nonlinear system (6.13) admits a parametrization in terms of Ω and Ω', namely

$$\begin{aligned} \tilde{U} &= 2\dot{\Omega}/\Omega, & \tilde{\Omega}' &= 2\dot{\Omega}'/\Omega, \\ \tilde{V} &= \tfrac{1}{2} - \Omega^{-2}, & \tilde{V}' &= \tfrac{1}{2} - \Omega'^{-2}, \\ \tilde{H} &= c_1\Omega^{-4}, & \tilde{H}' &= c_2\Omega'^{-4} \end{aligned} \tag{6.16}$$

where, remarkably, Ω and Ω' are governed by a Ermakov system originally derived in, [23], namely

$$\begin{aligned} \ddot{\Omega} + (\tfrac{1}{4} + \tfrac{\alpha}{\Omega'^4})\Omega &= \tfrac{\beta}{\Omega^3}, \\ \ddot{\Omega}' + (\tfrac{1}{4} + \tfrac{\gamma}{\Omega^4})\Omega' &= \tfrac{\delta}{\Omega'^3}, \end{aligned} \tag{6.17}$$

where

$$\begin{aligned} \alpha &= \tfrac{2\rho'}{\rho}c_2, & \beta &= 1 - c_1 \\ \gamma &= 1 - \beta, & \delta &= 1 - c_2 \end{aligned} \tag{6.18}$$

and c_1, c_2 are arbitrary constants of integration. Ermakov system of the type

$$\begin{aligned} \ddot{\Omega} + w^2(t)\Omega &= \tfrac{\beta}{\Omega^3} - \alpha\tfrac{\Omega}{\Omega'^4}, \\ \ddot{\Omega}' + w^2(t)\Omega' &= \tfrac{\delta}{\Omega'^3} - \gamma\tfrac{\Omega'}{\Omega^4}, \end{aligned} \tag{6.19}$$

have since been extensively investigated by Athorne [29]. In particular, it has been recently noted in [60] that application of the Douglas criterion [61] shows that the system is Hamiltonian if $\alpha = 3\delta, \gamma = 3\beta$.

The application of the preceding Lie group analysis to derive Ball-type theorems and physically important solutions remains. In this regard, it is noted that the application of Ball-type theorems in elliptic warm-core eddy theory is of current interest (Kirwan and Liu [62]). This has led to the recent discovery of pulsrodon solutions and underlying integrable Hamiltonian structure. (Rogers [63], Holm [64], [65]).

References

1. Ermakov, V.P. 1880, Univ. Isv. Kiev **20**, 1.
2. Lewis, H.R. Jr. 1967, Phys. Rev. Lett. **18**, 510.
3. Lewis, H.R. Jr. 1968, J. Math. Phys. **9**, 1976.
4. Lewis, J.R. Jr. and Riesenfeld, W.B. 1969, J. Math. Phys. **10**, 1458.
5. Khandekur, D.C. and Lawande, S.V. 1975, J. Math. Phys. **16**, 384.
6. Ray, J.R. and Reid, J.L. 1979, Phys. Lett. **A71**, 317.
7. Korsch, H.J. 1979, Phys. Lett. **A74**, 294.
8. Ray, J.R. 1980, Phys. Lett. **A78**, 4.
9. Wollenberg, L.S. 1980, Phys. Lett. **A79**, 269.
10. Lutsky, M. 1980, Phys, Lett. **A79**, 301.
11. Lutsky, M. 1980, J. Math. Phys. **21**, 1370.
12. Ray, J.R. and Reid, J.L. 1981, J. Math. Phys. **22**, 91.
13. Ray, J.R., Reid, J.L. and Lutsky, M. 1981, Phys. Lett. **A84**, 42.
14. Ray, J.R. 1981, Prog. Theor. Phys. **65**, 877.
15. Sarlet, W. and Ray, J.R. 1981, J. Math. Phys. **22**, 2504.
16. Sarlet, W. 1981, Phys. Lett. **A82**, 161.
17. Ray, J.L. 1981, Lett. Nuovo Cimento. **30**, 372.

18. Reid, J.L. and Ray, J.R. 1982, J. Phys. **A15**, 2751.
19. Saermark, K. 1982, Phys. Lett. **A90**, 5.
20. Reid, J.L. and Ray, J.R 1982, J. Math. Phys. **23**, 503.
21. Reid, J.L. and Ray, J.R. 1983, J. Math. Phys. **24**, 2433.
22. Goedert, J. 1989, Phys. Lett. **A136** 391.
23. Rogers, C. 1989, Research Report, Department of Mathematical Sciences, Loughborough University of Technology, U.K.
24. Athorne, C., Rogers, C, Ramgulam U. and Osbaldestin, A. 1990, Phys. Lett. **A143**, 207.
25. Athorne, C. 1990, J. Phys. A. Math. Gen **23** L137.
26. Athorne, C. 1990, Phys. Lett. **A151** 407.
27. Athorne, C. 1991, J. Phys. A. Math. Gen. **24** 945.
28. Leach, P.G.L. 1991, Phys, Lett. **A158** 102.
29. Athorne, C. J. Diff. Eqns. to be published.
30. Rogers, C. Hoenselaers, C. and Ray, J.R. 1992, J. Phys. A submitted.
31. Shahinpoor, M. and Nowinski, J.L. (1971), Int. J. Nonlin. Mech. **6**, 193.
32. Rogers, C. and Ames, W.F. (1989) Nonlinear Boundary Value Problems in Science and Engineering, Academic Press, New York.
33. Lee, R.A. (1984), J.Phys. **A17**, 535.
34. Cervero, J.M. and Lejareta, J.D. (1990) Quantum. Opt. **2** 333.
35. Goncharenko, A.M. and Kukushkin, V.G. (1990) Vetsi Akad. BSSR. Ser. Fiz. **6**, 32.
36. Cervero, J.M. and Lejareta, J.D. (1991), Phys. Lett. **A156**, 201.
37. Goncharenko, A.M., Logvin, Yu. A., Samson, A.M., Shapovatov, P.S. and Turovets, S.I. (1991) Phys. Lett. **A160**, 138.
38. Stoker, J.J. (1957), Water Waves Interscience Publ. New York.
39. Gibbon, J.D., Freeman, N.C. and Davey, A. (1978), J. Phys. A **11**, L93.
40. Gibbon, J.D., Freeman, N.C. and Johnson, R.S. (1978) Phys. Lett. **A65**, 380.
41. Lou, S. and Guang, G 1989 J. Math. Phys. **30**, 1614.
42. Lou, S. and Chen, W. 1991, Phys. Lett. **A156**, 260.
43. Leibbrandt, G. 1978 Phys. Rev. Lett. **41**, 435.
44. Rogers, C. and Shadwick, W.F. 1982, Bäcklund Transformations and Their Applications, Academic Press, New York.
45. Konopelchenko, B.G. and Rogers, C. 1991, Phys. Lett. **A158**, 397.
46. Konopelchenko, B.G. and Rogers, C. 1993, J. Math. Phys. to be published.
47. Nimmo, J.J.C. 1992, Phys. Lett. **A168**, 113.
48. Konopelchenko, B.G. and Dubrovsky, V.G. 1992, Budker Inst. Nuclear, Phys, Novosibirsk Preprint 92-1.
49. Konopelchenko, B.G., Schief, W. and Rogers, C. 1992 Phys. Lett. A to be published.
50. Schief, W. 1992, J. Phys. A: Math Gen **25**, L1351.
51. Ramgulam, U. 1991, Ph.D. Thesis Loughborough University of Technology, U.K.
52. Carrier, G.F. and Greenspan, H.P. 1958, J. Fluid. Mech. **1** 97.
53. Liu, P.F., Synolakis, C.E. and Yeh, H.H. 1991, J. Fluid. Mech. **229** 675.
54. Rogers, C. 1989, Proceedings IMACS Meeting, Paris, 69.
55. Levi, D., Nucci, M.C., Rogers, C. and Winternitz, P. 1989, J. Phys. A: Math. Gen. **22** 4743.
56. Ball, F.K. 1963, J. Fluid. Mech. **17**, 240.
57. Ball, F.K. 1963, Proceedings of the first Australasian Conference on Hydraulics and Fluid Mechanics 293 Pergamon Press, London.
58. Rogers, C and Ramgulam, U 1989 Int. J. Nonlin, Mech. **24**, 229.
59. Keller, J.B. 1948, Comm. Appl. Math. **1** 323.
60. Athorne, C. 1991, Phys. Lett. **A159** 375.
61. Douglas, J. 1941 Trans. Am. Math. Soc. **50** 71.
62. Kirwan, A.D. Jr. and Liu, J. 1991, Proc. Int. School of Physics "Enrico Fermi" Varenna, Italy, 99.
63. Rogers, C. 1989, Phys. Lett. **A138**, 267.
64. Holm, D 1991, Proc. Int. School of Physics "Enrico Fermi" Varenna, Italy 175.
65. Holm, D. 1991, J. Fluid Mech. **227**, 393.

SYMMETRIES OF SECOND-ORDER DIFFERENTIAL EQUATIONS AND DECOUPLING

W. SARLET
Instituut voor Theoretische Mechanica, Universiteit Gent
Krijgslaan 281, B-9000 Gent, Belgium

and

E. MARTÍNEZ
Departamento de Matemática Aplicada, Universidad de Zaragoza
Avda. María de Luna 3, E-50015 Zaragoza, Spain

Abstract. Necessary and sufficient conditions are discussed, which characterize complete separability of second-order ordinary differential equations via symmetry properties of the system.

1. Introduction

When considering systems of second-order ordinary differential equations, it very rarely happens that special coordinates can be found in which the system will decouple into completely separate equations. Intuitively, one expects that the existence of such coordinates may be related to symmetry properties of the given system. Clearly, however, separability cannot be explained by a standard group action, originating from Lie point symmetries of the system. As a matter of fact, one could easily compose an example by mixing up a number of separate equations which do not have point symmetries. Also, the mere availability of a sufficient number of more general symmetries will not be enough. To illustrate this point, consider for example the well-known Hénon-Heiles equations

$$\begin{aligned}\ddot{q}_1 &= b\,{q_1}^2 - a\,{q_2}^2 \\ \ddot{q}_2 &= -2a\,q_1 q_2\end{aligned}$$

and let us agree that, for simplicity, all considerations of this paper will be restricted to a strictly time-independent framework. This entails, in relation to standard works on symmetries and differential equations such as [8, 2, 9], that generators of symmetries will always appear in what is sometimes called their 'evolutionary form'. In this sense, insisting that a and b should not be zero, the above system has no point symmetries. As a next step in an algorithmic process, one can look for generators whose leading coefficients depend linearly on the first derivatives. Then, there always is the trivial symmetry $X_1 = \dot{q}_1\,(\partial/\partial q_1) + \dot{q}_2\,(\partial/\partial q_2)$ (equivalent to time translation invariance). A second generator is detected only for two special parameter combinations: for $b = -a$ we have $X_2 = \dot{q}_2\,(\partial/\partial q_1) + \dot{q}_1\,(\partial/\partial q_2)$, whereas for $b = -6a$ one finds $X_2 = q_2\dot{q}_2\,(\partial/\partial q_1) + (q_2\dot{q}_1 - 2q_1\dot{q}_2)\,(\partial/\partial q_2)$. The system turns out to decouple only in case 1.

N. H. Ibragimov et al. (eds.),
Modern Group Analysis: Advanced Analytical and Computational Methods in Mathematical Physics, 329–335.

Our objective is to prove necessary and sufficient criteria about symmetries, which will characterize complete separability of a second-order system. Two recent contributions which are relevant for this matter will briefly be reviewed in the next two sections.

2. Calculus on TM and submersive systems

The standard differential geometric framework for studying second-order equations is a tangent bundle TM, which has a natural type (1,1) tensor field S (called the 'vertical endomorphism') (see *e.g.* [1]). In local coordinates (q, v), we have $S = dq^i \otimes (\partial/\partial v^i)$, whereas a general second-order system $\ddot{q}^i = f^i(q, \dot{q})$ is represented by the vector field

$$\Gamma = v^i \frac{\partial}{\partial q^i} + f^i(q, v) \frac{\partial}{\partial v^i}.$$

Important in the calculus on TM is the tensor field $\mathcal{L}_\Gamma S$, which through its eigenspaces defines a splitting of the tangent space at each point of TM, *i.e.* determines a (non-linear) *connection* (see also [3]).

Kossowski and Thompson [4] recently studied so-called *submersive systems*. Submersiveness roughly refers to the existence of coordinates in which part of the system separates from the rest. Their main theorem states that submersive systems are characterized by the existence of an integrable distribution E on TM, which is *S-regular* and *invariant under $\mathcal{L}_\Gamma S$ and $\mathcal{L}_\Gamma^2 S$*. Here, S-regularity means that E is obtained from *complete* and *vertical* lifts of a distribution on M.

The authors also investigate the influence of symmetries and find as sufficient condition for submersiveness the existence of an algebra $\mathcal{G}$ of Lie point symmetries of Γ, such that for each $X \in \mathcal{G}$, the vector field $[X^V, \Gamma] - X^c$ on TM is tangent to $\mathcal{G}^V$. A further result which will give us a clue for an even more restrictive situation states: if the system is Lagrangian and $X \in \mathcal{X}(M)$ determines a Noether (point) symmetry, then there is a corresponding submersion iff $[X^V, \Gamma] - X^c = \mu X^V$ for some $\mu \in C^\infty(TM)$.

It may be worthwhile to recall the following coordinate expressions: for any $X \in \mathcal{X}(M)$ of the form $X = X^i(q)(\partial/\partial q^i)$, we have $X^V = X^i(q)(\partial/\partial v^i) \in \mathcal{X}(TM)$ and the complete lift (or prolongation) is given by

$$X^c = X^i(q) \frac{\partial}{\partial q^i} + v^j \frac{\partial X^i}{\partial q^j} \frac{\partial}{\partial v^i}.$$

Observe further that the coordinate expression of a non-point symmetry of Γ bears some resemblance to the latter:

$$J_\Gamma X = X^i(q, v) \frac{\partial}{\partial q^i} + \Gamma(X^i) \frac{\partial}{\partial v^i}.$$

At present, it suffices to regard the notation $J_\Gamma X$ as a way of indicating that such a symmetry is entirely determined by the object $X = X^i(q, v)(\partial/\partial q^i)$. The correct geometrical interpretation of this object is that X is a *vector field along the tangent bundle projection* $\tau : TM \to M$, for which we use the notation: $X \in \mathcal{X}(\tau)$.

3. Calculus along τ and separability

There are many more concepts of interest on TM, whose essential part is a corresponding object along τ. In [5] and [6], we have extensively studied the theory and classification of derivations of scalar and vector-valued forms along τ. An essential ingredient for this theory is a connection on $\tau : TM \to M$. With such a connection, one can horizontally lift vector fields along τ. In coordinates, for $X \in \mathcal{X}(\tau)$:

$$X^H = X^i H_i \in \mathcal{X}(TM), \quad H_i = \frac{\partial}{\partial q^i} - \Gamma_i^j \frac{\partial}{\partial v^j},$$

where the functions $\Gamma_i^j(q, v)$ are the connection coefficients. The vertical lift construction extends to $\mathcal{X}(\tau)$ and every $Z \in \mathcal{X}(TM)$ has a unique decomposition $Z = X^H + Y^V$, with $X, Y \in \mathcal{X}(\tau)$. Important derivations in our theory are the *horizontal* and *vertical covariant derivative*. They appear, for example, in the formula

$$[X^H, Y^V] = (\mathrm{D}_X^H Y)^V - (\mathrm{D}_Y^V X)^H,$$

and their action extends to forms along τ by duality.

As indicated before, each given Γ comes with its own connection, with coefficients $\Gamma_j^i = -\frac{1}{2}(\partial f^i / \partial v^j)$. Specifically for this case, two operations are most important: the *dynamical covariant derivative* ∇ and the *Jacobi endomorphism* Φ. They can be defined *e.g.* by the decomposition $\mathcal{L}_\Gamma X^H = (\nabla X)^H + (\Phi(X))^V$ and are determined in coordinates by

$$\nabla F = \Gamma(F), \text{ for functions } F \in C^\infty(TM)$$

$$\nabla(\partial/\partial q^j) = \Gamma_j^i(\partial/\partial q^i), \quad \nabla(dq^k) = -\Gamma_\ell^k \, dq^\ell$$

$$\Phi = \left(-\frac{\partial f^i}{\partial q^j} - \Gamma_k^i \Gamma_j^k - \Gamma(\Gamma_j^i)\right) dq^j \otimes \frac{\partial}{\partial q^i}.$$

To see the importance of these operations, observe *e.g.* that the determining equations of a (generalized) symmetry $X = X^i(q, v)\partial/\partial q^i \in \mathcal{X}(\tau)$ of Γ read $\nabla\nabla X + \Phi(X) = 0$. Also, the *curvature* of the connection

$$R = \frac{1}{2} R_{jk}^i \, dq^j \wedge dq^k \otimes \frac{\partial}{\partial q^i}, \quad R_{jk}^i = H_k(\Gamma_j^i) - H_j(\Gamma_k^i),$$

is obtained from Φ via a kind of vertical exterior derivative: $d^V\Phi = 3R$. For a link with the preceding section, note that Φ is related to the tensor field $\mathcal{L}_\Gamma^2 S$ on TM.

An interesting decomposition of ∇ is given by

$$\nabla = \mathrm{D}_{\mathbf{T}}^H + \mathrm{D}_{\nabla\mathbf{T}}^V + \mu_{\mathbf{t}}.$$

Here $\mathbf{T} = v^i \partial/\partial q^i$ is the canonical vector field along τ, μ refers to some well defined derivation of algebraic type, and

$$\mathbf{t} = \left(\Gamma_j^i - \frac{\partial \Gamma_j^i}{\partial v^k} v^k\right) dq^j \otimes \frac{\partial}{\partial q^i}$$

is the so-called *tension* of the connection.

We have used this calculus for a comprehensive study of complete separability of second-order equations [7]. In order to formulate the main theorem of that paper, we need one more definition: for each (1,1) tensor field U along τ,

$$C_U^V(X,Y) = [\mathrm{D}_X^V U, U](Y) \qquad \forall X,Y \in \mathcal{X}(\tau).$$

When C_U^V is zero for a diagonalizable tensor U, it means that all eigendistributions can be spanned by vector fields on M.

Theorem. *There exist coordinates in which a given* Γ *separates into n decoupled single second-order equations iff*

Φ *is diagonalizable,* $\quad R = 0$

$[\nabla\Phi, \Phi] = 0, \quad C_\Phi^V = 0$

$\mathbf{t}$ *is diagonalizable,* $\quad C_{\mathbf{t}}^V = 0.$

It is important to observe that all conditions of this theorem are algebraic and can (with computer algebra assistance) in principle be checked.

4. Symmetry properties characterizing full decoupling

Let us first try to understand the meaning of the conditions of Kossowski and Thompson in terms of the calculus along τ. Observe that for $X \in \mathcal{X}(M)$, we have $X^c = X^H + (\nabla X)^V$, $\mathcal{L}_\Gamma S(X^c) = -X^H + (\nabla X)^V$ and $\mathcal{L}_\Gamma^2 S(X^H) = -2(\Phi(X))^V$. It follows that the content of their main theorem essentially says that we have an integrable distribution on M which, regarded as distribution along τ is invariant under ∇ and Φ.

Still for basic X, we have the identity $[X^V, \Gamma] - X^c = -2(\nabla X)^V$, so that the condition for the Noether symmetry in [4] expresses that ∇X should be proportional to X.

As explained before, point symmetries cannot be sufficient for our purposes and a natural generalization of an S-regular distribution on TM consists of a distribution spanned by $\{X^V, J_\Gamma X = X^H + (\nabla X)^V\}$, for some distribution span$\{X\}$ along τ. Next, since we want an extreme case of 'submersiveness', the natural extension of the condition in the Noether case seems appropriate and a strong form of integrability will have to be imposed, in which *e.g.* all symmetries commute. It will now be clear that the formulation of the theorem below is directly inspired by the results in [4]. The proof, on the other hand, will be a matter of showing that all conditions of our separability theorem are verified and will thus heavily rely on the techniques developed in [7].

Theorem. *A* Γ *with n degrees of freedom is completely separable iff there exist n independent vector fields* X_i *along* τ*, such that:*

1. $\nabla X_i = \mu_i X_i$ *for some* $\mu_i \in C^\infty(TM)$
2. $J_\Gamma X_i$ *is a symmetry of* Γ *for all* i
3. $[J_\Gamma X_i, J_\Gamma X_j] = 0$ *and* $[J_\Gamma X_i, X_j^V] = 0, \quad i \neq j$
4. $S([J_\Gamma X_i, X_i^V])$ *is proportional to* X_i^V.

PROOF. From 1 and 2, it follows that $\Phi(X_i) = \lambda_i X_i$, with $\lambda_i = -\nabla\mu_i - {\mu_i}^2$. Hence Φ is diagonalizable and condition 1 further implies that the eigendistributions are

∇-invariant, so that $[\nabla\Phi, \Phi] = 0$. To proceed further, we need the following relations for general $X, Y \in \mathcal{X}(\tau)$:

$$\begin{aligned}
[J_\Gamma X, Y^V] &= -(\mathrm{D}^V_Y X)^H + (\mathrm{D}^H_X Y + \mathrm{D}^V_{\nabla X} Y - \mathrm{D}^V_Y \nabla X)^V \\
[J_\Gamma X, J_\Gamma Y] &= (\mathrm{D}^H_X Y - \mathrm{D}^H_Y X + \mathrm{D}^V_{\nabla X} Y - \mathrm{D}^V_{\nabla Y} X)^H \\
&\quad + (R(X,Y) + \mathrm{D}^H_X \nabla Y - \mathrm{D}^H_Y \nabla X + \mathrm{D}^V_{\nabla X} \nabla Y - \mathrm{D}^V_{\nabla Y} \nabla X)^V.
\end{aligned}$$

They are easy to obtain from general bracket relations in [7]. It follows from the first of these and condition 3 that $\mathrm{D}^V_{X_j} X_i = 0$ $(j \neq i)$ and

$$\mathrm{D}^H_{X_i} X_j - (\mathrm{D}^V_{X_j} \mu_i) X_i = 0 \quad (*).$$

It will be clear by now that the summation convention is being used only where appropriate.

The second relation and condition 3 next imply two further properties: the horizontal part, using $(*)$, gives rise to $\mathrm{D}^V_{X_j} \mu_i = 0$ and $\mathrm{D}^H_{X_j} X_i = 0$; the vertical part subsequently results in

$$R(X_i, X_j) + (\mathrm{D}^H_{X_i} \mu_j) X_j - (\mathrm{D}^H_{X_j} \mu_i) X_i = 0 \quad (**).$$

We next wish to exploit the relationship between the Jacobi endomorphism and the curvature. First of all, we have

$$d^V \Phi(X_i, X_j) = (\mathrm{D}^V_{X_i} \Phi)(X_j) - (\mathrm{D}^V_{X_j} \Phi)(X_i) = (\mathrm{D}^V_{X_i} \lambda_j) X_j - (\mathrm{D}^V_{X_j} \lambda_i) X_i\,.$$

Now $\mathrm{D}^V_{X_j} \lambda_i = -\mathrm{D}^H_{X_j} \mu_i$. This follows from the relation between the λ_i and μ_i, using the property $\mathrm{D}^V_{X_j} \mu_i = 0$ and the general commutator relation

$$[\mathrm{D}^V_X, \nabla] = \mathrm{D}^H_X - \mathrm{D}^V_{\nabla X} \quad (***)$$

established in [7]. We thus obtain:

$$3\, R(X_i, X_j) = d^V \Phi(X_i, X_j) = -(\mathrm{D}^H_{X_i} \mu_j) X_j + (\mathrm{D}^H_{X_j} \mu_i) X_i = R(X_i, X_j),$$

where the last equality follows from $(**)$. Since the X_i form a local basis of $\mathcal{X}(\tau)$, we conclude that $R = 0$.

We finally turn to condition 4. It implies, from the horizontal part of $[J_\Gamma X_i, X_i^V]$ that $\mathrm{D}^V_{X_i} X_i$ is proportional to X_i, which together with $\mathrm{D}^V_{X_j} X_i = 0$ $(i \neq j)$ means that each eigendistribution is basic, *i.e.* $C^V_\Phi = 0$. The commutator relation $(***)$ subsequently shows that also $\mathrm{D}^H_{X_i} X_i$ is proportional to X_i. Using the decomposition of the operator ∇ and knowing that $\mathbf{t}(X) = \mu_{\mathbf{t}}(X)$, one fairly easily obtains that the tension is also diagonal with respect to the basis X_i and since we already know that this is a basic distribution, it follows that $C^V_{\mathbf{t}} = 0$. All conditions of the separability theorem thus turn out to be satisfied.

The converse is trivial: if the system is separable, it suffices to think of each of the X_i as coming from a (generalized) symmetry of each separate equation. This completes the proof. □

5. Concluding remarks

Our goal was to characterize separability by symmetry properties of the equations. It may look a bit unsatisfactory that the conditions of the preceding theorem are so strong, particularly since, in the separating coordinates, the symmetries under consideration will be symmetries of each of the individual equations. One can formulate conditions which look much milder, but given the fact that differential equations locally always have (generalized) symmetries, translating the milder conditions to those symmetries will make the stronger conditions hold true anyway. To be specific, a milder looking theorem, which contains the minimal requirements for separability, says the following.

Theorem. *For complete separability of a second-order system, it is necessary and sufficient that n independent $X_i \in \mathcal{X}(\tau)$ exist, such that:*

1. $\nabla X_i = \mu_i X_i$
2. $\Phi(X_i) = \lambda_i X_i$
3. $\mathrm{D}^V_Z X_i = \sigma_i(Z)\, X_i, \quad \forall Z \in \mathcal{X}(\tau)$, with $\sigma_i \in C^\infty(TM)$.

For a sketch of the proof, observe first that conditions 1 and 3, using (***), imply that also $\mathrm{D}^H_Z X_i$ is proportional to X_i for all $Z \in \mathcal{X}(\tau)$. As a result (see the argumentation in [7]), the basic vector fields which span the eigendistributions of Φ constitute a number of distributions on M which are simultaneously integrable. Therefore, there exist new coordinates with respect to which each of the X_i is in fact a multiple of $\partial/\partial q^i$. Condition 1 then implies $\Gamma^i_j = 0$ for $i \neq j$ and condition 2 subsequently tells us that also $\partial f^i/\partial q^j = 0$ for $i \neq j$. The complete separation of Γ follows. □

With conditions 1 and 2 of this result, $J_\Gamma X_i$ need not necessarily be a symmetry, but we can always pass to one via multiplication by a suitable function. Having done this, condition 3 again looks milder, because it does not impose that these symmetries must in the end be the ones corresponding to separate equations. One can easily show, however, that they can at most differ from those by a factor which is then necessarily a first integral. Hence, if multiplicative first integrals are factored out, the stronger integrability conditions of the former theorem will hold true.

Concerning examples where one happens to know a sufficient number of independent symmetries, a strategy for checking separability could start by looking for a suitable set of linear combinations which will make the ∇X_i proportional to X_i. Applying this idea to the Hénon-Heiles example, it turns out that for $b = -a$, $X_1 + X_2$ and $X_1 - X_2$ are the only such combinations. The subsequent conditions appear to be satisfied. For the case $b = -6a$ on the other hand, no combination can be obtained which verifies the first requirement.

Recall, finally, that the Hénon-Heiles system, for $b = -6a$, is known to be separable in the sense of Hamilton-Jacobi. For future developments, it is therefore hoped that our approach may provide new insight, at the level of symmetry properties of the equations, concerning the difference between equations which completely decouple, are separable in the sense of Hamilton-Jacobi or are completely integrable.

Acknowledgements

W.S. is indebted to the Belgian National Fund for Scientific Research for a grant.

References

1. M. Crampin, Tangent bundle geometry for Lagrangian dynamics, *J. Phys. A: Math. Gen.* **16** (1983) 3755–3772.
2. G.W. Bluman and S. Kumei, *Symmetries and differential equations*, Appl. Math. Sciences **81** (Springer-Verlag, New York, 1989).
3. J. Grifone, Structure presque tangente et connexions I, *Ann. Inst. Fourier* **22** No 1 (1972) 287–334.
4. M. Kossowski and G. Thompson, Submersive second order ordinary differential equations, *Math. Proc. Camb. Phil. Soc.* **110** (1991) 207–224.
5. E. Martínez, J.F. Cariñena and W. Sarlet, Derivations of differential forms along the tangent bundle projection, *Diff. Geometry and its Applications* **2** (1992) 17–43.
6. E. Martínez, J.F. Cariñena and W. Sarlet, Derivations of differential forms along the tangent bundle projection. Part II, *Diff. Geometry and its Applications* **2** (1993), to appear.
7. E. Martínez, J.F. Cariñena and W. Sarlet, Geometric characterization of separable second-order differential equations, *Math. Proc. Camb. Phil. Soc.* (1992), to appear.
8. P.J. Olver, *Applications of Lie groups to differential equations* (Springer-Verlag, New York, 1985).
9. H. Stephani, *Differential equations: their solution using symmetries* (Cambridge University Press, Cambridge, 1989).

ALGORITHMIC METHODS FOR LIE PSEUDOGROUPS

JOACHIM SCHÜ, WERNER M. SEILER* and JACQUES CALMET
Institut für Algorithmen und Kognitive Systeme
Universität Karlsruhe, W-7500 Karlsruhe 1, Germany
Email: kg04@dkauni2.bitnet

Abstract. We present an algorithm to complete any given system of differential equations to an involutive system as needed e.g. for concrete applications of Lie pseudogroups. It is based on jet bundle formalism and formal theory. An implementation in the computer algebra system AXIOM is described.

1. Introduction

A central tool to study non-linear partial differential equations is provided by symmetry analysis [1, 8]. The usual procedure starts with setting up the determining equations for the infinitesimal symmetry generators (usually with one of the many programs dveloped for this purpose). Then this overdetermined system of linear partial differential equations has to be solved. To perform symmetry reductions, differential invariants must be constructed. This again requires to solve linear partial differential equations.

The ultimate goal of our work is to avoid solving differential equations during symmetry analysis. That this may be possible was recently indicated by Reid [12], who demonstrated that one can obtain the structure of the symmetry algebra without solving determining equations. For symmetry reductions this is, however, only of limited use, because there explicit expressions for the generators are needed. We hope that it will be possible to overcome some of these problems using Lie pseudogroups. They can be informally described as groups of transformations given as solutions of a system of partial differential equations [9]. These equations are, however, never explicitely solved!

As Pommaret [9] pointed out, it is very important for the application of Lie pseudogroups that both the system to be analysed and the system defining the group are involutive. Thus as a first step we need an algorithm to complete any given system of partial differential equations to an equivalent involutive one. In this paper, we will present such an algorithm and its implementation in the computer algebra system AXIOM.

The paper is organized as follows: The next section shortly introduces the notion of an involutive system and outlines the algorithm. Section 3 deals with some of the occuring computational problems, while Section 4 describes the implementation. After two examples in Section 5, finally some conclusions are given.

* Present address: Centre de recherches mathématiques, Université de Montréal, C.P. 6128 Succ. A, Montréal H3C 3J7, Canada

N. H. Ibragimov et al. (eds.),
Modern Group Analysis: Advanced Analytical and Computational Methods in Mathematical Physics, 337–344.

2. Involutive Systems

There exist several notions of involution in different frameworks. We will use a geometric approach to differential equations and define them as fibred submanifolds in a jet bundle. This formalism is the one usually used in symmetry theory. Other approaches involve e.g. differential forms (Cartan-Kähler theory [2]) or differential algebra (Riquier-Janet theory [5]). We can not give here an introduction to the underlying theory but must refer for all details to the literature [9].

A differential equation $\mathcal{R}_q$ is given in local coordinates by a set of equations $\Phi^\tau(x^i, u^\alpha, p^\alpha_\mu) = 0$ where the derivatives p^α_μ are of order less or equal q. Its prolongation $\mathcal{R}_{q+1}$ is obtained by formally differentiating the equations with respect to the independent variables x^i. It is well-known that during prolongation integrability conditions can arise, i.e. the projected system $\mathcal{R}^{(1)}_q$ has a smaller dimension than $\mathcal{R}_q$. This can happen at any prolongation order. A system without integrability conditions is called *formally integrable*, because it is possible to construct order by order a formal power series solution.

A formally integrable system is, however, not yet necessarily *involutive*. We request further that it has an involutive *symbol*. The symbol $\mathcal{M}_q$ of $\mathcal{R}_q$ is a system of linear (algebraic, not differential!) equations in some unknowns v^α_μ defined by:

$$\mathcal{M}_q : \sum_{\alpha,|\mu|=q} \frac{\partial \Phi^\tau}{\partial p^\alpha_\mu} v^\alpha_\mu = 0 . \tag{1}$$

Thus, for a quasi-linear equation the symbol is essentially the highest order part of the equation.

The intrinsic definition of an involutive symbol makes use of the Spencer cohomology [3, 9]. It is, however, possible to construct an algorithmic criterion to decide involution directly from the equations (1) using multiplicative variables as known from Riquier-Janet theory: $\mathcal{M}_q$ is involutive, if

$$\#(\text{multiplicative variables of } \mathcal{M}_q) = \operatorname{rank} \mathcal{M}_{q+1} . \tag{2}$$

The Cartan-Kuranishi theorem states that every differential equation $\mathcal{R}_q$ becomes involutive after a finite number of prolongations and projections (i.e. addition of integrability conditions). In other words: there exist two integers r, s such that $\mathcal{R}^{(s)}_{q+r}$ is an involutive system. The proof of this theorem yields an algorithm to construct r, s and thus the involutive system. It consists of two nested loops. The inner one prolongs till the symbol becomes involutive; the outer one adds the integrability conditions.

Two fundamental theorems form the basis of this algorithm: (*i*) Every symbol becomes involutive after a finite number of prolongations; (*ii*) $\mathcal{R}_q$ is involutive, iff $\mathcal{M}_q$ is involutive and $\mathcal{R}^{(1)}_q = \mathcal{R}_q$. Latter condition can be easily checked comparing the dimensions of the two submanifolds. The dimension of the projected submanifold is given by:

$$\dim \mathcal{R}^{(1)}_q = \dim \mathcal{R}_{q+1} - \dim \mathcal{M}_{q+1} . \tag{3}$$

Equations (2) and (3) lead to the simple algorithm depicted in Figure 1. The loop in [4.1] prolongs, until the symbol is involutive; [4.2] checks for integrability

[1]	r $\leftarrow$ 0; s $\leftarrow$ 0	
[2]	compute $\mathcal{R}_{q+1}$	{*prolong*}
[3]	compute $\mathcal{M}_q$,$\mathcal{M}_{q+1}$	{*extract symbols*}
[4]	**until** $\mathcal{R}^{(s)}_{q+r}$ involutive **repeat**	
[4.1]	**while** $\#multVar(\mathcal{M}^{(s)}_{q+r}) < \operatorname{rank} \mathcal{M}^{(s)}_{q+r+1}$ **repeat**	
[4.1.1]	r $\leftarrow$ r+1	{*counter for prolongations*}
[4.1.2]	compute $\mathcal{R}^{(s)}_{q+r+1}$	{*prolong*}
[4.1.3]	compute $\mathcal{M}^{(s)}_{q+r+1}$	{*extract symbol*}
[4.2]	**if** $\dim \mathcal{R}^{(s)}_{q+r+1} - \dim \mathcal{M}^{(s)}_{q+r+1} < \dim \mathcal{R}^{(s)}_{q+r}$ **then**	
[4.2.1]	s $\leftarrow$ s+1	{*counter for projections*}
[4.2.2]	compute $\mathcal{R}^{(s)}_{q+r}$	{*add integrability conditions*}
[4.2.3]	compute $\mathcal{R}^{(s)}_{q+r+1}$	{*prolong*}
[4.2.4]	compute $\mathcal{M}^{(s)}_{q+r}$,$\mathcal{M}^{(s)}_{q+r+1}$	{*extract symbols*}
[5]	**return** $\mathcal{R}^{(s)}_{q+r}$	

Fig. 1. Completion algorithm from the Cartan-Kuranishi theorem.

conditions. Their construction in step [4.2.2] is a purely linear operation. Equation (3) shows that they can only occur, if $\mathcal{M}_{q+1}$ has not maximal rank. The same linear combinations that yield zero rows in the symbol applied to the full system generate the integrability conditions, because a zero row in the symbol denotes a cancelling of the leading terms of the full equation.

3. Computational Problems

Our algorithm contains mostly simple operations: taking derivatives and computing ranks of matrices. The only non-trivial operation is the determination of the dimensions of the submanifolds $\mathcal{R}_{q+r}$. But already the ranks make trouble. The symbol $\mathcal{M}_{q+r}$ is a matrix whose entries can again be viewed as differential equations. Hence its rank may change depending on whether some additional equations are satisfied or not. In general, a whole tree of such case distinctions may exist, which must all be treated separately.

Similar problems can arise during prolongations. We have assumed throughout this paper, that we start with a regular system $\mathcal{R}_q$, i.e. all prolongations are again fibred submanifolds. Usually, however, only a suitable chosen restriction of $\mathcal{R}_q$ satisfies this condition. Thus we should check after each prolongation, whether the prolonged equations still define a variety of constant dimension.

This brings us back to the problem of determining the dimension of a submanifold defined by a set of equations. There exists no general algorithm for this purpose. One can design, however, many different algorithms for special classes of equations. The first approach would be to compute the rank of the Jacobi matrix. This rank has, however, to be evaluated *on* the submanifold. In general, this is only possible, if the equations are in solved form (e.g. quasi-linear equations).

For algebraic equations one could use Gröbner bases. The huge dimension of higher order jet bundles makes this not very promising. For homogeneous equations, i.e. equations which vanish, if the dependent variables and all derivatives are set equal to zero, a much simpler approach exists. There, $\mathcal{R}_{q+r}$ always contains the origin. The dimension of the submanifold can be computed as the dimension of the tangent space at the origin, i.e. through the rank of a linear system with *numerical* entries.

We omit here for lack of space the question of δ-regularity of the coordinate system [9]. In our algorithm it affects the criterion (2) for an involutive symbol, but it seems to be a more general problem (see e.g. [4]). It can be handled algorithmically, but only at the prize of a considerably higher complexity of the calculation.

A practical problem comes from the combinatorical explosion of the size of the occuring matrices. If several prolongations are necessary, the computations become quite lengthy. Consequent simplification of the equations can partially counter this effect. To what extent we can simplify depends on the class of equations treated. Linear and quasi-linear systems offer of course the most possibilities.

One of the most expensive operations in the algorithm is the prolongation of the equations. In principle it is only necessary for the determination of the dimension. The symbol can be prolonged independently using pure index manipulations, as one can easily see from its definition (1). Only when integrability conditions arise, we need explicitely the concerned equations. An alternative method to compute the dimension would thus lead to a considerable optimization.

The complexity of the algorithm is unknown. There exists only a bound for the number of prolongations necessary to get an involutive symbol [9]. It yields, however, usually much too high values (10^3 – 10^6 even for small examples!).

4. Implementation

We implemented the algorithm in the computer algebra system AXIOM [6]. Its highly modular and object-oriented design allows to imitate very closely the mathematical structure in the program. We have implemented general data structures and procedures for computations within the geometric theory of differential equations. This allows for an easy extension of the program for many other applications.

Figure 2 shows the present structure of the program. A more detailed description can be found in [14]. Italics denote some planned extensions; the final goal will be the implementation of the category and the domains in the dashed box.

C,D,P abbreviate the AXIOM structures category, domain and package: A *package* contains just a collection of procedures. A *category* defines an abstract datatype, i.e. it defines what operations are allowed in this type. A *domain* is a concrete implementation of such a category. In our programm e.g. **JetBundleFunctionCategory** contains basically only the signatures of operations like formal differentiation. The actual implementation of these procedures depends usually on the representation of the concrete chosen class of functions (linear, quasi-linear etc.). Thus it is contained in the corresponding domain.

We see that the domain **DifferentialEquation** and consequently the package **CartanKuranishi** are based only on the *category* **JetBundleFunctionCategory**. They are independent of the concrete implementation of any specific class (domain)

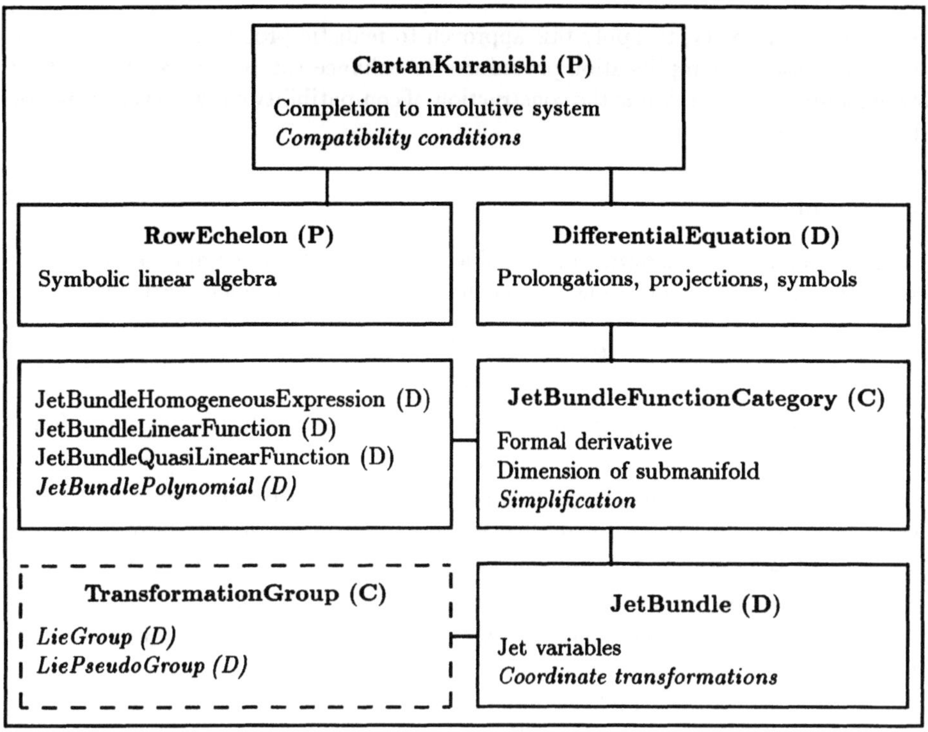

Fig. 2. Overview of the structure of the program.

of differential equations and can be left unchanged, if one adds new domains or changes old ones within `JetBundleFunctionCategory`. As they use of course many operations which are implemented in these domains, their performance can change considerably when applied to different ones.

So far we have implemented three domains within `JetBundleFunctionCategory`. Homogeneous, linear and quasi-linear functions. The main reason is the before mentioned problem of computing the dimension of submanifolds. For the homogeneous functions we use the above outlined tangent space method, as it appears to be the most efficient one. Linear and quasi-linear systems provide of course no problems.

The procedures in the package `RowEchelon` compute all necessary information to detect case distinctions. They use a fairly primitive approach, namely applying Gaußian elimination and storing the pivots. Sit [17] pointed recently out, that this leads often to many redundant distinctions and described an alternative algorithm based on methods from algebraic geometry which might be more effective in detecting the degenerate cases. Presently, the completion procedure in `CartanKuranishi` ignores them anyway restricting itself to the generic case.

Due to its general ansatz, many extensions of the program are possible. Besides the application of an infinitesimal generator to the system, nearly all ingredients for a symmetry package are already there. Then the completion procedure could be used to analyse determining equations, e.g. to calculate the size of the symmetry group (see below). However, to apply this approach to realistic problems will request the implementation of simplification procedures to enhance the performance. Another straight-forward extension is the construction of compatibility conditions at least for linear systems.

5. Examples

First, we look at a standard example due to Janet [10]. It comprises two linear equations of second order in one dependent and three independent variables:

$$\mathcal{R}_2 : \begin{cases} u_{zz} + y\, u_{xx} &= 0\,, \\ \quad u_{yy} &= 0\,. \end{cases} \tag{4}$$

Applying our algorithm we notice, that $\mathcal{M}_2$ is not involutive but $\mathcal{M}_3$. Since $\dim \mathcal{R}_3^{(1)} = 11 < \dim \mathcal{R}_3 = 12$, we find an integrability condition: $u_{xxy} = 0$. Because $\mathcal{M}_3^{(1)}$ is not involutive, we must prolong once to get the involutive symbol $\mathcal{M}_4^{(1)}$. But again an integrability condition occurs, namely $u_{xxxx} = 0$. Thus we have arrived at $\mathcal{R}_4^{(2)}$, who has, however, a non-involutive symbol. $\mathcal{M}_5^{(2)}$ vanishes and is therefore trivially involutive. As $\dim \mathcal{R}_5^{(3)} = 12 = \dim \mathcal{R}_5^{(2)}$, we are finished.

Because of the vanishing symbol, $\mathcal{R}_5^{(2)}$ is a finite type system and has a 12-dimensional solution space. This can also be seen by direct integration, which is easily performed in this simple example. $\mathcal{R}_4^{(2)}$ provides one of the few known examples of a formally integrable system that is not involutive.

To demonstrate some further possibilities of formal analysis, we consider the following linear system that arises in the theory of the KP equation [13].

$$\mathcal{R}_2 : \begin{cases} u_{xy} - u_t &= 0\,, \\ u_{xx} - u_y &= 0\,. \end{cases} \tag{5}$$

This second-order system is not yet involutive. We have to add the integrability condition $u_{yy} - u_{xt} = 0$ to get the involutive system $\mathcal{R}_2^{(1)}$.

A closer investigation of the symbol yields further information on the solution space [16]. We get the Cartan characters directly from its row echelon form. For (5) we find $\alpha_1 = 3, \alpha_2 = \alpha_3 = 0$. They allow to determine the arbitrariness of the general solution. Here it can be expressed by three functions each of one variable.

In our implementation, all these information are computed automatically at the end of the completion process. Figure 3 contains the input and the important part of the output of an AXIOM session analysing system (5). As AXIOM is a strongly typed system, it is sometimes necessary to declare types. Thus we introduce at the beginning abbreviations for some domains and packages. We use the ordering $x > y > t$. The **`setOutput`** command controls the amount of output generated.

6. Conclusions

We have presented an algorithm to complete a given system of partial differential equations to an involutive one within the framework of formal theory. There exist already several other, similiar algorithms using different approaches.

```
jbl := JetBundleLinearFunction('x,'u,'p,3,1)
lde := DifferentialEquation('x,'u,'p,3,1,jbl)
ck := CartanKuranishi('x,'u,'p,3,1,jbl)
eq1:jbl := P([0,0,2]) - P([0,1,0])
eq2:jbl := P([0,1,1]) - P([1,0,0])
sys:lde := generateSystem [eq1,eq2]
setOutput 12
complete sys
```

System $R_2^{(1)}$ involutive!

System without prolonged equations. Dimension: 7

$$\begin{aligned} -p_{[0,1,0]} + p_{[0,0,2]} &= 0, \\ -p_{[1,0,0]} + p_{[0,1,1]} &= 0, \\ p_{[1,0,1]} - p_{[0,2,0]} &= 0, \end{aligned}$$

Cartan characters: 3, 0, 0

Hilbert polynomial: 3

Number of arbitrary functions: 3, 0, 0

Fig. 3. In- and output for system (5).

Reid [11] implemented in Maple an algorithm to transform quasi-linear systems into a standard form based on Riquier-Janet theory. This algorithm yields in general only a formally integrable system. Another restriction is the necessity to have a quasi-linear systems in order to be able to solve for the leading derivatives. (The same is true for other implementations of Riquier-Janet theory [15, 18]). This allows on the other hand simplifications which improve the efficiency considerably and even lead to the definition of a standard form.

Mansfield [7] introduced differential Gröbner bases analogously to algebraic equations. The primary goal there is to obtain a normal form for bases of differential ideals. This requests to construct all integrability conditions of the system. They can be viewed as the differential analog to the S-polynomials in the Buchberger algorithm. Thus every differential Gröbner basis forms a formally integrable system. The converse does not hold in general, however, as the bases possess additional properties.

Hartley and Tucker [4] describe a REDUCE implementation of Cartan-Kähler theory, where they construct regular chains of involutive integral elements for exterior differential systems. This yields automatically the Cartan characters. Presently the program works only for involutive systems, as it does not yet contain a completion algorithm.

It is difficult to compare the different approaches, as for each one the class of equations for which it is applicable, the generated output and the complexity are different. For some applications, it might be more important to obtain a standard form; in others, one might prefer to get the Cartan characters. Our algorithm seems at present time to be the only one that is able to construct an equivalent involutive system for non-linear systems that are not in solved form. But it has e.g. compared with the algorithm of Reid a much higher complexity.

7. Acknowledgements

This work was partially supported by DFG. We thank J.-F. Pommaret for many helpful discussions. WMS is supported by Studienstiftung des deutschen Volkes and thanks P. Winternitz and the CRM for their hospitality.

References

1. G.W. Bluman, S. Kumei: *Symmetries and Differential Equations*, Springer, New York 1989.
2. R.L. Bryant, S.S. Chern, R.B. Gardner, H.L. Goldschmidt, P.A. Griffiths: *Exterior Differential Systems*, Springer, New York 1991.
3. H.L. Goldschmidt, J. Diff. Geom. 1 (1969) 269–307.
4. D. Hartley, R.W. Tucker, J. Symb. Comp. 12 (1991) 655–667.
5. M. Janet: *Leçons sur les Systèmes d'Équations aux Dérivées Partielles*, Gauthier-Villars, Paris 1929.
6. D. Jenks, R.S. Sutor: *AXIOM – The Scientific Computation System*, Springer, New York 1992.
7. E. Mansfield, E.D. Fackerell: *Differential Gröbner Bases*, Preprint 92-108, Macquarie University, Sydney 1992.
8. P.J. Olver: *Applications of Lie Groups to Differential Equations*, Springer, New York 1986.
9. J.F. Pommaret: *Systems of Partial Differential Equations and Lie Pseudogroups*, Gordon&Breach, London 1978.
10. J.F. Pommaret, A. Haddak: *Effective Methods for Algebraic Partial Differential Equations*, in: Proc. MEGA '90, T. Mora, C. Traverso (eds.), Birkhäuser, Boston 1991, pp. 411–426.
11. G.J. Reid, Eur. J. Appl. Math. 2 (1991) 293–318.
12. G.J. Reid, Eur. J. Appl. Math. 2 (1991) 319–340.
13. C. Rogers, W. Oevel, P.J. Vassiliou: *Invariance under Reciprocal Transformations in 1+1 and 2+1 Dimensions*, in: Proc. Int. Meeting on Nonlinear Diffusion Phenomena, Indian Institute of Science 1992.
14. J. Schü: *Implementierung des Cartan-Kuranishi-Theorems in AXIOM*, diploma thesis (in german), Karlsruhe 1992.
15. F. Schwarz: *Symmetries and Involution Systems: Some Experiments in Computer Algebra*, in: Proc. Oberwolfach Meeting on Nonlinear Evolution Equations, M. Ablowitz, B. Fuchssteiner, M. Kruskal (eds.), World Science Press, Singapore 1987.
16. W.M. Seiler: *On the Arbitrariness of the General Solution of an Involutive Partial Differential Equation*, Preprint CRM–1873, Montréal 1993.
17. W.Y. Sit, J. Symb. Comp. 13 (1992) 353–394.
18. V.L. Topunov, Acta Appl. Math. 16 (1989) 191–206.

A SPECIAL CLASS OF BÄCKLUND TRANSFORMATIONS FOR CERTAIN NONLINEAR PARTIAL DIFFERENTIAL EQUATIONS

C. SOPHOCLEOUS*
Department of Mathematics and Statistics University of Edinburgh
King's Buildings
Edinburgh EH9 3JZ
SCOTLAND

Abstract. The Bäcklund transformation of the special class $u = F(u', u'_x, u'_y)$ is considered for equations of the form $u_{xy} = f(u, u_x)$ and $u_y + u_{xxx} + f(u, u_x) = 0$. The nonlinear forms of $f(u, u_x)$ that admit such transformations are completely classified. Motivated by these results Bäcklund transformations of a general class are also obtained for the same equations.

Introduction

The Miura transformation [1] $u = u'_x - u'^2$ relates the Korteweg-de Vries (KdV) equation $u_y + u_{xxx} + 6uu_x = 0$ and the modified KdV equation $u'_y + u'_{xxx} - 6u'^2 u'_x = 0$. It is a relation between two nonlinear equations, neither of which can be solved in general. Nevertheless, this transformation is the key to prove the existence of an infinite number of conservation laws for the KdV equation [2]. It also provides the starting point for the inverse scattering method which provides exact solution for the same equation [2]. We generalize this well-known result by completely classifying the range of functions $f(u, u_x)$ and $f'(u', u'_x)$ for which the partial differential equations

$$u_y + u_{xxx} + f(u, u_x) = 0, \tag{1}$$

$$u'_y + u'_{xxx} + f'(u', u'_x) = 0 \tag{2}$$

are related by the Bäcklund transformation (BT) of the special class

$$u = F(u', u'_x, u'_y). \tag{3}$$

Miura's result is analogous to the Hopf-Cole transformation of Burgers' equation to the linear diffusion equation [3,4] and to the Nakamura transformation [5] $u = u' - \sin^{-1}(au'_x)$ which relates the sine-Gordon equation $u_{xy} = \sin u$ and the modified sine-Gordon equation $u'_{xy} = \sin u' \sqrt{1 - (au'_x)^2}$. A generalization of the Hopf-Cole transformation is presented by Sachdev [6]. We also generalize Nakamura's result by determining the complete range of functions $f(u, u_x)$ *and* $f'(u', u'_x)$ for which the equations

$$u_{xy} = f(u, u_x), \tag{4}$$

$$u'_{xy} = f'(u', u'_x) \tag{5}$$

* Present address: Department of Mathematics and Statistics, University of Cyprus, Nicosia, P.O. Box 537, CYPRUS

N. H. Ibragimov et al. (eds.),
Modern Group Analysis: Advanced Analytical and Computational Methods in Mathematical Physics, 345–352.

are related by transformations of the class (3). In the following analysis we will assume that at least one of the equations (1) and (2) and at least one of the equations (4) and (5) is nonlinear.

Using Nakamura's transformation, the BTs of the general class

$$u'_x = -\frac{1}{a}\sin(u - u'), \tag{6}$$

$$u'_y = u_y + a\sin u' \tag{7}$$

which relate the sine-Gordon and the modified sine-Gordon equations, are obtained. Therefore, motivated by the results which are obtained previously for equations of the form (1) and (4) we can derive BTs of the general class for the same equations.

The calculations involved in arriving at the conclusions of this work have been greatly facilitated by the computer algebraic manipulation package REDUCE [7].

1. Special Bäcklund Transformations for KdV type equations

We consider the BTs of the special class

$$u = F(u', u'_x, u'_y), \tag{8}$$

which links the two partial differential equations

$$u_y + u_{xxx} + f(u, u_x) = 0, \tag{9}$$

$$u'_y + u'_{xxx} + f'(u', u'_x) = 0. \tag{10}$$

In the subsequent analysis we determine the range of nonlinear equations of the form (9) and (10) that admit BTs of the class (8). In other words we derive the form of the functions $f(u, u_x), f'(u', u'_x)$ and $F(u', u'_x, u'_y)$.

Upon differentiating (8) with respect x and y respectively, we obtain

$$u_x = u'_x F_{u'} + u'_{xx} F_{u'_x} + u'_{xy} F_{u'_y}, \tag{11}$$

$$u_y = u'_y F_{u'} + u'_{xy} F_{u'_x} + u'_{yy} F_{u'_y}. \tag{12}$$

Remembering that $u'_{xxx} = -u'_y - f'$, from (10), and differentiating (11) twice with respect to x a similar expression for u_{xxx} is obtained. Upon substitution of this expression for u_{xxx} and (12) into equation (9) we obtain

$$u_y + u_{xxx} + f = E(u', u'_x, u'_y, u'_{xx}, u'_{xy}, u'_{yy}) = 0, \tag{13}$$

for some function E which may be calculated explicity in terms of F, its derivatives, f', its derivatives, $u', u'_x, u'_y, u'_{xx}, u'_{xy}$ and u'_{yy}. In view of (9) E must be identically zero with $u', u'_x, u'_y, u'_{xx}, u'_{xy}$ and u'_{yy} regarded as independent variables.

Without giving any computations, which can be found in ref. [8], we state that the functions F and f take the forms

$$F = \lambda u'_x + g(u'), \qquad f = L(u)u_x + M(u),$$

respectively, where g, L and M are functions of the indicated variables, to be determined and

$$\lambda L = \lambda f'_{u'_x} - 3u'_x g_{u'u'}, \tag{14}$$

$$u'_x g_{u'} L + M = f' g_{u'} - u'^3_x g_{u'u'u'} + \lambda u'_x f'_{u'}. \tag{15}$$

These two identities impose restrictions on the functions L, M, f' and g and enable the transformations to be derived.

The results partition naturally into two mutually exclusive cases: (*i*) $g_{u'u'} \neq 0$ and (*ii*) $g_{u'u'} = 0$.

(*i*) If $g_{u'u'} \neq 0$ equations (14) and (15) lead to the following theorem:

Theorem 1. The pair of partial differential equations

$$u_y + u_{xxx} + \frac{1}{\lambda^2}[\frac{3}{2}\mu(u - \mu\tau)^2 - 3\tau(1 - \mu^2)(u - \theta) - \frac{3}{2}(\phi^2 + \mu\theta^2)]u_x = 0,$$

$$u'_y + u'_{xxx} + \frac{1}{2}\mu u'^3_x - \frac{3}{2\lambda^2} g^2_{u'} u'_x = 0$$

are related by the special class of BT

$$u = \lambda u'_x + g(u'),$$

where g is a function of u' which satisfy the ordinary differential equation

$$\frac{d^2 g}{du'^2} + \mu g = \tau, \tag{16}$$

where $\lambda, \mu, \tau, \phi, \theta$ are all constants, $\mu = 0, -1, 1$ and ϕ, θ are constants of integration obtained by solving equation (16). These constants may be fixed without loss of generality.

We note that on setting $\mu = \phi = \theta = 0, \lambda = 1$ and $\tau = -2$, we obtain the Miura transformation, $u = u'_x - u'^2$, which links the KdV equation and the modified KdV equation. The Gardner's transformation [2] can also easily deduced from theorem 1. Also when $\mu = \pm 1$ we obtain the well known relationship to an equation that was first introduced by Calogero and Degasperis [9].

(*ii*) If $g = \nu u' + \tau$, where ν, τ are constants, we derive the following two trivial transformations:

(*a*) $u = \lambda u'_x + \nu u'$ relates the equations $u_y + u_{xxx} + \frac{1}{\nu}[\lambda M'(u) + \mu]u_x + M(u) = 0$ and $u'_y + u'_{xxx} + \frac{1}{\nu}[M(\lambda u'_x + \nu u') + \mu u'_x] = 0$, where M is an arbitrary function, M' is its derived function, μ is a constant and ν is non-zero.

(*b*) $u = \lambda u'_x + \tau$ relates the equations $u_y + u_{xxx} + J'(\frac{u-\tau}{\lambda})u_x + \mu u - \mu\tau = 0$ and $u'_y + u'_{xxx} + \mu u' + J(u'_x) = 0$, where J is an arbitrary function and J' is its derived. From the subcase (*b*) we deduce that on setting $\mu = \tau = 0$ and $J = -2\lambda^2 u'^3_x$ that the KdV equation and the equation $u'_y + u'_{xxx} - 2\lambda^2 u'^3_x = 0$ are related by the transformation $u = \lambda u'_x$.

2. Special BTs for equation of the type $u_{xy} = f(u, u_x)$

In this section we present some results of the work of Sophocleous and Kingston [10, 11], who classified the range of functions $f(u, u_x)$ and $f'(u', u'_x)$ for which the equations

$$u_{xy} = f(u, u_x), \tag{17}$$

$$u'_{xy} = f(u', u'_x) \tag{18}$$

admit BTs of the special class (8). Equations of this form appear in the study of pseudospherical surfaces. In what follows λ_1, λ_2, μ and τ are all constants.

(1)

$$u_{xy} = uu_x - \lambda_1\lambda_2 \quad \text{and} \quad u'_{xy} = \lambda_2 e^{-u'}u'_x + \lambda_1 e^{u'}$$

are related by the BT $u = u'_y + \lambda_2 e^{-u'}$.

(2)

$$u_{xy} = [u - g(h^{-1}(u_x))]h'(h^{-1}(u_x)) \quad \text{and} \quad u'_{xy} = -g'(u')u'_x + h(u')$$

are related by the BT $u = u'_y + g(u')$, where g and h are arbitrary functions and g', h' are the derived functions of g and h respectively.

(3)

$$u_{xy} = uu_x \quad \text{and} \quad u'_{xy} = e^{\lambda_1 u'}\sqrt{\tau^2 - (\lambda_1 u'_x)^2}$$

are related by the BT $u = J(u'_x)e^{\lambda_1 u'} + \lambda_1 u'_y$ where $J = \dfrac{\lambda_2^2 u'_x}{\tau + \sqrt{\tau^2 - (\lambda_2 u'_x)^2}}$.

(4)

$$u_{xy} = uu_x \quad \text{and} \quad u'_{xy} = \lambda_1 e^{u'}\sqrt{u'^2_x - \tau^2}$$

are related by the BT $u = u'_y + \lambda_1 e^{u'}$.

(5)

$$u_{xy} = \mu(\lambda_1 e^u + \lambda_2 e^{-u}) \quad \text{and} \quad u'_{xy} = (\lambda_1 e^{u'} + \lambda_2 e^{-u'})\sqrt{\mu^2 + u'^2_x}$$

are related by the BT $u = \sinh^{-1}\dfrac{u'_x}{\mu} + u'$.

(6)

$$u_{xy} = \Phi'(u)u_x + \lambda_1\Phi(u) - \lambda_2 u \quad \text{and} \quad u'_{xy} = \Phi(\lambda_1 u' + u'_x) - \lambda_2 u'$$

are related by the BT $u = u'_x + \lambda_1 u'$, where Φ is an arbitrary function and Φ' its derived function.

(7)

$$u_{xy} = u\sqrt{u_x^2 - 4\lambda_1\lambda_2\mu^2} \quad \text{and} \quad u'_{xy} = (\lambda_1 e^{u'} + \lambda_2 e^{-u'})\sqrt{u'^2_x + \mu^2}$$

are related by the BT $u = u'_y + \lambda_1 e^{u'} + \lambda_2 e^{-u'}$.

(8)

$$u_{xy} = Q'(Q^{-1}(u_x))[\lambda_1 u_x - \lambda_2 Q^{-1}(u_x) + u] \quad \text{and} \quad u'_{xy} = -\lambda_2 u'_x + Q(\lambda_1 u'_x + u')$$

are related by the BT $u = u'_y + \lambda_2 u'$, where Q is an arbitrary function and Q' is its derived.

In Ref. [12] it is shown that the modified Liouville equation $u_{xy} = e^u\sqrt{1-(au_x)^2}$, the modified sine-Gordon equation $u_{xy} = \sin u\sqrt{1-(au_x)^2}$ and the modified sinh-Gordon equation $u_{xy} = \sinh u\sqrt{1-(au_x)^2}$ admit auto-BTs of a general class. Here, in addition to other results, we have stated that the modified Liouville equation can also be related to equation $u_{xy} = uu_x$ by two different special BTs [Results (3) and (4)] and to the standard Liouville equation [Result (5)]. The latter result being a generalization of Nakamura's transformation [5]. Also we have presented two different special BTs for each of the modified sine-Gordon and the modified sinh-Gordon equations. One of these is the Nakamura transformation and the other one is given by result (7).

3. Bäcklund Transformations of a general class

We consider the KdV equation, the modified KdV equation and the BTs of the form

$$u_x = \Psi(u, u'), \qquad u_y = \Phi(u, u', u'_x, u'_{xx}). \tag{19}$$

Using the method of Clairin [13], it is shown [14] that the above partial differential equations are related by the BTs

$$u_x = \pm(u' + u^2), \tag{20}$$

$$u_y = \mp u'_{xx} - 2(uu'_x + u'u_x). \tag{21}$$

In other words, the forms of the functions Ψ and Φ are classified in order to derive BTs of the class (19) for the KdV and the modified KdV equations.

We note that (21) is the Miura transformation. Hence motivated by the latter transformation, BTs of a general class are also obtained for the same equations. In the same spirit we derive BTs of the form (19) for the equations which appear in the theorem 1. Hence, employment of the transformation $u_x = \frac{1}{\lambda}[u' - g(u)]$ (see theorem 1) and $u_y = \Phi(u, u', u'_x, u'_{xx})$, the method of Clairin lead to the theorem 2.

Theorem 2. The BTs

$$u_x = \frac{1}{\lambda}[u' - g(u)], \quad u_y = -\frac{1}{\lambda}u'_{xx} + \frac{1}{\lambda^2}u'_x g_u - \frac{1}{2}\mu u_x^3 + \frac{1}{2\lambda^2}u_x g_u^2 + \frac{1}{\lambda}u_x^2 g_{uu}$$

relate the partial differential equations

$$u_y + u_{xxx} + \frac{1}{2}\mu u_x^3 - \frac{3}{2\lambda^2}g_u^2 u_x = 0, \tag{22}$$

$$u'_y + u'_{xxx} + \frac{1}{\lambda^2}[\frac{3}{2}\mu(u' - \mu\tau)^2 - 3\tau(1-\mu^2)(u' - \theta) - \frac{3}{2}(\phi^2 + \mu\theta^2)]u'_x = 0, \tag{23}$$

where the functions $g(u)$ satisfy the differential equation $\frac{d^2g}{du^2}+\mu g = \tau$. The constants are as defined in theorem 1.

Choosing $\mu = \phi = \theta = 0, \lambda = 1$ and $\tau = -2$ in the above theorem gives the BTs (20) and (21). Auto-BTs for equations of the form (22) can be found in ref. [15, 16].
Similarly, using the transformations in the section 2, case (ii) we derive the following two results:

(a) The BTs

$$u_x = \frac{1}{\lambda}(u' - \nu u)$$

and

$$u_y = -\frac{1}{\lambda}u'_{xx} + \frac{\nu}{\lambda^2}u'_x - \frac{\nu^2}{\lambda^3}u' + \frac{\nu^3}{\lambda^3}u - \frac{1}{\nu}M(u') - \frac{\mu}{\lambda\nu}u' + \frac{\mu}{\lambda}u$$

relate the equations

$$u_y + u_{xxx} + \frac{1}{\nu}[M(\lambda u_x + \nu u) + \mu u_x] = 0$$

and

$$u'_y + u'_{xxx} + \frac{1}{\nu}[\lambda M'(u') + \mu]u'_x + M(u') = 0.$$

(b) The BTs

$$u_x = \frac{1}{\lambda}(u' - \tau) \quad \text{and} \quad u_y = -\frac{1}{\lambda}u'_{xx} - \mu u - J(\frac{u' - \tau}{\lambda})$$

relate the equations

$$u_y + u_{xxx} + \mu u + J(u_x) = 0$$

and

$$u'_y + u'_{xxx} + J'(\frac{u' - \tau}{\lambda})u'_x + \mu u' - \tau\mu = 0.$$

Also, employment of Clairin's method and the special BTs in results (4), (5), (6), (7), and (8) in section 3 lead to the following BTs of the general class:

(1) The BTs

$$u_x = \frac{1}{2}e^{-u}u'_x + \frac{1}{2}\frac{\tau^2 e^u}{u'_x} \quad \text{and} \quad u_y = u' - \lambda_1 e^u$$

relate the equations

$$u_{xy} = \lambda_1 e^u \sqrt{u_x'^2 - \tau^2} \quad \text{and} \quad u'_{xy} = u'u'_x.$$

(2) The BTs

$$u_x = \sinh\,[\mu(u' - u)] \quad \text{and} \quad u_y = u'_y - \lambda_1 e^u - \lambda_2 e^{-u}$$

relate the equations

$$u_{xy} = (\lambda_1 e^u + \lambda_2 e^{-u})\sqrt{\mu^2 + u_x^2} \quad \text{and} \quad u'_{xy} = \mu(\lambda_1 e^{u'} + \lambda_2 e^{-u'}).$$

(3) The BTs

$$u_x = u' - \lambda_1 u \quad \text{and} \quad u_y = \frac{1}{\lambda_1}[u'_y - \Phi(u') + \lambda_2 u]$$

relate the equations

$$u_{xy} = \Phi(\lambda_1 u + u_x) - \lambda_2 u \quad \text{and} \quad u'_{xy} = \Phi'(u')u'_x + \lambda_1 \Phi(u') - \lambda_2 u'.$$

(4) The BTs

$$u_x = -u'_x(\lambda_1 e^u - \lambda_2 e^{-u}) + (\lambda_1 e^u + \lambda_2 e^{-u})\sqrt{u_x'^2 + \mu^2}.$$

and

$$u_y = u' - \lambda_1 e^u - \lambda_2 e^{-u}$$

relate the equations

$$u_{xy} = (\lambda_1 e^u + \lambda_2 e^{-u})\sqrt{u_x^2 + \mu^2} \quad \text{and} \quad u'_{xy} = u'\sqrt{u_x'^2 - 4\lambda_1\lambda_2\mu^2}.$$

(5) The BTs

$$u_x = Q^{-1}(u'_x) - u - \lambda_2 \quad \text{and} \quad u_y = u' - \lambda_2 u$$

relate the equations

$$u_{xy} = -\lambda_2 u_x + Q(\lambda_1 u_x + u) \quad \text{and}$$

$$u'_{xy} = Q'(Q^{-1}(u'_x))[\lambda_1 u'_x - \lambda_2 Q^{-1}(u'_x) + u'].$$

For a number of partial differential equations, it is an impossible task to derive BTs of the general class even with use of today's computer algebra systems. A possible alternative is to attempt to obtain BTs of the special class (3). These transformations then, if they do not depend explicitly on both u'_x and u'_y, can usually be used to generate BTs of the general class. For example, it is not easy to classify all the functions $f(u, u_x)$ for equations of the form (4) which admit BTs of the general class.

Acknowledgement

The author would like to express his gratitude to Dr. John Kingston and Prof. David Parker for their support and encouragement. He is also grateful to SERC (UK) for financial support.

References

1. Miura R.M. 1968 J. Math. Phys. **9** 1202
2. Miura R.M. 1976 SIAM Review **18** 412
3. Hopf E. 1950 Commun. Pure Appl. Math. **3** 201
4. Cole J.D. 1951 Quart Appl. Math. **9** 225
5. Nakamura A. 1980 J. Phys. Soc. Jpn **49** 1167
6. Sachdev P. L. 1978 J. Appl. Math. (ZAMP) **29** 1963
7. Hearn A.C. 1987 REDUCE User's Manual, Version 3.3 (Santa Monica: Rand Corporation)
8. Sophocleous C. 1992 J. Phys. A:Math. Gen. **25** L89
9. Calogero F. and Degasperis A. 1981 J. Math. Phys. **22** 23
10. Sophocleous C. and Kingston J.G. 1991 J. Math. Phys. **32** 3176
11. Sophocleous C. 1991 PhD Thesis, Univ. of Nottingham
12. Beals R., Rabelo M. and Tenenblat K. 1989 Stud. Appl. Math **81** 125
13. Clairin M.J. 1903 Ann. Toulouse 2[e] Ser. **5** 437
14. Lamb G.L. 1974 J. Math. Phys. **15** 2157
15. Hlavatý L. 1985 Czech. J. Phys. B **35** 370
16. Hlavatý L. 1985 J. Phys. A: Math. Gen. **18** 1933

SYMMETRY GROUPS OF BALANCE EQUATIONS

ERDOĞAN S. ŞUHUBI
Department of Engineering Sciences, Faculty of Science,
Istanbul Technical University,
Maslak 80626, Istanbul, Turkey

Abstract. Explicit forms of the determining equations for the infinitesimal generators associated with the symmetry group of a general system of balance equations of finite arbitrary order involving n independent and N dependent variables are given by employing exterior calculus. The infinitesimal generators are identified as components of isovector fields of a closed ideal generated by certain exterior differential forms corresponding to a system of first order equations which are equivalent to the original balance equations. Two examples from nonlinear elastodynamics involving only two independent variables are considered to expose the efficiency of the approach and some similarity solutions corresponding to particular symmetry groups and satisfying ordinary differential equations are provided.

1. Introduction

The global balance equations on an n-dimensional manifold R are expressible as

$$\int_{\partial R} \Sigma^{\alpha i} n_i \, da + \int_R \bar{\Sigma}^{\alpha} \, dv = 0, \quad i = 1, \ldots, n, \quad \alpha = 1, \ldots, N$$

where $\Sigma^{\alpha i}$ represent fluxes whereas $\bar{\Sigma}^{\alpha}$ denote sources.They have to be prescribed by appropriate *constitutive relations.* In classical physics we have of course $n = 4$ and R is a tube along time axis with a three-dimensional cross-section. If we assume that the foregoing balance laws are valid in every subregions of R and the integrands are continuous we obtain the local balance equations as

$$\Sigma^{\alpha i}{}_{;i} + \bar{\Sigma}^{\alpha} = 0$$

where a semicolon denotes covariant differentiation in a curvilinear coordinate system. This set of equations can obviously be transformed to the form

$$\frac{\partial \Sigma^{\alpha i}}{\partial x^i} + \Sigma^{\alpha} = 0, \quad i = 1, \ldots, n, \quad \alpha = 1, \ldots, \tag{1.1}$$

We further assume that $\Sigma^{\alpha i}$ and Σ^{α} are given smooth functions of independent variables x^i, dependent variables u^{α} and partial derivatives $u^{\alpha}_{,i_1}, \ldots, u^{\alpha}_{,i_1 i_2 \cdots i_m}$ up to the m-th order. Hence the explicit form of the balance equations are

$$\sum_{r=0}^{m} \frac{\partial \Sigma^{\alpha i}}{\partial u^{\beta}_{,i_1 i_2 \cdots i_r}} u^{\beta}_{,i_1 i_2 \cdots i_r i} + \frac{\partial \Sigma^{\alpha i}}{\partial x^i} + \Sigma^{\alpha} = 0 \tag{1.2}$$

N. H. Ibragimov et al. (eds.),
Modern Group Analysis: Advanced Analytical and Computational Methods in Mathematical Physics, 353–365.

These are a quasilinear system of equations. Our aim is to obtain the determining equations for the infinitesimal generators of the symmetry group associated with the balance equations (1.2). Our approach will be based on the geometrical formulation of partial differential equations in terms of exterior differential forms due to Cartan [1]. Within the framework of exterior calculus, infinitesimal generators of symmetry groups associated with a given system of partial differential equations are identified as components of isovector fields of the ideal constructed by certain exterior forms generated by a system of first order partial differential equations which are equivalent to the original system. Isovector fields are defined in turn as such vector fields whose orbits induces transformations under which the ideal remains invariant. This scheme has been explored, in principle, in [2]. Detailed analysis of what we shall present here can be found in [3].

2. Isovector fields

We reduce Eqs. (1.2) to a first order system by introducing new variables

$$v^{\alpha}_{i_1\cdots i_r} = u^{\alpha}_{,i_1\cdots i_r}, \quad r = 0,1,\ldots,m \quad (v^{\alpha}_{i_0} = u^{\alpha}) \tag{2.1}$$

where $v^{\alpha}_{i_1\cdots i_r}$ are completely symmetric in lower indices. We thus have

$$\sum_{r=0}^{m} \frac{\partial \Sigma^{\alpha i}}{\partial v^{\beta}_{i_1 i_2\cdots i_r}} v^{\beta}_{i_1 i_2\cdots i_r,i} + \frac{\partial \Sigma^{\alpha i}}{\partial x^i} + \Sigma^{\alpha} = 0 \tag{2.2}$$

commas representing partial derivatives. Let us now define

$$\begin{aligned} M &= R^n, \text{ coordinate space with the coordinate cover } \{x^i\}, \\ G &= R^n \times R^N, \text{ graph space with the coordinate cover } \{x^i, u^{\alpha}\}, \\ K &= R^n \times R^N \times \cdots \times R^{n^m N}, \text{ kinematic space with the coordinate} \\ &\qquad \text{cover } \{x^i, u^{\alpha}, \ldots, v^{\alpha}_{i_1}, v^{\alpha}_{i_1\cdots i_m}\}. \end{aligned}$$

The dimension of the manifold K may be very large. In fact

$$\dim K = n + \frac{n^{m+1}-1}{n-1} N$$

However, it is obvious that we actually dwell on a submanifold of significantly lower dimension due to the symmetries imposed by (2.1).

Let $\Lambda(K)$ denote the graded exterior algebra over the manifold K. We define contact 1-forms by

$$\sigma^{\alpha}_{i_1\cdots i_r} = dv^{\alpha}_{i_1\cdots i_r} - v^{\alpha}_{i_1\cdots i_r i}\, dx^i \in \Lambda^1(K), \quad r = 0,1,\ldots,m-1 \tag{2.3}$$

and balance n-forms by

$$\omega^{\alpha} = d\Sigma^{\alpha i} \wedge \mu_i + \Sigma^{\alpha}\mu \in \Lambda^n(K) \tag{2.4}$$

where μ is the volume form in M, i.e.,

$$\mu = dx^1 \wedge dx^2 \wedge \cdots \wedge dx^n \in \Lambda^n(K) \subset \Lambda^n(K)$$

and denoting the interior product by $\rfloor$

$$\mu_i = \frac{\partial}{\partial x^i} \rfloor \mu \in \Lambda^{n-1}(M)$$

We have the identity [2] $dx^i \wedge \mu_j = \delta^i_j \mu$. Consider the map $\phi : M \to K$ such that $v^\alpha_{i_1\cdots i_r} = v^\alpha_{i_1\cdots i_r}(x^i)$, $r = 0, 1, \ldots, m$. ϕ induces a map $\phi^* : \Lambda(K) \to \Lambda(M)$ yielding

$$\phi^* \sigma^\alpha_{i_1\cdots i_r} = (v^\alpha_{i_1 i_2\cdots i_r, i} - v^\alpha_{i_1 i_2 \cdots i_r i}) dx^i, \quad r = 0, 1, \ldots, m-1$$

$$\phi^* \omega^\alpha = \frac{\partial \Sigma^{\alpha i}}{\partial x^j} dx^j \wedge \mu_i + \Sigma^\alpha \mu = \left(\frac{\partial \Sigma^{\alpha i}}{\partial x^i} + \Sigma^\alpha \right) \mu$$

Therefore the map ϕ annihilating the contact and balance forms corresponds to the solution of (1.2).

Let us consider now an ideal I of $\Lambda(K)$ generated by ω^α, $\sigma^\alpha_{i_1\cdots i_r}$ and $d\sigma^\alpha_{i_1\cdots i_r} = -dv^\alpha_{i_1\cdots i_r i} \wedge dx^i \in \Lambda^2(K)$. It can easily be shown that $I(\omega^\alpha, \boldsymbol{\sigma}^\alpha, d\boldsymbol{\sigma}^\alpha)$ is closed, i.e., $\forall \alpha \in I \Rightarrow d\alpha \in I$. A vector field V in the tangent space $T(K)$ is *an isovector field of* I if and only if $\forall \alpha \in I \Rightarrow \mathcal{L}_V \alpha \in I$ where $\mathcal{L}_V$ denotes the Lie derivative with respect to V. Hence the necessary and sufficient conditions for a vector field V given by

$$V = V^i \frac{\partial}{\partial x^i} + \sum_{r=0}^{m} V^\alpha_{i_1 i_2 \cdots i_r} \frac{\partial}{\partial v^\alpha_{i_1 i_2 \cdots i_r}}, \quad V^i, V^\alpha_{i_1 i_2 \cdots i_r} \in \Lambda^0(K)$$

to be an isovector field of I are [2]

$$\mathcal{L}_V \sigma^\alpha_{i_1\cdots i_k} = \sum_{r=0}^{m-1} \lambda^{\alpha j_1 \cdots j_r}_{\beta i_1 \cdots i_k} \sigma^\beta_{j_1 \cdots j_r}, \quad \lambda^{\alpha j_1 \cdots j_r}_{\beta i_1 \cdots i_k} \in \Lambda^0(K)$$

$$\mathcal{L}_V \omega^\alpha = \lambda^\alpha_\beta \omega^\beta + \sum_{r=0}^{m-1} \gamma^{\alpha i_1 \cdots i_r}_\beta \wedge \sigma^\beta_{i_1\cdots i_r} + \sum_{r=0}^{m-1} \Gamma^{\alpha i_1 \cdots i_r}_\beta \wedge d\sigma^\beta_{i_1 \cdots i_r} \tag{2.5}$$

$$\lambda^\alpha_\beta \in \Lambda^0(K), \quad \gamma^{\alpha i_1 \cdots i_r}_\beta \in \Lambda^{n-1}(K), \quad \Gamma^{\alpha i_1 \cdots i_r}_\beta \subset \Lambda^{n-2}(K)$$

Since $\mathcal{L}_V\, d\sigma^\alpha_{i_1\cdots i_k} = d\mathcal{L}_V \sigma^\alpha_{i_1 \cdots i_k}$ we deduce that $\mathcal{L}_V\, d\boldsymbol{\sigma}^\alpha \in I$ if $\mathcal{L}_V \boldsymbol{\sigma}^\alpha \in I$.

If we employ the well-known relation

$$\mathcal{L}_V \alpha = V \rfloor d\alpha + d(V \rfloor \alpha), \quad \forall \alpha \in \Lambda(K)$$

in $(2.5)_1$ and compare the coefficients of the independent differentials we obtain

$$\frac{\partial V^\alpha_{i_1\cdots i_k}}{\partial x^i} - v^\alpha_{i_1\cdots i_k j} \frac{\partial V^j}{\partial x^i} - V^\alpha_{i_1 \cdots i_k i} = \sum_{r=0}^{m-1} \lambda^{\alpha j_1 \cdots j_r}_{\beta i_1 \cdots i_k} v^\beta_{j_1 \cdots j_r i}$$

$$\lambda^{\alpha j_1 \cdots j_r}_{\beta i_1 \cdots i_k} = \frac{\partial V^\alpha_{i_1 \cdots i_k}}{\partial v^\beta_{j_1 \cdots j_r}} - v^\alpha_{i_1 \cdots i_k i} \frac{\partial V^i}{\partial v^\beta_{j_1 \cdots j_r}} \tag{2.6}$$

$$\frac{\partial V^\alpha_{i_1 \cdots i_k}}{\partial v^\beta_{j_1 \cdots j_m}} - v^\alpha_{i_1 \cdots i_k i} \frac{\partial V^i}{\partial v^\beta_{j_1 \cdots j_m}} = 0, \quad k, r = 0, 1, \ldots, m-1$$

Introducing $(2.6)_2$ into $(2.6)_1$ and working on the resulting expressions together with $(2.6)_3$ recursively we find after some laborious calculations that if $N > 1$

$$V^i = \omega^i(\boldsymbol{x}, \boldsymbol{u}), \qquad V^\alpha = \Omega^\alpha(\boldsymbol{x}, \boldsymbol{u}) \tag{2.7}$$

and remaining components of the isovector field satisfy the following recurrence relations

$$V^\alpha_{i_1\cdots i_k i} = D_i^{(k)} V^\alpha_{i_1\cdots i_k} - v^\alpha_{i_1\cdots i_k j}\, D_i^{(k)} \omega^j, \quad k = 0, 1, \ldots, m-1 \tag{2.8}$$

where the total derivative is defined as

$$D_i^{(k)} = \frac{\partial}{\partial x^i} + v_i^\alpha \frac{\partial}{\partial u^\alpha} + \sum_{r=1}^{k} v^\alpha_{i_1\cdots i_r i} \frac{\partial}{\partial v^\alpha_{i_1 i_2 \cdots i_r}}$$

For detail we refer to [3]. It is easy to see that these results are also valid for any contact ideal of *finite dimension*. Hence, in this case, all symmetries are Lie symmetries (point transformations) [4, 5, 6] and higher order components of the isovector field are found by usual prolongation formulas [7].

If we have a single balance equation, namely if $N = 1$, certain types of generalized symmetries may be admitted. We can indeed find, omitting the greek superscripts which are all equal to 1, that

$$V^i = \omega^i = -\frac{\partial \omega}{\partial v_i}, V = \Omega = \omega - v_i \frac{\partial \omega}{\partial v_i} \tag{2.9}$$

and the recurrence relations (2.8) become

$$V_{i_1\cdots i_k i} = D_i^{(k)} V_{i_1\cdots i_k} + v_{i_1\cdots i_k j} D_i^{(k)} \frac{\partial \omega}{\partial v_i} + v_{ki} v_{i_1\cdots i_k j} \frac{\partial^2 \omega}{\partial v_k \partial v_j} \tag{2.10}$$

where $\omega = \omega(\boldsymbol{x}, \boldsymbol{u}, \boldsymbol{v})$.

We shall now investigate restrictions imposed by $(2.5)_2$ on the isovector field. Making use of the general formula to evaluate the Lie derivative of balance forms we find that, for $N > 1$

$$\mathcal{L}_V \omega^\alpha = V \rfloor d\omega^\alpha - d(V \rfloor \omega^\alpha), \quad V \rfloor d\omega^\alpha = V(\Sigma^\alpha)\mu - V^i\, d\Sigma^\alpha \wedge \mu_i$$

$$V \rfloor \omega^\alpha = V(\Sigma^{\alpha i})\mu_i - V^j d\Sigma^{\alpha i} \wedge \mu_{ji} + V^i \Sigma^\alpha \mu_i$$

where μ_{ij} is defined by

$$\mu_{ij} = -\mu_{ji} = \frac{\partial}{\partial x^i} \rfloor \mu_j = \frac{\partial}{\partial x^i} \rfloor \frac{\partial}{\partial x^j} \rfloor \mu \in \Lambda^{n-2}(M)$$

satisfying the identity $dx^k \wedge \mu_{ji} = \delta_j^k \mu_i - \delta_i^k \mu_j$ [2] and for all $f \in \Lambda^0(K)$

$$V(f) = V \rfloor f = V^i \frac{\partial f}{\partial x^i} + \sum_{r=0}^{m} V^\alpha_{i_1 i_2 \cdots i_r} \frac{\partial f}{\partial v^\alpha_{i_1 i_2 \cdots i_r}}$$

We thus get

$$\mathcal{L}_V\omega^\alpha = V(\Sigma^\alpha)\mu + [dV(\Sigma^{\alpha i}) + \Sigma^\alpha dV^i]\wedge\mu_i - dV^j\wedge d\Sigma^{\alpha i}\wedge\mu_{ji}$$

or

$$\mathcal{L}_V\omega^\alpha = A^\alpha\mu + A^{\alpha i}_\beta\, du^\beta\wedge\mu_i + A^{\alpha ij}_{\beta\gamma}\, du^\beta\wedge du^\gamma\wedge\mu_{ji}$$
$$+\sum_{r=0}^{m} A^{\alpha i i_1\cdots i_r}_\beta\, dv^\beta_{i_1\cdots i_r}\wedge\mu_i + \sum_{r=1}^{m} A^{\alpha i j i_1\cdots i_r}_{\beta\gamma}\, dv^\beta_{i_1\cdots i_r}\wedge du^\gamma\wedge\mu_{ji} \tag{2.11}$$

In the above expressions all coefficients are smooth functions on the manifold K and are given by

$$A^\alpha = V(\Sigma^\alpha) + \frac{\partial V(\Sigma^{\alpha i})}{\partial x^i} + \Sigma^\alpha\frac{\partial\omega^i}{\partial x^i} + \frac{\partial\Sigma^{\alpha i}}{\partial x^i}\frac{\partial\omega^j}{\partial x^j} - \frac{\partial\Sigma^{\alpha i}}{\partial x^j}\frac{\partial\omega^j}{\partial x^i}$$

$$A^{\alpha i}_\beta = \frac{\partial V(\Sigma^{\alpha i})}{\partial u^\beta} + \Sigma^\alpha\frac{\partial\omega^i}{\partial u^\beta} + \frac{\partial\Sigma^{\alpha j}}{\partial x^j}\frac{\partial\omega^i}{\partial u^\beta} - \frac{\partial\Sigma^{\alpha i}}{\partial x^j}\frac{\partial\omega^j}{\partial u^\beta} + \frac{\partial\Sigma^{\alpha i}}{\partial u^\beta}\frac{\partial\omega^j}{\partial x^j} - \frac{\partial\Sigma^{\alpha j}}{\partial u^\beta}\frac{\partial\omega^i}{\partial x^j}$$

$$A^{\alpha ij}_{\beta\gamma} = -A^{\alpha ji}_{\beta\gamma} = -A^{\alpha ij}_{\gamma\beta} = \frac{\partial\Sigma^{\alpha[i}}{\partial u^{[\beta}}\frac{\partial\omega^{j]}}{\partial u^{\gamma]}} \tag{2.12}$$

$$A^{\alpha i i_1\cdots i_r}_\beta = \frac{\partial V(\Sigma^{\alpha i})}{\partial v^\beta_{i_1\cdots i_r}} + \frac{\partial\Sigma^{\alpha i}}{\partial v^\beta_{i_1\cdots i_r}}\frac{\partial\omega^j}{\partial x^j} - \frac{\partial\Sigma^{\alpha j}}{\partial v^\beta_{i_1\cdots i_r}}\frac{\partial\omega^i}{\partial x^j}$$

$$A^{\alpha i j i_1\cdots i_r}_{\beta\gamma} = -A^{\alpha j i i_1\cdots i_r}_{\beta\gamma} = \frac{\partial\Sigma^{\alpha[i}}{\partial v^\beta_{i_1\cdots i_r}}\frac{\partial\omega^{j]}}{\partial u^\gamma}, \quad r = 1, 2, \ldots, m$$

where the bracketed upper or lower indices represent the antisymmetric parts.

Substituting $d\sigma^\alpha$ from (2.3) into (2.4) we can transform it to

$$\omega^\alpha = \omega'^\alpha + \sum_{r=0}^{m-1}\frac{\partial\Sigma^{\alpha i}}{\partial v^\beta_{i_1\cdots i_r}}\sigma^\beta_{i_1\cdots i_r}\wedge\mu_i$$

where n-forms ω'^α are given by

$$\omega'^\alpha = \left(\frac{\partial\Sigma^{\alpha i}}{\partial x^i} + \sum_{r=0}^{m-1}\frac{\partial\Sigma^{\alpha i}}{\partial v^\beta_{i_1\cdots i_r}}v^\beta_{i_1\cdots i_r i} + \Sigma^\alpha\right)\mu + \frac{\partial\Sigma^{\alpha i}}{\partial v^\beta_{i_1\cdots i_m}}dv^\beta_{i_1\cdots i_m}\wedge\mu_i$$

It is a simple exercise to show that the forms ω^α, $\boldsymbol{\sigma}^\alpha$, $d\boldsymbol{\sigma}^\alpha$ and ω'^α, $\boldsymbol{\sigma}^\alpha$, $d\boldsymbol{\sigma}^\alpha$ generate the same ideal. Therefore the isovector condition may be replaced by

$$\mathcal{L}_V\omega'^\alpha = \lambda^\alpha_\beta\omega'^\beta + \sum_{r=0}^{m-1}\bar{\gamma}^{\alpha i_1\cdots i_r}_\beta\wedge\sigma^\beta_{i_1\cdots i_r} + \sum_{r=0}^{m-1}\bar{\Gamma}^{\alpha i_1\cdots i_r}_\beta\wedge d\sigma^\beta_{i_1\cdots i_r} \tag{2.13}$$

It now follows directly from (2.11) that

$$\mathcal{L}_V\omega'^\alpha = \left\{A^\alpha + \sum_{r=0}^{m-1}\left(A^{\alpha i i_1\cdots i_r}_\beta + 2A^{\alpha i j i_1\cdots i_r}_{\beta\gamma}v^\gamma_j\right)v^\beta_{i_1\cdots i_r i}\right\}\mu$$
$$+\left(A^{\alpha i i_1\cdots i_m}_\beta + 2A^{\alpha i j i_1\cdots i_m}_{\beta\gamma}v^\gamma_j\right)dv^\beta_{i_1\cdots i_m}\wedge\mu_i \tag{2.14}$$

By carefully comparing (2.13) and (2.14) we find that [3]

$$\bar{\gamma}^{\alpha}_{\beta} = 0; \quad \bar{\gamma}^{\alpha i_1 \cdots i_r}_{\beta} = -\bar{\Gamma}^{\alpha i_1 \cdots i_r}_{\beta} \wedge dx^i, \qquad r = 0, 1, \ldots, m-2$$

$$\bar{\Gamma}^{\alpha i_1 \cdots i_{m-1}}_{\beta} = \gamma^{\alpha i_1 \cdots i_{m-1} ij}_{\beta} \mu_{ji}, \quad \gamma^{\alpha i_1 \cdots i_{m-1} ij}_{\beta} = -\gamma^{\alpha i_1 \cdots i_{m-1} ji}_{\beta} \in \Lambda^0(K)$$

$$4\gamma^{\alpha i_1 \cdots i_{m-1} i_m i}_{\beta} = A^{\alpha i i_1 \cdots i_{m-1} i_m}_{\beta} - A^{\alpha i_m i_1 \cdots i_{m-1} i}_{\beta} + 2\Big(A^{\alpha i j i_1 \cdots i_{m-1} i_m}_{\beta\gamma} - A^{\alpha i_m j i_1 \cdots i_{m-1} i}_{\beta\gamma}\Big) v^{\gamma}_{j} - \lambda^{\alpha}_{\gamma} \left(\frac{\partial \Sigma^{\gamma i}}{\partial v^{\beta}_{i_1 \cdots i_m}} - \frac{\partial \Sigma^{\gamma i_m}}{\partial v^{\beta}_{i_1 \cdots i_{m-1} i}} \right)$$

and

$$\lambda^{\alpha}_{\beta} \left(\frac{\partial \Sigma^{\beta i}}{\partial x^i} + \sum_{r=0}^{m-1} \frac{\partial \Sigma^{\beta i}}{\partial v^{\gamma}_{i_1 \cdots i_r}} v^{\gamma}_{i_1 \cdots i_r i} + \Sigma^{\beta} \right) = A^{\alpha} + \textstyle\sum_{r=0}^{m-1} \big(A^{\alpha i i_1 \cdots i_r}_{\beta} + 2A^{\alpha i j i_1 \cdots i_r}_{\beta\gamma} v^{\gamma}_{j}\big) v^{\beta}_{i_1 \cdots i_r i}$$

$$\lambda^{\alpha}_{\gamma} \left(\frac{\partial \Sigma^{\gamma i}}{\partial v^{\beta}_{i_1 \cdots i_m}} + \frac{\partial \Sigma^{\gamma i_m}}{\partial v^{\beta}_{i_1 \cdots i_{m-1} i}} \right) = A^{\alpha i i_1 \cdots i_{m-1} i_m}_{\beta} + A^{\alpha i_m i_1 \cdots i_{m-1} i}_{\beta} + 2\Big(A^{\alpha i j i_1 \cdots i_{m-1} i_m}_{\beta\gamma} + A^{\alpha i_m j i_1 \cdots i_{m-1} i}_{\beta\gamma}\Big) v^{\gamma}_{j} \tag{2.15}$$

Eqs. (2.15) are determining equations for the infinitesimal generators (isovector components) of the symmetry group associated with the balance equations since they constitute, together with (2.7) and (2.8), an overdetermined system of equations to evaluate λ^{α}_{β}, $\omega^i(\boldsymbol{x}, \boldsymbol{u})$, $\Omega^{\alpha}(\boldsymbol{x}, \boldsymbol{u})$.

If we have a single balance equation, i.e., if $N = 1$ the above analysis should be modified to some extent. We just record here only the final results. The detailed analysis can be found in [3]. The Lie derivative of the balance form ω is written as

$$\mathcal{L}_V \omega = A\mu + \sum_{r=0}^{m} A^{i i_1 \cdots i_r} dv_{i_1 \cdots i_r} \wedge \mu_i + \sum_{r=1}^{m} B^{i j i_1 \cdots i_r} dv_{i_1 \cdots i_r} \wedge \mu_{ji} + \sum_{r=1}^{m} C^{i j k i_1 \cdots i_r} dv_{i_1 \cdots i_r} \wedge dv_k \wedge \mu_{ji}$$

where, in lieu of (2.12), we define for $m > 1$

$$A = A^1, \quad A^i = A^{1i}_1 \ (r = 0), \quad A^{ij} = A^{1ij}_1 + \Sigma \frac{\partial \omega^i}{\partial v_j} + \frac{\partial \Sigma^k}{\partial x^k} \frac{\partial \omega^i}{\partial v_j} - \frac{\partial \Sigma^i}{\partial x^k} \frac{\partial \omega^k}{\partial v_j} \quad (r = 1)$$

$$A^{i i_1 \cdots i_r} = A^{1 i i_1 \cdots i_r}_1, \quad r = 2, 3, \ldots, m$$

$$B^{ijk} = -B^{jik} = A^{1ijk}_{11} + \frac{1}{2} \left(\frac{\partial \Sigma^j}{\partial u} \frac{\partial \omega^i}{\partial v_k} - \frac{\partial \Sigma^i}{\partial u} \frac{\partial \omega^j}{\partial v_k} \right) \quad (r = 1)$$

$$B^{iji_1\cdots i_r} = -B^{jii_1\cdots i_r} = A_{11}^{1iji_1\cdots i_r}, \quad r = 2,3,\ldots,m \tag{2.16}$$

$$C^{ijkl} = -C^{jikl} = -C^{ijlk} = \frac{\partial \Sigma^{[i}}{\partial v_{[l}} \frac{\partial \omega^{j]}}{\partial v_{k]}} \quad (r=1)$$

$$C^{ijki_1\cdots i_r} = -C^{jiki_1\cdots i_r} = \frac{\partial \Sigma^{[i}}{\partial v_{i_1\cdots i_r}} \frac{\partial \omega^{j]}}{\partial v_k}, \quad r = 2,3,\ldots,m$$

These relations should be supplemented by (2.9) and (2.10). The determining equations then become

$$\lambda\left(\frac{\partial \Sigma^i}{\partial x^i} + \sum_{r=0}^{m-1} \frac{\partial \Sigma^i}{\partial v_{i_1\cdots i_r}} v_{i_1\cdots i_r i} + \Sigma\right) =$$

$$A + A^i v_i + \sum_{r=0}^{m-1} \left(A^{ii_1\cdots i_r} + 2B^{iji_1\cdots i_r} v_j + 2C^{ijki_1\cdots i_r} v_{kj}\right) v_{i_1\cdots i_r i} \tag{2.17}$$

$$\lambda\left(\frac{\partial \Sigma^i}{\partial v_{i_1\cdots i_m}} + \frac{\partial \Sigma^{i_m}}{\partial v_{i_1\cdots i_{m-1} i}}\right) = A^{ii_1\cdots i_{m-1} i_m} + A^{i_m i_1\cdots i_{m-1} i} + 2\Big(B^{iji_1\cdots i_{m-1} i_m}$$

$$+ B^{i_m j i_1\cdots i_{m-1} i}\Big) v_j + 2\Big(C^{ijki_1\cdots i_{m-1} i_m} + C^{i_m jki_1\cdots i_{m-1} i}\Big) v_{kj}$$

for $\lambda \in \Lambda^0(K)$ and $\omega(x^i, u, v_i)$.

The case $m = 1$ should be treated separately. In the same fashion we find that

$$\lambda\left(\frac{\partial \Sigma^i}{\partial x^i} + \frac{\partial \Sigma^i}{\partial u} v_i + \Sigma\right) = A + A^i v_i$$

$$\lambda\left(\frac{\partial \Sigma^i}{\partial v_j} + \frac{\partial \Sigma^j}{\partial v_i}\right) = A^{ij} + A^{ji} - 2\Big(B^{kij} + B^{kji}\Big) v_k \tag{2.18}$$

$$2C^{i(jk)l} = C^{ijkl} + C^{ikjl} = 0$$

Expressions (2.15) or (2.17) or (2.18) together with the appropriate recurrence relations for isovector components yield a number of equations to determine the infinitesimal generators of the symmetry group directly from the balance equations.

Similarity solutions

These are solutions which remain invariant under a particular symmetry transformation. Hence the corresponding isovector field must be a characteristic vector field of the ideal. Therefore the following restrictions should be imposed on the solving map:

$$\phi^*(V \rfloor \boldsymbol{\sigma}^\alpha) = 0, \quad \phi^*(V \rfloor \boldsymbol{\sigma}^\alpha) = 0, \quad \phi^*(V \rfloor \omega^\alpha) = 0$$

One can easily prove that only the first set of conditions for $r = 0$ (σ^α) are essential [3]. Hence, for $N > 1$, the form of similarity solutions is found by solving the quasilinear system of equations corresponding to a given isovector field as follows

$$\phi^*(V \rfloor \sigma^\alpha) = \omega^i(\boldsymbol{x}, \boldsymbol{u}) \frac{\partial u^\alpha}{\partial x^i} - \Omega^\alpha(\boldsymbol{x}, \boldsymbol{u}) = 0 \tag{2.19}$$

If there is a single balance equation (2.19) becomes a nonlinear equation

$$\omega^i(x^j, u, u_{,j})\frac{\partial u}{\partial x^i} - \Omega(x^j, u, u_{,j}) = 0$$

In the next two sections we shall try to show the efficiency of our approach by applying it to two problems from nonlinear elastodynamics.

3. Plane motion of a hyperelastic solid

Plane motion of a hyperelastic solid in the absence of body forces is governed by the following equations in material coordinates

$$\frac{\partial}{\partial X}\left(\frac{\partial \Sigma}{\partial p_k}\right) - \rho_0(X)\frac{\partial v_k}{\partial t} = 0, \quad k = 1, 2, 3 \tag{3.1}$$

where $\Sigma = \Sigma(p_k, X)$ is the given strain energy function of the material, $\boldsymbol{uu}(X, t)$ is the displacement vector,

$$p_k = \frac{\partial u_k}{\partial X}, \qquad v_k = \frac{\partial u_k}{\partial t} \tag{3.2}$$

are the components of the displacement gradient and velocity vector, respectively, and $\rho_0(X)$ is the density of the medium in the reference configuration. The system defined by (3.1) and (3.2) is characterized by $m = 1$, $n = 2$, $N = 3$. We shall now take $x^1 = X$, $x^2 = t$ and interchange the roman and greek indices for convenience in formulas recorded in Sec. 2. We thus have to take

$$\Sigma_{k1} = \frac{\partial \Sigma}{\partial p_k}, \qquad \Sigma_{k2} = -\rho_0 v_k, \qquad \Sigma_k = 0$$

The isovector field is to be expressed as

$$V = \phi\frac{\partial}{\partial X} + \psi\frac{\partial}{\partial t} + \omega_k\frac{\partial}{\partial u_k} + P_k\frac{\partial}{\partial p_k} + V_k\frac{\partial}{\partial v_k}$$

where ϕ, ψ and ω_k should be functions of X, t, $\boldsymbol{u}$ and $\boldsymbol{P}$, $\boldsymbol{V}$ are given by

$$\begin{aligned} P_k &= \frac{\partial \omega_k}{\partial X} - \frac{\partial \phi}{\partial X}p_k - \frac{\partial \psi}{\partial X}v_k + \frac{\partial \omega_k}{\partial u_l}p_l - \frac{\partial \phi}{\partial u_l}p_k p_l - \frac{\partial \psi}{\partial u_l}v_k p_l \\ V_k &= \frac{\partial \omega_k}{\partial t} - \frac{\partial \phi}{\partial t}p_k - \frac{\partial \psi}{\partial t}v_k + \frac{\partial \omega_k}{\partial u_l}v_l - \frac{\partial \phi}{\partial u_l}p_k v_l - \frac{\partial \psi}{\partial u_l}v_k v_l \end{aligned} \tag{3.3}$$

It follows then from (2.12) that

$$A_k = \frac{\partial B_k}{\partial X} - \rho_0\frac{\partial V_k}{\partial t} + C_k\frac{\partial \psi}{\partial t}$$

$$A_{1kl} = \frac{\partial B_k}{\partial u_l}, \quad A_{2kl} = -\rho_0\frac{\partial V_k}{\partial u_l} + C_k\frac{\partial \psi}{\partial u_l}$$

$$A_{11kl} = \frac{\partial B_k}{\partial p_l} + C_{kl}\frac{\partial \psi}{\partial t}, \quad A_{12kl} = \frac{\partial B_k}{\partial v_l} + \rho_0\frac{\partial \phi}{\partial t}\delta_{kl}$$

$$A_{21kl} = -\rho_0 \frac{\partial V_k}{\partial p_l} - C_{kl}\frac{\partial \psi}{\partial X}, \quad A_{22kl} = -\rho_0 \frac{\partial V_k}{\partial v_l} - \left(\rho_0 \frac{\partial \phi}{\partial X} + \rho_0' \phi\right)\delta_{kl}$$

$$A_{12klm} = 0, \quad 2A_{121klm} = -2A_{211klm} = C_{kl}\frac{\partial \psi}{\partial u_m}$$

$$2A_{122klm} = -2A_{212klm} = \rho_0 \frac{\partial \phi}{\partial u_m}\delta_{kl}$$

where we defined

$$B_k = C_k\phi + C_{kl}P_l, \quad C_{kl} = \frac{\partial^2 \Sigma}{\partial p_k \partial p_l}, \quad C_k = \frac{\partial^2 \Sigma}{\partial p_k \partial X}$$

Due to (2.15) the determining equations become

$$\begin{aligned}
\lambda_{kl}C_l &= \frac{\partial B_k}{\partial X} - \rho_0 \frac{\partial V_k}{\partial t} + C_k\frac{\partial \psi}{\partial t} + \frac{\partial B_k}{\partial u_l}p_l - \left(\rho_0 \frac{\partial V_k}{\partial u_l}p_l - C_k\frac{\partial \psi}{\partial u_l}\right)v_l \\
\lambda_{kl}C_{lm} &= \frac{\partial B_k}{\partial p_m} + C_{km}\left(\frac{\partial \psi}{\partial t} + \frac{\partial \psi}{\partial u_l}v_l\right) \\
-\lambda_{kl}\rho_0\delta_{lm} &= -\rho_0\lambda_{km} = -\rho_0\frac{\partial V_k}{\partial v_m} - \left(\frac{\partial}{\partial X}(\rho_0\phi) + \rho_0\frac{\partial \phi}{\partial u_l}p_l\right)\delta_{km} \\
0 &= \frac{\partial B_k}{\partial v_m} + \rho_0\frac{\partial \phi}{\partial t}\delta_{km} - \rho_0\frac{\partial V_k}{\partial p_m} - C_{km}\frac{\partial \psi}{\partial X} - C_{km}\frac{\partial \psi}{\partial u_l}p_l + \delta_{km}\rho_0\frac{\partial \phi}{\partial u_l}v_l
\end{aligned} \tag{3.4}$$

In the above expressions $(3.4)_3$ yields the coefficients λ_{kl}. Introducing these into other relations in (3.4) we obtain 21 equations to determine five unknown functions ϕ, ψ and ω_k. We refer [8] for the details about the solution of that overdetermined set of equations. We just give here the final result for strictly nonlinear materials:

$$\begin{aligned}
\phi = \phi(X) &= \frac{c_1 R_0(X) + c_2}{R_0'(X)}, \qquad R_0(X) = \int \rho_0(X)\,dX \\
\psi &= \psi(t) = b_1 t^2 + 2b_2 t + b_3 \\
\omega_k &= [(a_0 + b_2 + b_1 t)\delta_{kl} + e_{klm}a_m]u_l + \lambda_k(X)t + \mu_k(X)
\end{aligned} \tag{3.5}$$

Here e_{klm} is the permutation symbol, a_0, a_1, a_2, a_3, b_1, b_2, b_3, c_1, c_2 are arbitrary constants and $\lambda_k(X)$, $\mu_k(X)$ are arbitrary functions. One finds that the strain energy function has to satisfy simultaneously the following equations:

$$\begin{gathered}
(b_1 p_k + \lambda_k')\frac{\partial \Sigma}{\partial p_k} + 2b_1\Sigma = 0 \\
[(a_0 + b_2 - \phi')p_k + e_{klm}a_m p_l + \mu_k']\frac{\partial \Sigma}{\partial p_k} + \phi\frac{\partial \Sigma}{\partial X} + (\phi' + 2b_2 - 2a_0 - c_1)\Sigma = 0
\end{gathered} \tag{3.6}$$

As a special case, let us take $b_1 = a_k = \lambda_k = 0$, write

$$\phi = c_1\frac{R_0(X) + \beta}{R_0'(X)}, \quad \psi = 2b_2(t + \gamma), \quad \omega_k = a u_k + \mu_k(X) \tag{3.7}$$

and define $a = a_0 + b_2$, $b = 2b_2 - 2a_0 - c_1$. Then $(3.6)_1$ is satisfied identically and the second equation yields for the admissible form of Σ

$$\Sigma(p, X) = \frac{R_0'}{c_1(R_0+\beta)^{(b/c_1)+1}}\widetilde{\Sigma}(\pi),\ \pi_k = \frac{c_1 p_k}{R_0'(R_0+\beta)^{(a/c_1)-1}} - \int \frac{\mu_k'}{(R_0+\beta)^{a/c_1}}\,dX$$

The similarity solution associated with the symmetry group generated by (3.7) follows from

$$c_1\frac{R_0(X)+\beta}{R_0'(X)}\frac{\partial u_k}{\partial X} + 2b_2(t+\gamma)\frac{\partial u_k}{\partial t} - a u_k - \mu_k(X) = 0$$

whose solution is given by

$$u_k = (R_0+\beta)^{a/c_1}U_k(\xi) + \frac{1}{c_1}(R_0+\beta)^{a/c_1}\int \frac{R_0'\mu_k}{(R_0+\beta)^{(b/c_1)+1}}\,dX \tag{3.8}$$

where the similarity variable ξ is

$$\xi = \frac{(R_0+\beta)^\alpha}{t+\gamma}, \qquad \alpha = \frac{2b_2}{c_1} \tag{3.9}$$

Inserting (3.8) into (3.1) we find that U_k satisfy the following ordinary differential equations

$$2b_2\xi\left(\frac{\partial\widetilde{\Sigma}}{\partial\pi_k}\right)' + (a_0 - 3b_2 + c_1)\frac{\partial\widetilde{\Sigma}}{\partial\pi_k} = c_1\xi^3(\xi U_k'' + 2U_k')$$

where it is easily found that $\pi_k(\xi) = \alpha c_1\xi U_k' + aU_k$.

4. Radial motion of hyperelastic spheres and cylinders

Radial motion of an isotropic, heterogeneous, compressible hyperelastic cylinder or sphere is governed in the material description by the equation [9]

$$\frac{\partial\Sigma_p}{\partial R} - \rho_0(R)\frac{\partial v}{\partial t} + \frac{s\Sigma_p - \Sigma_q}{R} = 0 \tag{4.1}$$

where $\Sigma = \Sigma(p, q, R)$ is the given strain energy function, $r = r(R, t)$ denotes the radial motion, $\rho_0(R)$ is the initial density distribution, $s = 1$ for a cylinder, $s = 2$ for a sphere and

$$p = \frac{\partial r}{\partial R}, \qquad q = \frac{r}{R}, \qquad v = \frac{\partial r}{\partial t}$$

We now take $x^1 = R$, $x^2 = t$ and $u = r$. Since $n = 2$, $m = 1$, $N = 1$ we shall have to employ (2.16) and (2.18). The relevant coefficients are

$$\Sigma^1 = \Sigma_p, \qquad \Sigma^2 = -\rho_0(R)v, \qquad \Sigma = \frac{1}{R}(s\Sigma_p - \Sigma_q)$$

Therefore we immediately deduce from (2.16) that

$$\begin{aligned}
A &= \frac{s\Sigma_p - \Sigma_q}{R^2}(\omega_p - R\omega_{pR} - R\omega_{vt}) - \frac{s\Sigma_{pR} - \Sigma_{qR}}{R}\omega_p \\
&+ \frac{s\Sigma_{pr} - \Sigma_{qr}}{R}(\omega - p\omega_p - v\omega_v) + \frac{s\Sigma_{pp} - \Sigma_{qp}}{R}(\omega_R + p\omega_r) - \Sigma_{pR}(\omega_{pR} + \omega_{vt}) \\
&- \Sigma_{pRR}\omega_p + \Sigma_{pr}(\omega_R - p\omega_{pR} - v\omega_{vR}) + \Sigma_{prR}(\omega - p\omega_p - v\omega_v) \\
&+ \Sigma_{pp}(\omega_{RR} + p\omega_{rR}) + \Sigma_{ppR}(\omega_R + p\omega_r) - \rho_0(\omega_{tt} + v\omega_{rt}) \\
A^1 &= -[\Sigma_{pR} + \frac{1}{R}(s\Sigma_p - \Sigma_q)]\omega_{pr} - \Sigma_{pRr}\omega_p + \Sigma_{pr}(\omega_r - p\omega_{pr} - v\omega_{vr} - \omega_{vt}) \\
&+ \Sigma_{pp}(\omega_{Rr} + p\omega_{rr}) + \Sigma_{ppr}(\omega_R + p\omega_r) + \Sigma_{prr}(\omega - p\omega_p - v\omega_v) \\
A^2 &= -\rho_0(\omega_{tr} + v\omega_{rr}) - [\Sigma_{pR} - \Sigma_{pr} + \frac{1}{R}(s\Sigma_p - \Sigma_q)]\omega_{vr} \\
A^{11} &= -\Sigma_{pR}\omega_{pp} - \Sigma_{ppR}\omega_p - \Sigma_{pr}(p\omega_{pp} + v\omega_{vp}) + \Sigma_{ppr}(\omega - p\omega_p - v\omega_v) \qquad (4.2) \\
&- \frac{1}{R}(s\Sigma_p - \Sigma_q)\omega_{pp} + \Sigma_{pp}(\omega_{Rp} + \omega_r + p\omega_{rp} - \omega_{vt}) + \Sigma_{ppp}(\omega_R + p\omega_r) \\
A^{12} &= -\Sigma_{pR}\omega_{pv} - \Sigma_{pr}(p\omega_{pv} + v\omega_{vv}) + \Sigma_{pp}(\omega_{Rv} + p\omega_{rv}) \\
&- \frac{1}{R}(s\Sigma_p - \Sigma_q)\omega_{pv} - \rho_0\omega_{pt} \\
A^{21} &= -\rho_0(\omega_{tp} + v\omega_{rp}) - \frac{1}{R}(s\Sigma_p - \Sigma_q)\omega_{vp} + \Sigma_{pp}\omega_{vR} - \Sigma_{pR}\omega_{vp} \\
A^{22} &= \rho_0'\omega_p - \rho_0(\omega_{tv} + \omega_r + v\omega_{rv} - \omega_{pR}) - \frac{1}{R}(s\Sigma_p - \Sigma_q)\omega_{vv} - \Sigma_{pR}\omega_{vv}
\end{aligned}$$

$$B^{121} = \frac{1}{2}(-\Sigma_{pp}\omega_{vr} + \Sigma_{pr}\omega_{vp}), \quad B^{122} = \frac{1}{2}(-\rho_0\omega_{pr} + \Sigma_{pr}\omega_{vv})$$

$$C^{1212} = \frac{1}{4}(-\rho_0\omega_{pp} + \Sigma_{pp}\omega_{vv})$$

Then (2.18) yields

$$\lambda[\Sigma_{pR} + \Sigma_{pr}p + \frac{1}{R}(s\Sigma_p - \Sigma_q)] = A + A^1 p + A^2 v$$

$$\lambda\Sigma_{pp} = A^{11} + 2B^{121}v, \qquad -\rho_0\lambda = A^{22} - 2B^{122}p \qquad (4.3)$$

$$0 = A^{12} + A^{21} - 2B^{121}p + 2B^{122}v, \qquad C^{1212} + C^{1122} = C^{1212} = 0$$

Eliminating λ through $(4.3)_3$, employing (4.2) in the resulting equations and noting that $q_r = 1/R$, $q_R = -q/R$ we get

$$
\begin{aligned}
&\Sigma_{pp}\omega_{vv} - \rho_0\omega_{pp} = 0\\
&\Sigma_{pp}[\omega_{vR} + \frac{1}{R}(p-q)\omega_{vq}] - \frac{1}{R}[(p-q)\Sigma_{pq} + s\Sigma_p - \Sigma_q + R\Sigma_{pR}]\omega_{pv}\\
&\qquad - \rho_0\omega_{pt} - \frac{1}{R}\rho_0 v\omega_{pq} = 0\\
&\Sigma_{ppp}[\omega_R + \frac{1}{R}(p-q)\omega_q] + \frac{1}{R}\Sigma_{ppq}[\omega - (p-q)\omega_p - v\omega_v] + \left(\frac{\rho_0'}{\rho_0}\Sigma_{pp} - \Sigma_{ppR}\right)\omega_p\\
&\quad + 2\Sigma_{pp}\Big\{\omega_{pR} + \frac{1}{R}(p-q)\omega_{pq} - \omega_{vt} - \frac{v}{R}\omega_{vq}\\
&\quad - \frac{1}{\rho_0 R}[(p-q)\Sigma_{pq} + s\Sigma_p - \Sigma_q + R\Sigma_{pR}]\omega_{vv}\Big\} = 0\\
&\frac{1}{R}(p-q)\Big\{\Sigma_{ppq}[\omega_R + \frac{1}{R}(p-q)\omega_q] + \frac{1}{R}\Sigma_{pqq}[\omega - (p-q)\omega_p - v\omega_v]\Big\}\\
&\quad + \Sigma_{ppR}[\omega_R + \frac{1}{R}(p-q)\omega_q] + \frac{1}{R}\Sigma_{pqR}[\omega - 2(p-q)\omega_p - v\omega_v] - \Sigma_{pRR}\omega_p\\
&\quad + \Sigma_{pp}[\omega_{RR} + \frac{2}{R}(p-q)(\omega_{qR} - \frac{1}{R}\omega_q) + \frac{1}{R^2}(p-q)^2\omega_{qq}] \qquad (4.4)\\
&\quad + \frac{1}{R^2}\Sigma_{pq}\Big\{2(p-q)(\omega_p - R\omega_{vt} - v\omega_{vq}) + R\omega_R + v\omega_v - \omega + \frac{R\rho_0'}{\rho_0}(p-q)\omega_p\Big\}\\
&\quad - \frac{1}{\rho_0 R^2}[(p-q)\Sigma_{pq} + s\Sigma_p - \Sigma_q + R\Sigma_{pR}]^2\omega_{vv} + \frac{1}{R}(s\Sigma_{pp} - \Sigma_{pq})[\omega_R + \frac{1}{R}(p-q)\omega_q]\\
&\quad + \frac{1}{R^2}(s\Sigma_{pq} - \Sigma_{qq})[\omega - (p-q)\omega_p - v\omega_v] + \frac{1}{R^2}(s\Sigma_p - \Sigma_q)\Big(\omega_p + \frac{R\rho_0'}{\rho_0}\omega_p - \omega_q\\
&\quad - 2v\omega_{qv} - 2R\omega_{vt}\Big) - \frac{1}{R}\Sigma_{pR}\left(\omega_q - \frac{R\rho_0'}{\rho_0}\omega_p + 2v\omega_{qv} + 2R\omega_{vt}\right) - \frac{1}{R}(s\Sigma_{pR} - \Sigma_{qR})\omega_p\\
&\qquad - 2\rho_0\frac{v}{R}\omega_{qt} - \rho_0\omega_{tt} - \rho_0\frac{v^2}{R^2}\omega_{qq} = 0
\end{aligned}
$$

The four equations listed above should be solved to obtain $\omega = \omega(R, t, q, p, v)$. If we denote the isovector field by

$$
\mathcal{V} = \phi\frac{\partial}{\partial R} + \psi\frac{\partial}{\partial t} + \Omega\frac{\partial}{\partial r} + P\frac{\partial}{\partial p} + V\frac{\partial}{\partial v}, \qquad (4.5)
$$

its components are determined by the relations

$$
\begin{aligned}
&\phi = -\omega_p, \qquad \psi = -\omega_v, \qquad \Omega = \omega - p\omega_p - v\omega_v\\
&P = \omega_R + p\omega_r = \omega_R + \frac{p-q}{R}\omega_q, \qquad V = \omega_t + v\omega_r = \omega_t + \frac{v}{R}\omega_q
\end{aligned} \qquad (4.6)
$$

Eqs. (4.4) supply the generalized symmetries of the balance equation (4.1). If we are only interested in point symmetries then ω should have the form

$$\omega = \alpha(R,t,q)p + \beta(R,t,q)v + \gamma(R,t,q) \tag{4.7}$$

Assuming that $\Sigma_{ppp} \neq 0$ we can show that

$$\begin{gathered}\alpha(R) = \frac{1}{R^s \rho_0}\left(c_1 \int R^s \rho_0 \, dR + 2\int R^{s-1}\nu\rho_0 \, dR + c_2\right), \beta(t) = 2b_1 t + b_2 \\ \gamma = [-b_1 R + \nu(R)]q + \mu(R)\end{gathered} \tag{4.8}$$

where b_1, b_2, c_1, c_2 are arbitrary constants and $\nu(R)$, $\mu(R)$ are arbitrary functions. It is seen that the strain energy function should have the form

$$\Sigma = \widetilde{\Sigma} + (p-q)G_q + RG_R + (s+1)G \tag{4.9}$$

where $G(R,q)$ is an arbitrary function and $\widetilde{\Sigma}$ satisfies

$$\begin{gathered}[(\alpha' - b_1 + \frac{\nu}{R})p + (\nu' - \frac{\nu}{R})q + \mu']\widetilde{\Sigma}_p + [(\frac{\alpha+\nu}{R} - b_1)q + \frac{\mu}{R}]\widetilde{\Sigma}_q \\ -\alpha\widetilde{\Sigma}_R - 2\left(b_1 + \frac{\nu}{R} - \frac{\rho_0'\alpha}{2\rho_0}\right)\widetilde{\Sigma} = 0\end{gathered} \tag{4.10}$$

It can be shown that Ko material, which represent fairly accurately certain class of elastomers, with modulus and density varying according to arbitrary power laws falls into this category. Hence a similarity solution may be obtainable for this kind of materials.

References

1. Cartan, E. (1945) Les Systèmes Différentiels Extérieurs et Leurs Applications Géometriques, Hermann, Paris.
2. Edelen, D. G. B. (1985) Applied Exterior Calculus, Wiley, New York.
3. Şuhubi, E. S. (1991) 'Isovector fields and similarity solutions for general balance equations', Int. J. Engng Sci. 29, 133–150.
4. Ovsiannikov, L. V. (1982) Group Analysis of Differential Equations, Academic Press, New York.
5. Ibragimov, N. H. (1985) Transformation Groups Applied to Mathematical Physics, D. Reidel Pub. Com., Dordrecht.
6. Anderson, R. L, Ibragimov, N. H. (1979) Lie-Bäcklund Transformations in Applications, SIAM, Philadelphia.
7. Olver, P. J. (1986) Applications of Lie Groups to Differential Equations, Springer, New York.
8. Şuhubi, E. S. (1992) 'Similarity solutions for plane waves in hyperelastic materials', Int. J. Engng Sci. 30, 701–715.
9. Şuhubi, E. S. (1991) 'A class of similarity solutions for radial motions of compressible hyperelastic spheres and cylinders', J. of Elasticity 25, 259–287.

ON EQUIVALENCE TRANSFORMATIONS APPLIED TO A NON-LINEAR WAVE EQUATION

M. TORRISI, R. TRACINÀ and A. VALENTI
Dipartimento di Matematica
Università di Catania
Viale A. Doria, 6
95125 Catania (Italy)

Abstract. The equivalence algebra for the non-linear wave equations $u_{xx} - u_{tt} = f(u, u_t, u_x)$ is obtained. Some algorithms are performed in order to extend the principal Lie algebra.

1. Introduction

We consider the non linear wave equations

$$u_{tt} - u_{xx} = f(u, u_t, u_x), \tag{1.1}$$

where f is an arbitrary function of its arguments.

Equations like (1.1) arise in several problems of mathematical physics [1].

An equivalence transformation for a given differential equation of the form (1.1) is a non degenerate change of independent and dependent variables which leaves invariant the structure of the equation under consideration. It means that any equation of the form (1.1) is mapped by equivalence transformations into an equation of the same form but, generally speaking, with different $f(u, u_t, u_x)$ [2–6].

Many equivalence transformations are used in the theory of ordinary and partial differential equations. However the problem how to find equivalence transformations as been considered only recently.

Ovsiannikov [2] observed that it is possible to construct the continuous group of equivalence transformations by applying the Lie criterion of infinitesimal invariance.

In this paper first, by using the aforementioned criterion, we construct the equivalence algebra for eqs. (1.1). After this we generalize some propositions stated in [4–5] which allow us to obtain the principal Lie algebra and, when the physical problem gives a specialization of the form of the function f, we are able to construct easily extensions of it.

Finally we apply the method of preliminary group classification [3, 4] in order to obtain extensions of the principal Lie algebra. This method reduces the problem of classification to the algebraic problem of constructing the optimal systems of subalgebras of the equivalence Lie algebra.

We must observe that these results give only a partial classification but often by simpler calculations. Instead the group classification of equations (1.1) by Lie method for symmetries leads usually to a very complicated system of determining equations.

N. H. Ibragimov et al. (eds.),
Modern Group Analysis: Advanced Analytical and Computational Methods in Mathematical Physics, 367–375.

2. Equivalence generators

In order to obtain the equivalence generator of eqs. (1.1) we consider the arbitrary function f as a dependent variable and apply the method of infinitesimal invariance transformations [2, 4] to the following system

$$\begin{aligned} &u_{tt} - u_{xx} = f, \\ &f_t = f_x = 0, \end{aligned} \tag{2.1}$$

where the last two equations of (2.1) give the independence of f on t and x.

The generator of the equivalence group is taken of the form

$$Y = \xi^1 \frac{\partial}{\partial t} + \xi^2 \frac{\partial}{\partial x} + \eta \frac{\partial}{\partial u} + \mu \frac{\partial}{\partial f} + \zeta_1 \frac{\partial}{\partial u_t} + \zeta_2 \frac{\partial}{\partial u_x}, \tag{2.2}$$

where the coordinates ξ^1, ξ^2 and η are thought as functions of t, x and u while the coordinate μ depends on t, x, u, u_t, u_x and f, and

$$\zeta_1 = D_t(\eta) - u_t D_t(\xi^1) - u_x D_t(\xi^2), \tag{2.3}$$

$$\zeta_2 = D_x(\eta) - u_t D_x(\xi^1) - u_x D_x(\xi^2). \tag{2.4}$$

It is worthwhile observing that the projection of operator Y on the space t, x and u

$$\mathrm{Pr}_{(t,x,u)} Y = X = \xi^1 \frac{\partial}{\partial t} + \xi^2 \frac{\partial}{\partial x} + \eta \frac{\partial}{\partial u} \tag{2.5}$$

is a symmetry operator.

The prolongation of operator (2.2) which we need is

$$\widetilde{Y} = Y + \zeta_{11} \frac{\partial}{\partial u_{tt}} + \zeta_{22} \frac{\partial}{\partial u_{xx}} + \omega_1 \frac{\partial}{\partial f_t} + \omega_2 \frac{\partial}{\partial f_x}, \tag{2.6}$$

where

$$\begin{aligned} &\zeta_{11} = D_t(\zeta_1) - u_{tt} D_t(\xi^1) - u_{tx} D_t(\xi^2), \\ &\zeta_{22} = D_x(\zeta_2) - u_{tx} D_x(\xi^1) - u_{xx} D_x(\xi^2), \\ &\omega_1 = \widetilde{D}_t(\mu) - f_u \widetilde{D}_t(\eta) - f_{u_t} \widetilde{D}_t(\zeta_1) - f_{u_x} \widetilde{D}_t(\zeta_2), \\ &\omega_2 = \widetilde{D}_x(\mu) - f_u \widetilde{D}_x(\eta) - f_{u_t} \widetilde{D}_x(\zeta_1) - f_{u_x} \widetilde{D}_x(\zeta_2). \end{aligned} \tag{2.7}$$

The operators D_t and D_x denote the total derivatives with respect to t and x:

$$\begin{aligned} D_t &= \frac{\partial}{\partial t} + u_t \frac{\partial}{\partial u} + u_{tt} \frac{\partial}{\partial u_t} + u_{tx} \frac{\partial}{\partial u_x} + \cdots, \\ D_x &= \frac{\partial}{\partial x} + u_x \frac{\partial}{\partial u} + u_{tx} \frac{\partial}{\partial u_t} + u_{xx} \frac{\partial}{\partial u_x} + \cdots, \end{aligned} \tag{2.8}$$

and

$$\begin{aligned} \widetilde{D}_t &= \frac{\partial}{\partial t} + f_t \frac{\partial}{\partial f}, \\ \widetilde{D}_x &= \frac{\partial}{\partial x} + f_x \frac{\partial}{\partial f}, \end{aligned} \tag{2.9}$$

which, in view of eqs. (2.1) become

$$\widetilde{D}_t = \frac{\partial}{\partial t}, \qquad \widetilde{D}_x = \frac{\partial}{\partial x}. \tag{2.10}$$

Applying the operator (2.6) to the system (2.1) and following the well known algorithm we obtain

$$\begin{aligned} Y &= (c_3 t + c_0 x + c_1)\frac{\partial}{\partial t} + (c_0 t + c_3 x + c_2)\frac{\partial}{\partial x} + \\ &+ \varphi(u)\frac{\partial}{\partial u} + [(\varphi' - 2c_3)f + \varphi''(u_t^2 - u_x^2)]\frac{\partial}{\partial f} + \\ &+ [u_t(\varphi' - c_3) - u_x c_0]\frac{\partial}{\partial u_t} + [u_x(\varphi' - c_3) - u_t c_0]\frac{\partial}{\partial u_x}. \end{aligned} \tag{2.11}$$

where c_0, c_1, c_2, c_3 are arbitrary constants and φ is an arbitrary function of u. The equivalence algebra $L_\mathcal{E}$ is infinite-dimensional and is spanned by

$$\begin{aligned} Y_0 &= x\frac{\partial}{\partial t} + t\frac{\partial}{\partial x} - u_x\frac{\partial}{\partial t} - u_t\frac{\partial}{\partial x}, \\ Y_1 &= \frac{\partial}{\partial t}, \\ Y_2 &= \frac{\partial}{\partial x}, \\ Y_3 &= t\frac{\partial}{\partial t} + x\frac{\partial}{\partial x} - 2f\frac{\partial}{\partial f} - u_t\frac{\partial}{\partial u_t} - u_x\frac{\partial}{\partial u_x}, \\ Y_\varphi &= \varphi\frac{\partial}{\partial u} + [\varphi' f + \varphi''(u_t^2 - u_x^2)]\frac{\partial}{\partial f} + \varphi' u_t\frac{\partial}{\partial u_t} + \varphi' u_x\frac{\partial}{\partial u_x}. \end{aligned} \tag{2.12}$$

3. Applications of equivalence algebra

The most general symmetry algebra of equation (1.1) for arbitrary function f is said Principal Lie Algebra $L_\mathcal{P}$ [4]. This algebra can be obtained from equivalence algebra $L_\mathcal{E}$.

Generalizing proposition 1 of [5] we can state:

PROPOSITION I: *The projection of operator Y on the space t, x and u is the operator X of the principal Lie Algebra $L_\mathcal{P}$ of eqs. (1.1) if and only if the projection of operator Y on the space u, u_t, u_x and f*

$$\begin{aligned} \mathrm{Pr}_{(u,u_t,u_x,f)} Y &= \varphi(u)\frac{\partial}{\partial u} + [(\varphi' - 2c_3)f + \varphi''(u_t^2 - u_x^2)]\frac{\partial}{\partial f} + \\ &+ [u_t(\varphi' - c_3) - u_x c_0]\frac{\partial}{\partial u_t} + [u_x(\varphi' - c_3) - u_t c_0]\frac{\partial}{\partial u_x} \end{aligned} \tag{3.1}$$

is identically zero.

Thus from (3.1) we get $c_0 = c_3 = 0$ and $\varphi = 0$.
Therefore the symmetry operator of $L_{\mathcal{P}}$ is

$$X = c_1 \frac{\partial}{\partial t} + c_2 \frac{\partial}{\partial x} \tag{3.2}$$

and the principal Lie algebra is two-dimensional spanned by

$$X_1 = \frac{\partial}{\partial t}, \qquad X_2 = \frac{\partial}{\partial x}. \tag{3.3}$$

In many applications of group analysis several extensions of the principal Lie algebra admitted by the equations under consideration, are taken from the equivalence algebra $L_{\mathcal{E}}$. These extensions are called $\mathcal{E}$-extensions of the principal Lie algebra [4]. Here these extensions are obtained through the following proposition:

PROPOSITION II: *The operator*

$$X = \Pr_{(t,x,u)} Y \tag{3.4}$$

is a symmetry operator for the equation (1.1) if and only if the function f is invariant with respect to the operator

$$Z = \Pr_{(u,u_t,u_x,f)} Y. \tag{3.5}$$

The last proposition allows us to give two classes of simple but interesting applications.

A) Taking a specialization Y^* of the equivalence operator (2.11), from

$$Z^* (f - \hat{f}(u, u_t, u_x))\Big|_{f=\hat{f}(u,u_t,u_x)} = 0 \tag{3.6}$$

we can look for the form of functions f which admit $\Pr_{(t,x,u)} Y^*$ as symmetry operator.

For istance, we look for the equations (1.1) admitting the following specialization of equivalence operator

$$Y^* = c_1 \frac{\partial}{\partial t} + c_2 \frac{\partial}{\partial x} + c_4 e^u \left\{ \frac{\partial}{\partial u} + [f + (u_t^2 - u_x^2)] \frac{\partial}{\partial f} + u_t \frac{\partial}{\partial u_t} + u_x \frac{\partial}{\partial u_x} \right\}. \tag{3.7}$$

Then the operator $Z^* = \Pr_{(u,u_t,u_x,f)} Y^*$ is

$$Z^* = \frac{\partial}{\partial u} + [f + (u_t^2 - u_x^2)] \frac{\partial}{\partial f} + u_t \frac{\partial}{\partial u_t} + u_x \frac{\partial}{\partial u_x}. \tag{3.8}$$

From (3.6) we obtain

$$\frac{\partial}{\partial u} \hat{f} + u_t \frac{\partial}{\partial u_t} \hat{f} + u_x \frac{\partial}{\partial u_x} \hat{f} = \hat{f} + (u_t^2 - u_x^2) \tag{3.9}$$

whose solution is

$$\hat{f} = u_x F\left(\frac{e^u}{u_x}, \frac{u_t}{u_x}\right) + (u_t^2 - u_x^2) \tag{3.10}$$

where F is an arbitrary function.

Thus the equations

$$u_{tt} - u_{xx} = u_x F\left(\frac{e^u}{u_x}, \frac{u_t}{u_x}\right) + (u_t^2 - u_x^2) \tag{3.11}$$

admit the symmetry operator

$$X^* = c_1 \frac{\partial}{\partial t} + c_2 \frac{\partial}{\partial x} + c_4 e^u \frac{\partial}{\partial u}.$$

B) In contrast to A) we can specialize $f = \hat{f}(u, u_t, u_x)$ and from

$$Z\big(f - \hat{f}(u, u_t, u_x)\big)\Big|_{f=\hat{f}(u,u_t,u_x)} = 0 \tag{3.12}$$

we can look for some possible additional operators which extend $L_{\mathcal{P}}$.

As an example we consider

$$\hat{f} = F(u)(u_t^2 - u_x^2). \tag{3.13}$$

So equation (3.12) becomes

$$Z\big(f - F(u)(u_t^2 - u_x^2)\big)\Big|_{f=F(u)(u_t^2-u_x^2)} = 0 \tag{3.14}$$

which, taking (3.5) and (3.1) into account, is satisfied iff

$$\varphi' = (F\varphi)'. \tag{3.15}$$

From (3.15) we have

$$\varphi = e^{\int F(u)\,du}\left[c_4 + c_5 \int e^{-\int F(u)\,du} du\right] \tag{3.16}$$

with c_4 and c_5 arbitrary constants.

Hence the symmetry operator of equations

$$u_{tt} - u_{xx} = F(u)(u_t^2 - u_x^2) \tag{3.17}$$

which extend $L_{\mathcal{P}}$ is

$$\begin{aligned} X = (c_3 t + c_0 x + c_1)\frac{\partial}{\partial t} + (c_0 t + c_3 x + c_2)\frac{\partial}{\partial x} + \\ + e^{\int F(u)\,du}\left[c_4 + c_5 \int e^{-\int F(u)\,du} du\right] \end{aligned} \tag{3.18}$$

Then the $\mathcal{E}$-extended algebra of $L_{\mathcal{P}}$ is six-dimensional.

If we choose

$$F(u) = ku^{-1} \tag{3.19}$$

we obtain, for $k = 1$,

$$X = (c_3 t + c_0 x + c_1)\frac{\partial}{\partial t} + (c_0 t + c_3 x + c_2)\frac{\partial}{\partial x} + u(c_4 + c_5 \log u)\frac{\partial}{\partial u} \tag{3.20}$$

and, for $k \neq 1$,

$$X = (c_3 t + c_0 x + c_1)\frac{\partial}{\partial t} + (c_0 t + c_3 x + c_2)\frac{\partial}{\partial x} + (c_4 u^k + \frac{c_5}{1-k} u)\frac{\partial}{\partial u}. \tag{3.21}$$

Both applications could be useful in the practice when we need some special forms of invariant solutions compatible with $\mathrm{Pr}_{(t,x,u)} Y$, or when we know a priori the form of the function f and we wish to look quickly for some additional operators.

4. On the preliminary group classification

The method of the preliminary group classification was introduced first in [3]. This method consists in the classification of all non-equivalent equations, with respect to a given equivalence group $G_{\mathcal{E}}$ admitting $\mathcal{E}$-extensions of the principal Lie algebra. $G_{\mathcal{E}}$ is not necessary the largest equivalence group, but it can be any subgroup of the group of all equivalence transformations.

Here, bearing in mind to apply this method to a finite-dimensional algebra and for the sake of simplicity, we select the L_6 subalgebra of $L_{\mathcal{E}}$ spanned by

$$\begin{aligned}
Y_1 &= \frac{\partial}{\partial t}, \\
Y_2 &= \frac{\partial}{\partial x}, \\
Y_3 &= t\frac{\partial}{\partial t} + x\frac{\partial}{\partial x} - 2f\frac{\partial}{\partial f} - u_t\frac{\partial}{\partial u_t} - u_x\frac{\partial}{\partial u_x}, \\
Y_4 &= \frac{\partial}{\partial u}, \\
Y_5 &= u\frac{\partial}{\partial u} + f\frac{\partial}{\partial f} + u_t\frac{\partial}{\partial u_t} + u_x\frac{\partial}{\partial u_x}, \\
Y_6 &= \frac{u^2}{2}\frac{\partial}{\partial u} + [uf + (u_t^2 - u_x^2)]\frac{\partial}{\partial f} + uu_t\frac{\partial}{\partial u_t} + uu_x\frac{\partial}{\partial u_x}.
\end{aligned} \tag{4.1}$$

The operators Y_4, Y_5 and Y_6 correspond to the specialization of φ as: 1, u and $u^2/2$ respectively in (2.12).

Their non-zero projections in the space u, u_t, u_x and f are:

$$\begin{aligned}
Z_1 &= \mathrm{Pr}\, Y_3 = 2f\frac{\partial}{\partial f} + u_t\frac{\partial}{\partial u_t} + u_x\frac{\partial}{\partial u_x}, \\
Z_2 &= \mathrm{Pr}\, Y_4 = \frac{\partial}{\partial u}, \\
Z_3 &= \mathrm{Pr}\, Y_5 = u\frac{\partial}{\partial u} + f\frac{\partial}{\partial f} + u_t\frac{\partial}{\partial u_t} + u_x\frac{\partial}{\partial u_x}, \\
Z_4 &= \mathrm{Pr}\, Y_6 = \frac{u^2}{2}\frac{\partial}{\partial u} + [uf + (u_t^2 - u_x^2)]\frac{\partial}{\partial f} + uu_t\frac{\partial}{\partial u_t} + uu_x\frac{\partial}{\partial u_x}.
\end{aligned} \tag{4.2}$$

We denote by L_4 the Lie algebra spanned by operators (4.2) whose table of commutators is TABLE I.

Let L_m and L'_m $(m \leq 3)$ be m-dimensional similar subalgebras of the algebra L_4 [2, 7], the equations (1.1) and

$$u_{tt} - u_{xx} = f'(u, u_t, u_x) \tag{4.3}$$

constructed according to Proposition II respectively, are equivalent with respect to the finite-dimensional subalgebra L_6.

The problem of preliminary group classification of eqs. (1.1) is to construct the non-similar subalgebras of L_4, or optimal systems of subalgebras.

5. One-dimensional optimal system and equations admitting an extension by one of $L_\mathcal{E}$

We are interested in the construction of non-similar one-dimensional subalgebras of L_4.

According to [2, 7] we need the adjoint algebra L_4 and the corresponding adjoint group. Fixed $\alpha = 1, 2, 3$ or 4, the elements A of the adjoint algebra L_4 are

$$A_\alpha = \sum_{\gamma=1}^{4} [Z_\alpha, Z_\gamma] \frac{\partial}{\partial Z_\gamma} \tag{5.1}$$

which generate the following one-parameter groups of linear transformations

$$Z'_\gamma = Z_\gamma + a_\alpha [Z_\alpha, Z_\gamma], \qquad \gamma = 1, 2, 3, 4. \tag{5.2}$$

Using the table of commutators, we find the matrices $M_\alpha(a_\alpha)$ associated to the linear transformations (5.2), which we do not write for the sake of brevity, being interested in their product

$$M = \prod_\alpha M_\alpha(a_\alpha) = \begin{bmatrix} 1 & 0 & 0 & 0 \\ 0 & 1 - a_3 & a_4(1 - a_3) & 0 \\ 0 & a_2(1 - a_3) & a_2 a_4 (1 - a_3) + 1 & a_4 \\ 0 & 0 & a_2 & 1 + a_3 + a_2 a_4 \end{bmatrix}, \tag{5.3}$$

where a_2, a_3 and a_4 are real numbers and $1 - a_3^2 \neq 0$.

The linear trasformations associated to the matrix M allow us to obtain the collection of pairwise nonconjugate one-dimensional subalgebras of L_4.

The components e^i of the vector $Z \in L_4$ in the basis (4.2) are transformed by means of the transposed matrix M^T of M as follows

$$\begin{aligned} \bar{e}^1 &= e^1 \\ \bar{e}^2 &= (1 - a_3)[e^2 + a_2 e^3] \\ \bar{e}^3 &= a_4(1 - a_3)e^2 + [1 + a_2 a_4 (1 - a_3)]e^3 + a_2 e^4 \\ \bar{e}^4 &= a_4 e^3 + [1 + a_3 + a_2 a_4] e^4 \end{aligned} \tag{5.4}$$

So, the construction of the optimal system of one-dimensional subalgebras of L_4 follows in a very natural way. Obviously we choose as representative of each non equivalent class those which have the simplest form. After some calculations we obtain the following optimal system of one-dimensional subalgebra of L_4:

$$Z_1, \qquad Z_2 + \beta Z_1, \qquad Z_3 + \beta Z_1, \qquad Z_4 + \beta Z_1 \tag{5.5}$$

where β is an arbitrary real number.

Then the non-equivalent equations which admit an extension by one of $L_{\mathcal{P}}$ are obtained by applying Proposition II to operators (5.5). The results are summarized in TABLE II.

Acknowledgements

This work was partially supported by C.N.R. through G.N.F.M. and by M.U.R.S.T. through "Fondi 40% e 60%".

References

1. Fushchych W., Shtelen W., Serov M., Symmetry Analysis and Exact Solutions of non Linear Equations in Mathematical Physics, Naukova Dumka, Kiev (1989) and Kluwer, Dordrecht (1993)
2. Ovsiannikov L. V., Group Analysis of Differential Equations, Academic Press, New York (1982)
3. Akhatov I. Sh., Gazizov R. K., Ibragimov N. H., Nonlocal Symmetries. Heuristic Approach, J. Sov. Math. 55, 1401–1450 (1991)
4. Ibragimov N. H., Torrisi M., Valenti A., Preliminary Group Classification of Equations $v_{tt} = f(x, v_x)v_{xx} + g(x, v_x)$, J. Math. Phys. 32, 11, 2988–2995 (1991)
5. Ibragimov N. H., Torrisi M., A Simple Method for Group Analysis and its Applications to a Model of Detonation, J. Math. Phys. 33, 11, 3931–3937 (1992)
6. Lisle I. G., Equivalence Transformation for Classes of Differential Equations, Ph. D. Thesis, University of British Columbia (1992)
7. Ibragimov N. H., Transformation Groups Applied to Mathematical Physics, Reidel, Dordrecht (1985)

TABLE I

	Z_1	Z_2	Z_3	Z_4
Z_1	0	0	0	0
Z_2	0	0	Z_2	Z_3
Z_3	0	$-Z_2$	0	Z_4
Z_4	0	$-Z_3$	$-Z_4$	0

TABLE II

Z	Equation	Additional operator
Z_1	$u_{tt}-u_{xx}=u_t^2F\left(u,\frac{u_t}{u_x}\right)$	$X_3=t\frac{\partial}{\partial t}+x\frac{\partial}{\partial x}$
$Z_2+\beta Z_1$ ($\beta=0$)	$u_{tt}-u_{xx}=F(u_t,u_x)$	$X_3=\frac{\partial}{\partial u}$
$Z_2+\beta Z_1$ ($\beta\neq 0$)	$u_{tt}-u_{xx}=e^{2\beta u}F\left(\frac{e^{\beta u}}{u_x},\frac{u_t}{u_x}\right)$	$X_3=\beta t\frac{\partial}{\partial t}+\beta x\frac{\partial}{\partial x}+\frac{\partial}{\partial u}$
$Z_3+\beta Z_1$ ($\beta=-1$)	$u_{tt}-u_{xx}=u^{-1}F(u_t,u_x)$	$X_3=-t\frac{\partial}{\partial t}-x\frac{\partial}{\partial x}+u\frac{\partial}{\partial u}$
$Z_3+\beta Z_1$ ($\beta\neq -1$)	$u_{tt}-u_{xx}=u^{2\beta+1}F\left(\frac{u^{\beta+1}}{u_x},\frac{u_t}{u_x}\right)$	$X_3=\beta t\frac{\partial}{\partial t}+\beta x\frac{\partial}{\partial x}+u\frac{\partial}{\partial u}$
$Z_4+\beta Z_1$	$u_{tt}-u_{xx}=u^2e^{-\frac{4\beta}{u}}F\left(\frac{u^2e^{-\frac{2\beta}{u}}}{u_x},\frac{u_t}{u_x}\right)+$ $+2\frac{u_t^2-u_x^2}{u}$	$X_3=\beta t\frac{\partial}{\partial t}+\beta x\frac{\partial}{\partial x}+\frac{u^2}{2}\frac{\partial}{\partial u}$

AN EFFICIENCY IMPROVED PROGRAM LIEPDE FOR DETERMINING LIE-SYMMETRIES OF PDES

THOMAS WOLF
School of Mathematical Sciences
Queen Mary and Westfield College
University of London, London E1 4NS,
T.Wolf@maths.qmw.ac.uk

Abstract. A REDUCE program `LIEPDE` to formulate and automatically solve conditions for determining Lie-symmetries of PDE-systems is described and examples with timings are given. Mainly responsible for the improved efficiency are two modules of the REDUCE package `CRACK`, which is used to solve the symmetry conditions. These modules integrate exact PDEs and perform a generalized separation.

1. Introduction

So far the REDUCE program `CRACK` has been applied to investigate a number of aspects of ordinary differential equations (ODEs) (infinitesimal symmetries, first integrals, equivalent variational principles and differential factorizability) [38]. In each case a relatively short program formulates the application-dependent conditions in the form of an overdetermined system of partial differential equations (PDEs) which were then investigated with `CRACK`. Along these lines the new program `LIEPDE` allows the investigation of point symmetries of systems of PDEs or contact symmetries for single PDEs. New features of `LIEPDE` aim at a high efficiency partly through an alternation of determining some of the symmetry conditions and solving them before deriving the next longer conditions and partly through an as automatic as possible integration strategy provided by `CRACK` which actually does the hard work by solving the overdetermined PDE-system of symmetry conditions. For a detailed description see the manual [38]. Two of its modules which distinguish it from other programs are described in the next chapter. In the third chapter the operation of the program `LIEPDE` is described together with two examples.

2. Solving overdetermined systems with `CRACK`

The main application of programs for solving overdetermined systems is to solve symmetry conditions for DEs. An extensive overview of software for formulating and solving symmetry conditions is given in [9]. The main references for [2], [3], [4], [5], [10], [15], [18], [21], [22], [23], [26], [28], [29], [31], [32], [36]. Common to most programs is that they have a module for adding integrability conditions to the system and solving algebraically highest derivatives of the enlarged system to do further substitutions and to get new integrability conditions. The corresponding procedure is algorithmic and finite and stops when an involutive system in its standard form is

N. H. Ibragimov et al. (eds.),
Modern Group Analysis: Advanced Analytical and Computational Methods in Mathematical Physics, 377–385.

reached. The theory dates back to C. Riquier [25], M.J. Janet [11, 12] and Thomas [34]. Main references of programs which implement these or related algorithms are [1], [6], [7], [8], [20], [21], [22], [23], [24], [27], [30], [35], [39].

CRACK also has such a module, but its distinctive features are an integration and a separation algorithm, now in a generalized version [37], [38]. In the next two subsections both algorithms are sketched.

2.1. Integrating exact PDEs

The purpose of the integration module in **CRACK** is to decide whether a given differential expression D which involves unknown functions $f^i(x^j)$, $1 \le i \le m$ of independent variables x^j, $1 \le j \le n$ is a total derivative of another expression I w.r.t. any variable x^k, $1 \le k \le n$

$$D(x^i, f^j, f^j{}_{,p}, f^j{}_{,pq}, \ldots) = \frac{dI(x^i, f^j, f^j{}_{,p}, f^j{}_{,pq}, \ldots)}{dx^k}$$

and to invert the total derivative i.e. to find I. The index k is reserved in the following for the integration variable x^k. Because the appropriate constant or function of integration which depends on all variables except x^k is added to I, a replacement of $0 = D$ by $0 = I$ in a system of equations is no loss of generality.

Of course there always exists a function I with a total derivative equal to D but the question is whether for arbitrary f^i the integral I is functionally dependent only on the f^i and their derivatives, and not on integrals of f^i.

Preconditions:

D is a polynomial in the f^i and their derivatives. The number of functions and variables is free. For deciding the existence of I only, the explicit occurrence of the variables x^i is arbitrary. To actually calculate I explicitly, all those terms in D which contain x^k only explicitly must be formally integrable w.r.t. x^k. That means only a special explicit occurrence of x^k can prevent the calculation of I if it exists and furthermore only in those terms which do not contain any unknown function of x^k. If such terms occur in D and I exists then I can still be expressed as a polynomial in the $f^i, f^i{}_{,j}, \ldots$ and terms containing indefinite integrals with integrands explicit in x^k.

Algorithm:

Successive partial integration of the term with the highest x^k-derivative of any f^i. By that the differential order w.r.t. x^k is reduced successively. This procedure is always applicable because steps involve only differentiations and the polynomial integration ($\int h^n \frac{\partial h}{\partial x} dx = h^{n+1}/(n+1)$) where h is a partial derivative of some function f^i. For a more detailed description see [37].

Stop:

Iteration stops if no term with any x^k-derivative of any f^i is left. If in the remaining un-integrated terms any $f^i(x^k)$ itself occurs, then I is not expressible with f^i and its derivatives only. In case no $f^i(x^k)$ occurs then any remaining terms can contain x^k only explicitly. Whether they can be integrated depends on their formal integrability. For their integration the REDUCE integrator is applied.

Example :

The above algorithm applied to

$$D := 2f_{,y}\,g' + 2f_{,xy}\,g + gg'^3 + xg'^4 + 3xgg'^2g'' = 0 \tag{1}$$

with $f = f(x,y)$, $g = g(x)$, $' \equiv d/dx$ provides

$$I := 2fg + xygg'^3 + c_1(x) + c_2(y) = 0.$$

Generalization:
If after applying the above basic algorithm, terms are left which contain functions of x^k but each of these functions depends only on a subset of all x^i, $1 \leq i \leq n$, then a generalized version of the above algorithm can still provide a formal expression for the integral I (see [37]). The price consists of additional differential conditions, but they are equations in less variables than occur in the integrated equation. Integrating for example

$$\tilde{D} = D + g^2(y^2 + x\sin y + x^2e^y) \tag{2}$$

gives for the integral

$$\tilde{I} = 2fg + xygg'^3 + c_1(x) + c_2(y) + \frac{1}{3}y^3c_3'' - \cos y(xc_3'' - c_3) + e^y(x^2c_3'' - 2xc_3' + 2c_3) \tag{3}$$

with $c_3 = c_3(x)$ and the single additional condition $g^2 = c_3'''$. The integration of the new terms of (2) is achieved by partial integration again.

Characterization:
This algorithm is a decision algorithm which does not involve any heuristic. It is fast because time and memory requirements grow only linearly with the number of terms in D and with the order of the highest derivative. After integration the new equation is still a polynomial in f^i and in the new constant or function of integration,. Therefore the algorithms for bringing the system into standard form can still be applied to the PDE-system after the equation $D = 0$ is replaced by $I = 0$.

The complexity of algorithms for bringing a PDE-system into a standard form depends nonlinearly on the order of these equations because of the nonlinear increase of the number of different leading derivatives and by that the number of equations generated intermediately by such an algorithm. It therefore in general pays off to integrate equations (with linear expenses) during such a standard form algorithm. The increase in the number of unknown constants/functions through integration does not play a role for standard form algorithms because these functions have less variables and do not increase the number of possible leading derivatives and therefore not the number of equations to be maintained during execution of the standard form algorithm.

If an f^i, which depends on all variables, can be eliminated after an integration, then depending on its length it is in general helpful to substitute f^i in other equations and to reduce the number of equations and functions by one. This is especially profitable if the replaced expression is short and contains only functions of less variables than f^i.

2.2. Separation of PDEs

As a result of repeated integrations the functions in the remaining equations have less and less variables. A direct separation in the case that at least one variable occurs only explicitly and not as an argument of any unknown function is incorporated in all similar programs of other authors. The generalization implemented in **CRACK** concerns a more general class of equations.

Preconditions:

D is a polynomial in the f^i and their derivatives. No unknown function depends on all variables but each independent variable occurs in at least one unknown function as in

$$0 = f(x)g(y) - \frac{1}{2}xf'(x) - g'(y) - (1 + x^2)y. \tag{4}$$

The number of functions and variables is free as well as the explicit occurrence of the variables.

Algorithm:

The idea is to do a sequence of divisions and differentiations to be able to separate directly w.r.t. one variable. To have, for example, x occur only explicitly in (4) we must get rid of all terms which contain $f(x)$ and for that we differentiate w.r.t. y, to cancel the second term, divide by g and differentiate again w.r.t. y to cancel the first term.

Because problems such as (4) are nonlinear, book-keeping is necessary in the case that the equations can be factorized with the **REDUCE** factorizer and the factors have to be set to zero individually. A separate investigation is also necessary whenever a division is to be done where the divisor is set to zero. For more details see the manual [38].

Result:

For equation (4) **CRACK** obtains the general solution $f = 1 + x^2, \quad g = 1 + y$ and $f = -1 - x^2, \quad g = 1 - y$. The special case $g = 0$ leads to a contradiction and no further solution.

Characterization:

To split one PDE in n variables into a number of shorter PDEs/ODEs in less variables is a major simplification step.

3. Determining symmetries of PDE-systems iteratively

The standard method to simplify or solve nonlinear differential equations is to use Sophus Lie's theory and investigate their infinitesimal symmetries. In this section we describe a program **LIEPDE** which investigates point symmetries for a PDE-system and contact symmetries for single PDEs.

To obey symmetries, differential equations

$$H_A = 0 \tag{5}$$

for unknown functions u^α of independent variables x^i must be forminvariant against infinitesimal transformations

$$\tilde{x}^i = x^i + \varepsilon\xi^i, \quad \tilde{u}^\alpha = u^\alpha + \varepsilon\eta^\alpha \tag{6}$$

in first order of ε. To transform the equations (5) by (6), derivatives of u^α must be transformed, i.e. the part linear in ε must be determined. The corresponding formulas are (see e.g. [19], [33])

$$\begin{aligned} \tilde{u}^\alpha_{j_1\ldots j_k} &= u^\alpha_{j_1\ldots j_k} + \varepsilon\eta^\alpha_{j_1\ldots j_k} + O(\varepsilon^2) \\ \eta^\alpha_{j_1\ldots j_{k-1}j_k} &= \frac{D\eta^\alpha_{j_1\ldots j_k}}{Dx^k} - u^\alpha_{ij_1\ldots j_{k-1}}\frac{D\xi^i}{Dx^k} \end{aligned} \tag{7}$$

and the complete symmetry condition then takes the form

$$XH_A = 0 \quad \mathrm{mod} \quad H_A = 0 \tag{8}$$

$$X = \xi^i\frac{\partial}{\partial x^i} + \eta^\alpha\frac{\partial}{\partial u^\alpha} + \eta^\alpha_m\frac{\partial}{\partial u^\alpha_m} + \eta^\alpha_{mn}\frac{\partial}{\partial u^\alpha_{mn}} + \ldots + \eta^\alpha_{mn\ldots p}\frac{\partial}{\partial u^\alpha_{mn\ldots p}}. \tag{9}$$

where mod $H_A = 0$ means that the original PDE-system is used to replace some partial derivatives of u^α to reduce the number of independent variables, because the symmetry condition (8) must be fulfilled identically in x^i, u^α and all partial derivatives of u^α.

For point symmetries ξ^i, η^α are functions of x^j, u^β and for contact symmetries they depend on x^j, u^β and u^β_k.

Usually the determining equations for symmetries of PDEs are simpler to solve than those for ODEs. The reason is that the more variables a function has the higher is the number of different derivatives of a given total order in which the symmetry condition has to be satisfied identically. It therefore can be separated w.r.t. more different partial derivatives, splitting the whole condition into a larger number of smaller equations, such that the resulting overdetermined system is simpler to solve. The price for this simplicity in the PDE case is that symmetries are less useful for solving PDEs than for solving ODEs. Given a point symmetry for a PDE, it only allows one to lower the number of variables at the price of losing generality. In contrast a point symmetry of an ODE allows one lower the order of the ODE in full generality. As a consequence of this relative simplicity, the integration module and generalized separation module in most practical cases allow a quick automatic solution of the determining equation as can be seen from the CPU-times given in the examples below.

The difficulty of determining symmetries of PDEs lies more in the computational effort for formulating the conditions (8) if the number of x^i and u^α is high. This can partially be avoided if at first only a few conditions are formulated and solved such that the formulation of the remaining ones becomes shorter.

A first step in this direction is to investigate one PDE $H_A = 0$ after another, as done in [4]. Two methods to partition the conditions for a single PDE are described by Bocharov/Bronstein [2] and Stephani [33].

In the first method only those terms of the symmetry condition $XH_A = 0$ are calculated which contain at least a derivative of u^α of a minimal order m. Setting coefficients of these u-derivatives to zero provides symmetry conditions. Lowering the minimal order m successively then gradually provides all symmetry conditions.

The second method is even more selective. If H_A is of order n then only terms of the symmetry condition $XH_A = 0$ are generated which contain n'th order derivatives of u^α. Furthermore these derivatives must not occur in H_A itself. They can therefore occur in the symmetry condition (8) only in $\eta^\alpha_{j_1\dots j_n}$, i.e. in the terms

$$\eta^\alpha_{j_1\dots j_n}\frac{\partial H_A}{\partial u^\alpha_{j_1\dots j_n}}.$$

If only coefficients of n'th order derivatives of u^α need to be accurate then from the total derivatives to be taken in (8) only that part is performed which differentiates w.r.t. the highest u^α-derivatives. This means, for example, to form only $u^\alpha_{mnp}\partial/\partial u^\alpha_{mn}$ if the expression, which is to be differentiated totally, contains at most second order derivatives of u^α.

3.1. Examples

All reported calculations below have been performed on a PC 486/33 MHz with 4 MByte. Times are given in the form 'CPU time'/'garbage collection time' in seconds. During calculation no intermediate results were printed. Because the given intermediate times are rounded upwards, the total time may be less than the sum of the partial times.

One commonly used PDE to test software for the determination of symmetries is the Kadomtsev-Petviashvili equation [30]. It reads for a function $u(t,x,y)$:

$$4u_{,tx} + u_{,xxxx} + 6uu_{,xx} + 6u_{,x} + 3u_{,yy} = 0.$$

After 1.3/0 sec the program has formulated the two conditions

$$0 = \xi^t{}_{,u} \qquad \text{and} \qquad 0 = \xi^y{}_{,u}$$

by using the second of the above described selective methods. These conditions are solved to

$$\xi^t = c_1(t,x,y) \qquad \text{and} \qquad \xi^y = c_2(t,x,y)$$

within 0.5/0 sec. After a further 8.1/0.1 sec the full symmetry condition (8), as yet unseparated (i.e. as a single equation), is derived which is then solved by CRACK within 53.5/0.7 sec to give the symmetry generators

$$\xi^t = c_3 \quad \xi^x = -\frac{2}{3}c_4'y - c_5 \quad \xi^y = c_4 \quad \eta^u = -\frac{4}{9}c_4''y - \frac{2}{3}c_5'$$

where $c_4 = c_4(t)$, $c_5 = c_5(t)$ and c_3 =constant are free. The total time is 63.2/0.8 sec. The corresponding three symmetry-generators are

$$\partial_t, \qquad -\frac{2}{3}c_4'y\partial_x + c_4\partial_y - \frac{4}{9}c_4''y\partial_u \qquad \text{and} \qquad -c_5\partial_x - \frac{2}{3}c_5'\partial_u.$$

A more difficult system of PDEs are the Karpman equations [14] which play a role in plasma physics. In their real form, which is used to determine symmetries, they are ([4])

$$\rho_{,t} + w_1\rho_{,z} + \frac{1}{2}\left[s_1(2\rho_{,x}\phi_{,x} + 2\rho_{,y}\phi_{,y} + \rho\phi_{,xx} + \rho\phi_{,yy}) + s_2(2\rho_{,z}\phi_{,z} + \rho\phi_{,zz})\right] = 0$$

$$\phi_{,t} + w_1\phi_{,z} - \frac{1}{2}\left[s_1\left(\frac{\rho_{,xx}}{\rho} + \frac{\rho_{,yy}}{\rho} - \phi_{,x}^2 - \phi_{,y}^2\right) + s_2\left(\frac{\rho_{,zz}}{\rho} - \phi_{,z}^2\right)\right] + a_1\nu = 0$$

$$\nu_{,tt} - (w_2)^2(\nu_{,xx} + \nu_{,yy} + \nu_{,zz}) - 2a_2\rho(\rho_{,xx} + \rho_{,yy} + \rho_{,zz}) - 2a_2(\rho_{,x}^2 + \rho_{,y}^2 + \rho_{,z}^2) = 0$$

with the three functions ρ, ϕ, ν of four variables t, x, y, z and with constant parameters a_i, s_i, w_i.

At first, preliminary symmetry conditions are derived and solved successively for each of the three equations. To initialize all internal variables and to formulate preliminary conditions for the first equation takes 7.7/0.1 sec and 20.4/0.1 sec to solve them. The solution requires 31 integrations and states that all ξ^i are functions of t, x, y, z only. The preliminary conditions of the second equation do not provide new constraints which to realize takes the program 4.8/0 sec. To formulate the preliminary conditions for the third equation costs 1.8/0 sec and 0.3/0 sec to solve them with the result that ν is a function of t, x, y, z only. This needs 4 integrations. After 7.1/0.1 sec the full conditions for the first equation are derived which to solve takes 179.2/2.6 sec and requires further 77 integrations. To formulate the remaining full conditions for the second equation takes 7.6/0.1 sec and 3.8/0.1 sec to solve them which requires 6 integrations. Finally, the full conditions for the last equation are formulated in 10/0.2 sec and solved with three more integrations after 26.9/0.2 sec.

The general solution for the symmetry generators then reads

$$\begin{aligned} \xi^x &= -yc_1 + c_2 & \eta^\rho &= 0 \\ \xi^y &= xc_1 + c_3 & \eta^\phi &= a_1 c_6 t^2 + a_1 c_7 t + c_8 \\ \xi^z &= c_4 & \eta^\nu &= -c_6 t - c_7 \\ \xi^t &= c_5 \end{aligned}$$

with constants $c_1, \ldots, c_8$. The corresponding 8 symmetry generators are

$$\begin{aligned} X_1 &= \partial_x & X_5 &= y\partial_x - x\partial_y \\ X_2 &= \partial_y & X_6 &= \partial_\phi \\ X_3 &= \partial_z & X_7 &= a_1 t\partial_\phi - \partial_\nu \\ X_4 &= \partial_t & X_8 &= a_1 t^2\partial_\phi - 2t\partial_\nu. \end{aligned}$$

The total time is 260/3.2 seconds, i.e. less than 5 min on a PC 486. For comparison, the same calculation reported in [4], also split into 6 parts but with integrations done by hand, took more than 2 hours CPU on a VAX 8600 to formulate and simplify the symmetry conditions.

3.2. INPUT

The call of **LIEPDE** is similar to that of other application programs of the package **CRACK** which are included in the REDUCE network library. For the Kadomtsev-Petviashvili equation the input to determine point symmetries would be

```
depend u,t,x,y;
de := {{4*df(u,t,x)+6*df(u,x)+6*u*df(u,x,2)+df(u,x,4)+3*df(u,y,2)},
      {u},{t,x,y}}$
```

```
mo := {0,nil,nil};
LIEPDE(de,mo);
nodepend u,t,x,y;
end$
```

The first argument to **LIEPDE** is a list which includes the list of equations, the list of unknown functions and the list of independent variables. The second argument to **LIEPDE** is a list `mo = {od, lp, fl}` where

- `od` is the order of the ansatz for ξ, η ($= 0$ for point symmetries and $= 1$ for contact symmetries),
- if `lp` is *nil* then the standard ansatz for ξ^i, η^α is taken, otherwise `lp` is the ansatz for ξ^i, η^α and must have the form {`xi_`$x1$ `= ..., ..., eta_`$u1$ `= ..., ...`} where `xi_`, `eta_` are fixed and $x1, \ldots, u1$ are to be replaced by the actual name of the variables and functions.
- `fl` is the list of free functions in the ansatz in case `lp` is not *nil*.

3.3. SUMMARY

To summarize, the described integration and separation modules allow a very efficient automatic solution of simple PDEs like those which result from symmetry conditions of PDE-systems. A successive iteration consisting of formulating and solving symmetry conditions cuts further down the complexity of the calculations, which e.g. makes it possible to run difficult problems on a PC.

References

1. E.A. Arais, V.P. Shapeev and N.N. Janenko, DAN SSSR 214, 296 (1974)
2. A.V. Bocharov, M.L. Bronstein, Efficiently implementing two methods of the geometrical theory of differential equations: An experience in algorithm and software design, Acta Appl. Math. 16, 143-166 (1988)
3. J. Carminati, J.S. Devitt and G.J.Fee, Isogroups of differential equations using algebraic computing, J. Sym. Comp. 14, 103-120 (1992)
4. B. Champagne, W. Hereman and P. Winternitz, The computer calculation of Lie point symmetries of large systems of differential equation, Comp. Phys. Comm. 66, 319-340 (1991)
5. W.I. Fushchych, V.V. Kornyak, Computer algebra application for determining Lie and Lie-Bäcklund symmetries of differential equations, J. Symb. Comp. 7, no 6, 611-619 (1989)
6. V.G. Ganja, S.V. Meleshko, F.A. Murzin, V.P. Shapeev and N.N. Janenko, DAN SSSR 261, 1044 (1981)
7. V.G. Ganzha and N.N. Yanenko, Compatibility Analysis for Systems of Differential Equations on a Computer, ITPM Preprint # 20-82 (ITPM, Novosibirsk, USSR, in Russian) (1982)
8. V.P. Gerdt and A.Yu. Zharkov, Computer generation of necessary integrability conditions for polynomial-nonlinear evolution systems, in: Proc. ISSAC '90, Tokyo, Japan, Eds.: S. Watenabe and M. Nagata (ACM Press, New York) 250-254 (1990)
9. W. Hereman, Review of Symbolic Software for the Computation of Lie Symmetries of Differential Equations, to appear in Euromath Bulletin 2, no 1 (1993)
10. S. Herod, MathSym: A Mathematica program for computing Lie symmetries, Preprint, Program in Applied Mathematics (The University of Colorado, Boulder, Colorado, 1992) submitted to the Mathematica Journal (1993)
11. M. Janet, J. Math. 3, 65-151 (1920)
12. M. Janet, Lecons sur les Systemes d'Equations aux Derivees (Gauthier-Villars, Paris, 1929); Equations aux Derivees Partielles (CNRS, Paris, 1956)
13. E. Kamke, Differentialgleichungen, Lösungsmethoden und Lösungen, Band 1, Gewöhnliche Differentialgleichungen, Chelsea Publishing Company, New York, (1959)

14. V.I. Karpman, Phys. Lett. A 136, 216 (1989)
15. V.V. Kornyak and W.I. Fushchych, A program for symmetry analysis of differential equations, Proc. of the 1991 International symposium on Symbolic and Algebraic Computation, (S.Watt, ed.), ACM Press, Bonn, 315-316 (1991)
16. M.A.H. MacCallum, An Ordinary Differential Equation Solver for REDUCE, Proc. ISAAC'88, Springer Lect. Notes in Comp Sci. 358, 196–205.
17. M.A.H. MacCallum, F.J. Wright, Algebraic Computing with REDUCE, Clarendon Press, Oxford (1991).
18. M.C. Nucci, Interactive REDUCE programs for calculating, classical, non-classical, and Lie-Bäcklund symmetries of differential equations, Preprint, School of Mathematics, Georgia Institute of Technology, Atlanta, (1990)
19. P.J. Olver, Applications of Lie Groups to Differential Equations, Springer-Verlag New York (1986)
20. G.J. Reid, J. Phys. A: Math. Gen. 23, L853-L859 (1990)
21. G.J. Reid, Algorithmic determination of Lie symmetry algebras of differential equations, Lie Theory, Differential Equations and Representation Theory, (V. Hussin, ed.) Les Publications de Centres de Recherches Mathematiques, Montreal, Canada, 363-372 (1990)
22. G.J. Reid, A triangularization algorithm which determines the Lie symmetry algebra of any system of PDEs, J. Phys. A: Math. Gen. 23, L853-L859 (1990)
23. G.J. Reid, Finding symmetries of differential equations without integrating determining equations, Europ. J. Appl. Math. 2, 293-340 (1991)
24. G.J. Reid and A. Boulton, Reduction of systems of differential equations to standard form and their integration using directed graphs, in: Proc. ISSAC '91, Bonn, Germany, Ed.: M. Watt (ACM Press, New York), 308-312 (1991)
25. C. Riquier, Les systemes d'equations aux derives partielles, Paris Gauthier-Villars (1910)
26. W. Sarlet and J. Vanden Bonne, REDUCE-procedures for the study of adjoint symmetries of second-order differential equations, J. Sym. Comp. 13, 683-693 (1992)
27. F. Schwarz, Physica D 11, 243-251 (1984)
28. F. Schwarz, Automatically determining symmetries of partial differential equations, Computing 34, 91-106 (1985); Addendum: Computing 36, 279-80 (1986)
29. F. Schwarz, Symmetries of differential equations from Sophus Lie to computer algebra, SIAM Review 30, 450-81 (1988)
30. F. Schwarz, An Algorithm for Determining the Size of Symmetry Groups, Computing 49, 95-115 (1992)
31. J. Sherring and G. Prince, DIMSYM - Symmetry determination and linear differential equations package, Report, LaTrobe University, Bundoora, Australia (1992)
32. S. Steinberg, Symmetries of differential equations, MACSYMA Newsletter 7, 3-7 (1990)
33. H. Stephani, Differential Equations, Their solution using symmetries, Edit. M.A.H. MacCallum, Cambridge Univ. Press (1989)
34. J. Thomas, Differential Systems, Colloquium Publications 21, AMS, New York (1937)
35. V.L. Topunov, Acta Appl. Math. 16, 191-206 (1989)
36. P. Vafeades, PDELIE: A partial differential equation solver, Macsyma Newsletter 9, 1-13 (1992)
37. T. Wolf, The Symbolic Integration of Exact PDEs, preprint (1991).
38. T. Wolf, A. Brand, The Computer Algebra Package CRACK for Investigating PDEs, preprint, manual for the package in the REDUCE network library and in Proceedings of ERCIM School on Partial Differential Equations and Group Theory, April 1992 in Bonn, GMD Bonn.
39. T. Wolf, An analytic algorithm for decoupling and integrating systems of nonlinear partial differential equations, J. of Comp. Phys. 60, no 3, 437-46 (1985)

CONDITIONAL SYMMETRIES AND THE DIRECT REDUCTION OF PARTIAL DIFFERENTIAL EQUATIONS

S. ZIDOWITZ
Institut für Mathematische Physik,
TU-Braunschweig,
Mendelssohnstr. 3,
D-3300 Braunschweig, Germany

Abstract. In the 'classical' symmetry reduction of PDEs one looks for solutions, that are invariant under some chosen subgroup of the symmetry group of the given PDEs. An attempt to generalize this method leads to the concept of conditional symmetries [1, 3, 4, 5]. It seems to be connected to the 'direct' method suggested by Clarkson and Kruskal [2]. By giving a – as far as we know new – definition of the conditional symmetries, which corresponds to the algorithm of Levi and Winternitz, we are able to prove the equivalence between the reduction by conditional symmetries and the 'direct' method in the case of reducing a locally solvable system of PDEs by one dimension.

Further we show, that the methods where also manageable for (simple) systems of PDEs: a hydrodynamic model for a plasma with hot electrons is discussed – it forms a system of PDEs with two independent and three dependent variables.

1. Reduction of partial differential equations

We consider a system $\Theta(x,u,u^{(k)}) = (\Theta_1(x,u,u^{(k)}),\ldots,\Theta_l(x,u,u^{(k)}))$ of PDEs given by

$$\Theta(x,u,u^{(k)}) = 0$$

$x := (x^1,\ldots,x^n) \in X \subset I\!R^n$ ($u := (u^1,\ldots,u^m) \in U \subset I\!R^m$) stands for the independent (for the dependent) variables. $u^{(k)}$ means the partial derivatives of the u^σ with respect to the x^i up to order k. To reduce this equations, let us consider an invertible transformation T over $M \subset X \times U$:

$$T : M \to I\!R^{n+m}$$

$$z^1 = \gamma^1(x,u)\ ,\ldots,\ z^{n-r} = \gamma^{n-r}(x,u) \qquad s^1 = S^1(x)\ ,\ldots,\ s^r = S^r(x)$$
$$w^1 = \gamma^{n+1-r}(x,u)\ ,\ldots,\ w^{n+m-r} = \gamma^{n+m-r}(x,u)$$

with the inversion: $\quad x = \mathcal{X}(s,z,w) \quad u = \mathcal{U}(s,z,w)$

Using the chain rule it is possible to prolongate this transformation to an invertible transformation over $M^{(k)}$, so one gets:

$$u^{(k)} = \mathcal{U}^{(k)}(s,z,w,w^{(k)})$$

Putting this into the system $\Theta(x,u,u^{(k)})$ leads for $w(s,z)$ to the transformed system

$$\widetilde{\Theta}'_\nu(s,z,w,w^{(k)}) = 0 \qquad \nu = 1,\ldots,l$$

Demanding w should be independent of the parametric variables $s = (s^1,\ldots,s^r)$, one is left with two cases:

N. H. Ibragimov et al. (eds.),
Modern Group Analysis: Advanced Analytical and Computational Methods in Mathematical Physics, 387–393.

1) The transformed system is algebraic equivalent to a system, which contains only the reduction variables $z, w, w^{(k_z)}$ (the derivatives of the w^σ with respect to the z^i):
$$\widetilde{\Theta}'(s,z,w,w^{(k)})\Big|_{\substack{w_{s^i}=0\\ i\in\{1,\dots,r\}}} = A(s,z)\cdot\widetilde{\Theta}(z,w,w^{(k_z)}) \qquad (1)$$
with a regular $l\times l$ matrix $A(s,z)$. In this case a solvable system $\Theta(x,u,u^{(k)})$ is transformed into a solvable system $\widetilde{\Theta}(z,w,w^{(k_z)})$ for $w(z)$, containing less independent variables.

2) In general it is not possible to get rid of the parametric variables s. In this case the expansion of $\widetilde{\Theta}'(s,z,w,w^{(k)})$ with $w_{s^i}=0$ $(i=1,\dots,r)$ for instance in powers of the s^i leads to an overdetermined set of differential equations for $w(z)$. If these equations aren't compatible, then there is no solution of the original system $\Theta(x,u,u^{(k)})$ with $u=\mathcal{U}(s,z,w(z))$.

The question is, how to find systematic approaches for calculating transformations leading to the form (1).

2. The direct reduction

An answer to this question is the direct reduction established by Clarkson and Kruskal [2]. In this one takes an arbitrary transformation
$$u=\mathcal{U}(x,w(z)) \qquad z=z(x) \qquad (z\in\mathbb{R}^{n-r})$$
(Note that $\mathcal{U}$ depends on the original independent variables x.) By the demand, that substituting this transformation into the differential equation $\Theta(x,u,u^{(k)})$ should lead to a differential equation only depending on $z,w,w^{(k_z)}$ in the sense of (1) one is be able to calculate proper transformations. Making the calculations manageable, $A(s,z)$ should be restricted to diagonal form. For practical calculations further the ansatz $\mathcal{U}^\sigma(x,w(z))=\alpha^\sigma(x)+\beta^\sigma(x)\cdot w^\sigma(z)$ $(\sigma=1,\dots,m)$ is convenient. In many cases this means no loss of generality.

3. Conditional symmetries

In generalizing the 'classical' Lie formalism of similarity reduction, the concept of conditional symmetries is discussed in literature for calculating proper transformations [1, 3, 4, 5]. A careful investigation of this 'working instructions' leads us to following definition:

Definition 1 *Let G be a connected (local) Lie group of transformations on the set $M\subset X\times U=\mathbb{R}^{n+m}$, $(\mathbf{v}_{(1)},\dots,\mathbf{v}_{(r)})$ a basis of the corresponding Lie algebra $\mathcal{G}$ with characteristics $Q_{(1)},\dots,Q_{(r)}$.*[1] *Let $\Theta(x,u,u^{(k)})=0$ be a system of differential*

[1] The characteristic $Q=(Q^1,\dots,Q^m)$ of an infinitesimal generator $\mathbf{v}=\sum_{j=1}^{n}\xi^j\frac{\partial}{\partial x^j}+\sum_{\sigma=1}^{m}\eta^\sigma\frac{\partial}{\partial u^\sigma}$ is defined by $Q^\sigma(x,u,u^{(1)}):=\eta^\sigma(x,u)-\sum_{j=1}^{n}\xi^j(x,u)\frac{\partial u^\sigma}{\partial x^j}$ $(\sigma=1,\dots,m)$

equations of order k, Λ the on $M^{(k)}$ prolongated combined system:

$$\Lambda := \{\Theta_\nu \,,\ D_{J_{(i)}} Q^\sigma_{(i)} \mid \nu = 1,\ldots,l \,,\ \sigma = 1,\ldots,m \,,\ i = 1,\ldots,r \,, \\ |J_{(i)}| \leq k-1 \text{ for } \textstyle\sum_{j=1}^n (\xi^j_{(i)})^2 \neq 0 \,,\ |J_{(i)}| \leq k \text{ for } \sum_{j=1}^n (\xi^j_{(i)})^2 = 0\}$$

(D_J stands for the total differential characterised by J, where $|J|$ means the order of differentation).

G is a **conditional symmetry** *of the original system $\Theta = 0$, if the set $\mathcal{S}_\Lambda$, which is given by $\Lambda = 0$ in $M^{(k)}$, is invariant under the prolongated action of G on $M^{(k)}$.*

It is important, that definition 1 means a stronger condition then the demand, that G should be a 'classical' symmetry of the combined system Λ (which leads to the 'weak symmetries' of Olver and Rosenau [5]). Here not only the points of locally solvability of $\mathcal{S}_\Lambda$ are taken into account. This 'geometric' symmetry condition enforces the compatibility of the combined system Λ:

Theorem 1 *Let $\Theta(x, u, u^{(k)}) = 0$ be a analytic, locally solvable system of PDEs with maximal rank on $M^{(k)}$. Let G be a local, projectable Lie transformation group with a complete set of globally defined functionally independent invariants $z^1 = \gamma^1(x), \ldots, z^{n-r} = \gamma^{n-r}(x)$, $w^1 = \gamma^{n+1-r}(x,u), \ldots, w^{n+m-r} = \gamma^{n+m-r}(x,u)$, providing local coordinates on an open subset $M/G \subset Z \times W = I\!R^{n-r} \times I\!R^m$. Let $Q_{(1)}, \ldots, Q_{(r)}$ be the characeristics for a basis of the corresponding Lie algebra.*

G is a conditional symmetry of Θ if and only if the prolongated combined system Λ is locally solvable in every point of the set $\mathcal{S}_\Lambda \subset M^{(k)}$.

Proof: For $\mathcal{S}_\Lambda = \emptyset$ the equivalence is trivial. Hence we assume $\mathcal{S}_\Lambda \neq \emptyset$.

Let a conditional symmetry be given. We start by completing the invariants to a full set of coordinates on M. For this the solutions of

$$\mathbf{v}_{(i)}(s^j(x)) = \delta_{ij} \qquad i = 1,\ldots,r \qquad j = 1,\ldots,r$$

are taken as parametric variables s. The independence of the $\mathbf{v}_{(i)}$ ensures the existence of r functionally independent solutions $s^j(x)$ with $j = 1,\ldots,r$. The transformation into the new variables $\gamma^1,\ldots,\gamma^{n+m-r}$ and $s^1,\ldots,s^r$ is due to this functionally independency invertible. In the new variables we get for Λ the equivalent system:

$$\tilde{\Lambda} := \{\tilde{\Theta}_\nu(s,z,w,w^{(k)}) \,,\ D_{J_{s^i}} w^\sigma(s,z) \mid \nu = 1,\ldots,l \,,\ \sigma = 1,\ldots,m \,, \\ i = 1,\ldots,r \,,\ |J| \leq k-1\}$$

$\tilde{\Theta}_\nu(s,z,w,w^{(k)}) = 0$ is again locally solvable. Moreover the infinitesimal symmetry condition for Λ is preserved under an invertible transformation:

$$\frac{\partial}{\partial s^i} \tilde{\Theta}_\nu(s,z,w,w^{(k)})|_{\tilde{\Lambda}=0} = 0 \qquad |J| \leq k-1 \qquad i = 1,\ldots,r$$

So $\tilde{\Theta}(s,z,w,w^{(k)})$ is on the set $\mathcal{S}_{\tilde{\Lambda}}$ (given by $\tilde{\Lambda} = 0$) independent of the parametric variables s. The solutions of $\tilde{\Theta}(s,z,w,w^{(k)})$ with initial conditions on $\mathcal{S}_{\tilde{\Lambda}}$, which exists by assumption, are therefore independent of s and solves $D_{J_{s^i}} w = 0$. Hence $\tilde{\Lambda} = 0$ is locally solvable and so is its transformed Λ.

The other direction of the equivalence is easy to prove: Any solution of $\Lambda = 0$ solves especially the characteristics $Q_{(1)}, \ldots, Q_{(r)}$. Hence it is an invariant function of G. The locally solvability of Λ ensures, that the invariance of the set $\mathcal{S}_\Lambda$ is equivalent to the invariance of the set of solutions. So G is a conditional symmetry of Θ if Λ is locally solvable. □

4. On the equivalence of conditional symmetries and the direct reduction

Restricting us to the (practical) case of reduction by one dimension and projectable transformations we are able to prove equivalence in the following sense:

Theorem 2 *Let $\Theta(x, u, u^{(k)}) = 0$ be a analytic, locally solvable system of PDEs with maximal rank on $M^{(k)}$. Let the ansatz*

$$\begin{aligned} u^\sigma &= \mathcal{U}^\sigma(x, w(z)) && \sigma = 1, \ldots, m \\ z^i &= \mathcal{Z}^i(x) && i = 1, \ldots, n-1 \\ \textit{with inversion} \quad w^\sigma &= \mathcal{W}^\sigma(x, u) && \sigma = 1, \ldots, m \end{aligned}$$

and

$$\operatorname{rang}\left(\frac{\partial \mathcal{W}^\sigma(x,u)}{\partial u^\beta}\right) = m \quad \operatorname{rang}\left(\frac{\partial \mathcal{Z}^i(x)}{\partial x^j}\right) = n-1 \tag{2}$$

transform $\Theta(x, u, u^{(k)}) = 0$ in a locally solvable system $\widetilde{\Theta}(z, w, w^{(k_z)}) = 0$ for $w(z)$. Then there is a (projectable) conditional symmetry of $\Theta = 0$, for which $w = \mathcal{W}(x, u)$ und $z = \mathcal{Z}(x)$ are the corresponding invariants.

Proof: The invariance conditions:

$$\sum_{j=1}^{n} \xi^j(x,u) \frac{\partial \mathcal{Z}^i(x)}{\partial x^j} = 0 \qquad i = 1, \ldots, n-1$$

$$\sum_{j=1}^{n} \xi^j(x,u) \frac{\partial \mathcal{W}(x,u)}{\partial x^j} + \sum_{\sigma=1}^{m} \eta^\sigma(x,u) \frac{\partial \mathcal{W}(x,u)}{\partial u^\sigma} = 0 \qquad \sigma = 1, \ldots, m$$

gives the infinitesimal generator $\mathbf{v} = \sum_{j=1}^n \xi^j \frac{\partial}{\partial x^j} + \sum_{\sigma=1}^m \eta^\sigma \frac{\partial}{\partial u^\sigma}$ connected with $\mathcal{W}$ and $\mathcal{Z}$. Due to (2) the coefficients are settled up to multiplication by a free function. Further gives $z = \mathcal{Z}(x)$ a projectable infinitesimal generator.
Following theorem 1 we are left with the proof, that the system

$$\Lambda := \{\Theta_\nu = 0 \ , \ D_J Q^\sigma = 0 \mid \nu = 1, \ldots, l \ , \ \sigma = 1, \ldots, m \ , \ |J| \leq k-1\}$$

is locally solvable. For this we transform on the reduction variables:

$$\begin{aligned} u^\sigma &= \mathcal{U}^\sigma(x, w)|_{x = \mathcal{X}(z,s)} && \sigma = 1, \ldots, m \\ x^j &= \mathcal{X}^i(z, s) && i = 1, \ldots, n-1 \end{aligned} \tag{3}$$

($\mathcal{X}$ means the inversion of $\mathcal{Z}(x)$ completed by a parameter variable s, which solves $\mathbf{v}(s) = 1$.) Substituting this into Λ we get

$$\tilde{\Lambda} = \{\widetilde{\Theta}'_\nu(s, z, w, w^{(k)}) = 0 \ , \ w^\sigma_{J_s} = 0 \mid \nu = 1, \ldots, l \ , \ \sigma = 1, \ldots, m \ , \ |J| \leq k-1\}$$

as equivalent to the transformed system. By assumption $\tilde{\Lambda}$ is equivalent to a locally solvable system $\widetilde{\Theta}(z, w, w^{(k_s)}) = 0$ for $w = w(z)$. Since locally solvability is preserved under invertible transformations, Λ is locally solvable too. This proves the theorem. □

In the other direction one gets:

Theorem 3 *Let $\Theta(x, u, u^{(k)}) = 0$ be a analytic, locally solvable system of PDEs with maximal rank on $M^{(k)}$. Let $\mathbf{v} = \sum_{j=1}^{n} \xi^j(x)\frac{\partial}{\partial x^j} + \sum_{\sigma=1}^{m} \eta^\sigma(x, u)\frac{\partial}{\partial u^\sigma}$ with $\sum_{j=1}^{n}(\xi^j(x))^2 \neq 0$ be the generator for a conditional symmetry of $\Theta = 0$ with a complete set of globally defined, functionally independent invariants*

$$\begin{aligned} w^\sigma &= \mathcal{W}^\sigma(x, u) \qquad & \sigma &= 1, \dots, m \\ z^i &= \mathcal{Z}^i(x) & i &= 1, \dots, n-1 \\ \textit{with inversion} \qquad u^\sigma &= \mathcal{U}^\sigma(x, w) & \sigma &= 1, \dots, m \end{aligned} \tag{4}$$

Then transition to the invariants gives a locally solvable system $\widetilde{\Theta}(z, w, w^{(k_s)}) = 0$ for w=w(z).

Proof: By transforming into the room of invariants following (3) the system Λ which is locally solvable by asumption and theorem 1 is maped into the system $\tilde{\Lambda} = \{\widetilde{\Theta}'_\nu(s, z, w, w^{(k)}) = 0\ ,\ w^\sigma_{J_s} = 0 \mid \nu = 1, \dots, l\ ,\ \sigma = 1, \dots, m\ ,\ |J| \leq k-1\}$ Hence this system is also locally solvable. Thus $\widetilde{\Theta}'(s, z, w, w^{(k)}) = 0$, into which Θ is transformed by ansatz (4), is locally solvable on $\mathcal{S}_{\tilde{\Lambda}}$. For this $\widetilde{\Theta}'$ must be equivalent to a locally solvable system $\widetilde{\Theta}(z, w, w^{(k_s)}) = 0$ for $w(z)$. So ansatz (4) leads to a direct reduction, which proves the theorem. □

The above mentioned restrictions are more technical than of principle nature. Projectability (to be exact only the fraction $\frac{\xi^j}{\xi^\alpha}$ for $j = 1, \dots n$ and a chosen α must be independent of the dependent variables) could be weakend. Comparing with non-projectable symmetries one had to calculate 'non-projectable' direct reductions of the form $u = \mathcal{U}(x, w(z))$ with $z = \mathcal{Z}(x, u)$. The extension to reductions by more than one dimension is obvious

5. Reduction of the hydrodynamic sytem for a hot electron plasma

Since applications of both methods to single differential equations are detailed discussed in literature, we would like to show the direct reduction working on a sytem of PDEs. Let's consider the following system describing hot electron with an immovable backround of ions (n stands for the electron mass-density, v for the electron velocity and ϕ for the electrical potential):

$$\begin{aligned} \frac{\partial n}{\partial t} + \frac{\partial}{\partial x}(nv) &= 0 \\ \frac{\partial v}{\partial t} + v\frac{\partial v}{\partial x} &= -\frac{1}{n}\frac{\partial n}{\partial x} + \frac{\partial \phi}{\partial x} \\ \frac{\partial^2 \phi}{\partial x^2} &= n - 1 \end{aligned} \tag{5}$$

The 'classical' symmetry algebra of this system is spanned by

$$\mathbf{v} = \vartheta(t)\frac{\partial}{\partial x} + a_1\frac{\partial}{\partial t} + \vartheta_t(t)\frac{\partial}{\partial v} + (\varepsilon(t) + \vartheta_{tt}(t)x)\frac{\partial}{\partial \phi} \tag{6}$$

with a free parameter a_1 and two free functions $\vartheta(t)$ and $\varepsilon(t)$.

For the direct reduction we make the ansatz:

$$\begin{aligned} n &= \alpha^n(x,t) + \beta^n(x,t)w^n(z) \\ v &= \alpha^v(x,t) + \beta^v(x,t)w^v(z) \qquad z = z(x,t) \\ \phi &= \alpha^\phi(x,t) + \beta^\phi(x,t)w^\phi(z) \end{aligned} \tag{7}$$

Substitution into (5) gives

$$\begin{aligned} 0 = {} & w^v w^n_z(\beta^v\beta^n z_x) + w^n w^v_z(\beta^v\beta^n z_x) + w^n_z(\alpha^v\beta^n z_x + \beta^n z_t) + w^v_z(\alpha^n\beta^v z_x) \\ & + w^n w^v(\beta^v\beta^n_x + \beta^n\beta^v_x) + w^n(\alpha^v\beta^n_x + \beta^n\alpha^v_x + \beta^n_t) \\ & + w^v(\beta^v\alpha^n_x + \alpha^n\beta^v_x) + (\alpha^n_t + \alpha^v\alpha^n_x + \alpha^n\alpha^v_x) \end{aligned}$$

$$\begin{aligned} 0 = {} & w^v w^v_z(\alpha^n + \beta^n w^n)((\beta^v)^2 z_x) + w^v_z(\alpha^n + \beta^n w^n)(\alpha^v\beta^v z_x + \beta^v z_t) \\ & + (w^v)^2(\alpha^n + \beta^n w^n)(\beta^v\beta^v_x) - w^\phi_z(\alpha^n + \beta^n w^n)(\beta^\phi z_x) \\ & - w^\phi(\alpha^n + \beta^n w^n)(\beta^\phi_x) + w^n w^v(\beta^n\beta^v\alpha^n_x + \alpha^v\beta^n\beta^v_x + \beta^n\beta^v_t) \\ & + w^v(\alpha^n\beta^v\alpha^v_x + \alpha^v\alpha^n\beta^v_x + \alpha^n\beta^v_t) + w^n_z(\beta^n z_x) \\ & + w^n(\alpha^v\beta^n\alpha^v_x - \beta^n\alpha^\phi_x + \beta^n\alpha^v_t + \beta^n_x) \\ & + (\alpha^n\alpha^v_t + \alpha^n\alpha^v\alpha^v_x + \alpha^n_x - \alpha^n\alpha^\phi_x) \end{aligned}$$

$$\begin{aligned} 0 = {} & w^\phi_{zz}(\beta^\phi z_x^2) + w^\phi_z(\beta^\phi z_{xx} + 2\beta^\phi_x z_x) - w^n(\beta^n) + w^\phi(\beta^\phi_{xx}) \\ & + (\alpha^\phi_{xx} - \alpha^n + 1) \end{aligned}$$

Demanding this equations to be equivalent to ordinary differential equations for $w^n(z)$, $w^v(z)$ and $w^\phi(z)$ gives with

$\mathbf{z_x \neq 0}$:

$$\begin{aligned} n &= -\left(\frac{1}{s_1}\right)^2 w^n(z) \\ v &= s_1 f^b_t(t) + w^v(z) \qquad z = -\frac{x}{s_1} + f^b(t) \\ \phi &= f^c(t) + s_1 f^b_{tt} x + w^\phi(z) \end{aligned}$$

with a free parameter s_1 an two free functions $f^b(t)$ and $f^c(t)$. System (5) is transformed to:

$$\begin{aligned} w^n_z w^v + w^n w^v_z &= 0 \\ w^v w^v_z - w^\phi_z + \frac{w^n_z}{w^n} &= 0 \\ w^\phi_{zz} + w^n + (s_1)^2 &= 0 \end{aligned} \tag{8}$$

Easy calculations show, that this reduction is a similarty reduction to the 'classical' symmetry (6). After direct integration of the first and second equation system (8) could be solved via Bernoulli equation.

$\mathbf{z_x = 0}$: Here we get two solutions:

1. $$\begin{aligned} n &= w^n(z) \\ v &= f^d(t)x + w^v(z) \qquad z = t \\ \phi &= \tfrac{1}{2}((f^d(t))^2 + f^d_t(t))x^2 + f^e(t)x + (f^g(t) + c_3 x)w^\phi(z) \end{aligned}$$

with three free functions $f^d(t)$, $f^e(t)$ and $f^g(t)$ gives the reduced system:

$$\begin{aligned} w^n_z + f^d(z)w^n w^v &= 0 \\ w^v_z + f^d(z)w^v - c_3 w^\phi - f^e(z) &= 0 \\ (f^d(z))^2 + f^d_z(z) - w^n + 1 &= 0 \end{aligned}$$

2. $n = w^n(z)$
$v = -f^p_t(t) + (x + f^p(t))w^v(z) \qquad z = t$
$\phi = -f^p_{tt}(t)x + (c_4 x^2 + 2c_4 f^p(t)x + f^q(t))w^\phi(z)$
with two free functions $f^p(t)$ and $f^q(t)$ and a free parameter c_4 leads to

$$\begin{aligned} w^n_z + w^n w^v &= 0 \\ w^v_z + (w^v)^2 - 2c_4 w^\phi &= 0 \\ 2c_4 w^\phi - w^n + 1 &= 0 \end{aligned}$$

In both cases the reduction variables are no compatible sets of invariants for the 'classical' symmetry (6). So for $z_x = 0$ we are left with new reductions.

The calculation of the general conditional symmetry for system (5) is hard. But it is easy to prove, that the infinitesimal generators connected with the direct reductions for $z_x = 0$ are generators of conditional symmetries.

6. Conclusion

Stated by theorem 1 the conditions characterising conditional symmetries means the compatibility conditions for the combination of the original system Θ with the invariance equations. Based on this observation the equivalence between both methods – the method of conditional symmetries and the direct reduction – seems to be obvious.

While the conditional symmetries could be calculated in a more algorithmic manner, the direct reduction has its own right in being independent of symmetry methods. In practical calculations during both methods one is faced with the problem to solve nonlinear determining equations. In many cases one is left with solutions, which could also be found by the 'classical' similarity reduction.

For reductions by more than one dimension these extensions may become interesting. Here hidden symmetries – meaning symmetries being 'classical' symmetries of an intermediat step of reduction but not of the original equations – could be taken into account in an one step calculation.

References

1. G.W. Bluman, J.D. Cole. The general similarity solution of the heat equation. *J. Math. Mech.*, 18:1025–1042, 1969.
2. P.A. Clarkson, M.D. Kruskal. New similarity reduction of the Bousinesq equation. *J. Math. Phys.*, 30(10):2201–2213, 1989.
3. V.I. Fushchich. Conditional symmetry of the equations of nonlinear mathematical physics. *Ukrainskii Mathematicheskii Zhurnal*, Vol. 43, No. 11:1456–1470, 1991.
4. D. Levi, P. Winternitz. Non-classical symmetry reduction: example of the Bousinesq equation. *J. Phys. A.: Math. Gen.*, 22:2915–2924, 1989.
5. P.J. Olver, P. Rosenau. Group-invariant solutions of differential equations. *SIAM Journal of Applied Mathematics*, 47(2):263–278, 1987.

MIX
Papier aus verantwortungsvollen Quellen
Paper from responsible sources
FSC® C105338

If you have any concerns about our products,
you can contact us on
ProductSafety@springernature.com

In case Publisher is established outside the EU,
the EU authorized representative is:
Springer Nature Customer Service Center GmbH
Europaplatz 3, 69115 Heidelberg, Germany

Printed by Libri Plureos GmbH
in Hamburg, Germany